AF775942

SOURCES OF CONTAMINATION IN MEDICINAL PRODUCTS AND MEDICAL DEVICES

SOURCES OF CONTAMINATION IN MEDICINAL PRODUCTS AND MEDICAL DEVICES

DENISE BOHRER
Federal University of Santa Maria
Santa Maria, Brazil

WILEY

A JOHN WILEY & SONS, INC., PUBLICATION

Copyright © 2013 by John Wiley & Sons, Inc. All rights reserved

Published by John Wiley & Sons, Inc., Hoboken, New Jersey
Published simultaneously in Canada

No part of this publication may be reproduced, stored in a retrieval system, or transmitted in any form or by any means, electronic, mechanical, photocopying, recording, scanning, or otherwise, except as permitted under Section 107 or 108 of the 1976 United States Copyright Act, without either the prior written permission of the Publisher, or authorization through payment of the appropriate per-copy fee to the Copyright Clearance Center, Inc., 222 Rosewood Drive, Danvers, MA 01923, (978) 750-8400, fax (978) 750-4470, or on the web at www.copyright.com. Requests to the Publisher for permission should be addressed to the Permissions Department, John Wiley & Sons, Inc., 111 River Street, Hoboken, NJ 07030, (201) 748-6011, fax (201) 748-6008, or online at http://www.wiley.com/go/permissions.

Limit of Liability/Disclaimer of Warranty: While the publisher and author have used their best efforts in preparing this book, they make no representations or warranties with respect to the accuracy or completeness of the contents of this book and specifically disclaim any implied warranties of merchantability or fitness for a particular purpose. No warranty may be created or extended by sales representatives or written sales materials. The advice and strategies contained herein may not be suitable for your situation. You should consult with a professional where appropriate. Neither the publisher nor author shall be liable for any loss of profit or any other commercial damages, including but not limited to special, incidental, consequential, or other damages.

For general information on our other products and services or for technical support, please contact our Customer Care Department within the United States at (800) 762-2974, outside the United States at (317) 572-3993 or fax (317) 572-4002.

Wiley also publishes its books in a variety of electronic formats. Some content that appears in print may not be available in electronic formats. For more information about Wiley products, visit our web site at www.wiley.com.

Library of Congress Cataloging-in-Publication Data:

Bohrer, Denise.
Sources of contamination in medicinal products and medical devices / Denise Bohrer.
p. cm.
Includes bibliographical references and index.
ISBN 978-0-470-48750-1 (cloth)
1. Microbial contamination. 2. Pharmaceutical technology–Standards. 3. Pharmaceutical industry–Quality control. I. Title.
RS192.B64 2012
338.4'76151–dc23

2012017302

Printed in the United States of America

ISBN: 9780470487501

10 9 8 7 6 5 4 3 2 1

A man will turn over half a library to make one book.

Samuel Johnson (English writer, 1709–1784)

This book is dedicated to my sons, Daniel and Marcelo,
and to the memory of my mother.

CONTENTS

PREFACE

This book was designed to discuss products that can carry unexpected foreign species into the human body. Therefore, two categories of products were considered: medicinal products and medical devices. The differentiation between medicinal products and medical devices, although obvious, can, in some cases, be difficult to establish, as, for example, when products incorporate a device used to administer a drug.

Since the purpose of this book is to contribute to the understanding of how unforeseen species can enter a human body along with a specific product or treatment, only those medicinal products and medical devices that may lead to exposure to contaminants that pose a significant risk to human health have been included.

There are distinctive ways to classify contaminants. A primary classification could be related to their nature: chemical, microbiological, or physical (particulate matter). Another basic division could be related to their origin: extrinsic or intrinsic. Beyond these two main divisions, others, related to the nature of the product, sterile or nonsterile, for example, could also be used. In fact, any sort of classification, although necessary, is not fully satisfactory. Chemical, microbiological, and physical contaminants appear as the main divisions in this book, and include drugs and medical devices.

Under chemical contaminants, besides the customary residual solvents, and extractables and leachables, diagnostic imaging agents, including radiopharmaceuticals and contrast agents, are considered. Because of the great variety among these agents, mainly among radiopharmaceuticals, only the most employed—technetium-, iodine-, and fluoride-based—diagnostic imaging agents are discussed. These radionuclides are obtained either by fission of uranium or by neutron activation (the bombardment of an atom nucleus to produce the desired radioactive species). Therefore, major contaminants are other radioactive species concomitantly produced, species derived from the interaction of the radiation with matter, and species formed through radionuclide decay. Contrast agents dealt with in this book are iodine- and gadolinium-based formulations, and barium sulfate. Gadolinium (Gd) belongs to the so-called rare earth metals, a group of 17 elements, which occur collectively in nature. Rare earth metals are chemically very similar, which makes their separation a challenge. Since Gd comprises less than 4% of the rare earth metals, the others are potential impurities in Gd-based contrast agents.

Catalyst residuals and genotoxic impurities in active pharmaceutical ingredients (APIs) are also discussed in individual sections. Both classes are relevant impurities recently issued by regulatory bodies. Synthetic starting materials, reagents, and intermediates are reactive by design and may occur as impurities in the final APIs. This chemical reactivity translates into subsequent biological reactivity and may lead to DNA damage with mutagenic or carcinogenic consequences.

Although sources of microbial contamination are well-known, microbial contamination is not a solved problem. Evidently, its relevance is related to sterile products, where any form of microorganism should be absent. Although sterile products usually leave their production site as such, it is during handling for administration that they become contaminated. The major contribution comes from personnel due to nonadherence to aseptic practices. In fact, contamination occurs as a combination of microbial accessibility and drug formulation or device design, which enables proliferation. The anesthetic propofol, for example, is itself bactericidal; however, when formulated in a lipid vehicle, it becomes an excellent milieu for bacterial growth. Nevertheless, it is only when handled under conditions that allow microbials to enter the product that it becomes a threat.

Biofilms seem to be the most challenging form of microbial transmission since they harbor communities of microorganisms attached to a surface and are covered by a protective film formed by polymeric substances, which offer resistance to the penetration of antimicrobial agents. Indwelling medical devices are often primarily colonized by a single bacterial species, but as the biofilm develops, multispecies consortia quickly grow and remain practically irreversibly adhered.

It is not possible to discuss microbial contamination without considering nosocomial infections, although they do not seem to fall within the scope of this book. Nevertheless, because they may be caused by infected medical devices, they are briefly discussed from the point of view of microorganism sources.

Although sterilization is a necessary step in the production of sterile products, it is a source of contaminants as well. Radiation sterilization may lead to degradation of drugs and polymeric materials, and therefore the generated species are potential contaminants. On the other hand, sterilizing chemical agents may not be completely eliminated after the procedure, mainly upon reprocessing of medical devices. Residuals of sterilization agents appear in an individual chapter. Although they are chemical contaminants extrinsically introduced into an article, it would be difficult to locate this extensive chapter under chemical contaminants, mainly because they can contaminate both drugs and medical devices.

Medicinal gases are an important source of both chemical and microbiological contamination. Compressed air can carry all impurities present in atmospheric air when collected directly outdoors from a healthcare facility. The humid and dark environment that exists in a compressed air system is ideal for microbial growth.

Biopharmaceuticals are very sensitive drugs that present a distinct universe of contaminants. The challenge of purification increases when dealing with recombinant DNA technology products. The manufacture requires the isolation of the target product from the other components of the cell culture—DNA, proteins, viruses, and endotoxins—and also from others added to the medium such as antibiotics and preservatives. Moreover, protein drugs may aggregate, denaturize, or be fragmented, introducing new entities into the final product even after distribution, during shelf life.

Finally, there is a brief discussion of quality assurance with a focus on guidelines from regulatory agencies that contemplate the contamination issue.

This book includes an Appendix that deals with polymeric materials. In the form of tables, this appendix brings together general information on synthetic and natural polymers (structural units and additives), organic extractables from polymers used by the pharmaceutical industry, and degradation products of polymers submitted to ionizing radiation for sterilization purposes. Because polymeric materials have several applications (from containers to implants), this has been done in an attempt to locate the data of several chapters in one place.

This book was not written with the intention of raising suspicion about the quality of commercially available pharmaceutical products but, rather, to be a source of information on the origin of impurities and contaminants of a product when doubts about its integrity may endanger its actual therapeutic role.

Consulted sources include scientific journals and publications of official bodies and regulatory agencies. Unpublished data are difficult to find; therefore, information about research and findings that were not made public or information broadcasted in communication media will not be found in this book. Contaminants originated by counterfeit or adulteration are not included in this book.

Denise Bohrer

ACKNOWLEDGMENTS

My special thanks go to my husband (and colleague), Paulo, for his continuing support and for assuming several of my usual activities to give me the necessary time to write this book.

I am indebted to my students and to my colleagues. Not only did they provide me with encouragement and support but they also showed a remarkable tolerance for the lapses in my attention to my regular duties caused by this special activity.

I am especially grateful to Marlei Veiga, Marcella Schmidt, Lucas Bressan, and Sandra Ribeiro, my students, who assisted me in several tasks during the book preparation.

To Amy Pippi for her specialized support to improve text clarity and style.

To Jonathan Rose, my editor, whose insight was fundamental to the creation of this book.

D.B.

1

INTRODUCTION

Joseph Lister, Ignaz Philipp Semmelweis, Albert Woolley, and Cecil Roe are probably the most famous names linked to the history of contamination. While Lister and Semmelweis evidenced, in 1850–1860, the importance of asepsis in dealing with surgical instruments and patients to avoid infections, Woolley and Roe were victims of contamination with a chemical contaminant inadvertently infused during spinal anesthesia 100 years later, in 1947. In the late nineteenth century, practices such as hand washing with a solution of chlorinated lime to reduce the incidence of fatal childbed fever, and spraying instruments, surgical incisions, and dressings with a solution of phenol to reduce the incidence of gangrene were neither accepted nor recognized as means of avoiding the transmission of microorganisms. Microbial contamination was simply not an issue. In contrast, a chemical contaminant delivered to Woolley and Roe along with an anesthetic, which caused both to become paraplegic, was recognized as such, although the explanation for the case does not seem fully plausible and has not been totally elucidated even today. Curiously, it was the phenol solution, in which the ampoules of the anesthetic had been immersed for asepsis, that was the supposed contaminant! According to the trial conclusion, contamination occurred by the penetration of the contaminant through invisible cracks in the glass ampoules of the anesthetic.

Nowadays, microbial contamination is, with rare exceptions, well recognized through immediate and specific bodily reactions. Chemical contamination, on

---

Sources of Contamination in Medicinal Products and Medical Devices, First Edition. Denise Bohrer.
© 2013 by John Wiley & Sons, Inc. Published 2013 by John Wiley & Sons, Inc.

the other hand, is not so widely recognized, except when ill-fated episodes like the Woolley and Roe case occur. Chemical contamination rarely provokes an acute bodily reaction, and therefore, its manifestation is not promptly linked to the contaminant. This absence of an immediate response hampers the recognition of chemical contamination, making a substantial amount of evidence necessary in order for the effect of a chemical to be acknowledged and measures for its eradication to be taken. Aluminum was irrevocably recognized as causative of dementia dialytica in 1976, 3 years after clinical manifestations of the syndrome had emerged. The water used for hemodialysis was considered to be the primary source of aluminum, but it was only after 20 years that precautions to eliminate aluminum in the water were routine and the syndrome ceased to be a threat. In other situations, even when data indicate the presence of a contaminant and research has demonstrated that it could be harmful, risks are taken because no viable or satisfactory solution exists. Diethylhexyl phthalate, a polyvinyl chloride (PVC) additive that makes the polymer flexible and functional is recognized as causing infertility and endocrine disruption in rats. Despite all studies indicating this hazardous effect, it is still the chief plasticizer for the PVC used in medical devices such as infusion lines and catheters. A strong argument favoring its permanence is that, without plasticizers, PVC is useless and substitute candidates could even be more hazardous than the phthalate itself.

Physical contamination, a third type of contamination, entails the presence of solid particles suspended in liquid formulations. Particulate matter is a problem because the introduction of particulate matter into the bloodstream may result in phlebitis or cause damage to vital organs. The most common particulates in intravenous preparations are glass fragments, from the opening of glass ampoules, particles from rubber stoppers and intravenous equipment, and particles from plastic syringes.

Of all the recalls linked to contamination issued by the United States by the Food and Drug Administration over the last 5 years (2007–2011), 33% were due to particulate matter, demonstrating that, although it may be controlled through careful inspection, this is not a solved problem (Fig. 1.1).

Of the remaining recalls, 32% were related to microbial contamination and 29% were related to chemical contaminations. Contributing to increased rates of chemical contamination were the recalls of toothpaste (contaminated by diethylene glycol) in 2007, heparin (contaminated by oversulfated chondroitin sulfate) in 2008, and Tylenol (contaminated by 2,4,6-tribromoanisole) in 2010. The remaining 2% of chemical contamination involved iron in a lens care solution and a cross contamination of several drug products by penicillin.

With the advent of biopharmaceutical drugs, new modalities of contaminants have arisen. Since most of them are proteins, minimal changes in their conformational structure are sufficient to introduce a new entity in the formulation, which, able to trigger adverse reactions, is deemed to be an undesirable species and therefore a contaminant. Allergic and adverse reactions accounted for more than 40% of recalls over the last 5 years (2007–2011) (Fig. 1.2).

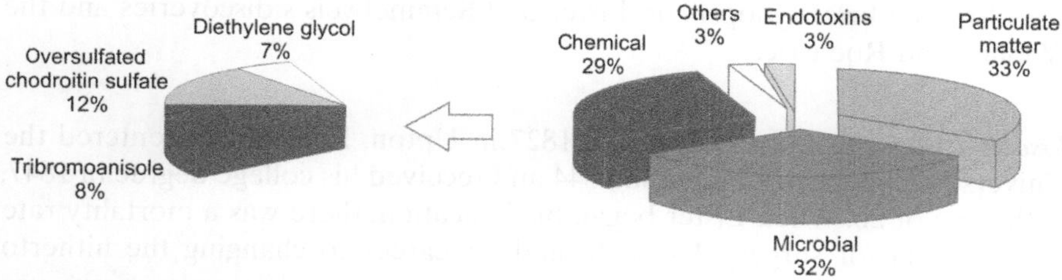

FIGURE 1.1 Food and Drug Administration drug recalls linked to contamination over the last 5 years (2007–2011).

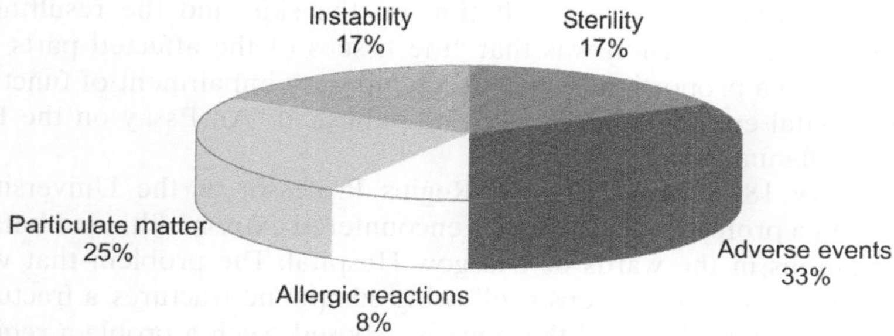

FIGURE 1.2 Food and Drug Administration biologic product recalls linked to contamination over the last 5 years (2007–2011).

The number of drug recalls due to contamination over the last 5 years clearly indicates that contamination of drug products is a topic that demands discussion (Fig. 1.3). Thus, the goal of this book is to gather together data regarding contamination sources associated with the production, storage, and delivery of pharmaceuticals.

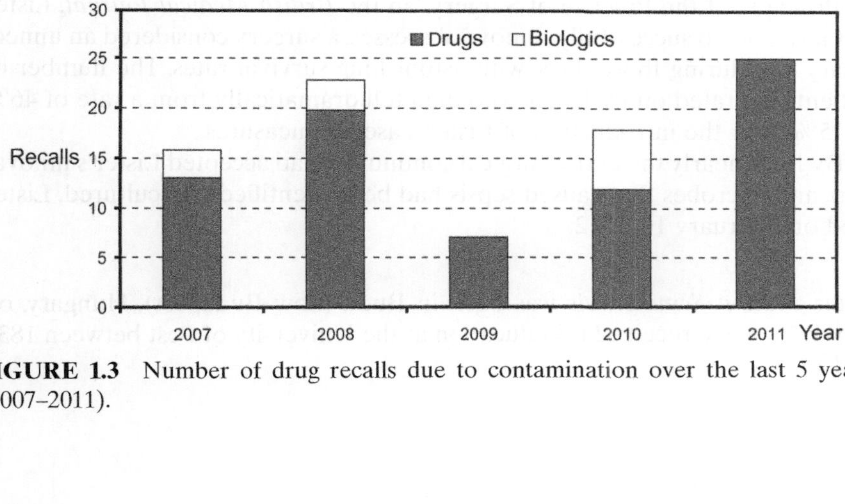

FIGURE 1.3 Number of drug recalls due to contamination over the last 5 years (2007–2011).

Below is a brief synopsis on Lister and Semmelweis's discoveries and the Woolley and Roe case.

Joseph Lister was born on April 5, 1827 in Upton, England. He entered the University College of London in 1844 and received his college degree in 1847, at the age of 20. When Lister began his education, there was a mortality rate of over 50% for surgery. He dedicated his career to changing the hitherto accepted conventions of surgery.

He began researching inflammation. Lister was aware that inflammation was the first stage of many postoperative conditions and, although many theories of inflammation existed, almost all of them were devoid of facts. Lister studied the varying effects of irritation on the skin and the resulting inflammation. His conclusion was that "the tissues of the affected parts have experienced to a proportionate extent a temporary impairment of functional activity or vital energy," and, in 1857, he published "An Essay on the Early Stages of Inflammation."

In January 1860, Lister became Regius Professor at the University of Glasgow. As a professor of surgery, he encountered extreme filth and unfavorable conditions in the wards of Glasgow Hospital. The problem that vexed Lister the most was that of sepsis following compound fractures, a fracture in which the skin is broken and the bone is exposed. Such a problem required surgery and had an extremely high mortality rate, especially when the individual remained in the hospital following the surgery.

In 1865, he read about the work done by Louis Pasteur on fermentation and microbes. Based on Pasteur's ideas, Lister deduced that wounds had to be thoroughly cleansed to avoid the entry of germs into the body. He tested spraying instruments, surgical incisions, and dressings with a solution of phenol, which at that time was used to deodorize sewage. He used it on a small boy with a compound fracture in his leg. The wound did not suppurate following surgery, and the only injury was that the acid burned the boy's skin. Lister explained the case and subsequent ones in a series of articles on the "Antiseptic Principle of the Practice of Surgery" in the *British Medical Journal*. Lister was also able to successfully remove abscesses, a surgery considered an unnecessary risk during those days, with astonishing survival rates. The number of patients operated on by Lister who died fell dramatically from a rate of 46% to 15% after the introduction of Lister's asepsis measures.

By 1890, nearly the entire surgical community had accepted Lister's innovation, and microbes that caused sepsis had been identified and cultured. Lister died on February 10, 1912.

Ignaz Philipp Semmelweis was born in Buda (now Budapest), Hungary, on July 1, 1818. He received his education at the University of Pest between 1835 and 1837.

In 1837, Semmelweis moved to Vienna and studied at the Second Vienna Medical School. He completed his studies in 1844 and remained in Vienna after graduation, becoming an assistant in the First Obstetrical Clinic of the university's teaching institution, the Vienna General Hospital (Wien Allgemeines Krankenhaus). In July 1846, Semmelweis became the titular house officer of the First Obstetrical Clinic, and his numerous duties included assistance with surgical procedures and clinical examinations. One of the most pressing problems he faced was the high maternal and neonatal mortality due to puerperal fever. Curiously, however, the Second Obstetrical Clinic in the same hospital exhibited a much lower mortality rate. The difference between them laid in their functions. The First Obstetrical Clinic was used for teaching medical students, while the Second Obstetrical Clinic was for the instruction of midwives. No clear explanation for the difference in mortality rates was forthcoming. Most women at the time delivered at home, but those who had to go to hospitals due to poverty, illegitimacy, or birth complications were exposed to high mortality rates. The disease was considered to be an inevitable aspect of contemporary hospital-based obstetrics, a product of unknown agency operating in conjunction with elusive atmospheric conditions. Semmelweis was severely disturbed that his First Clinic had a much higher mortality rate due to puerperal fever than the Second Clinic.

In 1847, Jakob Kolletschka, his friend and a professor of forensic medicine, died after being accidentally punctured with a scalpel while performing a postmortem examination. Kolletschka's own autopsy revealed a pathological situation similar to that of the women who were dying of puerperal fever. Semmelweis made a crucial association. He promptly connected the idea of cadaveric contamination with puerperal fever. He concluded that doctors and students carried the infecting particles on their hands from the autopsy room to the patients they examined during labor. This startling hypothesis led Semmelweis to devise a novel system of prophylaxis in May 1847.

Realizing that the cadaveric smell emanating from the hands of the dissectors reflected the presence of the incriminated matter, he instituted the use of a solution of chlorinated lime for washing hands between autopsy work and examination of patients. Despite protests, Semmelweis was able to enforce the new procedure vigorously and, in barely 1 month, the mortality from puerperal fever declined from 12% to 2% and remained low for the time his methodology was in practice.

In spite of the obvious conclusion, Semmelweis's observations conflicted with the established scientific and medical opinions of the time. Some doctors were offended at the suggestion that they should wash their hands, and Semmelweis could offer no acceptable scientific explanation for his findings. In 1861, Semmelweis published his discovery in the book *Die Ätiologie, der Begriff und die Prophylaxis des Kindbettfiebers* (*Etiology, Understanding and Preventing of Childbed Fever*), which received a number of unfavorable foreign reviews.

In July 1865, Semmelweis suffered what appeared to be a form of mental illness and was committed to an asylum, the Niederösterreichische Landesirrenanstalt, in Wien Döbling. He died there only 2 weeks later, on August 13, 1865.

On Monday, October 13, 1947, two patients, *Albert Woolley* and *Cecil Roe*, who were on the same operating list for a surgical procedure, developed permanent paraparesis following spinal anesthesia administered by the same anesthetist. Both patients sued the hospital and the anesthetist. At the trial, in October 1953, the court accepted evidence that the paralysis had been caused by the phenolic sterilizing solution seeping through invisible cracks in the glass ampoules of cinchocaine, the anesthetic. The court concluded that, because the anesthetist could not have been expected to know about this hypothetical risk, there had been no negligence.

An editorial in the *British Journal of Anaesthesia* at that time considered the sequence of events to be unlikely and thought it more probable that there had been contamination of the anesthetic with a different chemical. Dr. Malcolm Graham, the anesthetist, did not believe the "invisible crack theory" or the role of phenol. Phenol was known to be a chemical irritant, but no one was aware at that time of the effects of a solution of phenol in the subarachnoid space. Additionally, 1 year after the trial, the use of intrathecal phenol for the treatment of chronic pain was reported, which means that the neurological damage would be alleviated rather than caused by phenol.

In 1990, the case was critically reevaluated by Dr. Hutter [1]. His findings provided a more logical explanation for the events. He concluded that there is no doubt that the neurological damage was caused by a chemical contaminant, but that it was a mineral acid rather than phenol. Hydrochloric acid from a sterilizer could have been the contaminant. The ease with which contamination could happen, and the relatively small volume of acid that would have been required, makes this a realistic possibility. He hypothesized that the sterilizer would have been contaminated with acid on the Monday morning if, as a part of routine weekend maintenance, it were descaled (with the acidic solution) and the person undertaking this duty had forgotten to drain and wash out the acid. Needles and syringes placed into the sterilizer containing the acidic solution instead of ordinary water would have become contaminated and then used by the anesthetist.

While the reassessment conducted by Dr. Hutter absolved phenol, it continues to be accepted that the cause was some sort of chemical contaminant.

REFERENCE

1. Hutter CDD. The Woolley and Roe case. Anaesthesia 45, 859–864, 1990.

2

DIRECTIVES FOR CONTAMINATION CONTROL

The primary sources for the standardization of medicines are the pharmacopeias. They are nonprofit scientific organizations that set standards for the quality, purity, identity, and strength of medicines manufactured, distributed, and consumed worldwide.

Although editions of books dedicated to the management of drugs date back the sixteenth century (or even earlier), official pharmacopeias, as we know them today, were launched in the 1800s. The *United States Pharmacopeia* was issued in 1820, and the *British Pharmacopoeia* in 1864. The *Japanese Pharmacopeia* was first published in 1886, and the *German Pharmacopoeia* (Deutsches Arzneibuch) in 1872.

The creation of a worldwide, unified pharmacopeia, first envisioned in the 1940s, required the collaboration between national pharmacopeial commissions. The first edition of *The International Pharmacopoeia* was published in two volumes (1951 and 1955) and a supplement (1959), in English, French, and Spanish, and was also translated into German and Japanese. Today, the fourth edition of *The International Pharmacopoeia*, published by the World Health Organization (WHO), comprises volumes 1 and 2, published in 2006, the First Supplement, published in 2008, and the Second Supplement published in 2011.

In 1964, Belgium, France, Germany, Italy, Luxembourg, The Netherlands, Switzerland, and the United Kingdom signed a convention, drawn up under

Sources of Contamination in Medicinal Products and Medical Devices,
First Edition. Denise Bohrer.
© 2013 by John Wiley & Sons, Inc. Published 2013 by John Wiley & Sons, Inc.

the aegis of the Council of Europe, for the elaboration of a unified European Pharmacopoeia. The objectives were to harmonize specifications for medicinal substances of interest to the population of Europe. A protocol for this convention was signed in 1989 and it came into force in 1992. The European Pharmacopoeia currently has 37 European members, including the European Union (EU).

Although pharmacopeias define how medicinal products should be prepared and controlled, they are not regulatory bodies. Their publications are not regulated by law; therefore, the control is exerted by other bodies. The regulation and supervision of pharmaceutical manufacturing is carried out by governmental agencies such as the Food and Drug Administration (FDA) in the United States and the European Medicines Agency (EMEA) in the European Union, which are responsible for protecting and promoting public health.

The unification of the world's economic order through intensified international trade has led to an increased transnational exchange of material wealth, goods, and services. Consequently, the pharmaceutical industry, as many other industrial segments, has been built around production and marketing arrangements that include partners in several countries. Such arrangements have created the need for a unified technical and regulatory language.

The International Conference on Harmonisation of Technical Requirements for Registration of Pharmaceuticals for Human Use (ICH), created in 1990, congregates regulatory authorities and the pharmaceutical industry of Europe, Japan, and the United States in an answer to this increasingly global development. It is a joint initiative involving both regulators and research-based industry focusing on the worldwide unification of technical requirements for medicinal products.

Recently, there has been a growing interest in the use of ICH Guidelines from regions outside those that have already adopted them. In response, the ICH Global Cooperation Group (GCG) was established in 1999 as a subcommittee of the ICH Steering Committee (ICH-SC). Through the GCG, the ICH has begun to discuss scientific and technical aspects of drug registration. ICH's mission is to achieve an independent evaluation of medicinal products before they are allowed on the market.

Today, ICH is composed of representatives from six parties that represent the regulatory bodies and research-based industry:

- U.S. FDA
- EMEA
- Japanese Ministry of Health, Labour and Welfare (MHLW)
- Japan Pharmaceutical Manufacturers Association (JPMA)
- European Federation of Pharmaceutical Industries and Associations (EFPIA)
- Pharmaceutical Research and Manufacturers of America (PhRMA)

Other members include observers and representatives from non-ICH countries and regions. The ICH observers are the following:

- European Free Trade Association (EFTA)—currently represented by Swissmedic (Swiss Agency for Therapeutic Products)
- Health Canada
- WHO

Additionally, there are representatives of the GCG, the so-called Regional Harmonisation Initiatives (RHIs). They are formed by

- Asia-Pacific Economic Cooperation (APEC)
- Association of Southeast Asian Nations (ASEAN)
- Gulf Cooperation Council (GCC)
- Pan American Network on Drug Regulatory Harmonization (PANDRH)
- South African Development Community (SADC)

In 2005, the ICH-SC adopted a new system to ensure a clearer coding of ICH Guidelines. With the new codification, revisions to ICH Guidelines are shown as (R1), (R2), (R3) depending on the number of revisions. Annexes or Addenda to Guidelines have now been incorporated into the core guidelines and are indicated as revisions to the core guidelines (e.g., R1).

Although covering several aspects of drug production, ICH Guidelines do not encompass medical devices. Besides regulatory agencies, which define how medical devices are regulated, there are organizations that are responsible for normalizing material features and procedures related to medical devices. One of these organizations is the International Organization for Standardization (ISO).

The ISO is a nongovernmental international standard-setting body. Manufacturing medical devices and components for the pharmaceutical industry such as containers requires standards to ensure desirable characteristics of products such as quality and safety. In this area, guidelines are set by the ISO.

The Association for the Advancement of Medical Instrumentation (AAMI) is another organization responsible for regulating the use of medical instrumentation through standardization. The AAMI produces Standards, Recommended Practices, and Technical Information Reports for medical devices. AAMI standards are approved by the American National Standards Institute (ANSI) and AAMI also administers a number of international technical committees for the ISO.

Compiling into one short chapter the regulations, directives, and standards that regulate contaminants in medicinal products and medical devices is a challenging goal mainly due to transpositions and superimpositions of directives among the bodies. The Medicine and Healthcare Products Regulatory Agency (MHRA) in the United Kingdom regulates medical devices (Bulletin

No. 17) in accordance with the EMEA (Directive 93/42/EEC) by transposing EU directives into U.K. law. The EU, in its turn, implements ISO standardization for active implantable medical devices (2009/C 293/02). Similarly, there are ANSI/AAMI/ISO regulations for ethylene oxide sterilization residuals (10993-7:2008), to cite only a few examples. Therefore, a single directive may appear under different designations. Nevertheless, these apparent duplications spring from an awareness of the necessity to align national laws and regulations in order to promote safe exchange among internal markets.

Directives for the quality control of medicines and medical devices that fall within the scope of this book are presented in Table 2.1. The table shows the book chapters and the corresponding guidelines. This is to provide the readers with direct and summarized sources on regulation. The table includes some organizations not cited above. The identification of these organizations appears in the table's footnote.

TABLE 2.1 Reference and Title of Guidelines and Standards Regulating Impurities and Contaminants in Medicinal Products and Medical Devices

Contamination Source	Organization	Identification	Description
Chemical contamination			
Water	ANSI/AAMI/ISO	13959:2009	Water quality for hemodialysis and related therapies
Raw materials	FDA	CMC	New drug application: impurities in drug substances
	FDA	CMC	Drugs, biologics, and medical devices derived from bioengineered plants for use in humans and in animals
Inorganic impurities	ICH	Q4B Annex 1	Residue on ignition/sulfated ash general chapter (pharmacopeial text)
	ANSI/AAMI/ISO	11663:2009	Quality of dialysis fluid for hemodialysis and related therapy
Metallic impurities	ICH	Q3D	Impurities: guidelines for metal impurities
Organic impurities	ICH	Q3A	Impurities in new drug substances
	FDA	CMC	Development of new stereoisomeric drugs

TABLE 2.1 (*Continued*)

Contamination Source	Organization	Identification	Description
Genotoxic impurities	ICH	Q3B	Impurities in new drug products
Degradation products	ICH	Q3B	Impurities in new drug products
Excipients	PQRI		Testing and control strategies, used by excipient manufacturers, excipient distributors, and drug product manufacturers
Residual solvents	ICH	Q3C	Impurities: guideline for residual solvents
Medicinal gases and volatile anesthetics	PIC/S	PI 025-2	Inspection of medicinal gases
Medicinal gases	EMEA	ENTR/F/2/ AM/an D(2010) 3374	Manufacture of medicinal gases
Volatile anesthetics	FDA	21 C.F.R. PART 868	Anesthesiology devices
Diagnostic imaging agents	FDA		Developing medical imaging drug and biological products. Part 1: conducting safety assessments
Containers	FDA	CMC	Container closure systems for packaging human drugs and biologics
Plastic containers	EMEA	CVMP/204/04	Guideline on plastic immediate packaging materials
	ICH	Q4B Annex 2	Test for extractable volume of parenteral preparations general chapter
	EMEA	CVMP/204/04	Guideline on plastic immediate packaging materials
Metallic containers	EU	EC 1895/2005	On the restriction of use of certain epoxy derivatives in materials and articles intended to come into contact with food
Closures	EMEA	CVMP/204/04	Guideline on plastic immediate packaging materials

(*Continued*)

TABLE 2.1 (*Continued*)

Contamination Source	Organization	Identification	Description
	CEN	SCCP/0883/05	Opinion on 2-mercaptobenzothiazole
	FDA	CMC	Container closure systems for packaging human drugs and biologics
	ISO	EN ISO 8871-1	Elastomeric parts for parenterals and for devices for pharmaceutical use—part 1: extractable in aqueous autoclavates
	FDA	21CFR177.2600	Indirect food additives: polymers. Rubber articles intended for repeated use
Delivery systems and medical devices	EU	EN ISO 10993-55:2009	Biological evaluation of medical devices—part 9: framework for identification and quantification of potential degradation products
	ANSI/ AAMI/ ISO	10993-13:2009	Biological evaluation of medical devices—part 13: identification and quantification of degradation products from polymeric materials
	ANSI/ AAMI/ ISO	10993-17:2009	Biological evaluation of medical devices—part 17: establishment of allowable limits for leachable substances
Medical gloves	ASTM	D3578	Standard specification for rubber examination gloves
	FDA	Guidance documents (medical devices and radiation-emitting products)	Medical glove guidance manual

TABLE 2.1 (*Continued*)

Contamination Source	Organization	Identification	Description
	EU	SCCP/0883/05	2-Mercaptobenzothiazole (MBT)
Extracorporeal circuits	ANSI/ AAMI	RD47-2008	Reprocessing of hemodialyzers
Tissue substitutes	ISO	10993-9:2009	Biological evaluation of medical devices—part 9: framework for identification and quantification of potential degradation products
	ISO	10993-15:2009	Biological evaluation of medical devices—part 15: identification and quantification of degradation products from metals and alloys
	ASTM	F648-10a	Standard specification for ultra-high-molecular-weight polyethylene powder and fabricated form for surgical implants
	ISO	5834-1:2005	Implants for surgery—ultra-high-molecular-weight polyethylene. Part 1: powder form
	ISO	6474-1:2010	Implants for surgery—ceramic materials. Part 1: ceramic materials based on high purity alumina
	ISO	13781:1997	Poly(L-lactide) resins and fabricated forms for surgical implants—in vitro degradation testing
Devices for administration of aerosolized drugs	IPAC-RS and PQRI		Safety thresholds and best practices for extractables and leachables in orally inhaled and nasal drug products
	FDA	CMC	Nasal spray and inhalation solution, suspension, and drug products

(*Continued*)

TABLE 2.1 (*Continued*)

Contamination Source	Organization	Identification	Description
Particulate matter			
	ICH	Q4B Annex 3	Test for particulate contamination: subvisible particles general chapter
	ISO	ISO 8871-3:2003	Elastomeric parts for parenterals and for devices for pharmaceutical use—part 3: determination of released-particle count
Microbiological and endotoxin contamination			
	ICH	Q4B Annex 4A	Microbiological examination of nonsterile products: microbial enumeration tests general chapter
	ICH	Q4B Annex 14	Bacterial endotoxins general chapter
	EMEA	CVMP/815/00 Rev 1	Guideline on specifications: test procedures and acceptance criteria for herbal substances, 1 Herbal preparations, 2 Herbal medicinal products, and 3 Traditional herbal medicinal products
	ANSI/AAMI/ISO	11663:2009	Quality of dialysis fluid for hemodialysis and related therapy
	FDA	CMC	Container and closure system integrity testing in lieu of sterility testing as a component of the stability protocol for sterile products
Water	ANSI/AAMI/ISO	13959:2009	Water quality for hemodialysis and related therapies

TABLE 2.1 ***(Continued)***

Contamination Source	Organization	Identification	Description
Raw materials	ICH	Q4B Annex 4B	Microbiological examination of nonsterile products: tests for specified micro-organisms general chapter
	ICH	Q4B Annex 4C	Microbiological examination of nonsterile products: acceptance criteria for pharmaceutical preparations and substances for pharmaceutical use general chapter
Sterile formulations	ICH	Q4B Annex 8	Sterility test general chapter.
Contamination from sterilization procedures			
Heat sterilization	EU	EN 556-2:2003	Sterilization of medical devices—requirements for medical devices to be designate "sterile"—part 1: requirement for terminally sterilized medical devices
Residuals from chemical disinfection/ sterilization agents	FDA	Guidance documents (medical devices and radiation-emitting products)	Liquid chemical sterilants/ high level disinfectants
Ethylene oxide	ISO	10993-7: 2008	Ethylene oxide sterilization residuals
Formaldehyde	ANSI/ AAMI	RD47-2008	Reprocessing of hemodialyzers
Biotechnological products			
	FDA	CMC	Monoclonal Antibodies used as reagents in drug manufacturing
DNA and host cell protein residuals	FDA		Points to consider in the characterization of cell lines used to produce biologicals

(*Continued*)

TABLE 2.1 (*Continued*)

Contamination Source	Organization	Identification	Description
	ICH	Q5B	Quality of biotechnological products: analysis of the expression constructing cells used for production of recombinant DNA derived proteins
Viruses and mycoplasma	ICH	Q5A	Quality of biotechnological products: viral safety of biotechnology products derived from characterized cell lines of human or animal origin
Endotoxins	ICH	Q4B Annex 14	Bacterial endotoxins general chapter
Protein degradation	ICH	Q5C	Quality of biotechnological products: stability testing of biotechnological/ biological products
	FDA	CMC (draft 12/04/2009)	Assay development for immunogenicity testing of therapeutic proteins
Protein aggregation	ICH	Q5C	Quality of biotechnological products: stability testing of biotechnological/ biological products
	FDA	CMC (draft 12/04/2009)	Assay development for immunogenicity testing of therapeutic proteins

CMC, Chemistry, Manufacturing and Controls Guidances (FDA); CEN, European Commission Scientific Committee on Consumer Products; PIC/S, The Pharmaceutical Inspection Convention and Pharmaceutical Inspection Co-operation Scheme; IPAC-RS, International Pharmaceutical Aerosol Consortium on Regulation and Science; PQRI, Product Quality Research Institute; ASTM, American Society for Testing and Materials.

PART I

CHEMICAL CONTAMINATION

3

RAW MATERIALS

3.1 WATER

Water is the major component of the human body, and as a result, it is the most widely used vehicle for drug delivery. It is the primary raw material in pharmaceutical formulations. Water must be present even in drugs containing non-water-soluble substances. For example, lipophilic drugs are prepared as water/oil emulsions.

The amount and level of contaminants or impurities in water for pharmaceutical purposes depend on its use. Because water is used in all industries and scientific work, international and national standard authorities have established water quality parameters for all of its applications. Health-related water standards are established by organizations such as the World Health Organization (WHO) [1], the Environmental Protection Agency (EPA) [2], and the American Society for Testing and Materials Standards (ASTM) [3] in the United States, and the pharmacopeial compendia when the aim is specifically related to water for pharmaceutical products for human and veterinary consumption.

Standards for water quality are similar among pharmacopeias (*United States Pharmacopeia* [USP] [4], *British Pharmacopoeia* [BP] [5], *Deutsches Arzneibuch* [DAB] [6], *European Pharmacopoeia* [Ph. Eur.] [7], *The International Pharmacopoeia* [Ph. Int.] [8]), with only minor differences in the accepted

Sources of Contamination in Medicinal Products and Medical Devices,
First Edition. Denise Bohrer.
© 2013 by John Wiley & Sons, Inc. Published 2013 by John Wiley & Sons, Inc.

levels of chemical contaminants. Pharmacopeias classify water into three main categories: *purified water*, *highly purified water*, and *water for injection* (*WFI*). In all water purification processes, the raw material is always drinking water. Drinking water parameters, established by governmental regulatory agencies, include a large number of chemicals. These parameters comprise not only naturally occurring substances but also a series of chemicals that may be present from anthropogenic activities in natural waters, such as benzene, EDTA, and cadmium. Table 3.1 shows the guidelines for drinking water quality adopted by the WHO and the EPA.

TABLE 3.1 WHO Guideline Values for Chemicals in Drinking Water and EPA National Drinking Water Standards [1, 2]

Contaminant	EPA	WHO
Inorganic chemicals	Guideline Value (mg/L)	
Antimony	0.006	0.02
Arsenic	0.010	0.01
Asbestos (fibers > 10 μm)	7 million fibers/L	–
Barium	2.0	0.7
Beryllium	0.004	–
Boron	–	0.5
Cadmium	0.005	0.003
Chromium (total)	0.1	0.05
Cyanide (as free cyanide)	0.2	0.07
Cyanogen chloride	–	0.07
Lead	0.015	0.01
Mercury (inorganic)	0.002	0.006
Molybdenum	–	0.07
Nickel	–	0.07
Nitrate (measured as nitrogen)	10.0	50
Nitrite (measured as nitrogen)	1.0	3.0
Selenium	0.05	0.01
Thallium	0.002	–
Uranium	0.030	0.015
Organic chemicals	Guideline Value (mg/L)	
Acrylamide	–	0.0005
Alachlor	0.002	0.02
Aldicarb	–	0.01
Aldrin and dieldrin	–	0.00003
Atrazine	0.003	0.002
Benzene	0.005	0.01
Benzo(a)pyrene (PAHs)	0.0002	0.0007
Bromodichloromethane	–	0.06
Bromoform	–	0.1
Carbofuran	0.04	0.007
Carbon tetrachloride	0.005	0.004

TABLE 3.1 (*Continued*)

Contaminant	EPA	WHO
Chlordane	0.002	0.0002
Chlorobenzene	0.1	–
Chloroform	–	0.3
Chlorotoluron	–	0.03
Chlorpyrifos	–	0.03
Cyanazine	–	0.0006
Dalapon	0.2	–
DDT and metabolites	–	0.001
2,4-Dichlorophenoxyacetic acid	0.07	0.03
4-(2,4-Dichlorophenoxy)butyric acid	–	0.09
Dibromoacetonitrile	–	0.07
Dibromochloromethane	–	0.1
1,2-Dibromo-3-chloropropane (DBCP)	0.0002	0.001
1,2-Dibromoethane	–	0.0004
Dichloroacetate	–	0.05
Dichloroacetonitrile	–	0.02
o-Dichlorobenzene	0.6	1.0
p-Dichlorobenzene	0.075	0.3
1,2-Dichloroethane	0.005	0.03
1,1-Dichloroethene	0.007	–
cis-1,2-Dichloroethene	0.07	0.05[a]
trans-1,2-Dichloroethene	0.1	–
Dichloromethane	0.005	0.02
1,2-Dichloropropane	0.005	0.04
1,3-Dichloropropane	–	0.02
Dichlorprop	–	0.1
Di(2-ethylhexyl) adipate	0.4	–
Di(2-ethylhexyl) phthalate	0.006	0.008
Dimethoate	–	0.006
Dinoseb	0.007	–
1,4-Dioxane	–	0.05
Dioxin (2,3,7,8-TCDD)	0.00000003	–
Diquat	0.02	–
EDTA	–	0.6
Endothall	0.1	–
Endrin	0.002	0.0006
Epichlorohydrin	0.0004	0.0004
Ethylbenzene	0.7	0.3
Ethylene dibromide	0.00005	–
Fenoprop, 2,4,5-TP (Silvex)	0.05	0.009
Glyphosate	0.7	–
Heptachlor	0.0004	–
Heptachlor epoxide	0.0002	–
Hexachlorobenzene	0.001	–

(*Continued*)

TABLE 3.1 (*Continued*)

Contaminant	EPA	WHO
Hexachlorobutadiene	–	0.0006
Hexachlorocyclopentadiene	0.05	–
Isoproturon	–	0.009
Lindane	0.0002	0.002
2-Methyl-4-chlorophenoxyacetic acid (MCPA)	–	0.002
Mecoprop	–	0.01
Methoxychlor	0.04	0.02
Metolachlor	–	0.01
Microcystin-LR	–	0.001
Molinate	–	0.006
Monochloroacetate	–	0.02
Nitrilotriacetic acid (NTA)	–	0.2
N-Nitrosodimethylamine (NDMA)	–	0.1
Oxamyl (Vydate)	0.2	–
Pendimethalin	–	0.02
Pentachlorophenol	0.001	0.009
Permethrin	–	0.3
Picloram	0.5	–
Pyriproxyfen	–	0.3
Polychlorinated biphenyls (PCBs)	0.0005	–
Simazine	0.004	0.002
Sodium dichloroisocyanurate	–	50
Styrene	0.1	0.02
Tetrachloroethene	0.005	0.04
Terbuthylazine	–	0.007
Toluene	1.0	0.7
Toxaphene	0.003	–
2,4,5-T (2,4,5-Trichlorophenoxy)acetic acid	–	0.009
Trichloroacetate	–	0.2
1,2,4-Trichlorobenzene	0.07	–
1,1,1-Trichloroethane	0.2	–
1,1,2-Trichloroethane	0.005	–
Trichloroethene	0.005	0.02
2,4,6-Trichlorophenol	–	0.2
Trifluralin	–	0.02
Vinyl chloride	0.002	0.0003
Xylenes (total)	10.0	0.5
Disinfectants	Guideline Value (mg/L)	
Chloramines (as Cl_2)	4.0	3.0
Chlorine (as Cl_2)	4.0	5.0
Chlorine dioxide (as ClO_2)	0.8	–
Disinfection byproducts	Guideline Value (mg/L)	
Bromate	0.010	0.01

TABLE 3.1 (*Continued*)

Contaminant	EPA	WHO
Chlorate	–	0.7
Chlorite	1.0	0.7
Haloacetic acids (HAA5)	0.060	–
Total trihalomethanes (TTHMs)	0.080	1.0
EPA Secondary Standard	Guideline Value (mg/L)	
Aluminum	0.05–0.2	–
Chloride	250.0	–
Copper	1.0	2.0
Fluoride	2.0	1.5
Foaming agents	0.5	–
Iron	0.3	–
Manganese	0.05	0.4
Silver	0.10	–
Sulfate	250.0	–
Total dissolved solids	500.0	–
Zinc	5.0	–

Secondary Standards: nonenforceable guidelines regulating contaminants that may cause cosmetic or aesthetic effects in drinking water. EPA recommends secondary standards to water systems but does not require systems to comply.

The water purification process adopted by pharmaceutical industries must be able to furnish water with the quality parameters presented in Table 3.2. The ability to achieve a guideline value depends on a number of factors, including the following:

- concentration of the chemical in the raw water
- nature of the raw water
- treatment processes.

There are treatments available for purifying water, which should be chosen according to the degree of purity required. These treatments are listed in Table 3.3 according to their degree of complexity. The higher the ranking, the more complex the process.

Table 3.4 lists the quality parameters of water obtained by the purification processes listed in Table 3.3. The effectiveness of each treatment in removing the contaminants listed in Tables 3.1 and 3.2 are listed in Table 3.5.

As a result of anthropogenic activities, surface water systems are contaminated with chemicals that originate from wastewater discharges, agricultural activities, and atmospheric deposition. These chemicals may not be completely removed during drinking water treatments. Schriks et al. [10] evaluated 50 emerging contaminants relevant for drinking water based on their inherent

TABLE 3.2 Water Quality for Pharmaceutical Purposes (Pharmacopeial Standards)

	Water Grade		
Parameter	Purified Water	Highly Purified Water	Water for Injection
Conductivity at 20°C (μS/cm)	4.3	1.1	1.1
TOC (mg/L)	0.5	0.5	0.5
Nitrates (ppm)	0.2	0.2	0.2
Aluminum (μg/L)	10	10	10
Heavy metals (ppm)[a]	0.1	0.1	0.1
Chloride	Pass/fail	–	0.5 ppm
Sulfate	Pass/fail	–	Pass/fail
Ammonium (ppm)	0.2	–	0.2
Calcium/magnesium	Pass/fail	–	Pass/fail
Residue evaporation (mg/100 mL)	1	–	0.4 (volume ≤10 mL) 0.3 (volume >10 mL)

[a] Measured as lead.
TOC, total organic carbon.

TABLE 3.3 Ranking of Complexity of Water Treatment Processes for Chemicals

Ranking	Process
1	Distillation
2	Ion exchange
3	Reverse osmosis
4	Membrane filtration

difficulty to be removed by traditional treatment techniques. The n-octanol-water partition coefficient (log K_{ow}) was used as the criterion for selection. Compounds with log $K_{ow} > 3$ were excluded because they are less likely to pass through drinking water treatment plants. Table 3.6 lists these contaminants and classifies them according to the level at which they were found in drinking water samples. Although the levels of the contaminants are low, if these compounds are not satisfactorily removed on the treatment, they might not be eliminated in the purification step of the production of water for pharmaceutical purposes.

Even when water complies with quality parameters as a raw material, it can contain impurities after being turned into a pharmaceutical product. Table 3.7 lists the level of some contaminants found in water used for injections. Because the raw material passed the quality test, contaminants were either below the allowed concentration level or introduced after packaging. Contaminants

TABLE 3.4 Quality Parameters of Water after Different Purification Treatments

Purification Process	Conductivity at 20°C (μS/cm)	TOC (mg/L)	Heavy Metals (ppm)	Ammonium (ppm)	Calcium/ Magnesium (mg/L)	Residue Evap. (as Silicate) (mg/100 mL)
None (regular tap water)	240	0.6	1	1	35	1
Distillation (single)	2–10	0.1	0.5–1	0.01	1–3	0.5–1
Distillation (double)	1–2	<0.1	0.1–0.8	0.01	0.3–0.1	0.1–0.7
Ion exchange	2–30	–	<0.01	–	–	1
Reverse osmosis	10–25	0.015	<0.04	0.4	1.6	0.1
Membrane filtration	0.056	0.01	<0.01	<0.01	–	<0.01

Source: Millipore catalog.
–, not informed.

TABLE 3.5 Effectiveness of Each Treatment in Removing the Contaminants

	Treatment			
Contaminants	Distillation	Ion Exchange	Reverse Osmosis	Membrane Filtration
Ions	>70%	>80%	Monovalent >95% Polyvalent >97%	>80%
Organic compounds	>80%	Not efficient	>99%	>80%
Particles	>80%	Not efficient	>99%	>99%

Adapted from References 1 and 9.

introduced after packaging most likely originate from the packaging materials. Chapter 6 discusses containers as sources of contamination.

In summary, water can be a source of contaminants. If the raw material (drinking water) complies with the quality parameters established by authorities, contaminants that are still present can be eliminated by typical water purification processes available to the pharmaceutical industry. While distillation and reverse osmosis provide water with the required quality specifications for purified water and highly purified water, WFI is generally obtained by membrane filtration (associated with another purification process) because of chemical contamination and sterility requirements.

TABLE 3.6 Concentration in Drinking Water of Selected Emerging Contaminants

Concentration Level (μg/L)		
<0.1	0.1–1.0	>1.0
Benzothiazole[e]	1,4-Dioxane[e]	Bis(chloroisopropyl)ether[e]
Carbamazepine[b]	2,6-Dichlorobenzamide[a]	Diethylene triamine pentaacetic acid[e]
Diethyl toluamide[a]	Acetylsalicylate[b]	Ethylenediamine tetraacetic acid[e]
Diuron[a]	Amidotrizoic acid[b]	Methyl tert-butyl ether[c]
Iohexol[b]	Bentazone[a]	Metoprolol[b]
Iomeprol[b]	Benzene[e]	Trichloroethene[e]
4-Methylbenzene-sulfonamide[e]	Benzotriazole[e]	α-Amino-3-hydroxy-5-methyl-4-isoxazole propionic acid[a]
Iopromide[b]	Clofibric acid[b]	
Isoproturon[a]	Dichlorophenoxyacetic acid[a]	
n-Butylbrnzenesufonamide[e]	Diethylene glycol dimethyl ether[e]	
n-Nitrosodimethylamine[e]	Glyphosate[a]	
Perfluoroctane sulfonate[d]	Iopamidol[b]	
Phenazone[b]	Perfluorooctanoic acid[d]	
Simazine[a]	Triphenylphosphine oxide[e]	
Sulfamethoxazole[b]		

Chemical categories: [a]pesticide, [b]pharmaceutical, [c]gasoline additive, [d]perfluorinated compound, [e]miscellaneous.
Adapted from Reference 10.

TABLE 3.7 Contaminants Found in Water for Injection

Parameter	Sample	*n*	Content (μg/L)	Reference
Aluminum	Sterile water, Abbott	3	<5	[11]
	Sterile water, McGaw	2	<5	[11]
	Sterile water, Travenol	2	<5	[11]
	Aqua ad injectabilia, Braun, 50 mL	1	1	[12]
	Water for injection	36	24–450	[13]
Arsenic	Water for injection, EMS	1	39.3	[14]
	Water for injection, Geyer	1	30.9	[14]
Zinc	Sterile water	1	13.9	[15]
Silicate	Sterile water, Geyer	3	220–280	[16]
Silicate	Sterile water, SEM	1	292	[16]

3.2 INORGANIC IMPURITIES

Inorganic impurities are present in naturally occurring, nonsynthesized raw materials. They may either present toxic effects (as with arsenic) or be as harmless as chloride ions. An overview of usual concomitants and their limits cited in pharmacopeial compendia are listed in Table 3.8.

Although inevitable, the presence of concomitants in pharmaceutical products may not exactly be a problem. The presence of magnesium in calcium salts is very common due to the similarity between these cations. Both cations are alkaline earth metals with similar chemical behavior. As a result, most raw materials containing calcium also contain small amounts of magnesium. The same occurs with sodium and potassium. Thus, the limit for potassium as an impurity in sodium chloride (according to the BP monograph) is 550 ppm or 0.55 mg potassium per gram sodium salt. Species such as sodium, potassium, chloride, and sulfate are tolerable in a fairly large concentration range, whereas other toxic species such as arsenic and lead have narrower limits. For arsenic and lead, the limits in the pharmacopeial compendia are generally set between 1 and 10 ppm. In fact, these limits are high for such harmful species. There are a limited number of studies that show the presence of impurities in raw materials mainly because the quality of raw materials is certified by guarantee bulletins and products are supposed to be within the range of the quality attested by the certificate.

Metallic species can be introduced in active pharmaceutical ingredients (APIs) through various sources; they can naturally occur in plants or raw materials, they can be introduced as a concomitant impurity, or they can result from the manufacturing process. Reactors, pipes, and other equipment are

TABLE 3.8 Main Inorganic Impurities Listed in Pharmacopeial Compendia and Their Limits

Impurity	Limit	Impurity	Limit (ppm)
Sulfated ash	0.01–1.0%	Heavy metals	1.0–50.0
Chloride	10–500 ppm	Arsenic	1.0–4.0
Sulfate	50–400 ppm	Aluminum	0.2–1.0
	0.1–0.6%	Cadmium	5.0–10.0
Fluoride	3 ppm	Chromium	0.05–10.0
Bromide	50 ppm	Copper	0.1
Oxalate	100–350 ppm	Iron	2.0–100.0
Phosphate	25–400 ppm	Lead	0.1–50.0
Sulfite	15 ppm	Nickel	0.2–1.0
Ammonium	200 ppm	Silver	250.0
		Zinc	10.0–30.0

Obs: Some impurities are related to certain products only. For example, the limit for silver is in cysplatin and that for sulfite is in sugars.

TABLE 3.9 Concentration Limits for Metallic Impurities Used as Catalysts in the Production of a Pharmaceutical Substance Defined by EMEA [17]

Element	Oral Concentration Limit (ppm)	Parenteral Concentration Limit (ppm)
Pt, Pd, Ir, Rh, Ru, Os	5 (group)	0.5 (group)
Mo, V, Ni, Cr	10	1.0
Cu, Mn	15	1.5
Zn, Fe	20	2.0

potential sources. If a metal catalyst is used for the production of a pharmaceutical drug, residuals of the original catalyst or another species of the metal element altered by downstream chemical processes can be present in the product.

Metallic catalysts are not included in the general compendia tests for heavy metals; these species are considered in separate monographs. Limits for metal catalysts are already set by the European Medicines Agency (EMEA) [17] and will be adopted by the United States Pharmacopeia [18]. In the EMEA guidelines, metals are classified based on their toxicity, and their limits differ depending on the route of administration. The permitted daily exposure (PDE) is used as the key indicator of the maximum safe intake limit for individual elements. Fourteen elements were included in the classification, which are listed, along with their limits, in Table 3.9.

Table 3.10 lists the catalyst residuals in four drugs and seven APIs (and some intermediates). The metal residual catalysts listed in Table 3.9 were investigated in samples in which they were constituents of the employed catalysts. Most samples exhibited levels within the EMEA specifications. The only exception was synthetic vitamin E, in which the levels of Cr in both samples and Ni in one sample were higher than the permitted levels. In another sample [20], the API 2 intermediate, the metallic (Pd) catalyst residual, was also above the limit. However, the level was within the allowed range for the final drug.

Metallic impurities not associated with catalysts are presented in Table 3.11. The most investigated metals were Pb and Cd. A comparison of their levels demonstrates that synthetic drugs are the least contaminated, followed by natural products extracted from plants or animals and inorganic salts, such as calcium carbonate. Due to the wide use of calcium supplements, Table 3.12 summarizes the lead contamination in different calcium salts from different sources. The Pb level in different brands of calcium supplements varies by almost 300-fold. Calcium chelates and refined calcium carbonate were the only two categories with a mean Pb concentration below 1 μg/g. Although the Pb levels in other samples (natural sources of calcium) averaged at least 10 times higher, no significant differences among the Pb levels in these samples were found. These results indicate that natural sources of calcium are significant sources of Pb.

TABLE 3.10 Residuals of Some Metal Catalyst Impurities in Drug Products and Drug Substances

	Metal Range or Amount (μg/g)											
Drug	Pd	Pt	Rh	Cr	Fe	V	Zn	Ni	Mn	Mo	Cu	Reference
Enalapril maleate	0.077	–	–	–	–	0.38	4.3	4.9	<0.01	0.09	0.14	[19]
Calcium folinate	–	2.36	–	–	–	0.36	8.3	<0.01	<0.01	0.23	1.03	
Levadopa	–	–	0.027	–	–	0.47	1.5	<0.01	<0.01	0.08	<0.01	
API 1	n.d. – 2	–	–	–	–	–	–	–	–	–	–	[20]
API 1 intermediate (dicyclohexylamine salt)	n.d. – 4	–	–	–	–	–	–	–	–	–	–	
API 2	n.d. – 4	–	–	–	–	–	–	–	–	–	–	
API 2 intermediate (tryptophol)	100–130	–	–	–	–	–	–	–	–	–	–	
Synthetic vitamin E sample 1	–	–	–	200 ± 19	–	–	–	60 ± 3	–	–	–	[21]
Synthetic vitamin E sample 2	–	–	–	162 ± 14	–	–	–	n.d.	–	–	–	
API 3	0.34 ± 0.16	–	–	–	–	–	–	–	–	–	–	[22]
API 4	0.79 ± 0.06	–	–	–	–	–	–	–	–	–	–	
API 5	0.25 ± 0.09	–	–	–	–	–	–	–	–	–	–	
API 6	–	–	–	–	–	–	14.8 ± 1.1	–	–	–	–	[23]
API 7	–	–	–	–	6.0 ± 0.3	–	–	0.88 ± 0.04	–	–	–	

n.d., not detected.

TABLE 3.11 Metallic Impurities in Drug Products and Drug Substances

	Metal Range or Amount (μg/g)								
Drug	Pb	Sn	Tl	Cd	Co	Be	As	Hg	Reference
Enalapril maleate	0.16	0.7	<0.001	<0.001	<0.02	<0.01	–	–	[19]
Calcium folinate	0.13	<0.001	<0.001	<0.001	<0.07	<0.01	–	–	
Levadopa	<0.01	0.27	<0.01	<0.001	<0.01	<0.01	–	–	
Synthetic vitamin E sample 1	45 ± 3	9 ± 1	–	–	–	–	–	–	[21]
Synthetic vitamin E sample 2	n.d.	266 ± 24	–	–	–	–	–	–	
Zinc gluconate	–	–	–	0.33	–	–	–	–	[24]
Zinc gluconate	–	–	–	0.31	–	–	–	–	
Zinc gluconate	–	–	–	0.31	–	–	–	–	
Zinc sulfate	–	–	–	0.83	–	–	–	–	
Zinc citrate/chelate/picolinate	–	–	–	0.45	–	–	–	–	
Calcium carbonate (brand A)	3.39 ± 0.12	–	–	0.12 ± 0.09	–	–	–	–	[25]
Calcium carbonate (brand B)	3.56 ± 0.34	–	–	0.71 ± 0.06	–	–	–	–	
Calcium carbonate (brand C)	1.97 ± 0.29	–	–	3.55 ± 0.27	–	–	–	–	
Echinacea extract	0.02	–	–	0.036	–	–	0.134	0.080	[26]
Ginkgo biloba extract	0.063	–	–	0.010	–	–	0.022	0.080	
Chitosan	0.136	–	–	0.072	–	–	0.097	0.080	
Chitosan (plus vitamins)	0.074	–	–	0.018	–	–	0.010	0.080	
Soybean, lecithin (plus vitamins and minerals)	0.098	–	–	0.127	–	–	0.516	0.088	
Bovine cartilage	0.143	–	–	0.023	–	–	0.104	0.080	
Glucosamine (plus vitamins and minerals)	0.033	–	–	0.010	–	–	0.009	0.080	

TABLE 3.12 Lead Concentration in Different Calcium Supplements

Compound	Pb Concentration (μg/g)	Reference
Calcium folinate	0.13	[19]
Calcium carbonate (A)	3.39 ± 0.12	[25]
Calcium carbonate (B)	3.56 ± 0.34	
Calcium carbonate (C)	1.97 ± 0.29	
Calcium carbonate (D)	2.11 ± 1.33	[27]
Calcium carbonate (E)[a]	0.34 ± 0.24	
Calcium gluconate, calcium lactate, amino acid salts	0.26 ± 0.36	
Calcium magnesium carbonate	1.11 ± 0.71	
Calcium phosphate and hydroxyapatite	2.67 ± 2.74	

[a] Samples analyzed after laboratory refine.

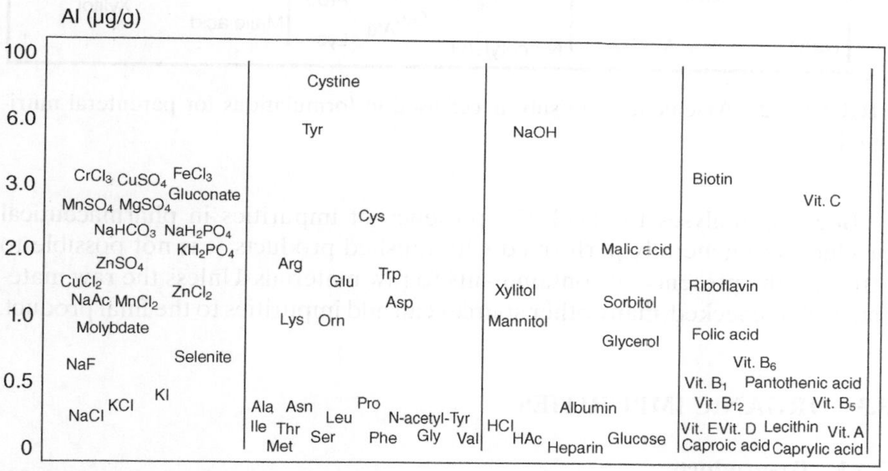

FIGURE 3.1 Aluminum present as impurity in substances used in parenteral nutrition [28].

Two studies were conducted on raw materials used to prepare parenteral formulations, which include several inorganic salts, to determine their aluminum and arsenic content [28, 29]. Figures 3.1 and 3.2 demonstrate that aluminum and arsenic were present in all investigated raw materials. There were also different levels of contamination among the substances. Salts, such as NaCl and KCl, presented low aluminum contaminations, whereas phosphates, gluconate, and malic acid were relatively contaminated. The authors attributed this difference to the affinity of aluminum to the latter substances. Arsenic showed a more uniform distribution of contamination. With the exception of the amino acid tyrosine, the arsenic level in all substances was below 1 μg/g and did not exceed the limits established by pharmacopeias.

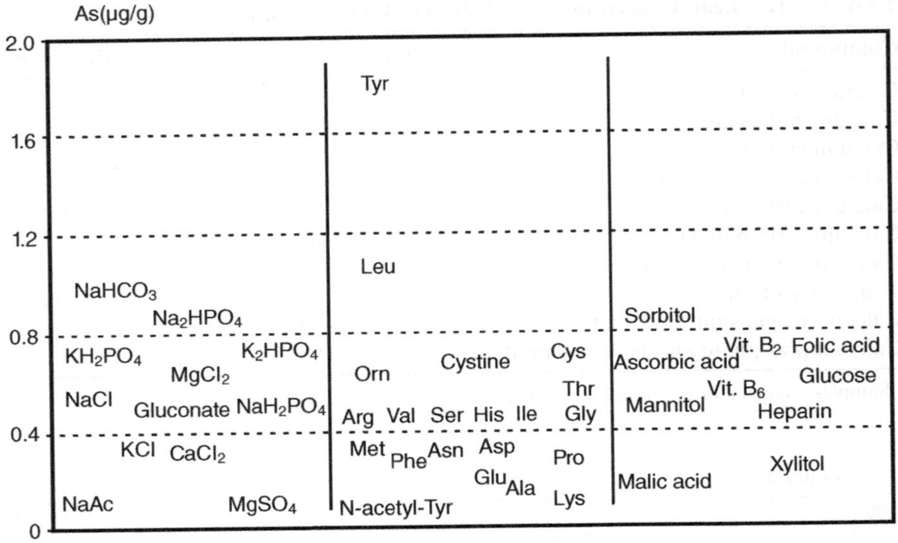

FIGURE 3.2 Arsenic level in substances used in formulations for parenteral nutrition [29].

Because analyses to check the presence of impurities in pharmaceutical products are generally performed with finished products, it is not possible to attribute the presence of contaminants to raw materials. Unless the raw material itself is checked, many other sources can add impurities to the final product.

3.3 ORGANIC IMPURITIES

3.3.1 By-products

By-products are perhaps the most difficult impurities to summarize because each drug has its own by-products that may appear as impurities. These by-products are synthesized along with active ingredients and are generally difficult to separate due to their similarity to the molecule of interest. Most are isomeric species, differing from each other by the presence of only a small group or the position of a hydrogen atom. Modern drugs that contain only one chiral isomer are even more difficult to separate and purify.

The control of organic impurities in new drug substances and drug products is based on the maximum daily dose (amount of drug substance administered per day) and the total daily intake (TDI) of impurities. Limits and impurity identification depend on the maximum daily dose. These limits were established by the International Conference on Harmonisation of Technical Requirements for Registration of Pharmaceuticals for Human Use (ICH) Q3A [30] and ICH Q3B [31] and are listed in Table 3.13.

TABLE 3.13 Limits for Impurities Established by the International Conference of Harmonization (ICH) for Drug Substances (Q3A) and for Drug Products (Q3B) [30, 31]

ICH	Maximum Daily Dose	Reporting Threshold: Limit above Which an Impurity Should Be Reported	Identification Threshold: Limit above Which an Impurity Should Be Identified	Qualification Threshold: Limit above Which an Impurity Should Be Qualified (Data That Establishes the Biological Safety of an Individual Impurity)
Q3A	≤2 g	0.05%	0.10% or 1.0 mg/day intake	0.15% or 1.0 mg/day intake
	>2 g	0.03%	0.05%	0.05%
Q3B	≤1 g	0.1%		
	>1 g	0.05%		
	<1 mg		1.0% or 5 μg TDI	
	<10 mg			1.0% or 50 μg TDI
	1–10 mg		0.5% or 20 μg TDI	
	10–100 mg			0.5% or 200 μg TDI
	>10 mg–2 g		0.2% or 2000 μg TDI	
	>100 mg–2 g			0.2% or 3000 μg TDI
	>2 g		0.10%	0.15%

A study conducted by Görög et al. [32] provides an example of how impurity profiles in drugs can be estimated. They surveyed impurities that were not related to a drug itself but rather to a functional group, such as the 3-oxo-group present in steroid drugs and mainly 4-ene-3-one or 1,4-diene-3-one moiety. The related impurities and their origins are listed in Table 3.14. In this study, the authors offered an analogy between the general structures of impurities and the genre of "Theme with Variations" in classical music, which requires that variations (harmony, melody, rhythm, and timbre) maintain the same duration and shape of the tune. With drugs, the general structure of impurities is usually the same, whereas variations are restricted to certain functional groups and their close neighbors. Some of the impurities listed in Table 3.14 are present in the chromatogram of mazipredone (Fig. 3.3), one of the steroids used in the study to outline the impurity profile. In this analysis, the 13 impurities detected in bulk mazipredone either above or below the ICH threshold of 0.1% were identified (Table 3.15).

TABLE 3.14 Estimation of the Profile of 3-Oxo-Related Impurities in 4-Ene-3-oxo Steroid and 3-Oxo-Related Impurities in 1,4-Diene-3-oxo-11β-hydroxy Steroids and Probable Pathway for Impurity Formation [32]

Impurity	Origin
Drug substance: 4-ene-3-oxo steroid	
4-Ene-3-one	
4,5-Dihydro	Birch reduction[a]
5(10)-Ene-3-one	Birch reduction/hydrolysis
3-Deoxo	Birch reduction
6α-Hydroxy	Oxidative degradation
6β-Hydroxy	Oxidative degradation
10β-Hydroxy	Oxidative degradation
4-Ene-3, 6-dione	Oxidative degradation
4,6-Diene-3-one	Oxidative degradation
4,8(14)-Diene-3-one	Total synthesis
4,6,8(14)-Triene-3-one	Total synthesis/degradation
3-Deoxo-3-ethynyl-3,5-diene	Ethynylation
1-Ene-3-one	Birch reduction
Drug substance: 1,4-diene-3-oxo-11β-hydroxy steroid	
11β-Hydroxy-1,4-diene-3-one	
1,2-Dihydro	Incomplete dehydrogenation
11α-Hydroxy	Fermentation
11-Oxo	Fermentation
11-Deoxy	Fermentation
8(9)-Ene	Synthetic steps
11-Deoxy-9(11)-ene	Synthetic steps

[a] It converts aromatic compounds having a benzenoid ring into a product, 1,4-cyclohexadienes, in which two hydrogen atoms have been attached on opposite ends of the molecule.

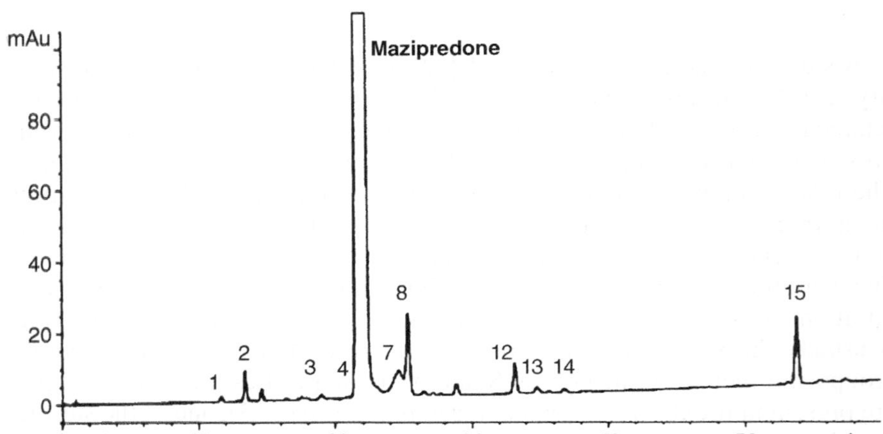

FIGURE 3.3 Chromatogram of a production batch of mazipredone. For the numbering and identification of the peaks, see Table 3.15 [33].

TABLE 3.15 Retention Data of Mazipredone and Its Impurities and Degradation Products

Derivative	Number	Retention Time (min)
17α-Hydroxy-17-oic acid	1	11.3
17α,20-Dihydroxy-21-oic acid	2	13.8
17-Deoxy-20-keto-21-hydroxy-21-(4-methyl-1-piperazinyl)	3	18.7
11-Keto-mazipredone	4	21.0
1,2-Dihydro-mazipredone	5	21.6
Mazipredone		21.6
Prednisolone	6	23.0
17α-Hydroxy-20-keto-21,21-dihydroxy	7	24.6
$\Delta^{9(11)}$-Dehydro-mazipredone	8	25.5
11-Deoxy-mazipredone	9	26.6
17-Ketone	10	27.1
17-Deoxy-20-keto-21,21-dihydroxy	11	32.8
Prednisolone-21-mesylate	12	33.0
20-Hydroxy-17(20)*E*-ene-21-al	13	34.7
20-Hydroxy-17(20)*Z*-ene-21-al	14	36.7
Piperazine-bridged mazipredone dimer	15	53.7

The numbers correspond to those in the chromatogram of Figure 3.3 [33].

3.3.2 Genotoxic Impurities (GTIs)

Genotoxics are species capable of inducing genetic mutations and chromosomal breaks or rearrangements. Due to their action on DNA, these species have the potential to cause cancer even in trace amount quantities. Thus, genotoxics must be avoided in the formulation of pharmaceutical products. Among the species known to be genotoxics are some solvents, alkylating agents, aflatoxins, and polycyclic aromatic hydrocarbons (PAHs) [34].

Solvents are classified in specific pharmacopeial monographs. Those known to present human carcinogen action are included in a distinct class of "solvents to be avoided" (see Section 3.5, Residual Solvents). Although PAHs have been deeply investigated in natural environments, such as atmospheric air and sediments, they are not considered an issue in pharmaceutical products even though they may be present. Among the products discussed in this book, only metered dose inhalers (MDIs) and solutions for parenteral nutrition were investigated for the presence of PAHs. In both cases, the source was rubber, which was a part of the aerosol device in MDIs and the cap of solution bottles. Rubber may contain PAHs due to the use of carbon black as a pigment or filler (see Chapter 7, Closures).

Aflatoxins are naturally occurring mycotoxins that are produced by the fungus *Aspergillus*. The habitat of *Aspergillus* is soil, hay, and grains. Therefore, aflatoxins are unlikely to be found as contaminants in pharmaceutical products.

Alkylating agents are among the reagents utilized in the synthesis of APIs and have been increasingly investigated as potential GTIs in drug substances. Aside from the use of reactive reagents, drug synthesis may produce reactive intermediates that may also be genotoxic. The problem is when these intermediate or by-products, even at low levels, remain in the final drug substance and drug products.

Regulation of GTIs in APIs was implemented in 2007 by the EMEA in Europe [35]. In the United States, the FDA released a draft guidance to address GTI in drug substances and products in 2008 [36]. The guide proposes the use of the "threshold of toxicological concern" (TTC) approach for determining the levels of GTI intake. For the determination of acceptable levels of exposure, GTI should be distinguished into the following two classes:

- genotoxic compounds with sufficient (experimental) evidence for a threshold-related mechanism
- genotoxic compounds without sufficient (experimental) evidence for a threshold-related mechanism.

The TTC has been developed to define a common exposure level for unstudied chemicals that do not pose significant risks of genotoxicity. Based on the analysis of 343 carcinogens and their carcinogenic potencies, a TTC level of 1.5 μg/person/day was established. Although the TTC is limited to species that do not pose significant risks, candidate species do not have the same degree of genotoxicity. Notably, Müller et al. [37] proposed a classification of API starting materials and intermediates based on published data on genotoxicity testing and comparative structural analysis to identify chemical functional moieties correlated with mutagenicity.

The categories and their genotoxic potential are summarized in Table 3.16. Table 3.16 also displays the results of a structure-based analysis of starting materials and intermediates that were found to fit within each category conducted by Dobo et al. [38]. Using a structure-based analysis (see below), the authors selected 272 starting materials and intermediates that were previously tested in the Ames assay (a test to access the mutagenic potential of chemical compounds) and defined the prevalence of the structural classes observed for each category. The results indicated that the two chemicals in category 1 were simple aromatic amines. Category 2 was primarily composed of nitro-aromatic compounds, aromatic amines, and secondary amines. The largest diversity of structural classes was observed among chemicals assigned to category 3, which included 16 different structural classes, with nitro-aromatic compounds being the most prevalent. Alkylating agents were the second most prevalent chemicals in this category. Category 4 included mostly secondary amines, aromatic amides, and aromatic amines. Chemicals in category 5 were primarily compounds with one or multiple aromatic and/or aliphatic rings, a large diversity

TABLE 3.16 Prevalence of Structural Classes in Starting Materials and Intermediates Evaluated in the Retrospective Analysis [37, 38]

Category	Definition	Structural Class	No Compounds
1	Precedent for mutagenicity and carcinogenicity		
		Aromatic amine	2
2	Mutagens with unknown carcinogenic potential or a "close-in" structural analog		
		Aromatic nitro	13
		Aromatic amine	4
		Secondary amine	4
		Alkylator	1
		α-Halo ether	1
		Carboxylic acid halide	1
		Hydrazine	1
3	Alerting structure—unique and unknown mutagenic potential		
		Aromatic nitro	14
		Alkylator	13
		Aromatic amide	12
		Aromatic amine	7
		Halogenated quinolone	4
		Michel acceptors	4
		Aliphatic nitro	2
		Halogenated heterocyclic	2
		Pyridine dialdehyde	2
		Acid halide	1
		Alkyl ester	1
		α-Halo ether	1
		Epoxide	1
		Hydrazine	1
		N-Methylol derivative	1
		Oxime	1
4	Alerting structure—nonunique and qualified in comparison to API		
		Secondary amine	27
		Aromatic amide	12
		Aromatic amine	8
		Aldehyde	4
		Alkyl aldehyde	4
		Gem dihalide	3
		Vinyl ketone	3
		Aliphatic carboxylic acid	2
		Hydroxamic acid	2
		Halogenated alkene	1
		Alkyl carbamate	1
5	No structural alerting features	–	111

of carbocyclic and heterocyclic ring structures, and substituted alkanes and alkynes.

This structure-based analysis was conducted with the aid of computer-based systems designed to qualitatively predict the toxic actions of compounds based on their chemical structures. These systems were able to perceive chemical substructures within molecules and to relate these to substructures of similar toxicity. Examples of the software used to conduct these analyses are DEREK [39] and MC4PC [40].

Based on the compound classes listed in Table 3.16 those with the highest concerns are the alkylating agents, especially esters derived from sulfonic acids. Although sulfonic esters have been known to be genotoxic for several years, the Viracept incident most likely triggered an incisive reaction from regulatory agencies that control GTI in pharmaceutical drugs. In 2007, patients reported having a bad smell and adverse reactions after taking Viracept tablets. A technical review [41] confirmed the presence of high levels of ethyl methane sulfonate in the Viracept tablets and led to the recall of the drug.

Alkylation is the transfer of an alkyl group from one molecule to another. An alkyl group may be transferred through several ways. Alkylating agents, which are widely used in organic chemistry, are classified according to their nucleophilic or electrophilic character. Nucleophilic alkylating agents provide the equivalent of an alkyl anion (carbanion). Examples include organometallic compounds such as Grignard (organomagnesium), organolithium, organocopper, and organosodium reagents. Electrophilic alkylating agents provide the equivalent of an alkyl cation. Examples include the use of alkyl halides with a Lewis acid catalyst to alkylate aromatic substrates in Friedel–Crafts reactions. Alkyl halides can also directly react with amines to form C–N bonds [42]. Electrophilic alkylating agents are able to alkylate DNA through a C–N bond by the addition of an alkyl group to the nitrogen (number 7) atom in the guanine base of DNA. As a result, alkylated DNA is unable to uncoil and separate, preventing DNA replication. Thus, the cells can no longer divide. This mechanism of genotoxicity of alkylating agents is precisely the same mechanism of toxicity that is responsible for the ability of some alkylating agents to perform as anticancer drugs.

Alcohols and phenols can be alkylated to yield alkyl ethers:

$$\text{R-OH} + \text{R}'\text{-X} \rightarrow \text{R-O-R}' + \text{H-X}. \tag{3.1}$$

Alcohols can also be alkylated via reactions with sulfonic acids:

$$\text{R-OH} + \text{R}'\text{-SO}_2\text{-OH} \rightarrow \text{R-O-SO}_2\text{-R}' + \text{H-OH}. \tag{3.2}$$

Sulfonic acids, which are widely used for salt formation during the synthesis of drug substances, may react with low-molecular-weight alcohols, such as methanol, ethanol, and isopropanol, to form the genotoxic alkyl esters. Alkyl sulfonate esters have the following general formula:

TABLE 3.17 Sulfonate Esters Most Commonly Formed in APIs and the Respective Acid and Alcohol from Which They Are Derived

Compound	Acid	Alcohol
Mesylates	Methane sulfonic	
Methyl methanesulfonate (MMS)		Methanol
Ethyl methanesulfonate (EMS)		Ethanol
Isopropyl methanesulfonate (IMS)		Isopropanol
Besylates	Benzene sulfonic	
Methyl benzenesulfonate (MBS)		Methanol
Ethyl benzenesulfonate (EBS)		Ethanol
Isopropyl benzenesulfonate (IBS)		Isopropanol
Tosylates	*p*-Toluene sulfonic	
Methyl *p*-toluenesulfonate (MpTS)		Methanol
Ethyl *p*-toluenesulfonate (EpTS)		Ethanol
Isopropyl *p*-toluenesulfonate (IpTS)		Isopropanol

Adapted from Reference 44.

$$R\text{-}SO_2\text{-}O\text{-}R'.$$

The development of APIs in the form of mesylate, tosylate, or besylate salts is one approach to increase drug solubility. As the synthesis of such compounds uses sulfonic acids and alcohols in the same milieu, genotoxic sulfonate esters may form [43]. Table 3.17 lists the most commonly formed sulfonate esters in APIs and the respective acids and alcohols from which they are derived.

Studies have demonstrated that, although viable, the formation of sulfonate esters does not occur at high levels. Table 3.18 lists the levels of sulfonate esters measured in APIs. Figure 3.4 shows the chromatograms of residual MMS found in APIs. MMS was determined either by ion (Fig. 3.4A) or gas chromatography (Fig. 3.4B) after derivatization with sodium thiocyanate. In both cases, the amount detected corresponds to less than 1.5 μg/g, which is the TTC established by EMEA.

Despite the toxicological concerns and regulatory constraints, it is unrealistic to avoid using the procedures and reagents that may generate GTI. These reactive species are indispensable for the synthesis of complex organic molecules, even though the properties that render these species useful are precisely those associated with genotoxicity.

3.3.3 Degradation Products

The main routes for drug substance decomposition are hydrolysis and oxidation. Functional groups with a propensity for each reaction are required for both reactions to occur.

TABLE 3.18 Levels of Sulfonate Esters and Solvents Measured in APIs and Some Drug Substances

Drug	Synthesis Solvents	Solvents Residuals	MMS Residuals (μg/g)	EMS Residuals (μg/g)	Other Sulfonic Esters	Residuals (μg/g)	Reference
Bismesylate API	Methanol	9.0	0.51	1.31	IMS	n.d.	[45]
	Ethanol	2.0					
	Propan-2-ol	10.8					
	Acetonitrile	n.d.					
	Isopropanol	12.8					
API	Methanol	<10	(see Fig. 3.4B)	n.d.	–	–	[46]
	Ethanol	<10					
API 1 (mesylate)	–	–	<5	<5	–	–	[47]
API 2 (tosylate)	–	–	<5	<5	–	–	
API 3 (besylate)	–	–	<5	<5	–	–	
API 4 (besylate)	–	–	<5	<5	–	–	
Different APIs	Methanol	–	<0.11	<0.03	–	–	[44]
	Ethanol						
	Isopropanol						
Lopinavir (six batches)	–	–	<185.7	<80	–	–	[48]
Ritonavir (six batches)	–	–	n.d.	n.d.	–	–	

n.d., not detected.

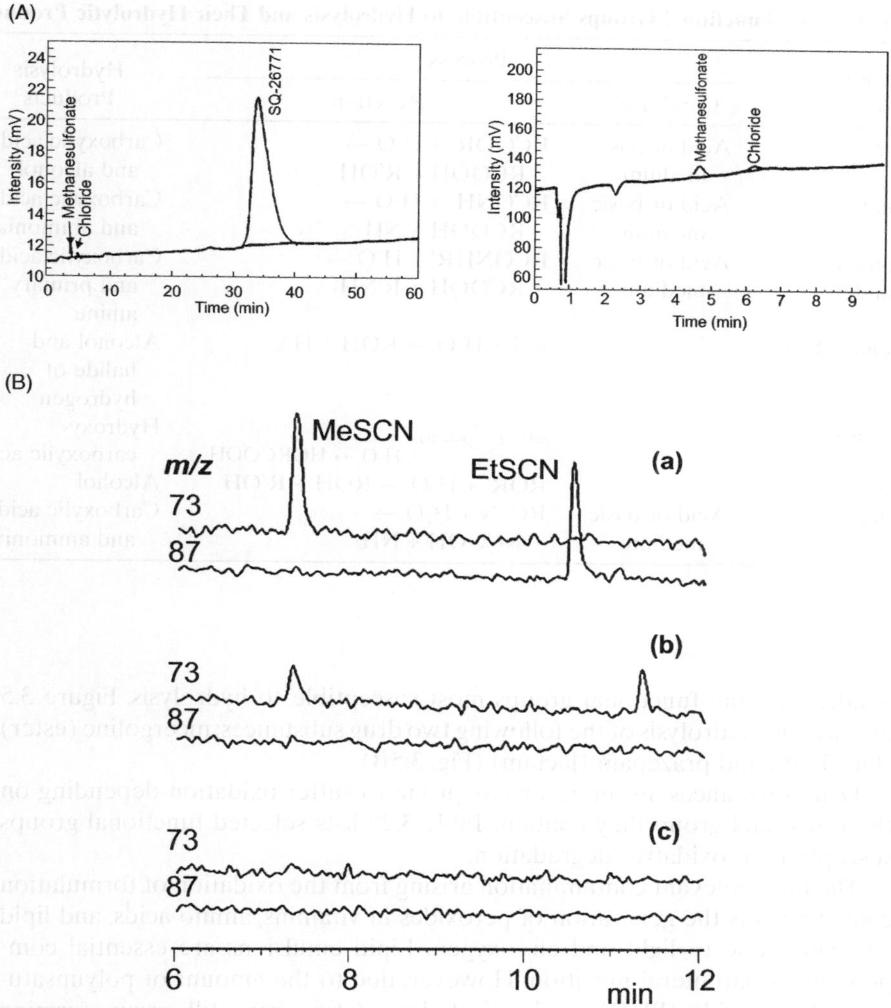

FIGURE 3.4 (A) Typical chromatogram of an API intermediate (SQ-26771) and the methanesulfonate residue present in 1 mg/mL drug solution [49]. (B) Selected ion recording chromatograms: (a) drug substance spiked with 1 μg/g methyl and ethyl mesylates, (b) drug substance alone, (c) reagent blank. Retention times (minute): methylthiocyanate 7.3, ethylthiocyanate 10.4. The small unknown peak at 11.3 minutes (chromatogram b, *m/z* 73) was not observed for all injections [46].

Hydrolysis reactions presuppose the presence of water, which can remain as a residual solvent from the manufacturing process, excipients, or even the environment (humidity). Usually, hydrolysis is catalyzed by both an acid and a base. Table 3.19 lists the functional groups susceptible to hydrolysis and their hydrolytic products. Esters, amides, lactones (cyclic esters), and lactams (cyclic

TABLE 3.19 Functional Groups Susceptible to Hydrolysis and Their Hydrolytic Products

Functional Group	Process: Conditions	Process: Reaction	Hydrolysis Products
Ester	Acid or basic medium	$RCOOR' + H_2O \rightarrow RCOOH + R'OH$	Carboxylic acid and alcohol
Primary amide	Acid or basic medium	$RCONH_2 + H_2O \rightarrow RCOOH + NH_3$	Carboxylic acid and ammonia
Secondary amide	Acid or basic medium	$RCONHR' + H_2O \rightarrow RCOOH + R'NH_3$	Carboxylic acid and primary amine
Halogenalkane		$RX + H_2O \rightarrow ROH + HX$	Alcohol and halide of hydrogen
Lactones		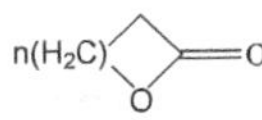 $+ H_2O \rightarrow HORCOOH$	Hydroxy-carboxylic acid
Ether		$ROR' + H_2O \rightarrow ROH + R'OH$	Alcohol
Nitrile	Acid or basic medium	$RC{\equiv}N + H_2O \rightarrow RCOOH + NH_4^+$	Carboxylic acid and ammonium

amides) are the functional groups most susceptible to hydrolysis. Figure 3.5 displays the hydrolysis of the following two drug substances: nicergoline (ester) (Fig. 3.5A) and prazepam (lactam) (Fig. 3.5B).

Drug substances are more or less prone to suffer oxidation depending on the functional group they contain. Table 3.20 lists selected functional groups susceptible to oxidative degradation.

The most relevant contamination arising from the oxidation of formulation constituents is the generation of peroxides in vitamins, amino acids, and lipid emulsions due to light and air–oxygen. Lipid emulsions are essential components of parenteral nutrition. However, due to the amount of polyunsaturated fatty acids (PUFAs), chemical degradation can still occur, forming hydroperoxides.

The exposure of lipid emulsions to room light and spotlight simulating phototherapy conditions in a neonatal intensive care unit for 24 hours can increase the level of hydroperoxides in these formulations by up to 60 times [52]. Lipid admixtures stored in ethylene vinyl acetate (EVA) bags formed a significant amount of lipid hydroperoxide. After 1 month under stressful conditions (gas-permeable container, 40°C), the peroxide value (PV = mmol peroxides/L) was 450 times higher than that of the controls (nitrogen-overlaid glass bottles) [53]. The results presented in Table 3.21 show that in contrast to lipids stored in closed glass containers gassed with nitrogen, significant and time-dependent peroxide formation occurred in all-in-one (AIO) plastic bags. Moreover, the lower PV of the lipid stored in V90 bags (polypropylene:

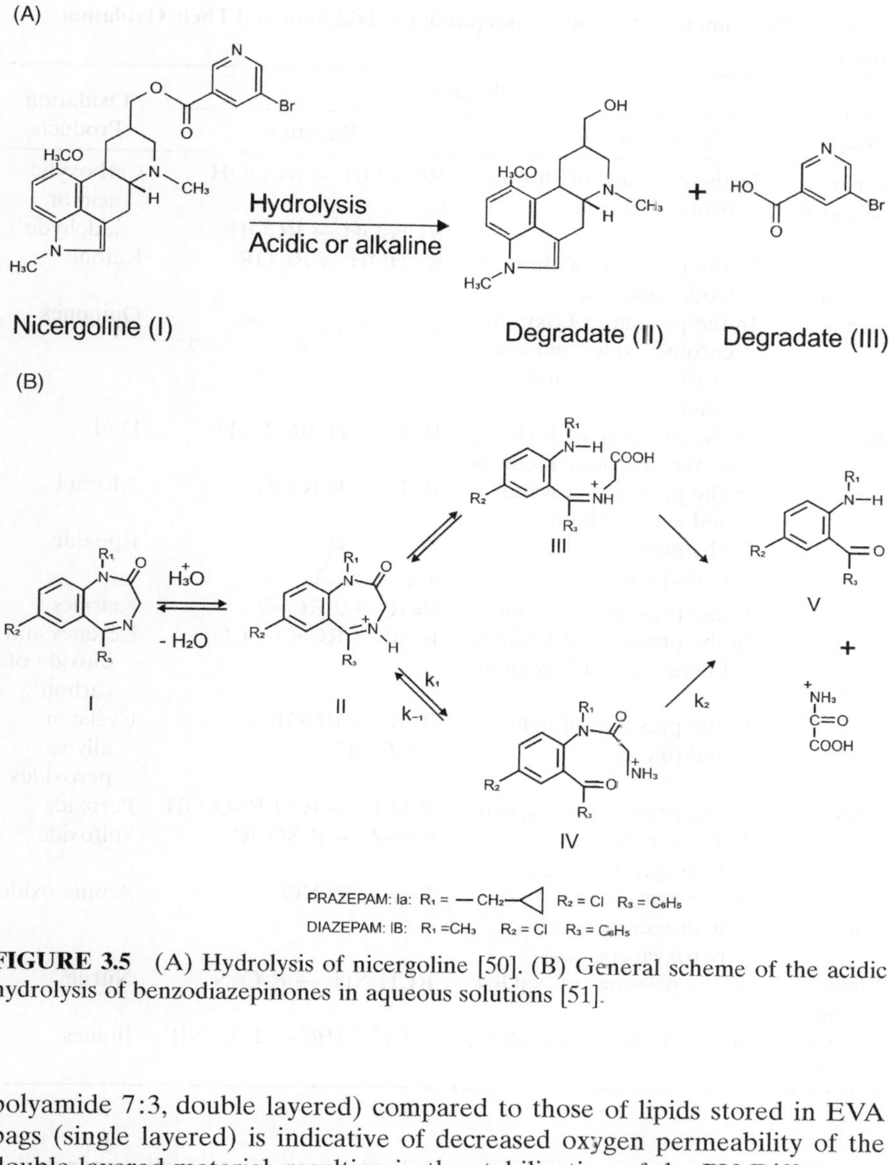

FIGURE 3.5 (A) Hydrolysis of nicergoline [50]. (B) General scheme of the acidic hydrolysis of benzodiazepinones in aqueous solutions [51].

polyamide 7:3, double layered) compared to those of lipids stored in EVA bags (single layered) is indicative of decreased oxygen permeability of the double-layered material, resulting in the stabilization of the PV. Differences were found between the PV of Intralipid (long-chain triglyceride [LCT]) and Lipofundin (medium-chain triglyceride [MCT]) samples. The initial PV of Intralipid (0.02 mmol/L) was lower than that of Lipofundin (0.10 mmol/L), whereas in the latter, the formation rate of PV was slower over time. This behavior is due to the double PUFA content of Intralipid.

PUFAs are oxidized by enzymatic and nonenzymatic pathways. Nonenzymatic oxidation is a free radical-mediated peroxidation. It is a chain reaction

TABLE 3.20 Functional Groups Susceptible to Oxidation and Their Oxidation Products

Functional Group	Process		Oxidation Products
	Conditions	Reaction	
Primary alcohol	In the presence of chrome oxides and acid	$RCH_2OH \rightarrow RCOOH$ or $RCH_2OH \rightarrow RCOH$	Carboxylic acid or aldehyde
Secondary alcohol	In the presence of chrome oxides and acid	$RCHOH \rightarrow RCOR$	Ketone
Phenol	In the presence of oxygen, chrome oxides and acid, or silver oxides and ether	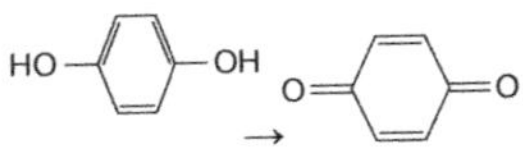	Quinones
Olefins	In the presence of $KMnO_4$, water, and basic medium	$R{=}R' \rightarrow HOR\text{-}R'OH$	Diol
	In the presence of water and acid medium	$R{=}R' \rightarrow R\text{-}R'OH$	Alcohol
	In the presence of peroxyacids	R=R' → 	Epoxide
	In the presence of ozone	$R{=}R' \rightarrow 2\ RC{=}O$	Ketones
	In the presence of $KMnO_4$ in basic or acid medium	$R{=}R' \rightarrow RC{=}O + CO_2$	Ketones and dioxide of carbon
	In the presence of light and oxygen	$R{=}R' \rightarrow ROOR'$ or $ROOH$	Cyclic or allylic peroxides
Ether	In the presence of oxygen	$R\text{-}O\text{-}R' \rightarrow R\text{-}O\text{-}R'\text{-}O\text{-}OH$	Peroxide
Thioether	In the presence of hydrogen peroxide	$R\text{-}S\text{-}R' \rightarrow R\text{-}SO\text{-}R'$	Sulfoxide
Tertiary amine	In the presence of hydrogen peroxide or peroxyacetic acid	$R_3N \rightarrow R_3\text{-}NO$	Amine oxide
Primary amine	In the presence of NaOCl	$RCH_2NH_2 \rightarrow R\text{-}C{\equiv}N$	Nitrile
Secondary amine	In the presence of $KMnO_4$	$R_2CH_2NHR' \rightarrow R_2C{=}NR'$	Imines

providing a continuous supply of free radicals that initiate further peroxidation. The whole process can be depicted as follows [54]:

$$RH + X^{\bullet} \rightarrow R^{\bullet} + XH \tag{3.3}$$

$$R^{\bullet} + O_2 \rightarrow ROO^{\bullet} \tag{3.4}$$

$$ROO^{\bullet} + RH \rightarrow ROOH + R^{\bullet}. \tag{3.5}$$

TABLE 3.21 Peroxidation of Intralipid 20% and Lipofundin MCT 20% during Storage: Influence of Container Material, Light Exposure, and Temperature [53]

Exposure Condition	Nutrimix 2/3 Bag (EVA) 20–27°C Daylight*	Nutrimix 2/3 Bag (EVA) 20–27°C Light Protection	3 – Chamber Bag (V90) 20–27°C Daylight*	Control Sample Closed Glass Bottle 20–27°C Daylight*
Intralipid 20%				
Time (day)				
1				0.02 ± 0.006[b]
5	0.47 ± 0.016[a]	0.06 ± 0.004[a,b]	0.33 ± 0.011	
8				0.02 ± 0.005
9	0.84 ± 0.010	0.05 ± 0.007	0.50 ± 0.017	
14	1.46 ± 0.020	0.17 ± 0.002	0.60 ± 0.014	
15				0.02 ± 0.001
19	1.87 ± 0.027	0.27 ± 0.017	0.76 ± 0.018	
22	2.48 ± 0.038	0.40 ± 0.013	0.92 ± 0.025	
28				0.02 ± 0.005
29	2.93 ± 0.052	0.52 ± 0.023	1.24 ± 0.032	
Lipofundin MCT 20%				
Exposure Condition	Nutrimix 2/3 Bag (EVA) 20–27°C Daylight*	Nutrimix 2/3 Bag (EVA) 20–27°C Light Protection	3 – Chamber Bag (V90) 20–27°C Daylight*	Control Sample Closed Glass Bottle 40.0 ± 0.5°C Light Protection
Time (day)				
1				0.11 ± 0.006[b]
5	0.57 ± 0.018[a]	0.19 ± 0.017[c,d]	0.64 ± 0.014	
8				0.10 ± 0.018
9	0.88 ± 0.024	0.35 ± 0.002	0.69 ± 0.010	
14	1.31 ± 0.048	0.54 ± 0.025	0.64 ± 0.021	
15				0.08 ± 0.003
19	1.59 ± 0.064	0.61 ± 0.018	0.67 ± 0.036	
22	1.99 ± 0.076	0.84 ± 0.007	0.70 ± 0.054	
28				0.10 ± 0.032
29	2.48 ± 0.040	0.99 ± 0.044	0.69 ± 0.032	

Daylight: 11 h/day; [a,b]$P < 0.001$ (Nutrimix 2/3 with light protection); [c,d]$P < 0.002$ (Nutrimix 2/3 with light protection vs. control samples). Peroxide values (mmol peroxides/L fat emulsion) are means of three samples ± standard error (SE).

* Daylight: 11h/day.

EVA, ethylene vinyl acetate; V90, polypropylene:polyamide 7:3.

The reaction is initiated by any existing free radical ($X^{\bullet}$), light, or metal ions that react with lipids (RH), generating peroxy radicals ($ROO^{\bullet}$), which eventually form lipid hydroperoxides (ROOH). Besides hydroperoxides, malondialdehyde (MDA), ethane, and pentane can also be formed if PUFA containing three or more double bonds (ω-3 PUFA and ω-6 PUFA, respectively) are present. The formation of pentane (2 μmol/L) and MDA (10 μmol/L) along with hydroperoxides was observed in a lipid emulsion (Intralipid 10%) [55]. MDA was also found in parenteral nutrition admixtures for newborn infants. In the 12 samples analyzed, MDA levels varied from 1632 to 14,679 nmol/L [56]. Pironi et al. [57] also detected MDA in fat emulsions. They compared three lipid emulsions (20%) containing different amounts of PUFA and α-tocoferol, 24 hours after admixing them to prepare AIO solutions. The authors concluded that lipid peroxide generation is directly related to PUFA content and inversely related to the α-tocoferol/PUFA ratio of the emulsion. Moreover, an increased MDA concentration was also measured in the soybean oil containing emulsions compared to that of the olive oil emulsion. Table 3.22 shows the PV and MDA measured in these samples.

Peroxide formation has also been attributed to the pro-oxidant action of vitamin E on PUFA, as vitamin E is another component present in lipid emulsions [52, 58]. Lipid oxidation occurring despite high concentrations of vitamin E (tocopherol) may appear to be surprising as the latter is generally regarded as the most efficient chain-breaking, lipid-soluble antioxidant [59]. However, studies have revealed that vitamin E requires a suitable "co-antioxidant," such as vitamin C, to be an efficient antioxidant in an isolated lipoprotein emulsion [60, 61]. The co-antioxidant eliminates the vitamin E radical formed in the interaction of the vitamin with initiating radical oxidants, which can have an adverse effect, promoting lipid peroxidation.

TABLE 3.22 Lipid Peroxide (LPX) and Malondialdehyde (MDA) Concentration in Lipid Emulsions [57]

	LPX (μM/L)				MDA (μM/L)			
Emulsions	Bottles (*n* = 6)	No Fat PN (*n* = 6)	AIO-T0 (*n* = 6)	AIO-T24 (*n* = 6)	Bottles (*n* = 6)	No Fat PN (*n* = 6)	AIO-T0 (*n* = 6)	AIO-T24 (*n* = 6)
Soybean oil	28.8 ± 29.9	<2	<2	11.3 ± 15.0	8.8 ± 3.1	4.9 ± 0.8	5.0 ± 0.3	6.9 ± 1.0
Soybean oil + MCT	1.7 ± 0.5	<2	<2	5.2 ± 8.8	3.7 ± 0.4	3.4 ± 0.4	4.1 ± 0.3	7.8 ± 1.8
Olive oil	4.1 ± 2.8	<2	<2	5.5 ± 6.4	0.7 ± 0.1	5.4 ± 0.3	5.3 ± 0.3	3.3 ± 0.4

Values are mean ± standard deviation of six bags (one sample per bag, analyzed in triplicate).
AIO, all-in-one mixture immediately after the addition of lipid emulsion; AIO-T24, all-in-one mixture 24 hours after the addition of lipid emulsion.

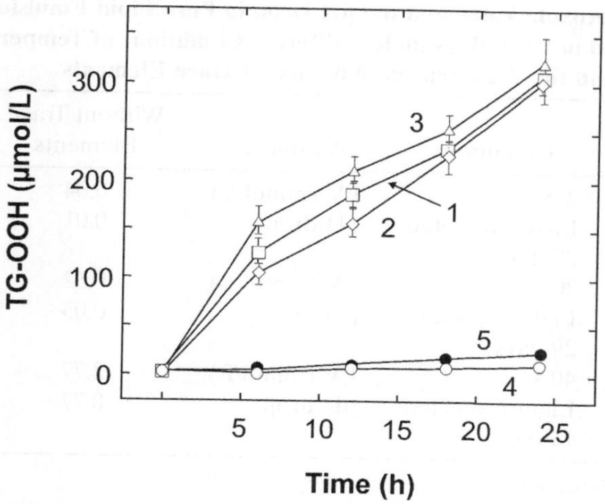

FIGURE 3.6 Kinetics of accumulation of triglyceride hydroperoxide in Intralipid exposed to phototherapy light under conditions that mimic the situation in a neonatal intensive care unit (NICU). A solution of 20% Intralipid was dispensed into a glass reservoir containing plastic tubing (line 3), plastic (line 1) or glass reservoir (line 2), plastic reservoir covered with aluminum foil (line 4), or a plastic reservoir after supplementation with 1 mmol/L sodium ascorbate (line 5) and exposed to phototherapy light for up to 24 hours. For all samples, the light fluxes measured were 11.7–25.9 μW/cm²/nm for 24 hours. At the time points indicated, aliquots were withdrawn and analyzed for the presence of triglyceride hydroperoxides [52].

Through the measurement of triglyceride hydroperoxides in Intralipid samples stored in glass and in plastic syringes exposed to direct light or wrapped in aluminum foil, it was possible to demonstrate that Intralipid is highly oxidizable under routine clinical conditions, in spite of its vitamin E content (≅60 γ-tocopherol and 20 μmol/L of the α- and δ-isomers).

The problem is particularly evident for neonates undergoing phototherapy. Despite the fact that lipids are used extensively to supply premature babies with calories, bags and extension sets are exposed to the intense radiation of spotlights for longer periods of time due to the low infusion rates of neonatal solutions. Figure 3.6 demonstrates the evolution of triglyceride hydroperoxide formation in a 20% Intralipid emulsion under different conditions of light exposure. Despite the high concentration of vitamin E present in the samples, lipid oxidation does occur. Decomposition can be prevented by either wrapping containers and tubing sets with aluminum foil or adding ascorbate to the infusate.

Lipid peroxides are also able to react with other components of parenteral nutrition admixtures (trace elements), causing a drop in the pH of the mixture

TABLE 3.23 Peroxide Value and the pH Drop in Pure Lipid Emulsion and AIO Admixture Stored in EVA Bags under Different Conditions of Temperature and Light Exposition in the Presence and Absence of Trace Elements

Sample	Conditions	Parameters	Without Trace Elements	With Trace Elements
AIO	2–8°C Light protected 29 days	PV (mmol/L) pH drop	0.04 0.01	0.19 0.02
AIO	20–30°C Light exposed 29 days	PV (mmol/L) pH drop	0.52 0.03	1.92 0.11
Intralipid 20%	40°C Light protected 14 days	PV (mmol/L) pH drop	2.77 0.77	18.04 1.54

Adapted from Reference 62.
AIO, all in one; PV, peroxide value.

and potential physicochemical instability [62]. Table 3.23 shows the PV and the pH drop in a pure lipid emulsion and a lipid-containing AIO admixture stored in EVA bags under different conditions of temperature and light exposure in the presence and absence of trace elements.

Peroxide formation has also been observed in multivitamin solutions for parenteral nutrition. Chessex et al. [63, 64] have studied the effects of light, air, and composition on the stability of multivitamin formulations and total parenteral nutrition (TPN) admixtures with or without vitamins and fatty acids. They analyzed the generation of peroxide in multivitamin solutions and TPN for adults and neonates. The analysis of multivitamin solutions for enteral use revealed the presence of peroxides at the initial opening of the bottle. The levels were higher in Poly-Vi-Sol (vitamin A, vitamin D, and vitamin C, vitamin B1, riboflavin, and niacinamide) than in Tri-Vi-Sol (vitamin A, vitamin D, and vitamin C) (Fig. 3.7). The authors attributed the difference in the peroxide content of the preparations to riboflavin. This vitamin catalyzes the photoinduced reaction between ascorbate and oxygen, which leads to hydrogen peroxide generation. The drop in hydrogen peroxide concentration over time observed with Poly-Vi-Sol is explained by the transformation of hydrogen peroxide into a more reactive species, which in turn could react with other components of the formulation. In contrast, the increase in catalase-resistant peroxides in the Tri-Vi-Sol preparation suggests a greater contribution of air in peroxide generation. In comparison to the formulation for parenteral use, the enteral multivitamin presented levels of 100 times higher, where catalase-resistant peroxides represented the bulk of peroxides. To study the effect of air and light on the generation of peroxides in multivitamins for TPN, the authors limited the experiment to fat-free TPN to prevent lipids from generating peroxides [64]. In solutions for neonates, the peroxide concentration

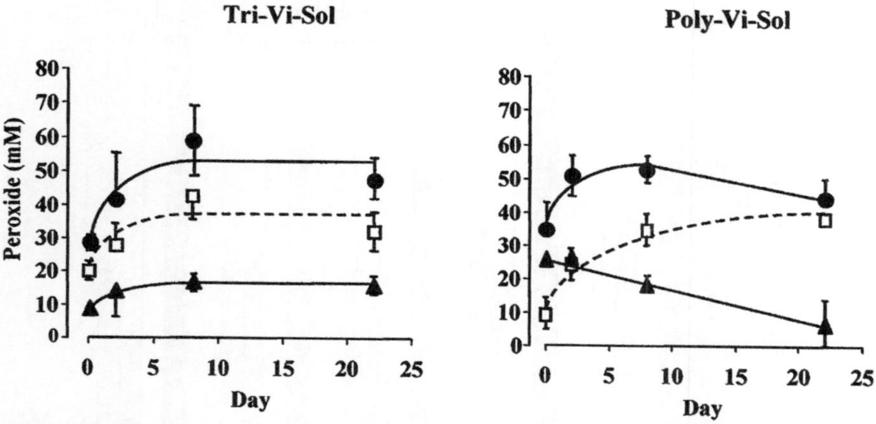

FIGURE 3.7 Total peroxides (circles), catalase-resistant peroxides (squares), and H_2O_2 as the difference (triangles) were measured in three lots of oral multivitamins without riboflavin (Tri-Vi-Sol) and three lots of oral multivitamins with 0.6 mg of riboflavin (Poly Vi-Sol) by time after the initial opening of the bottle. Compared with the preparation without riboflavin, the level of H_2O_2 was initially higher ($P < 0.01$) in Poly Vi-Sol, and that level dropped over time ($P < 0.05$). In both multivitamin preparations, the levels of catalase-resistant and total peroxides rose ($P < 0.05$) until day 8. Data represent the mean ± standard error of the mean. Variations too small relative to the symbol are not shown [63].

ranged from 190 to 300 μmol/L in sets unprotected from light, and sets protected from light ranged from 60 to 130 μmol/L. In the adult formulation, peroxide concentrations reached less than one-tenth of those in the neonatal preparation (Fig. 3.8). The authors attributed this difference to nutrient composition, as the adult formulation was four times lower in vitamins and higher in final amino acid and glucose concentrations. Amino acids and glucose contributed to a decrease in the concentration of peroxides [65]. Figure 3.8 demonstrates that protecting solutions from light and air decreases peroxide generation in these formulations. However, due to the high peroxide concentration in neonatal solutions, the protection from air during the infusion is not efficient.

The action of vitamins in the presence of lipids was also investigated by these researchers [66]. The generation of peroxide was compared among TPN preparations left in the dark and daylight for 6 hours. Formulations with vitamins and with vitamins with and without lipids were included in the experiment. The results demonstrated that the lipids had a significant but minor additive effect compared to vitamins in generating peroxides.

Based on the ability of ascorbate to reduce the photooxidation of lipid emulsions and multivitamin preparations (see Fig. 3.6), Lavoie et al. [67]

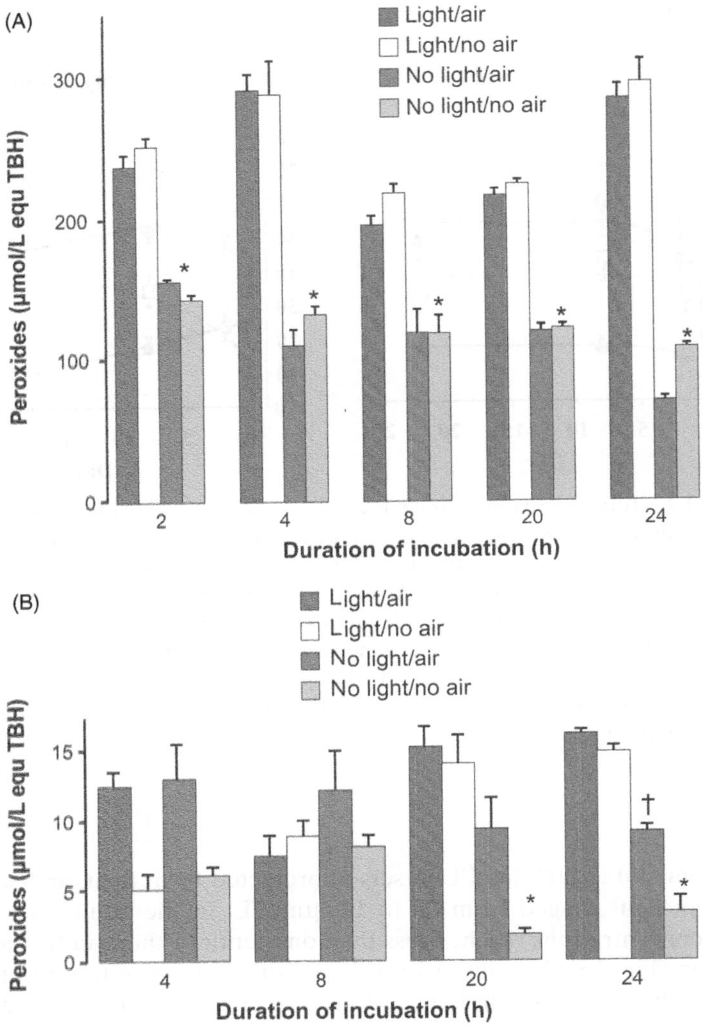

FIGURE 3.8 (A) Peroxide concentration in a neonatal fat-free solution of parenteral nutrition run through four different infusion sets: with or without air inlet, protected and unprotected from light (lux). Photoprotection was associated with a significantly lower peroxide content (indicated by *; $P < 0.05$). Protection from light resulted in a greater prevention against the generation of peroxides than did protection from air contamination. Results are expressed as mean ± SEM; $n = 3$. (B) Peroxide concentration in an adult solution of fat-free parenteral nutrition run through four different infusion sets: with or without air inlet, protected and unprotected from light (lux). The concentration of peroxides is 10-fold lower than in the neonatal solution. Protection from light and air was not associated with an overall effect on peroxide generation. Protection from light and air resulted in a significant decrease in peroxide generation at 20 and 24 hours of infusion (indicated by *; $P < 0.05$), whereas photoprotection alone resulted in a significant difference at 24 hours (indicated by +; $P < 0.05$). Results are expressed as mean ± SEM; $n = 3$. TBH, tert-butylhydroperoxide [64].

studied the formation of oxidative by-products of vitamin C in multivitamins exposed to light. The authors observed that the loss of ascorbic acid in photo-exposed multivitamin preparations was associated with the generation of products other than dehydroascorbate and 2,3-diketogulonic acid, which are the usual products of vitamin C oxidation. Furthermore, they showed that hydrogen peroxide at concentrations found in TPN solutions induced the transformation of dehydroascorbate into new, biologically active compounds with the potential to affect lipid metabolism. The authors speculated that these species have peroxide and aldehyde functions [68].

Because air (oxygen) is one of the factors responsible for peroxide generation in lipid emulsions and vitamin solutions, Balet et al. [69] compared multilayered versus single-layered (EVA) bags in terms of the oxidation of parenteral nutrition solutions. The authors measured PV, α-tocopherol, and ascorbic acid upon admixing, after 6 and 14 days in 24 parenteral solutions. Admixtures in multilayered bags exhibited less oxidation compared to those in EVA bags. No significant difference was observed in α-tocopherol content. However, after 6 days, ascorbic acid and dehydroascorbic acid disappeared in EVA bags.

Sodium metabisulfite is an antioxidant agent widely used in pharmaceutical preparations to reduce or prevent oxidation. However, there are studies that have shown that, under specific conditions, metabisulfite may have indirect oxidant properties. Baker et al. [70] demonstrated that a sulfite propofol emulsion and not an EDTA propofol emulsion underwent chemical changes during a simulated intravenous infusion. The compounds were identified as propofol oxidation products. The increase of propofol oxidation products demonstrated that sulfite from metabisulfite created a strong oxidant environment upon the introduction of air. Additionally, Lavoie et al. [71] showed that sulfite has been able to cause the oxidation of lipids. The reaction of sulfite with oxygen may potentially lead to several sulfite-derived oxidant species [72].

Aside from the oxidation of sulfite to sulfate ($SO_3^{2-} \rightarrow SO_4^{2-} + 2\ e^-$), sulfite may undergo a one-electron oxidation leading to the formation of a reactive sulfite radical ($SO_3^{2-} \rightarrow SO_3^{\bullet-} + e^-$). This reactive radical can react with oxygen forming the following two strong oxidant species: sulfite peroxyl ($SO_3^{\bullet-} + O_2 \rightarrow SO_3OO^{\bullet-}$) and sulfate radical ($SO_3^{\bullet-} + O_2 \rightarrow SO_4^{\bullet-}$). These species and the sulfite radical itself react as oxidants, which, depending on the circumstances, converts sodium metabisulfite into a pro-oxidant agent.

Peroxidation and free radical formation should be considered as important aspects of pharmaceutical stability and quality of parenteral nutrition and intravenous drugs. Peroxidation and free radical formation depend on environmental factors, such as storage conditions and container material, but they are also influenced by formulation components or additives, such as tocopherols and metabisulfite. Because the generation of these harmful species generally occurs at the time of use, manufacturing quality controls fail in demonstrating their existence.

3.4 ADDITIVES

Additives are all formulation constituents other than the active ingredient. Although additives could be classified into excipients (for solid preparations) and vehicles (for liquid preparations), there are several other agents used in pharmaceutical formulations with specific functions, such as preservatives, sweeteners, coatings, colorants, antioxidants, surfactants, emulsifying agents, and flavors. Because they comprise a vast amount of substances, this section will only deal with additives for compounding pharmaceutical products for internal use [73, 74].

Table 3.24 lists the additives and the possible impurities they may contain. The limits for these impurities were taken from the manufacturers' specification bulletins. Products of pharmacopeial grade (USP, BP, Ph. Eur., and DAB) were selected from qualified suppliers (Merck, Aldrich, Sigma, Fluka, EMS, and Riedel deHaën).

TABLE 3.24 Additives for Pharmaceutical Formulations and Impurities and Their Limits According to Manufacturers (Merck, Aldrich, Sigma, Fluka, EMS, Riedel deHaën)

Additive	Impurity	Limit
Solvents		
Ethanol	Acetone and isopropyl alcohol	≤0.01%
	Acetaldehyde and acetal	≤0.001%
	Benzene	≤0.0002%
	Methanol	≤0.01%
	Total of other impurities	≤0.03%
Glycerol	Chloride	≤0.0010%
	Sulfate	≤0.0010%
	Halogen compounds (as Cl)	≤0.0030%
	Heavy metals	≤0.0005%
	Arsenic	≤0.0001%
	Calcium	≤0.0001%
	Cadmium	≤0.0001%
	Mercury	≤0.0001%
	Ammonium	≤0.0005%
	Lead	≤0.0002%
	1,2,4-Butantriol	≤0.2%
	Residual solvents class 3[a]	≤0.5%
	Aldehydes	≤10 ppm
	Sulfated ash	≤0.01%
Polyethyleneglycol	Dioxane	10 ppm max
	Ethylene glycol and diethylene glycol	0.4% max
	Ethylene glycol (class 2)	620 ppm max
	Ethylene oxide	1 ppm max
	Formaldehyde (HCHO)	30 ppm max

TABLE 3.24 (*Continued*)

Additive	Impurity	Limit
	Heavy metals	20 ppm max
	Sulfated ash	0.2% max
Preservatives		
Benzyl alcohol	Peroxide value[a]	≤5
	Benzaldehye	≤0.15%
	Cyclohexylmethanol	≤0.10%
	Benzene	≤0.0002%
	Chlorobenzene	≤0.01%
	Toluene	≤0.01%
Benzoic acid	Sulfate	≤0.02%
	Heavy metals	≤0.001%
	Arsenic	≤0.0001%
	Cadmium	≤0.0010%
	Copper	≤0.0010%
	Mercury	≤0.0001%
	Lead	≤0.0005%
	Zinc	≤0.0010%
	Halogen compounds (as Cl)	≤0.01%
	Toluene	≤890 ppm
Chlorobutanol	Chloride	≤0.01%
	Chloroform	≤60 ppm
	Residual solvents class 3[a]	≤0.5%
	Sulfated ash	≤0.1%
Methylparaben	Heavy metals	≤0.001%
	Arsenic	≤0.0003%
	Cadmium	≤0.001%
	Copper	≤0.001%
	Mercury	≤0.0001%
	Lead	≤0.0005%
	Zinc	≤0.001%
	Methanol	≤0.3%
	Sulfated ash	≤0.05%
Methylparaben sodium salt	Chloride	≤0.03%
	Sulfate	≤0.03%
	Heavy metals	≤0.001%
	Arsenic	≤0.0003%
	Cadmium	≤0.001%
	Copper	≤0.001%
	Mercury	≤0.0001%
	Lead	≤0.0005%
	Zinc	≤0.001%
Propylparaben sodium salt	Chloride	≤0.03%
	Sulfate	≤0.03%

(*Continued*)

TABLE 3.24 (*Continued*)

Additive	Impurity	Limit
	Heavy metals	≤0.001%
	Arsenic	≤0.0003%
	Cadmium	≤0.001%
	Copper	≤0.001%
	Mercury	≤0.0001%
	Lead	≤0.0005%
	Zinc	≤0.001%
	1-Propylalcohol	≤0.5%
	Sulfated ash	34–36%
Propylparaben	Heavy metals	≤0.001%
	Arsenic	≤0.0003%
	Cadmium	≤0.001%
	Copper	≤0.001%
	Mercury	≤0.0001%
	Lead	≤0.0005%
	Zinc	≤0.001%
	Residual solvents class 3[a]	≤0.5%
	Sulfated ash	≤0.05%
Potassium sorbate	Arsenic	≤0.0003%
	Cadmium	≤0.001%
	Copper	≤0.001%
	Mercury	≤0.0001%
	Lead	≤0.0002%
	Zinc	≤0.001%
	Aldehydes (as acetaldehyde)	≤0.15%
	Residual solvents class 3[a]	<0.5%
Sodium benzoate	Chloride	≤0.02%
	Sulfate	≤0.01%
	Total chlorine	≤0.03%
	Heavy metals	≤0.001%
	Arsenic	≤0.0001%
	Cadmium	≤0.001%
	Copper	≤0.001%
	Mercury	≤0.0001%
	Lead	≤0.0002%
	Zinc	≤0.001%
Sorbic acid	Heavy metals	≤0.0010%
	Arsenic	≤0.0003%
	Cadmium	≤0.001%
	Copper	≤0.001%
	Mercury	≤0.0001%
	Lead	≤0.0002%
	Zinc	≤0.001%
	Aldehydes (as acetaldehyde)	≤0.15%
	Residual solvents class 3[a]	≤0.5%
	Sulfated ash	≤0.2%

TABLE 3.24 (*Continued*)

Additive	Impurity	Limit
Antioxidants		
Ascorbic acid	Heavy metals	≤0.001%
	Ign. residue	≤0.05% (as SO_4)
	Chloride	≤50 mg/kg
	Sulfate	≤20 mg/kg
	Copper	≤5 mg/kg
	Iron	≤2 mg/kg
L(+)ascorbyl palmitate	Heavy metals	≤0.001%
	Arsenic	≤0.0003%
	Copper	≤0.0025%
	Lead	≤0.001%
	Zinc	≤0.0025%
	Sulfated ash	≤0.1%
	Residual solvents class 3[a]	≤0.5%
Butylated hydroxyanisole	Arsenic	0.0003% max
	Heavy metals	0.001% max
	Lead	0.0005% max
	Mercury	0.0001% max
	3-tert-Butyl-4-methoxyphenol	10% max
	Hydroquinone	0.2% max
	Sulfated ash	0.01% max
Butylhydroxytoluene	Arsenic	0.0003% max
	Heavy metals	0.001% max
	Lead	0.0005% max
	Mercury	0.0001% max
	Residual solvents class 2 (MeOH)[a]	0.2% max
	Sulfated ash	0.002% max
Formaldehyde sulfoxylate sodium	Iron	0.0025% max
	Sodium Sulfite	5.0% max
	Residual solvents class 2 (MeOH)[a]	0.3% max
Phosphoric acid	Volatile acids (as CH_3COOH)	≤0.001%
	Chloride	≤0.0005%
	Fluoride	≤0.0010%
	Nitrate	≤0.0003%
	Phosphite and hypophosphite (as H_3PO_3)	≤0.02%
	Sulfate	≤0.005%
	Heavy metals	≤0.001%
	Arsenic	≤0.0002%
	Cadmium	≤0.00010%
	Copper	≤0.002%
	Iron	≤0.005%
	Mercury	≤0.0001%

(*Continued*)

TABLE 3.24 (*Continued*)

Additive	Impurity	Limit
	Potassium	≤0.005%
	Sodium	≤0.03%
	Lead	≤0.0010%
	Zinc	≤0.002%
Sodium bisulfite	Arsenic	0.001% max
	Heavy metals	0.003% max
	Iron	0.005% max
Sodium metabisulfite	Chloride	≤0.01%
	Heavy metals	≤0.001%
	Thiosulfate	≤0.02%
	Arsenic	≤0.0002%
	Iron	≤0.001%
	Mercury	≤0.0001%
	Lead	≤0.0005%
	Selenium	≤0.0006%
Sodium thiosulfate	Sulfate and sulfite (as SO_4)	≤0.2%
	Heavy metals	≤0.001%
Tocopherol	Heavy metals	≤0.001%
	Arsenic	≤0.0003%
	Copper	≤0.0025%
	Mercury	≤0.0001%
	Lead	≤0.0005%
	Zinc	≤0.0025%
	Methanol	≤3000 ppm
	Pyridine	≤200 ppm
	Toluene	≤890 ppm
	Sulfated ash	≤0.1%
Vehicles		
Cellulose powder	Ether-soluble substances	≤0.15%
	Water-soluble substances	≤1.0%
	Heavy metals	≤0.001%
	Sulfated ash	≤0.3%
Gelatin	Sulfur dioxide (SO_2)	≤0.004%
	Heavy metals	≤0.001%
	Arsenic	≤0.00008%
	Chromium	≤0.001%
	Iron	≤0.003%
	Zinc	≤0.003%
	Peroxide (as H_2O_2)	≤0.001%
	Ash	≤2.0%
Lactose	Heavy metals	≤0.0005%
	Arsenic	≤0.0001%
	Copper	≤0.0025%
	Lead	≤0.00005%
	Zinc	≤0.0025%
	Sulfated ash	≤0.1%

TABLE 3.24 (*Continued*)

Additive	Impurity	Limit
Starch	Reducing matter (as maltose)	Max 0.7%
	Sulfated ash	Max 0.4%
Sorbitol	Chloride	≤0.002%
	Sulfate	≤0.006%
	Heavy metals	≤0.0005%
	Arsenic	≤0.00013%
	Nickel	≤0.0001%
	Lead	≤0.00005%
	Related substances (mannitol)	≤2.0%
	Reducing sugars after hydrolysis/ total sugar (as glucose)	≤0.5%
	Reducing sugars (as glucose)	≤0.11%
	Sulfated ash	≤0.02%
Sucrose	Chloride	≤0.0035%
	Sulfate	≤0.005%
	Sulfite (as SO_2)	≤0.0010%
	Heavy metals	≤0.0005%
	Arsenic	≤0.0001%
	Lead	≤0.00005%
	Sulfated ash	≤0.02%
	Residual solvents class 3[a]	≤0.5%
Talc	Heavy metals	≤0.004%
	Aluminum	≤2.0%
	Arsenic	≤0.0003%
	Calcium	≤0.9%
	Iron	≤0.25%
	Lead	≤0.0005%
	Asbestos (according to Ph Eur)	Not detectable
Zinc oxide	Chloride	≤0.005%
	Sulfate	≤0.02%
	Arsenic	≤0.0005%
	Cadmium	≤0.001%
	Iron	≤0.001%
	Lead	≤0.005%
Chelating agents		
EDTA	Heavy metals	≤0.001%
	Calcium	≤0.001%
	Iron	≤0.001%
	Magnesium	≤5 ppm
	Nitrilotriacetic acid	≤0.1%
	Sulfated ash	≤0.1%
Buffers		
Boric acid	Sulfate	≤0.04%
	Heavy metals	≤0.0015%

(*Continued*)

TABLE 3.24 (*Continued*)

Additive	Impurity	Limit
Disodium hydrogenophosphate	Chloride	≤0.02%
	Fluoride	≤0.001%
	Sulfate	≤0.05%
	Heavy metals	≤0.001%
	Arsenic	≤0.0002%
	Cadmium	≤0.0001%
	Iron	≤0.002%
	Mercury	≤0.0001%
	Lead	≤0.0004%
Sodium dihydrogenophosphate	Chloride	≤0.005%
	Fluoride	≤0.001%
	Hydrogenphosphate (HPO_4)	≤0.5%
	Sulfate	≤0.01%
	Heavy metals	≤0.0005%
	Arsenic	≤0.0002%
	Cadmium	≤0.0001%
	Iron	≤0.001%
	Lead	≤0.0004%
	Mercury	≤0.0001%

[a] Residual solvents (see Section 3.5); Peroxide value: mmol peroxide/L.

3.5 RESIDUAL SOLVENTS

Residual solvents are volatile organic chemicals that remain in active substances, excipients and other pharmaceutical products after processing. Despite their toxic properties, solvents play an important role in the production of pharmaceuticals, during the synthesis, separation, or purification, and their use cannot be avoided. Solvents in this category do not include those used as excipients.

The ICH [75] has adopted Impurities Guidelines for Residual Solvents, which prescribes limits for the amount of residual solvent. They are divided into the following three classes: class 1 solvents are those known to cause toxic effects and should be avoided in the production of active substances and excipients; class 2 solvents present less severe toxicity than class 1; and class 3 solvents have such low toxic potentials that exposure limits are not necessary. Table 3.25 lists the general characteristics of the solvents in each class, and Table 3.26 lists the solvents and their concentration limit in pharmaceutical products.

Witschi et al. [76] reviewed this topic and listed several drugs and their residual solvents. Tables 3.27 and 3.28 list the results for encapsulated drugs and regular drugs, respectively. The data in Tables 3.27 and 3.28 show that most

TABLE 3.25 General Characteristics of Residual Solvents

Class	Action	Characteristics
1	Solvents to be avoided	Strongly suspected human carcinogens and environmental hazards
2	Solvents to be limited	Nongenotoxic animal carcinogens or possible causative agents of other irreversible toxicity such as neurotoxicity or teratogenicity
3	Solvents with low toxic potential	No health-based exposure limit is needed. Solvents with permitted daily exposure of 50 mg or less per day

According to ICH Guidelines [75].

TABLE 3.26 Solvents Included in Each Class and Their Concentration Limit in Pharmaceutical Products

Solvent	Concentration Limit (ppm)	Solvent
Class 1		Class 3
Benzene	2	Acetic acid
Carbon tetrachloride	4	Acetone
1,2-Dichloroethane	5	Anisole
1,1-Dichloroethane	8	1-Butanol
1,1,1-Trichloroethane	1500	2-Butanol
Class 2		Butyl acetate
Acetonitrile	410	tert-Butylmethyl ether
Chlorobenzene	360	Dimethyl sulfoxide
Chloroform	60	Ethanol
Cumene	70	Ethyl acetate
Cyclohexane	3880	Ethyl ether
1,2-Dichloroethene	1870	Ethyl formate
Dichloromethane	600	Formic acid
1,2-Dimethoxyethane	100	Heptane
N,N-Dimethylacetamide	1090	Isobutyl acetate
N,N-Dimethylformamide	880	Isopropyl acetate
1,4-Dioxan	380	Methyl acetate
2-Ethoxyethanol	160	3-Methyl-1-butanol
Ethylene glycol	620	Methyl ethyl ketone
Formamide	220	Methyl isobutyl ketone
Hexane	290	2-Methyl-1-propanol
Methanol	3000	Pentane
2-Methoxyethanol	50	1-Pentanol
Methyl butyl ketone	50	1-Propanol
Methylcyclohexane	1180	2-Propanol
N-methylpyrrolidone	530	Propyl acetate
Nitromethane	50	
Pyridine	200	
Sulfolane	160	
Tetrahydrofuran	720	
Tetralin	100	
Toluene	890	
1,1,2-Trichloroethene	80	
Xylene	2170	

According to ICH Guidelines [75].

TABLE 3.27 Residual Solvents in Encapsulated Drugs[a]

Product	Residual Solvent	Concentration (ppm)
Progesterone-loaded DL-PLA microspheres	Dichloromethane	18,000–47,000
Eudragit L	Isopropanol	44,000
Acetylsalicylic acid-loaded Eudragit RS100	Dichloromethane	n.d.
	mineral oil	≤1,000
DL-PLA and DL-PLGA microspheres	Silicone oil	2,000–5,000
Semisolid poly(ortoester)	Cyclohexanone	200
	Tetrahydrofuran	1,500
	Methanol	4,100
Ethyl cellulose	Dichloromethane	n.d. – 3,150
	Methanol	n.d.
	Acetone	n.d.
Ethyl cellulose and hydroxypropyl methylcellulose tablets	Methylene chloride	30–74
	Methanol	115–312 (per 760 mg tablet)
Cellulose acetophthalate tablets	Isopropanol	≤600–700
	Ethyl acetate	
Cisplatin-loaded DL-PLA and DL-PLGA microspheres	Dichloromethane	30,000
Cisplatin-loaded DL-PLA microspheres	Dichloromethane	260–37,500
BSA and drug-free DL-PLA and DL-PLGA microspheres	Dichloromethane	<100–43,200
	Ethyl acetate	20,700–84,800
	Hexane	24,900–55,100
Salmon calcitonin-loaded DL-PLGA microspheres	Dichloromethane	<10–129
TRH-loaded DL-PLGA microspheres	Dichloromethane	<20–3,100
	Acetonitrile	<20–4,100
Multilamellar liposomes with cisplatine derivative	*t*-Butyl alcohol	40–250
	Chloroform	40
Monomeric zinc phthalocyanine-loaded liposome	*t*-Butyl alcohol	30
	N-methyl pyrrolidone	40
PLGA film	Acetone	n.d. – 0.75%
Eudragit L100-55 film	Isopropanol	n.d. – 10%
Tetracosactide-loaded and drug-free PLA/PLGA microspheres	Dichloromethane	n.d. – 5,283
	Methanol	2.5–40
	Chloroform	934–6,000
Nitroglycerine-containing transdermal system	Petroleum spirit	

[a] Data compiled by Witschi et al. [76].
DL-PLA, DL- Poly(lactic acid); TRH, thyrotropin releasing hormone.

TABLE 3.28 Residual Solvents in Drugs and Raw Materials[a]

Product	Residual Solvent	Concentration (ppm)
Vegetable oils	Trichloroethylene	14–740
Soybean oil	Petroleum	n.d. – 1,000
Bacampicillin HCl	Butyl acetate	1,000–70,000
	Ethyl acetate	
Paracetamol	Dioxane	n.d. – 188
	Ethanol	1,495–2,501
Topiramate	Methanol	n.d.
	Ethanol	900
	Toluene	100
Antiangina API	*t*-Butyl methyl ether	1,500
	Toluene	n.d.
Triazolam API	Methyl ethyl ketone	n.d., 100–750
	Benzene	n.d., 6–30
	Isopropanol	n.d., 80–300
	Ethyl acetate	n.d. – 510
	Ethanol	n.d. – 100
Steroids	Acetone	210–10,760
CNS agents	Isooctane	330
Antibiotics	*n*-Butanol	800
	Diethyl ether	479–700
	Ethyl acetate	5,170
	Methanol	850
	Butyl acetate	110
Captopril tablets and raw material	Methanol	n.d. – 200
	Ethanol	n.d. – 2,600
	Acetone	n.d. – 2,200
	Butanol	n.d. – 90
	Dichloromethane	n.d. – 3,700
	2-Butanone	n.d. – 70
	Ethyl acetate	n.d. – 720
	Chloroform	n.d. – 3,150
	Butyl acetate	n.d. – 440
Phenobarbital	Ethanol	4,400
Lidocaine HCl	Acetone	2,700
Ca-pantothenate	Ethanol	1,100
Methyl nicotinate	Methanol	14.4
Na-ascorbate	Ethanol	127.1
Nicotinamide	Ethanol	52.8
Phenylbutazone	Methanol	50.9
	Ethanol	14.2
	Isopropanol	419.4
Streptomycin sulfate	Triethylamine	3,540–3,710
Methyl-acycline HCl	Triethylamine	1,170–1,250

(*Continued*)

TABLE 3.28 (*Continued*)

Product	Residual Solvent	Concentration (ppm)
Human gonadotrophin	Ethanol	6.67–16.0
Human chorionic gonadotrophin	Acetone	0.16–6.52
	Diethyl ether	0.6–14.2
Sulfamethazine	Acetone	0.2–1.0
Spironolactone	Benzene	<13
Steroid hormone	Ethyl acetate	160–660
	Ethanol	180–620
	Methyl ethyl ketone	1,050
	Acetone	120–2,020
	Methanol	150–620

[a] Data compiled by Witschi et al. [76].
CNS, central nervous system agents.

residual solvents belong to "class 3" and the limit of 0.5% was exceeded in only a few cases. In two products, benzene was found at levels above the permitted. The data also show that encapsulated drugs contain much higher levels of residual solvents than regular drugs. For example, 58% of the products listed in Table 3.27 contained at least one solvent with limits above the ICH compared to 28% in Table 3.28. One reason for this difference resides in the use of solvents for encapsulation. The process of encapsulation using poly(lactic-*co*-glycolic acid) (PLGA) microparticles includes emulsions, spray freezing, or spray-drying steps, which require the use of solvents. The most used is dichloromethane, which was found at highest levels in most encapsulated drugs. Although solvents are usual components in drug synthesis (encapsulated or not), solvent vaporization from polymeric materials is a more demanding task. Solvent evaporation from polymers is a process that depends on the volatility, vapor pressure, and diffusion coefficient of the solvent used. The internal diffusion resistance depends on the molecular branching of the solvent such that linear molecules have lower volatility but faster diffusion rate. Moreover, the removal of residual solvents from polymeric microparticles also depends on the polymer free volume: More compact matrices retain higher amounts of solvent.

Excipients may also be synthesized in the presence of organic solvents and may contain residuals. However, the analysis of typical pharmaceutical excipients revealed that they should not be considered as a source of residual solvents. The excipients listed in Table 3.29 did not present any class 1 solvent. Although methanol, ethyl acetate, and propanol were found in four products, their levels did not exceed those permitted by the ICH. However, del Barrio et al. [78] showed that formaldehyde and formic acid (only formic acid is enrolled in the ICH guideline as class 3 solvent) may be found in excipients. The data in Table 3.30 show that excipients are more contaminated with formic

TABLE 3.29 Residual Solvents in Typical Pharmaceutical Excipients [77]

Excipient	Class 1 Residual Solvent	Class 2 and 3 Residual Solvents	Amount (ppm)
Butylated hydroxyanisole	None	Methanol	140
Butylparaben sodium	None	None	
Carbomer	None	Ethyl acetate	1500
Citric acid	None	None	
Hydroxypropylcellulose	None	None	
Lactose	None	None	
Mannitol	None	None	
Microcrystalline cellulose	None	None	
Magnesium stearate	None	None	
Polyethylene glycol	None	None	
Poloxamer	None	None	
Polysorbates	None	None	
Propyl gallate	None	1-Propanol	80
Sucrose	None	None	
Tartaric acid	None	None	
Tromethamine	None	Methanol	200
		Acetone	Traces

acid than with formaldehyde. While almost all excipients contained formic acid above 1 ppm, the level of formaldehyde was below this value in the majority of the products. All samples of povidone contained significantly higher levels of formic acid compared with those of the other excipients analyzed. High levels of formic acid and formaldehyde were also found in polyethylene glycol 400 and hydroxypropyl methylcellulose.

CONCLUDING REMARKS

Inorganic impurities may naturally occur in nonsynthetic raw materials due to similar chemical behavior among the species. Inorganic impurities are also associated with the use of metallic catalysts in the synthesis of APIs. Usually, these impurities are transition metals with recognized toxicity. Because their presence is a matter of concern, regulatory agencies and pharmacopeias conceived limits for these metals in APIs.

Organic impurities are synthesized along with active ingredients and are generally difficult to separate due to their similarity to the molecule of interest. Most are isomeric species, differing from each other by the presence of only a small functional group or the position of a hydrogen atom. GTIs are among the organic impurities that should be strictly controlled. GTIs may be generated by the use of alkylating agents in synthetic routes leading to APIs. Problems occur when these agents remain in the final drug substance or drug

TABLE 3.30 Residuals of Formic Acid and Formaldehyde in Excipients [78]

Excipient	Manufacturer	Lot	Formic Acid (ppm)	Formaldehyde (ppm)
Lactose	A	1	1.0	<0.2
Powdered cellulose	B	1	2.9	0.4
Mannitol	C	1	0.0	0.0
Microcrystalline cellulose 50 μm	D	1	9.3	<0.2
Microcrystalline cellulose 50 μm	D	2	23.9	0.9
Microcrystalline cellulose 50 μm	D	3	11.8	1.0
Microcrystalline cellulose 50 μm	D	4	4.0	0.3
Starch 1500	E	1	3.0	<0.2
Starch	F	1	2.4	0.0
Starch	F	2	5.1	0.0
Dibasic calcium phosphate	G	1	1.5	0.0
Povidone K-25	H	1	3080.3	<0.2
Povidone K-25	I	1	1990.5	0.4
Povidone 90	H	2	630.7	0.4
Hydroxypropyl methylcellulose	J	1	58.3	11.1
Hydroxypropyl methylcellulose	J	2	86.4	15.7
Polyplasdone XL	I	1	10.8	0.0
Croscarmellose sodium	D	5	26.3	0.0
Sodium starch glycolate	G	2	1.9	0.9
Magnesium stearate	K	1	2.1	1.1
Stearic acid	L	1	0.0	0.0
Glycerol dibehenate	M	1	4.3	2.1
Polyethylene glycol 3500	N	1	2.3	0.3
Polyethylene glycol 3500	N	2	1.7	1.2
Polyethylene glycol 4000	O	1	1.7	0.9
Polyethylene glycol 4000	N	3	14.0	3.6
Polyethylene glycol 400	O	2	469.0	85.8
Sodium dodecyl sulfate	p	1	0.0	0.0

products. Organic impurities may also be generated by the decomposition of ingredients within a formulation. The main routes for drug substance decomposition are hydrolysis and oxidation. Notably, the most relevant contamination arising from the oxidation of formulation constituents is the generation of peroxides in vitamins, amino acids, and lipid emulsions. Peroxidation and free radical formation depend on environmental factors, such as incidence of light, storage conditions, and container material. However, formulation components or additives also influence them.

Although strictly regulated, residual solvents are potential contaminants in drug substances due to the limitations in solvent vaporization during drug manufacturing.

REFERENCES

1. World Health Organization. *Guidelines for Drinking-Water Quality*, Vol. 1, Recommendations, 3rd ed. World Health Organization, Geneva, 2008.
2. Water Programs, Environmental Protection Agency, National Interim Primary Drinking Water Regulations. USA 1985, Code of Federal Regulations, Part 141.
3. ASTM International. *ASTM Book of Standards*, Volume 11.02: *Water and Environmental Technology: Water (II)*. ASTM International, West Conshohocken, PA, 2005.
4. USP 30, *United States Pharmacopeia*, US Pharmacopeial Convention, Rockville, MD, 2007.
5. BP, British Pharmacopoeia Commission, London UK © Crown Copyright, 2003.
6. DAB, *Deutsche Arzneibuch*, Deutschen Apotheker Verlag, Stuttgart, 2005.
7. Ph. Eur., *European Pharmacopoeia*, 4th ed. European Directorate for the Quality of Medicines (EDQM), Strasbourg, 2004.
8. Ph. Int., *The International Pharmacopoeia*. 3rd ed. World Health Organization , Geneva, 2003.
9. Küster FW, Thiel A. *Rechentafeln für die Chemische Analytik*. Walter de Gruyter, Berlin, 1985, p. 258.
10. Schriks M, Heringa MB, van der Kooi MM, de Voogt P, van Wezel AP. Toxicological relevance of emerging contaminants for drinking water quality. Water Res 44(2), 461–476, 2010.
11. Koo WW, Kaplan LA, Horn J, Tsang R, Steichen J. Aluminum in parenteral nutrition solutions—sources and possible alternatives. JPEN J Parenter Enteral Nutr 10, 591–595, 1986.
12. Recknagel S, Brätter P, Chrissafidou A, Gramm H-J, Kotwas J, Rösick U. Parenteral aluminum loading in critical care medicine. Part I: aluminum content of infusion solutions and solutions for parenteral nutrition. Infusionsther Transfusionsmed 21, 266–273, 1994.
13. Fleming LW, Halls DJ, Henderson IS, Mactier RA, Taylor JE, Stewart WK. Aluminium content of water for injection used with recombinant human erythropoietin. Nephrol Dial Transplant 6, 857–861, 1991.
14. Bohrer D, do Nascimento PC, Becker E, de Carvalho LM, Dessuy M. Arsenic species in solutions for parenteral nutrition. JPEN J Parenter Enteral Nutr 29, 1–7, 2005.
15. Pluhator-Murton MM, Fedorak RN, Audette PJ, Marriage BJ, Yatscoff RW. Extent of trace-element contamination from simulated compounding of total parenteral nutrient solutions. Am J Health Syst Pharm 53, 2299–2303, 1996.
16. Bohrer D, Bortoluzzi F, Nascimento PC, Carvalho LM, Ramirez AG. Silicate release from glass for pharmaceutical preparations. Int J Pharm 355(1–2), 174–183, 2008.

17. European Medicines Agency (EMEA). Guidelines on the Specification Limits from Residues of Metal Catalysts and Metal Reagents (EMEA/CHMP/SWP/4446/2000—21 Feb. 2008).
18. USP, Pharmacopeial Forum, Elemental Impurities-Limits, Vol. 36(1), 2010.
19. Lasztity A, Kelko-Levai A, Varga I, Zih-Perenyi K, Bertalan E. Development of atomic spectrometric methods for trace metal analysis of pharmaceuticals. Microchem J 73(1-2), 59–63, 2002.
20. Wang T, Walden S, Egan R. Development and validation of a general non-digestive method for the determination of palladium in bulk pharmaceutical chemicals and their synthetic intermediates by graphite furnace atomic absorption spectroscopy. J Pharm Biomed Anal 15(5), 593–599, 1997.
21. Ponce De León CA, Montes Bayón M, Caruso JA. Trace element determination in vitamin E using ICP-MS. Anal Bioanal Chem 374(2), 230–234, 2002.
22. Marguí E, Van Meel K, Van Grieken R, Buendía A, Fontàs C, Hidalgo M, Queralt I. Method for the determination of Pd-catalyst residues in active pharmaceutical ingredients by means of high-energy polarized-beam energy dispersive X-ray fluorescence. Anal Chem 81(4), 1404–1410, 2009.
23. Marguí E, Fontas C, Buendia A, Hidalgo M, Queralt I. Determination of metal residues in active pharmaceutical ingredients according to European current legislation by using X-ray fluorescence spectrometry. J Anal Atom Spectrom 24(9), 1253–1257, 2009.
24. Krone CA, Wyse EJ, Ely JT. Cadmium in zinc-containing mineral supplements. Int J Food Sci Nutr 52(4), 379–382, 2001.
25. Bourgoin BP, Boomer D, Powell MJ, Willie S, Edgar D, Evans D. Instrumental comparison for the determination of cadmium and lead in calcium supplements and other calcium-rich matrices. Analyst 117(1), 19–22, 1992.
26. Dolan SP, Nortrup DA, Bolger PM, Capar SG. Analysis of dietary supplements for arsenic, cadmium, mercury, and lead using inductively coupled plasma mass spectrometry. J Agric Food Chem 51(5), 1307–1312, 2003.
27. Bourgoin BP, Evans DR, Cornett JR, Lingard SM, Quattrone AJ. Lead content in 70 brands of dietary calcium supplements. Am J Public Health 83(8), 1155–1160, 1993.
28. Bohrer D, do Nascimento PC, Binotto R, Becker E, Pomblum SG. Contribution of the raw material to the aluminum contamination in parenterals. JPEN J Parenter Enteral Nutr 26, 382–388, 2002.
29. Bohrer D, Becker E, do Nascimento PC, de Carvalho LM, Marques MS. Arsenic release from glass containers by action of intravenous nutrition formulation constituents. Int J Pharm 315, 24–29, 2006.
30. International Conference on Harmonisation—ICH Q3A—Impurities in New Drug Substances, 2008.
31. International Conference on Harmonisation—ICH Q3B—Impurities in New Drug Products, 2006.
32. Görög S, Babják M, Balogh G, Brlik J, Dravecz F, Gazdag M, Horváth P, Laukó A, Varga K. Estimation of impurity profiles of drugs and related materials. Part 19: theme with variations. Identification of impurities in 3-oxosteroids. J Pharm Biomed Anal 18(4–5), 511–525, 1998.

33. Gazdag M, Babják M, Brlik J, Mahó S, Tuba Z, Görög S. Estimation of impurity profiles of drugs and related materials. Part 18: impurities and degradation products of mazipredone. J Pharm Biomed Anal 17(6–7), 1029–1036, 1998.
34. Klaassen CD. *Toxicology: The Basic Science of Poisons*, 5th ed. McGraw-Hill, New York, 1996.
35. European Medicines Agency. Guideline on the Limits of Genotoxic Impurities (EMEA/CHMP/QWP/251344/2006), London, 2006.
36. FDA. Genotoxic and Carcinogenic Impurities in Drug Substances and Products: Recommended Approaches. FDA: The Food and Drug Administration (FDA), Center for Drug Evaluation and Research (CDER), Silver Spring, 2008.
37. Müller L, Mauthe RJ, Riley CM, Andino MM, Antonis DD, Beels C, DeGeorge J, De Knaep AG, Ellison D, Fagerland JA, Frank R, Fritschel B, Galloway S, Harpur E, Humfrey CD, Jacks AS, Jagota N, Mackinnon J, Mohan G, Ness DK, O'Donovan MR, Smith MD, Vudathala G, Yotti L. A rationale for determining, testing, and controlling specific impurities in pharmaceuticals that possess potential for genotoxicity. Regul Toxicol Pharmacol 44(3), 198–211, 2006.
38. Dobo KL, Greene N, Cyr MO, Caron S, Ku WW. The application of structure-based assessment to support safety and chemistry diligence to manage genotoxic impurities in active pharmaceutical ingredients during drug development. Regul Toxicol Pharmacol 44(3), 282–293, 2006.
39. MultiCASE Inc Softwares. MC4PC version v 2.00, Beachwood, OH. Available at http://www.multicase.com/products/prod01.htm.
40. Lhasa Limited Software & Services, Derek Nexus. Available at https://www.lhasalimited.org/derek_nexus, Leeds, UK.
41. Gerber C, Toelle HG. What happened: the chemistry side of the incident with EMS contamination in Viracept tablets. Toxicol Lett 190(3), 248–253, 2009.
42. Solomons TWG, Fryhle CB. *Organic Chemistry*, 8th ed. Wiley & Sons, Hoboken, NJ, 2003.
43. Elder DP, Teasdale A, Lipczynski AM. Control and analysis of alkyl esters of alkyl and aryl sulfonic acids in novel active pharmaceutical ingredients (APIs). J Pharm Biomed Anal 46(1), 1–8, 2008.
44. Alzaga R, Ryan RW, Taylor-Worth K, Lipczynski AM, Szucs R, Sandra P. A generic approach for the determination of residues of alkylating agents in active pharmaceutical ingredients by in situ derivatization-headspace-gas chromatography-mass spectrometry. J Pharm Biomed Anal 45(3), 472–479, 2007.
45. Ramjit HG, Singh MM, Coddington AB. Gas chromatographic/mass spectrometric analysis of methyl methanesulphonate and ethyl methanesulphonate in the bismesylate salt of DPI 201-106, a positive inotropic agent for the treatment of heart failure. J Mass Spectrom 31(8), 867–872, 1996.
46. Lee CR, Guivarch F, Van Dau CN, Tessier D, Krstulovic AM. Determination of polar alkylating agents as thiocyanate/isothiocyanate derivatives by reaction headspace gas chromatography. Analyst 128(7), 857–863, 2003.
47. Colón I, Richoll SM. Determination of methyl and ethyl esters of methanesulfonic, benzenesulfonic and p-toluenesulfonic acids in active pharmaceutical ingredients by solid-phase microextraction (SPME) coupled to GC/SIM-MS. J Pharm Biomed Anal 39(3–4), 477–485, 2005.

48. Kakadiya PR, Reddy BP, Singh V, Ganguly S, Chandrashekhar TG, Singh DK. Low level determinations of methyl methanesulfonate and ethyl methanesulfonate impurities in Lopinavir and Ritonavir active pharmaceutical ingredients by LC/MS/MS using electrospray ionization. J Pharm Biomed Anal 55(2), 379–384, 2011.

49. Jagota NK, Nair JB, Kurtulik PT. Ion chromatography of methanesulfonic acid in pharmaceuticals. J Pharm Biomed Anal 13(10), 1291–1295, 1995.

50. Ahmad AK, Kawy MA, Nebsen M. First derivative ratio spectrophotometric, HPTLC-densitometric, and HPLC determination of nicergoline in presence of its hydrolysis—induced degradation product. J Pharm Biomed Anal 30(3), 479–489, 2002.

51. Archontaki HA, Panderi IE, Gikas EE, Parissi-Poulou M. Kinetic study on the degradation of prazepam in acidic aqueous solutions by high-performance liquid chromatography and fourth-order derivative ultraviolet spectrophotometry. J Pharm Biomed Anal 17(4–5), 739–750, 1998.

52. Neuzil J, Darlow BA, Inder TE, Sluis KB, Winterbourn CC, Stocker R. Oxidation of parenteral lipid emulsion by ambient and phototherapy lights: potential toxicity of routine parenteral feeding. J Pediatr 126, 785–790, 1995.

53. Steger PJK, Mühlebach SF. In vitro oxidation of IV lipid emulsions in different all-in-one admixture bags assessed by an iodometric assay and gas-liquid chromatography. Nutrition 13, 133–140, 1997.

54. Murray RK, Granner DK, Mayes PA, Rodwe VW. *Harper's Biochemistry*, 25th ed. McGraw-Hill, New York, 2000, p. 169.

55. Helbock HJ, Motchnik PA, Ames BN. Toxic hydroperoxides in intravenous lipid emulsions used in preterm infants. Pediatrics 91, 83–87, 1993.

56. Picaud JC, Steghens JP, Auxenfans C, Barbieux A, Laborie S, Claris O. Lipid peroxidation assessment by malondialdehyde measurement in parenteral nutrition solutions for newborn infants: a pilot study. Acta Paediatr 93, 241–245, 2004.

57. Pironi L, Guidetti M, Zolezzi C, Fasano MC, Paganelli F, Merli C, Bersani G, Pizzoferrato A, Miglioni M. Peroxidation potential of lipid emulsions after compounding in all-in-one solutions. Nutrition 19, 784–788, 2003.

58. Steger PJK, Mühlebach SF. Lipid peroxidatoin of IV lipid emulsions in TPN bags: the influence of tocopherols. Nutrition 14, 179–185, 1998.

59. Burton GW, Ingold KU. Vitamin E as an in vitro and in vivo antioxidant. Ann N Y Acad Sci 570, 7–22, 1986.

60. Bowry VW, Ingold KU, Stocker R. Vitamin E in human low-density lipoprotein. When and how this antioxidant becomes a pro-oxidant. Biochem J 288, 341–344, 1992.

61. Bowry VW, Stocker R. Tocopherol-mediated peroxidation. The pro-oxidant effect of Vitamin E on the radical-initiated oxidation of human low-density lipoprotein. J Am Chem Soc 115, 6029–6040, 1993.

62. Steger PJK, Mühlebach SF. Lipid peroxidation of intravenous lipid emulsions and all-in-one admixtures in total parenteral nutrition bags: the influence of trace elements. JPEN J Parenter Enteral Nutr 24, 37–41, 2000.

63. Laborie S, Lavoie J-C, Rouleau T, Chessex P. Multivitamin solutions for enteral supplementation: a source of peroxides. Nutrition 18, 470–473, 2002.

64. Laborie S, Lavoie J-C, Pineaut M, Chessex P. Contribution of multivitamins, air and light in the generation of peroxides in adult and neonatal parenteral nutrition solutions. Ann Pharmacother 34, 440–445, 2000.
65. Helbock HJ, Motchnick PA, Ames BN. Toxic hydroperoxides in intravenous lipid emulsions used in preterm infants. Pediatrics 91, 83–88, 1993.
66. Lavoie J-C, Bélanger S, Spalinger M, Chessex P. Admixture of a multivitamin preparation to parenteral nutrition: the major contributor to in vitro generation of peroxides. Pediatrics 99, 61–70, 1997.
67. Lavoie J-C, Chessex P, Rouleau T, Migneault D, Comte B. Light-induced by products of vitamin C in multivitamin solutions. Clin Chem 50, 135–140, 2004.
68. Knafo L, Chessex P, Rouleau T, Lavoie J-C. Association between hydrogen peroxide-dependent by products of ascorbic acid and increased hepatic acetyl-CoA carboxylase activity. Clin Chem 51, 1462–1471, 2005.
69. Balet A, Cardona D, Jane S, Molins-Pujol AM, Sanchez-Quesada JL, Gich I, Mangues MA. Effects of multilayered bags vs. ethylvinyl-acetate bags on oxidation of parenteral nutrition. JPEN J Parenter Enteral Nutr 28, 85–91, 2004.
70. Baker M, Gregerson MS, Martin SM, Buettner GR. Free radical and drug oxidation products in an intensive care unit sedative: propofol with sulfite. Crit Care Med 31, 787–792, 2003.
71. Lavoie J-C, Lachance C, Chessex P. Antiperoxide activity of sodium metabisulfite. A double-edged sword. Biochem Pharmacol 47, 871–876, 1994.
72. Hayon E, Treinin A, Wilf J. Electronic spectra, photochemistry, and autoxidation: mechanism of the sulfite-bisulfite-pyrosulfite systems. The SO_2^-, SO_3^-, SO_4^- radicals. J Am Chem Soc 94, 47–57, 1972.
73. Thompson JE, Davidow L. *A Practical Guide to Contemporary Pharmacy Practice*. Lippincott Williams & Wilkins, Baltimore, 2003.
74. Wade A, Weller PJ. *Handbook of Excipients*, 2nd ed. The Pharmaceutical Press, London, 1994.
75. International Conference on Harmonisation—ICH Q3C(R5)—Impurities: Guidelines for Residual Solvents, 2010.
76. Witschi C, Doelker E. Residual solvents in pharmaceutical products: acceptable limits, influences on physicochemical properties, analytical methods and documented values. Eur J Pharm Biopharm 43(3), 215–242, 1997.
77. Fliszar K, Wiggins JM, Pignoli CM, Martin GP, Li Z. Analysis of organic volatile impurities in pharmaceutical excipients by static headspace capillary gas chromatography. J Chromatogr A 1027(1–2), 83–91, 2004.
78. del Barrio MA, Hu J, Zhou P, Cauchon N. Simultaneous determination of formic acid and formaldehyde in pharmaceutical excipients using headspace GC/MS. J Pharm Biomed Anal 41(3), 738–743, 2006.

4

MEDICINAL GASES AND VOLATILE ANESTHETICS

4.1 MEDICINAL GASES

Medicinal gases are classified as medicinal products and are included in pharmacopoeia compendia. Table 4.1 displays pharmacopeial purity specifications for medicinal gases and Table 4.2 shows the level of impurities in commercial gases for medicinal use. Comparing Tables 4.1 and 4.2, it can be seen that, in general, commercial gases fulfill compendia specifications, although some important items, such as hydrogen sulfide, are not mentioned by either of the suppliers listed in Table 4.2.

Medicinal gases and their impurities can by divided into two categories: gases that, even though presenting a helpful action, are toxic at some level and gases that pose no toxic hazards but behave as simple asphyxiants. In Table 4.3, gases are classified according to their toxicological data: the LC_{50}[1] and the Hazardous Materials Identification System (HMIS) number, which is a standardization of chemical compounds through a series of ratings (0–4). The items covered by this system are health, flammability, and physical hazards. The ratings are "0," no hazards; "1," slight; "2," moderate; "3," serious; and "4," severe. The ratings presented in Table 4.3 are related to effects on health.

[1] LC_{50} lethal concentration for gases, which kills 50% of the exposed animals.

Sources of Contamination in Medicinal Products and Medical Devices,
First Edition. Denise Bohrer.
© 2013 by John Wiley & Sons, Inc. Published 2013 by John Wiley & Sons, Inc.

TABLE 4.1 Pharmacopeial Specification of Gases for Medicinal Use

	Specifications	Medical Air		Synthetic Air	Oxygen O_2		Nitrogen N_2	Carbon Dioxide CO_2		Helium He		Nitrous Oxide N_2O		Nitric Oxide NO
	Formula	USP[a]	BP	BP	USP	BP	BP	USP	BP	USP	BP	USP	BP	BP
1	Purity (%) (v/v)			95.0–105.0	≥99.0	≥99.5	≥99.5	≥99.0	≥99.5	≥99.0	≥99.5	≥99.0	≥98.0	≥99.0
2	Water (vapor) ppm (v/v)	–	≤67	≤67	–	≤67	≤67	≤150 mg/m^3	≤67	–	≤67	≤150 mg/m^3	≤67	≤100
3	Oxygen (%) (v/v)	19.5–23.5	20.4–21.4	21.0–22.5	≥99.0	≥99.5	≤50 ppm	–	–	–	≤50 ppm	–	–	–
4	Air (%)	–	–	–	–	–	–	–	–	1.0%	–	1.0%	–	–
5	Carbon dioxide ppm (v/v)	≤500	≤500	–	0.03%	≤300	≤300	–	–	–	–	0.03%	≤300	≤3000
6	Carbon monoxide ppm (v/v)	≤10	≤5	–	0.001%	≤5	≤5	0.001%	≤5	0.001%	–	0.001%	≤5	–
7	Nitrogen (%) (v/v)	–	–	–	–	–	≥99.5	–	–	–	–	–	–	≤3000 ppm
8	Nitrogen oxide and nitrogen dioxide ppm (v/v)	≤2.5	≤2	–	–	–	–	≤2.5	≤2	–	–	≤1	≤2	≤400
9	Nitric oxide ppm (v/v)	–	–	–	–	–	–	≤2.5	–	–	–	≤1	–	–
10	Nitrous oxide ppm (v/v)	–	–	–	–	–	–	–	–	–	–	–	–	≤3000

(Continued)

TABLE 4.1 (*Continued*)

	Specifications	Medical Air		Synthetic Air	Oxygen		Nitrogen	Carbon Dioxide		Helium		Nitrous Oxide		Nitric Oxide
					O_2		N_2	CO_2		He		N_2O		NO
	Formula	USP[a]	BP	BP	USP	BP	BP	USP	BP	USP	BP	USP	BP	BP
11	Sulfur dioxide ppm (v/v)	≤5	≤1	–	–	–	–	≤5	≤2	–	–	–	–	–
12	Hydrogen sulfide ppm (v/v)	–		–	–	–	–	≤1	≤1	–	–	–	–	–
13	Carbonyl sulfide ppm (v/v)	–	–	–	–	–	–	–	–	–	–	–	–	–
14	Halogens ppm (v/v)	–	–	–	–	–	–	–	–	–	–	≤1	–	–
15	Methane ppm (v/v)	–	–	–	–	–	–	–	–	–	≤50	–	–	–
16	Ammonia ppm (v/v)	–	–	–	–	–	–	0.0025%	–	–	–	0.0025%	–	–
17	Nonvolatile residues ppm (m/m)	–	–	–	–	–	–	–	–	–	–	–	–	–
18	Inerts ppv (v/v)	–	–	–	–	–	–	–	–	–	–	–	–	–
19	Odor	n.d.	n.d.	n.d.	n.d.	n.d.	n.d.	–	–	n.d.	–	–	–	–
20	Oil (mg/m^3)	n.d.	0.1	–	–	–	–	–	–	–	–	–	–	–

[a] USP considers natural or synthetic air as a medical air.

n.d., not detectable.

TABLE 4.2 Specification of Commercial Gases for Medicinal Use

	Specifications[a]	Compressed Air	Synthetic Air	Oxygen	Nitrogen	Carbon Dioxide	Helium	Nitrous Oxide	Nitric Oxide
	Formula			O_2	N_2	CO_2	He	N_2O	NO
1	Purity (%)			≥99.5	≥99.5	99.9	≥99.5	99.0	99.9
2	Water (vapor) ppm (v/v)	≤67	≤67	≤67	≤5	32	≤67	150	<20
3	Oxygen ppm (v/v)	20.4–21.4	20.0–21.7		≤5	30	≤50	–	–
4	Air (%)	–	–	–	–	–	–	–	–
5	Carbon dioxide ppm (v/v)	≤500	≤500	≤300	≤300	–	–	300	<50
6	Carbon monoxide ppm (v/v)	≤5	≤10	≤5	≤5	10	≤10	10	–
7	Nitrogen	q.s.	q.s.	–	–	–	–	–	–
8	Nitrogen oxide ppm (v/v)	–	–	–	–	–	–	1	–
9	Nitrogen dioxide ppm (v/v)	≤2	≤2.5	–	–	5	–	1	<100
10	Nitrous oxide ppm (v/v)	–	–	–	–	–	–	–	<500
11	Sulfur dioxide ppm (v/v)	≤1	≤5	–	–	5	–	–	–
12	Hydrogen sulfide ppm (v/v)	–	–	–	–	–	–	–	–
13	Carbonyl sulfide ppm (v/v)	–	–	–	–	0.5	–	–	–
14	Halogens ppm (v/v)	–	–	–	–	–	–	–	–
15	Total hydrocarbons (as methane) ppm (v/v)	–	–	–	–	50	≤50	–	–
16	Ammonia ppm (v/v)	–	–	–	–	25	–	25	–
17	Nonvolatile residues ppm (m/m)	–	–	–	–	10	–	–	–
18	Inerts ppm (v/v)	–	–	–	–	1000	–	–	–
19	Odor	n.d.	n.d.	n.d.	n.d.	–	n.d.	–	–
20	Oil (mg/m^3)	≤0.1	≤0.1	–	–	–	–	–	–

[a] *Source*: Airgas, Spectra, and Linde Gases.

n.d., not detectable; q.s., quantum satis.

TABLE 4.3 Toxicity of Impurities Set in Pharmacopeial Compendia for Medicinal Gases

Toxic Gases			Nontoxic Gases		
Gas	HMIS[a]	LC_{50} Rat	Gas	HMIS[a]	Features
Carbon monoxide	2	3760 ppm/1 h	Oxygen	0	Toxic when pure
Nitric oxide	3	870 ppm/4 h	Nitrogen	0	–
Nitrogen dioxide	4	115 ppm/1 h	Carbon dioxide	0	LC_{50} rat >5000 ppm
Sulfur dioxide	3	2520 ppm/1 h	Nitrous oxide	1	–
Hydrogen sulfide	4	712 ppm/1 h	Methane	0	–
Ammonia	3	9500 ppm/1 h			

[a] Hazardous Materials Identification System (HMIS) health hazard classification: 0, no hazards; 1, slight; 2, moderate; 3, serious; and 4, severe.

Nitrogen dioxide and hydrogen sulfide are the most dangerous impurities with LC_{50} of 115 and 712 ppm for 1 hour of exposure, respectively, and a rating of 4 in HMIS.

Although *hydrogen sulfide* (H_2S) should be controlled in carbon dioxide (CO_2), none of the suppliers included in Table 4.2 presented H_2S limits for CO_2. By contrast, limits for *nitrogen dioxide* (NO_2), which should be controlled in medical air, CO_2, nitrous oxide (N_2O), and nitric oxide (NO) were, in fact, provided by all of the suppliers. Nitrogen dioxide may be present in both nitrous oxide (N_2O) and nitric oxide (NO) due to the facile transformation of one nitrogen oxide into another. It is present in compressed air and in CO_2 as a consequence of atmospheric pollution. NO_2 and NO are included together in the limits of compressed air in both the United States Pharmacopeia (USP) and the British Pharmacopoeia (BP).

Sulfur dioxide (SO_2) (HMIS rating 3) and *carbon monoxide* (CO) (HMIS rating 2) are also toxic contaminants that must be controlled in compressed air and CO_2 due to their presence in the atmospheric air as a consequence of the combustion of fuels. CO has to be additionally controlled in oxygen (O_2), helium (He), nitrogen (N_2), and nitrogen oxides. *Ammonia* (NH_3), although presenting the highest LC_{50}, is rated as 3 in the HMIS. Ammonia must be controlled in CO_2 and N_2O according to the USP. CO_2 as an impurity in the other gases and *methane* in He do not pose any toxic hazards.

Nitric oxide (NO) is an approved drug for the treatment of persistent pulmonary hypertension in full-term infants and presents potent antimicrobial activity and healing efficacy in the treatment of wounds despite its toxicity (HMIS rating 3). NO is corrosive and oxidizes in the presence of oxygen. NO is thermodynamically unstable and spontaneously decomposes into NO_2 and

TABLE 4.4 Characteristics of Nitrogen Oxides

Compound	Formula	Nitrogen Oxidation State	Features
Nitrous oxide	N_2O	+I	Gas, stable, colorless, neutral
Nitric oxide	NO	+II	Gas, unstable, colorless, neutral
Nitrogen sesquioxide	N_2O_3	+III	Liquid, stable (at low temperature), blue, acidic, reacts with water to form HNO_3 and NO
Nitrogen dioxide	NO_2	+IV	Gas, stable, red-brown, acidic
Dinitrogen pentoxide	N_2O_5	+V	Solid, unstable, colorless, acidic

N_2O when compressed (see equations below). Therefore, NO is susceptible to being contaminated with other nitrogen oxides.

Nitrous oxide (N_2O) has been used for anesthesia in dentistry and in hospitals. Inhalation of N_2O is frequently used to relieve pain associated with childbirth, trauma, oral surgery, and acute coronary syndrome. Nitrous oxide, although more stable than NO, may decompose explosively at high temperatures (650°C) into nitrogen and oxygen. Metallic surfaces, such as silver, copper, or nickel, act as catalysts causing the reaction to occur at lower temperatures.

Nitrous oxide is prepared by careful thermal decomposition of molten ammonium nitrate at about 170–240°C:

$$NH_4NO_3 \rightarrow N_2O + H_2O. \tag{4.1}$$

Nitrogen forms five different oxides with different oxidation states, which present different degrees of stability and toxicity. Table 4.4 [1] presents the general properties of nitrogen oxides.

The reactions involved in the transformation of one nitrogen oxide into another are presented below.

Nitric oxide reacts with oxygen to give NO_2 or nitrogen sesquioxide (N_2O_3) depending on the gas ratio. The reaction occurs instantly with pure oxygen (100%) and at a lower rate with atmospheric oxygen (21%):

$$2NO + O_2 \rightarrow 2NO_2 \tag{4.2}$$

$$4NO + O_2 \rightarrow 2N_2O_3. \tag{4.3}$$

Nitrogen sesquioxide (N_2O_3) is also formed by reacting NO and NO_2:

$$NO + NO_2 \rightarrow 2N_2O_3. \tag{4.4}$$

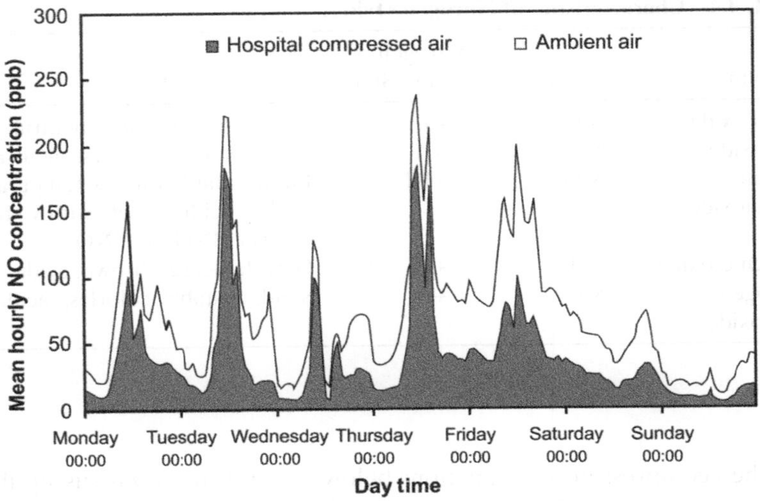

FIGURE 4.1 One-week nitrogen oxide monitoring in the compressed air delivered to patients and in the ambient air outside the hospital (300 m away) [2].

Nitrogen dioxide dimerizes to N_2O_4, which reacts with water to give a mixture of nitric and nitrous acids:

$$2NO_2 \leftrightarrow N_2O_4 \tag{4.5}$$

$$N_2O_4 + H_2O \rightarrow HNO_3 + HNO_2. \tag{4.6}$$

Therefore, the presence of water vapor in any of the nitrogen medicinal gases may lead to the formation of nitric acid and nitrous acid.

Compressed air is probably the medicinal gas most susceptible to contamination since it is drawn from the environment. In hospitals, compressed air is usually collected on-site using a compressor–reservoir system that draws fresh air from the local environment and delivers it to the patients.

Figure 4.1 displays the concentration of NO measured in the hospital's compressed air and in an environmental monitoring station 300 m away from the hospital. It is clear that the level of NO in the air delivered to patients is a reflection of its concentration in the environment. During the period of measurement, variations in the concentration of NO in the environmental air corresponded to the variations in the NO concentration in the hospital compressed air.

Nitric oxide and NO_2 have been found in compressed air and in oxygen mixtures in hospitals in different countries (Table 4.5). Although not exceeding the levels prescribed by pharmacopeial monographs, all samples presented NO or NO_2 at some level. In most cases, the authors correlated the levels of

TABLE 4.5 Nitric Oxide and Nitrogen Dioxide Concentration in Gas or Gas Mixtures Delivered to Patients in Different Hospitals

Impurity	Concentration (ppb) Compressed Air	Synthetic Air	N_2O	N_2	O_2	Reference
NO						
	<10–1270	–	–	–	–	[2]
	70	–	–	–	–	[3]
	2–550	–	–	n.d.	n.d.	[4]
	13–79	–	–	–	–	[5]
	12.4 ± 0.5	–	–	–	–	[6]
	5–550	–	–	n.d.	n.d.	[7]
	5.6	1.5	5.0	–	–	[8]
NO_2						
	<10–670	–	–	–	–	[2]

n.d., not detected.

NO in compressed air to environmental NO levels. Benzing et al. [2] observed that the highest levels occurred on workdays between 9 a.m. and 4 p.m. (Fig. 4.1), and Pinsky et al. [3] reported that inadvertent contamination occurred during a welding activity near the hospital intake port for fresh gas. The alteration in the NO level in compressed air due to this activity can easily be identified in the graph in Figure 4.2. Over 5 days, the NO concentration intermittently reached 5000–6000 ppb (corresponding to daytime) in the compressed air. The graph also shows that during weekends, when traffic and industrial activities were reduced, the level of NO was reduced.

Because compressed air delivered to the patients is a sample of atmospheric air, impurities existing in the environment will be inhaled by patients if no measures for its purification are taken. Table 4.6 lists the 33 relevant World Health Organization (WHO) air pollutants and their atmospheric concentrations in nonurban regions (background), in urban regions, and near industrial sites. Of course, the more industrialized the region, the heavier the pollution.

Air pollutants are classified according to their sources: Primary air pollutants are those that are emitted into the atmosphere from a variety of sources, and secondary air pollutants are those formed within the atmosphere itself through reactions of primary pollutants with natural components of the atmosphere [9]. The most important example of a secondary air pollutant is ozone.

Among the compounds listed in Table 4.6, the most relevant are NO_2, O_3, SO_2, and CO, whose levels in urban regions vary from 20 to more than 20,000 μg/m^3, although others such as benzene, formaldehyde, and toluene, whose toxicity is high, may be found in concentrations as high as 100 μg/m^3.

NO_2, O_3, and SO_2 are considered classical air pollutants, and their environmental concentration and heath effects have been extensively evaluated. Table

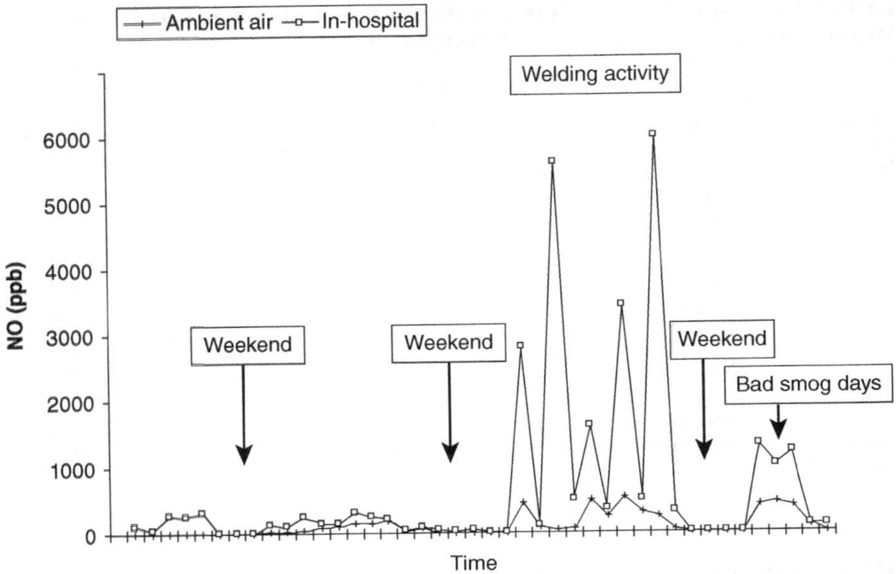

FIGURE 4.2 Relationship between time of day and NO levels in ambient air and in hospital compressed air. Twice daily NO levels for the 21-day study interval. Time lines represent 12-hour intervals. Note the decrease in NO contamination on weekends, which tended to follow both ambient air NO levels (adapted from Reference 3).

4.7 shows the concentration of these gases in global regions, measured in the period from 2000 to 2005. Ozone concentrations presented in Table 4.7 are not related to the ozone layer but are values measured at ground level. It is formed, along with other photochemical oxidants, by action of short-wavelength radiation from the sunlight on nitrogen oxides. Therefore, the highest levels of O_3 occur during periods of sunny weather.

Other problems associated to piped medicinal gases arise mainly from the following events [11]:

- contamination with an erroneous gas or solvent during installation or maintenance
- contamination at points of interconnection or cross connection of two pipelines
- contamination by backflow of a gas, so-called retropollution.

Although rare due to standards of monitoring, equipment design and alarm features, contamination with an erroneous gas or solvent may occur. Gilmour et al. [12] reported a case of oxygen supply contamination with trichloroethane, originated from the fluid used to clean the hose that connected the

TABLE 4.6 Relevant WHO Air Pollutants and Their Concentration Ranges (μg/m³) in Nonurban, Urban, and Industrial Regions [9]

Air Pollutants	Background Atmospheric Concentration	Urban Regions	Near Industrial Sites	Site
Classical				
Nitrogen dioxide	0.4–9.4	20–90	>940	Busy roads
Ozone	40–70	350	–	
Sulfur dioxide	–	20–60	1,000–2,000	
Carbon monoxide	–	60–140	20,000	Urban traffic
Hydrogen sulfide	0.14–0.4	–	–	
Carbon disulfide	–	–	10–1,500	Viscose rayon plants
Organic pollutants				
Acrylonitrile	–	<0.3	10	
Benzene	1	5–20	–	
Butadiene	–	2–20	–	
1,2-Dichloroethane	0.2	0.4–1.0	6.1	Petrol station
Dichloromethane	–	<5	50	
Formaldehyde	–	1–20	30–60	
Polycyclic aromatic hydrocarbons	<0.001	0.001–0.01	–	
Polychlorinated biphenyls	0.000003	0.003	–	
Polychlorinated dibenzodioxins and dibenzofurans	–	2 pg/m³	–	
Styrene	<1	20	–	
Tetrachloroethylene	<1	<5	1,000	Dry-cleaning operations
Toluene	<5	5–150	>150	
Trichloroethylene	<1	10	–	
Vinyl chloride	<1	–	10–100	PVC production facilities
Inorganic pollutants				
Arsenic	0.001–0.01	<0.03	1	Metal smelters, power plants
Asbestos	–	>100 fiber/m³	–	
Cadmium	–	–	–	
Chromium	–	0.005–0.2	–	
Fluoride	–		0.5–2	
Lead	<0.15	0.15–0.5	–	
Manganese	0.01–0.07	0.01–0.07	0.2–0.3	Foundries
Mercury	0.002–0.004	0.01	1	
Nickel	–	0.001–0.01	0.11–0.18	
Platimun	–	0.05–0.09 pg/m³	0.08–0.1	Platinum refineries
Vanadium	0.00002–0.0019	0.05–0.2 2 (in winter)	–	
Radon	–	20–100 Bq/m³	–	

TABLE 4.7 Ranges of Annual Average Concentrations (μg/m³) of Nitrogen Dioxide and Sulfur Dioxide and 1-Hour Average Maximum Concentrations of Ozone for Different Regions, Based on a Selection of Urban Data [10]

Region	Nitrogen Dioxide	Sulfur Dioxide	Ozone
Africa	35–65	10–100	120–300
Asia	20–75	6–65	100–250
Australia/New Zealand	11–28	3–17	120–310
Canada/United States	35–70	9–35	150–380
Europe	18–57	8–36	150–350
Latin America	30–82	40–70	200–600

delivery truck to the hospital tank. The trichloroethane concentration reached 400 ppm but did not lower the oxygen concentration enough to be detected by the security system.

Episodes of delivering hypoxic N_2O/O_2 gas mixtures due to failure in flow meter valves have been reported [13–16]. This is not a case of a true contamination but an erroneous proportion in the mixture of both gases.

4.2 VOLATILE ANESTHETICS

Volatile anesthetic agents are the basis of modern anesthetic practice. They are liquid at room temperature but evaporate easily for administration by inhalation. Early anesthetics, such as methoxyflurane, trichloroethylene, and diethylether have been replaced by halogenated anesthetics: halothane, enflurane, isoflurane, sevoflurane, and desflurane, which allow much faster induction and recovery times with lower hazard for human exposure and improved handling safety. They are chemically stable, nonflammable, and not explosive. Table 4.8 lists the chemical and commercial names of these volatile anesthetic agents as well as their structural forms.

Contamination associated with volatile anesthetics is a result of a combination of two parts: the anesthetic itself and the carbon dioxide absorber, necessary for inhalation anesthesia. Depending on the anesthetic/CO_2 absorber combination, degradation products can be formed to a greater or lesser degree.

The use of inhalation anesthesia machines allows for rebreathing of the anesthetic, which decreases its consumption and contributes to the maintenance of the patient's body temperature and airway humidity. Moreover, it prevents anesthetics to pollute the room. In order to be rebreathed, CO_2 must be removed from the exhaled air, which is achieved by the reaction of the gas with soda lime contained in an absorber canister inserted into the machine.

Soda lime is a mixture of chemicals, used in granular form and containing basically calcium hydroxide, water, and sodium hydroxide or potassium

TABLE 4.8 Chemical and Brand Names of Volatile Anesthetics

Chemical Name	Brand Names	Structure
2-Bromo-2-chloro-1,1,1-trifluoroethane	Halothane Fluothane Anestan Chalothane Fluktan Halan Halotan Halsan Narcotan Narcotane Narkotan Phthorothanum Rhodialothan	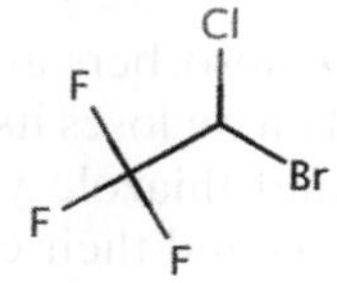
1,1,1,3,3,3-Hexafluoro-2-(fluoromethoxy) propane	Sevoflurane Ultane	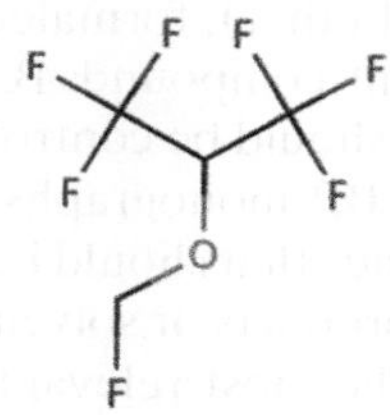
2-(Difluoromethoxy)-1,1,1,2-tetrafluoroethane	Desflurane Suprane	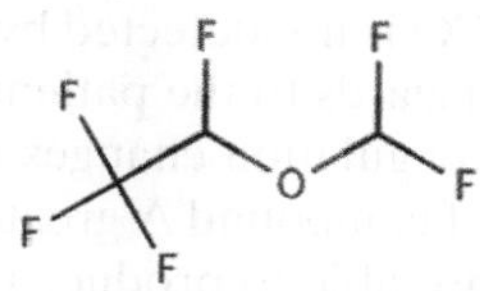
2-Chloro-1-(difluoromethoxy)-1,1,2-trifluoroethane	Enflurane Alyrane Efrane Ethrane	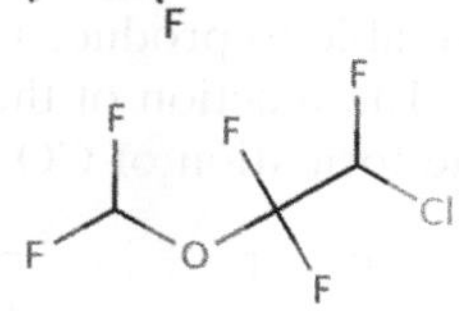
2-Chloro-2-(difluoromethoxy)-1,1,1-trifluoroethane	Isoflurane Aerrane Ethane Forane Forene	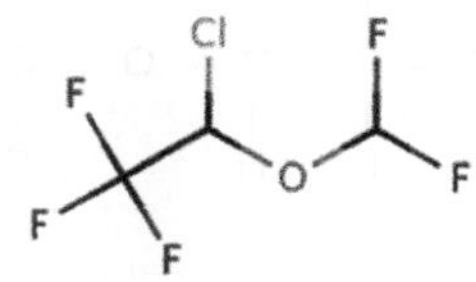

hydroxide, which undergo the following reactions when in contact with carbon dioxide:

$$1)\ CO_2 + H_2O \rightarrow H_2CO_{3(aq)} \quad (4.7)$$

$$2)\ H_2CO_3 + NaOH \rightarrow NaHCO_3 + H_2O \quad (4.8)$$

$$3)\ NaHCO_3 + Ca(OH)_2 \rightarrow CaCO_3 + H_2O + NaOH. \quad (4.9)$$

The overall reaction is

$$CO_2 + Ca(OH)_2 \rightarrow CaCO_3 + H_2O + \text{heat}. \tag{4.10}$$

All CO_2 absorbers are added with an indicating dye that changes color when the soda lime loses its carbon dioxide absorbing capacity. Usual dyes are ethyl violet and thiazole yellow. Table 4.9 lists some commercially available CO_2 absorbers and their characteristics.

Degradation may occur when halogenated anesthetics come into contact with the strong base present in soda lime. The degradation products depend on the anesthetic, the base and water content of the absorber, and the temperature. Carbon monoxide is produced by all halogenated anesthetics and *in vitro* experiments have shown that the instability follows the order desflurane > enflurane > isoflurane >> halothane = sevoflurane. Sevoflurane, besides CO, may also produce compound A (fluoromethy1-2,2-difluoro-1-(trifluoromethyl) vinyl ether), formaldehyde, methanol, and hydrofluoric acid, and, to a lesser extent, compounds B, C, D, and E (see Table 4.10). Table 4.10 lists the impurities that should be controlled in halogenated volatile anesthetics according to USP and BP monographs. Besides usual impurities, such as chloride, other compounds that should be controlled may arise from the synthesis process either as by-products or solvents.

The most relevant degradation products are CO and compound A. Since CO is not detected by anesthesia machines, this toxic by-product poses serious hazards to the patients. Carbon monoxide leads to severe hypoxia with blood coagulation changes and compound A is nephrotoxic. In rats, concentrations of compound A greater than 130 ppm are lethal, and those that are 25–50 ppm are able to produce renal injury [17].

The reaction of the -CHF_2 moiety with the hydroxyl ion is responsible for the formation of CO according to the following equations [18]:

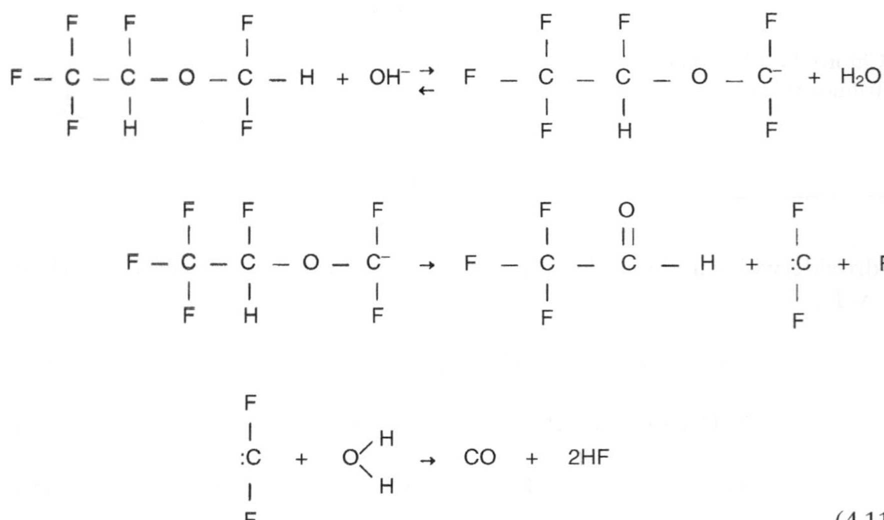

(4.11)

TABLE 4.9 Chemical Composition of Commercial Carbon Dioxide Absorbers

Component	Sodalime	Medisorb	Sodasorb	Spherasorb	Sofnolime	Agrisorb	Amsorb Plus	DrägerSorb 800	YabashiLime
$Ca(OH)_2$ (%)	94	70–80	76.5	75–80	>75	70–80	83	80	84
NaOH (%)	2–4	1–2	2.25	<2	3	<3.5%	–	2	–
KOH (%)	1–3	0.003	2.25	–	–	–	–	2	–
Humectant	–	–	–	–	–	–	$CaCl_2$ 1%	–	–
Hardener 1	–	–	–	Zeolite 4–5%	–	–	$CaSO_4$ 1%	–	Methylcellulose 0.3%
Hardener 2	–	–	–	–	–	–	Polyvinylpyrrolidine 1%	–	–
Water (%)	14–19	16–20	18.9	13.5–17.5	12–19	13–19	14.5	14	16
Size (mesh)	4–8	4–8	n.i.	5–7	4–8	4–8	4–8	4–8	n.i.
Indicator	Yes	Yes	Yes	Titan yellow <0.1%	Yes	Yes	Yes	Yes	n.i.
Company	Carlo Erba	GE Healthcare	W.R. Grace & Co.	Intersurgical Ltd.	Molecular, UK	GripoLab	Armstrong Medical	Dräger, Germany	Yabashi, Japan

n.i., not informed.

TABLE 4.10 Pharmacopeial Specifications for Volatile Anesthetics

		Impurity					
Anesthetic	Pharm.	Fluoride	Water	Chloride and/ or Bromide	Chlorine and Bromine	Nonvolatile Residue	Others
Ether	USP	–	<0.2%	–	–	1 mg/50 mL	Max 0.2% low-boiling hydrocarbons measured pentane Adehyde (pass/fail test) Peroxide <3 ppm
Halothane	USP	–	≤0.03%	Pass/fail test	–	1 mg/50 mL	0.008%–0.012% thymol Max 0.005% organic impurities measured as 1,1,2-trichloro-1,2,2-trifluoroethane
	BP	–	–	Pass/fail test	Pass/fail test	20 mg/L	0.008%–0.012% thymol Max 0.005% measured as trichlorotrifluoroethane (E)-1,1,1,4,4,4-Hexafluorobut-2-ene (EZ)-2-Chloro-1,1,1,4,4,4-hexafluorobut-2-ene (EZ)-2,3-Dichloro-1,1,1,4,4,4-hexafluorobut-2-ene (E)-2-Bromo-1,1,1,4,4,4-hexafluorobut-2-ene 2-Chloro-1,1,1-trifluoroethane 1,1,2-Trichloro-1,2,2-trifluoroethane 1-Bromo-1-chloro-2,2-difluoroethene 2,2-Dichloro-1,1,1-trifluoroethane 1-Bromo-1,1-dichloro-2,2,2-trifluoroethane 1,2-Dichloro-1,1- difluoroethane

Sevoflurane	USP	<2 mg/mL	<0.1%	–	–	1.0 mg/10 mL	<25 mg/g 2-(fluoromethoxy)-1,1,3,3,3-pentafluoro-1-propene (compound A) <100 mg/g 1-(methoxy)-2-(fluoromethoxy)-1,1,3,3,3-pentafluoropropane (compound B) <100 mg/g 1-(methoxy)-2-(fluoromethoxy)-1,1,3,3-tetrafluoro-2-propene (compound C) <0.22 mg/mL peroxide
Desflurane	USP	<0.001%	–	–	–	0.075%	<1.0% isoflurane <0.1% methylene chloride <0.1% 2-(fluoromethoxy)-1,1,3,3,3-pentafluoro-1-propene (compound A) <0.05% chloroform <0.025% trichlorotrifluoroethane <0.01% dichlorofluoroethane <0.01% trichlorofluoroethane <3 µg/g antimony
	BP	<10 ppm	–	–	–	100 mg/L	<0.1% 1,1′-oxybis(1,2,2,2-tetrafluoroethane) <0.2% isoflurane <0.01% methylene chloride <0.002% chloroform <0.01% 1,1,2-trichloro-1,2,2-trifluoroethane <0.01% dichlorofluoroethane <0.01% trichlorofluoroethane <0.01% acetone <3 ppm antimony

(Continued)

TABLE 4.10 (*Continued*)

		Impurity					
Anesthetic	Pharm.	Fluoride	Water	Chloride and/or Bromide	Chlorine and Bromine	Nonvolatile Residue	Others
Enflurane	USP	10 μg/mL	0.14%	0.9 mg/100 mL	–	2 mg/10 mL	–
Isoflurane	USP	10 ppm	<0.10%	10 ppm	–	2 mg/10 mL	<0.01% acetone <0.01% 2-(fluoromethoxy)-1,1,3,3,3-pentafluoro-1-propene (compound A) <0.007% 1-(methoxy)-2-(fluoromethoxy)-1,1,3,3,3-pentafluoropropane (compound B)
	BP	10 ppm	1 mg/mL	10 ppm	–	200 mg/L	<0.01% acetone <0.01% any other impurity: 2-(Chlorodifluoromethoxy)-1,1,1-trifluoroethane 2-(Difluoromethoxy)-1,1,1-trifluoroethane (2RS)-2-Chloro-2-(chlorodifluoromethoxy)-1,1,1-trifluoroethane 1,1-Dichloro-1-(difluoromethoxy)-2,2,2-trifluoroethane 1,1-Dichloro-1-(chlorodifluoromethoxy)-2,2,2-trifluoroethane

–, not specified.

The decomposition of sevoflurane to form compound A, methanol, and the less common compounds B, C, D, and E occurs according to following the equations [18]:

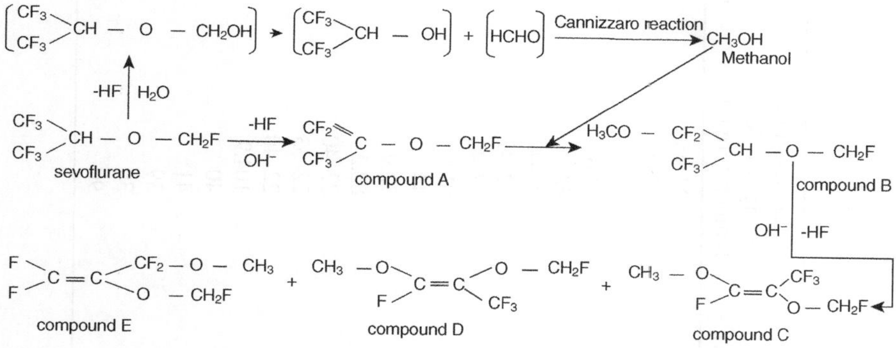

(4.12)

The presence of water in the absorber (14–16%) decreases direct contact of the anesthetic with hydroxides and therefore reduces the formation of degradation products. On the other hand, because the reactions for CO_2 absorption are exothermal, the temperature inside the absorber favors anesthetic degradation. Table 4.11 displays temperatures attained by absorbers under different conditions. The data show that, when using dry absorbers, temperatures are generally higher than when using fresh (wet) absorbers.

The amount of CO formed from halogenated anesthetics using different absorbers is shown in Table 4.12. Because there is a CO peak, generally at the beginning of the gas flow (see Fig. 4.4), the data in Table 4.12 correspond to the peak concentration. Results confirm that NaOH and KOH are the primary determinants of anesthetic degradation since Amsorb and DrägerSorb Free (containing no NaOH or KOH; see Table 4.9) caused negligible formation of CO. In contrast, for absorbers containing one or both bases, the amount of CO produced was higher than 45 ppm. Differences between wet and dry absorbers may also be seen in the records; the use of dry absorbers produced more CO than the use of hydrated ones.

The formation of compound A from sevoflurane is also determined by the base content of the CO_2 absorber. While using NaOH- or KOH-containing absorbers, the amount of compound A ranged from 20.4 to 38.3 ppm (Table 4.13). When using NaOH- or KOH-free absorbers, the amount decreased to 3.1 ppm or less. Methanol and formaldehyde were also found in one study and compound B in another [27]. Figure 4.3 displays the concentration of compounds A and B formed during the contact of sevoflurane with soda lime (Sodasorb) in a glass flask at different temperatures. While compound A was formed in all temperatures, compound B was not detected at 23 and 37°C and started to be formed only after 2 hours at 54°C. Differently from compound A, which presented a peak of concentration in approximately 1 hour and decreased thereafter, the concentration of compound B increased constantly after starting to be formed.

TABLE 4.11 Temperature Attained by Absorbers during Closed-Circuit Anesthesia

Anesthetic	Absorber	Water	Flow Rate (L/min)	Time Interval or Time to Peak Temperature (min)	Temperature Range or Peak Temperature (°C)	Reference
Desflurane	Sodasorb 1	Wet	3	0–45	22–45	[19]
	Sodasorb 2	Wet	3	0–45	22–45	
	Amsorb	Wet	3	0–45	22–41	
	DrägerSorb 800	Dry	2	12.6	51.8	[20]
	Sodasorb	Dry	6	32	42	[21]
Isoflurane	Sodasorb 1	Wet	3	0–45	22–42	[19]
	Sodasorb 2	Wet	3	0–45	22–42	
	Amsorb	Wet	3	0–45	22–42	
	DrägerSorb 800	Dry	2	12.5	81.9	[20]
	Sodasorb	Dry	6	35	44	[21]
Enflurane	DrägerSorb 800	Dry	2	7.3	72.9	[20]
Halothane	DrägerSorb 800	Dry	2	10.0	82.2	[20]
Sevoflurane	Sodalime			0–150	22–40	[23]
	Amsorb Plus			0–150	22–33	
	DrägerSorb Free			0–150	22–40	
	DrägerSorb 800		2	21.7	104.9	[20]
Sevoflurane	Sodasorb		0.16	60	40	[22]
	Sodasorb II		0.16	60	41	
	Sofnolime		0.16	60	39	
	Amsorb		0.16	60	39	
	Sodasorb	Dry	6	18	90	[21]

TABLE 4.12 Carbon Monoxide Formed during Closed-Circuit Anesthesia Using Different Halogenated Anesthetics and CO_2 Absorbers and Time to Reach CO Peak Concentration

Anesthetic	Absorber	Water	Flow Rate (L/min)	CO Peak Time (min)	CO Peak Conc. (ppm)	Reference
Desflurane	Sodasorb 1	Wet	3	10	780 ± 1,070	[19]
	Sodasorb 2	Wet	3	10	190 ± 180	
	Amsorb	Wet	3	–	n.d.	
	DrägerSorb Free	Dry	0.5	–	<10	[24]
	Amsorb Plus	Dry	0.5	–	<10	
	DrägerSorb 800	Dry	2	6.9	8,996 ± 673	[20]
Isoflurane	Sodasorb	Wet	3	5	58 ± 55	[19]
	Sodasorb	Wet	3	10	45 ± 42	
	Amsorb	Wet	3	–	3 ± 8	
	DrägerSorb 800	Dry	2	5.7	3,080 ± 183	[20]
Enflurane	DrägerSorb 800	Dry	2	11.4	10,106 ± 1,604	[20]
Halothane	DrägerSorb 800	Dry	2	8.8	423 ± 76	[20]
Sevoflurane	DrägerSorb 800	Dry	2	21	426 ± 11	[20]

n.d., not detected; –, not informed.

The influence of water and a strong base content of the absorber on the degradation of anesthetics can also be observed in Table 4.14 [28] and in Figure 4.4 [29]. These results were obtained by using a different approach: Absorbers were stored in sealed barrels and were maintained at a constant temperature, where the volatile anesthetics were allowed to flow through the barrels. Concentration of CO and compound A were measured in the outflow gas at regular time intervals. The data in Table 4.14 show that greater amounts of CO and compound A were formed in absorbers with higher contents of NaOH and KOH.

Figure 4.4 shows CO formed in desflurane, enflurane, and isoflurane degradation. While the amount of CO produced reached a peak early in the delivery of the anesthetic that flowed through dry soda lime, for soda lime containing 1.4% water, some degradation also occurred, but the concentration of CO was much lower.

Soda lime is also used as a CO_2 absorber in underwater breathing systems and hyperbaric facilities. In this case, the air is directed through a bed of soda

TABLE 4.13 Degradation Products from Sevoflurane during Closed-Circuit Anesthesia Using Different CO_2 Absorbers

Absorber	Water	Flow Rate (L/min)	Peak Time (min)	Compound A (ppm)	Methanol (ppm)	Formaldehyde (ppm)	Reference
Sodasorb 2	Wet	3	45	22	–	–	[19]
Sodasorb 1	Dry	3	30	42	–	–	
Amsorb	Wet	3	5	1	–	–	
Amsorb	Dry	3	5	1	–	–	
DrägerSorb Free	Wet	0.5	120	0.5	–	–	[24]
Amsorb Plus	Wet	0.5	240	0.3	–	–	
Medisorb	Wet	1	4 h	7	–	–	[25]
Medisorb	Wet	3	2 h	3	–	–	
Amsorb Plus	Wet	1 and 3	–	n.d.	–	–	
YabashiLime	Wet	1 and 3	–	n.d.	–	–	
Sodalime	Wet	0.5	150	20.4	0.4	7.6	[23]
Amsorb Plus	Wet	0.5	150	0.1	0.2	1.1	
DrägerSorb Free	Wet	0.5	150	0.5	0.2	1.9	
Sodasorb	Wet	0.16	60	23.3	–	–	[22]
Sodasorb II	Wet	0.16	30	30.1	–	–	
Sofnolime	Wet	0.16	60	38.3	–	–	
Amsorb	Wet	0.16	–	1.3	–	–	
Sodasorb	Wet	1	120	28.0 ± 10	–	–	[26]
Amsorb Plus	Wet	1	120	3.1 ± 0.5	–	–	
DrägerSorb Free	Wet	1	120	2.4 ± 0.8	–	–	

n.d., not detected; –, not measured.

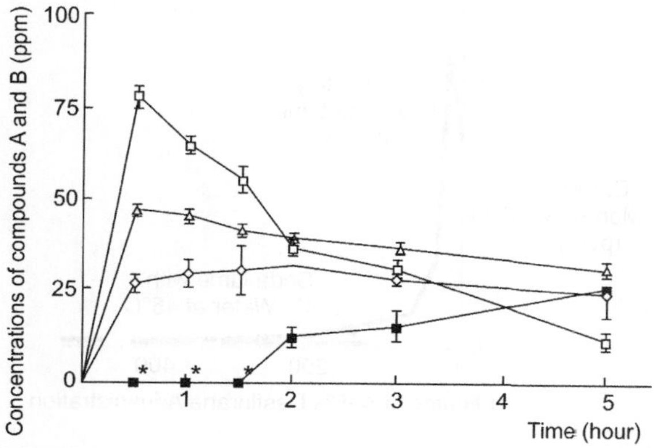

FIGURE 4.3 Concentrations of compounds A and B with Sodasorb in a glass flask. Open circles = compound A (23°C); open triangles = compound A (37°C); open squares = compound A (54°C); filled squares = compound B (54°C). *, not detected. Each point represents the mean ± SD of three samples. Compound B was not detected at 23 and 37°C [27].

TABLE 4.14 Amount of CO and Compound A Generated by CO_2 Absorbers Containing Different Base Compositions in 240 Minutes Kept at 45°C [28]

Absorber[a]	Base Content (%) $Ca(OH)_2$	NaOH	KOH	$Ba(OH)_2$	LiOH	CO Peak (ppm)	Compound A Peak (ppm)
Baralyme	72	0	5.3	10.5	0	36,100 ± 800	17.7 ± 0.7
Carbolime	83.2	3	0	0	0	37,200 ± 1,000	23.6 ± 0.4
DragerSorb 800	79.8	3	0	0	0	37,600 ± 1,600	32.6 ± 0.4
Amsorb	83.2	0	0	0	0	<10	4.5 ± 0.4
Grace 1M	80.7	1.6	2.6	0	0	27,100 ± 1,500	33.2 ± 0.5
Grace 1M	84.8	0	0.1	0	0	9,710 ± 1,300	23.6 ± 1.7
Grace 1M	81.1	3.2	0.1	0	0	22,000 ± 4,600	11.4 ± 0.5
LiOH	0	0	0	0	99	41.3 ± 0.8	0.7 ± 0.4

[a] All absorbents are commercially available. Baralyme was withdrawn from the market in 2004.

lime; the CO_2 reacts with the bed material; and the gas leaves the bed to be recirculated for rebreathing. Although not involving any other components than those present in the air, contamination of the gas that flowed through a CO_2 absorber was registered in the U.S. Navy [30]. The investigation started when an odor of ammonia was reported during diving. Two soda limes were in use, Sodasorb and Sofnolime, and were analyzed to identify the contaminants. Four major contaminants were detected: ammonia, diethyl amine, ethyl amine, and octane. Concentrations of these contaminants in the air containing 5% CO_2 flowing through a soda lime bed at 5 L/min are shown in Figure 4.5. Although not confirmed, the authors presumed that ammonia may have been a result of the breakdown of the dye indicator (ethyl violet) and the hydrocarbon from the material of which the bucket was constructed.

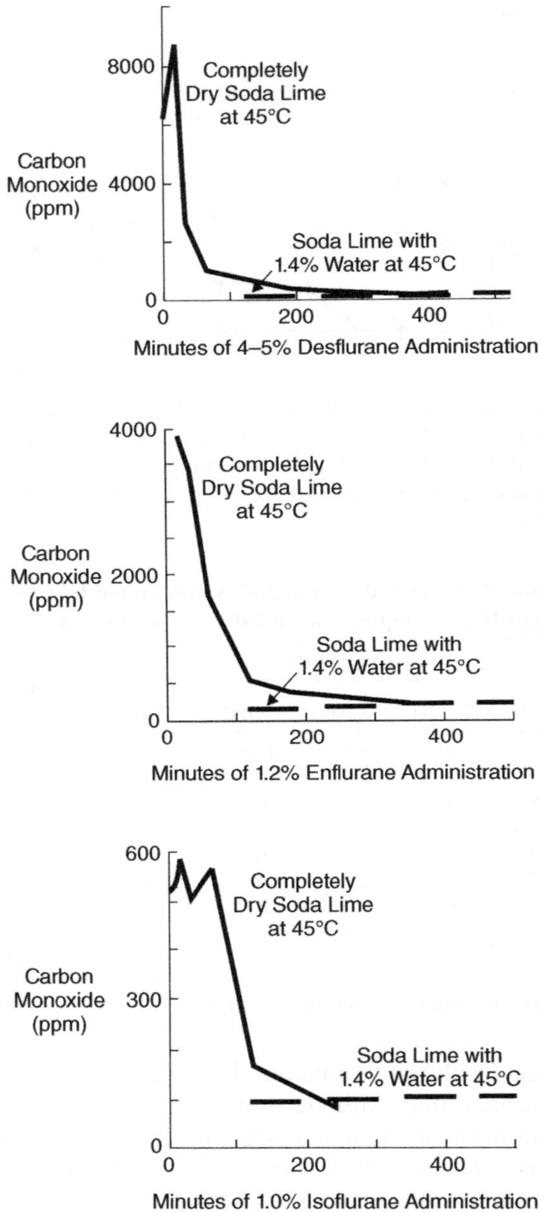

FIGURE 4.4 Carbon monoxide production from degradation of desflurane, enflurane, and isoflurane by dry and wet (1.4% water) soda lime at 45°C [29].

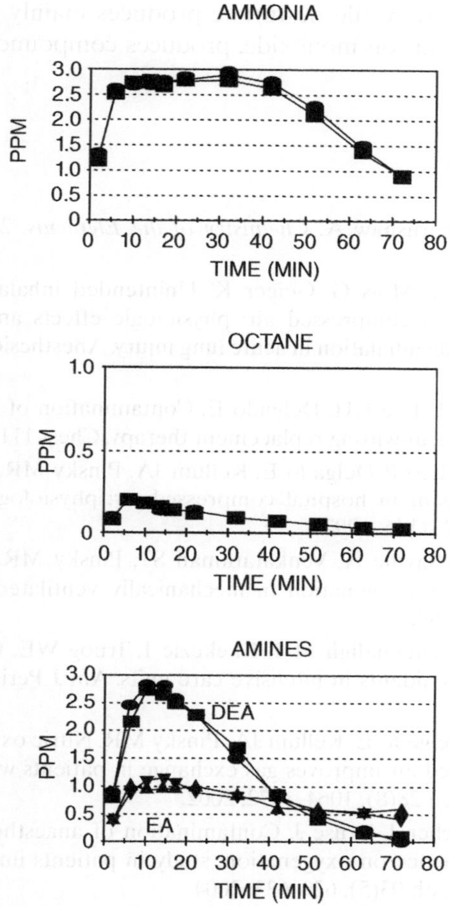

FIGURE 4.5 Representative contaminant washout curves for indicating Sofnolime based on Fourier transform infrared spectroscopy (FTIR) analysis. Replicate tests from the same bucket done the same day are plotted. DEA, diethyl amine; EA, ethyl amine [30].

CONCLUDING REMARKS

Among the gases used for medical purposes, compressed air is the most vulnerable to carrying impurities. Because compressed air is usually collected from the environment, the level of impurity reflects their concentration in the atmospheric air. Halogenated anesthetics, although liquid at room temperature, are administered in the form of gas. Degradation products, generated from the reaction of the anesthetic with soda lime (carbon dioxide absorber necessary for rebreathing the anesthetic), are the main contaminants in these products. The number and extent of degradation products formed depend on the anesthetic structure; while desfurane produces mainly carbon monoxide, sevoflurane, besides carbon monoxide, produces compound A, a nephrotoxic substance.

REFERENCES

1. Greenwood NN, Earnshaw A. *Chemistry of the Elements*, 2nd ed. Butterworth, Oxford, 2002.
2. Benzing A, Loop T, Mols G, Geiger K. Unintended inhalation of nitric oxide by contamination of compressed air: physiologic effects and interference with intended nitric oxide inhalation in acute lung injury. Anesthesiology 91(4), 945–950, 1999.
3. Pinsky MR, Genc F, Lee KH, Delgado E. Contamination of hospital compressed air with nitric oxide: unwitting replacement therapy. Chest 111(6), 1759–1763, 1997.
4. Lee KH, Tan PS, Rico P, Delgado E, Kellum JA, Pinsky MR. Low levels of nitric oxide as contaminant in hospital compressed air: physiologic significance? Crit Care Med 25, 1143–1146, 1997.
5. Lum LC, Tan PS, Saville A, Venkataraman ST, Pinsky MR. Occult nitric oxide inhalation improves oxygenation in mechanically ventilated children. J Pediatr 133(5), 613–616, 1998.
6. Thibeault DW, Rezaiekhaligh MH, Ekekezie I, Truog WE. Compressed air as a source of inhaled oxidants in intensive care units. Am J Perinatol 16(9), 497–501, 1999.
7. Tan PS, Genc F, Delgado E, Kellum JA, Pinsky MR. Nitric oxide contamination of hospital compressed air improves gas exchange in patients with acute lung injury. Intensive Care Med 28(8), 1064–1072, 2002.
8. Hess W, Kannmacher J, Kruse J. Contamination of anaesthetic gases with nitric oxide and its influence on oxygenation: study in patients undergoing open heart surgery. Br J Anaesth 93(5), 629–633, 2004.
9. World Health Organization Regional Office for Europe. *Air Quality Guidelines for Europe*, 2nd ed. (WHO Regional Publications, European Series, No. 91), WHO, Copenhagen, 2000.
10. World Health Organization. *Air Quality Guidelines for Particulate Matter, Ozone, Nitrogen Dioxide and Sulfur Dioxide—Global Update, Particulate Matter, Ozone, Nitrogen Dioxide and Sulfur Dioxide*. WHO, Genève, 2005.
11. Hay H. Contamination of piped medical gas supply with water. Eur J Anaesthesiol 17(8), 512–514, 2000.
12. Gilmour IJ, McComb RC, Palahniuk RJ. Contamination of a hospital oxygen supply. Anesth Analg 71(3), 302–304, 1990.
13. Lye A, Patrick R. Oxygen contamination of the nitrous oxide pipeline supply. Anaesth Intensive Care 6(2), 207–209, 1998.
14. Hay H. Delivery of an hypoxic gas mixture due to a defective rubber seal of a flowmeter control tube. Eur J Anaesthesiol 17(7), 456–458, 2000.
15. Herff H, Paal P, von Goedecke A, Lindner KH, Keller C, Wenzel V. Fatal errors in nitrous oxide delivery. Anaesthesia 62(12), 1202–1206, 2007.

16. Herff H, Paal P, Lindner KH, von Goedecke A, Keller C, Wenzel V. Fatalities due to nitrous oxide. Complications from mistakes in nitrous oxide supply. Anaesthesist 57(10), 1006–1010, 2008.
17. Anders MW. Formation and toxicity of anesthetic degradation products. Annu Rev Pharmacol Toxicol 45, 147–176, 2005.
18. Coppens MJ, Versichelen LFM, Rolly G, Mortier EP, Struys MMRF. The mechanisms of carbon monoxide production by inhalational agents. Anaesthesia 61(5), 462–468, 2006.
19. Kharasch ED, Powers KM, Artru AA. Comparison of Amsorb, Sodalime, and Baralyme degradation of volatile anesthetics and formation of carbon monoxide and compound A in swine in vivo. Anesthesiology 96(1), 173–182, 2002.
20. Wissing H, Kuhn I, Warnken U, Dudziak R. Carbon monoxide production from desflurane, enflurane, halothane, isoflurane, and sevoflurane with dry soda lime. Anesthesiology 95(5), 1205–1212, 2001.
21. Laster MJ, Eger EI, 2nd. Temperatures in soda lime during degradation of desflurane, isoflurane, and sevoflurane by desiccated soda lime. Anesth Analg 101(3), 753–757, 2005.
22. Versichelen LF, Bouche MP, Rolly G, Van Bocxlaer JF, Struys MM, De Leenheer AP, Mortier EP. Only carbon dioxide absorbents free of both NaOH and KOH do not generate compound A during in vitro closed-system sevoflurane: evaluation of five absorbents. Anesthesiology 95(3), 750–755, 2001.
23. Marini F, Bellugi I, Gambi D, Pacenti M, Dugheri S, Focardi L, Tulli G. Compound A, formaldehyde and methanol concentrations during low-flow sevoflurane anaesthesia: comparison of three carbon dioxide absorbers. Acta Anaesthesiol Scand 51(5), 625–632, 2007.
24. Struys MM, Bouche MP, Rolly G, Vandevivere YD, Dyzers D, Goeteyn W, Versichelen LF, Van Bocxlaer JF, Mortier EP. Production of compound A and carbon monoxide in circle systems: an in vitro comparison of two carbon dioxide absorbents. Anaesthesia 59(6), 584–589, 2004.
25. Yamakage M, Takahashi K, Takahashi M, Satoh JI, Namiki A. Performance of four carbon dioxide absorbents in experimental and clinical settings. Anaesthesia 64(3), 287–292, 2009.
26. Kobayashi S, Bito H, Obata Y, Katoh T, Sato S. Compound A concentration in the circle absorber system during low-flow sevoflurane anesthesia: comparison of Drägersorb Free, Amsorb, and Sodasorb II. J Clin Anesth 15(1), 33–37, 2003.
27. Morio M, Fujii K, Satoh N, Imai M, Kawakami U, Mizuno T, Kawai Y, Ogasawara Y, Tamura T, Negishi A, Kumagai Y, Toshikazu K. Reaction of sevoflurane and its degradation products with soda lime. Toxicity of the by-products. Anesthesiology 77(6), 1155–1164, 1992.
28. Stabernack CR, Brown R, Laster MJ, Dudziak R, Eger E. Absorbents differ enormously in their capacity to produce compound A and carbon monoxide. Anesth Analg 90(6), 1428–1435, 2000.
29. Fang ZX, Eger E, Laster MJ, Chortkoff BS, Kandel L, Ionescu P. Carbon monoxide production from degradation of desflurane, enflurane, isoflurane, halothane, and sevoflurane by soda lime and Baralyme. Anesth Analg 80(6), 1187–1193, 1995.
30. Lillo RS, Ruby A, Gummin DD, Porter WR, Caldwell JM. Chemical safety of U.S. Navy Fleet soda lime. Undersea and Hyperbaric Medical Society 43-53, 1994.

5

DIAGNOSTIC IMAGING AGENTS

Several types of contrast media (CM) and imaging agents are used for diagnostic purposes. They can be roughly classified based on the imaging modalities they provide: X-ray attenuation, magnetic resonance signal enhancement, and scintigraphy (gamma-radiation emission). Table 5.1 presents the most common imaging agents.

Since most of these elements are toxic species, even at low concentrations, they are usually present in the form of an insoluble compound, such as barium in barium sulfate, or in the form of an indissociable (nonionizable) chelate, such as technetium in Tc-sestamibi and gadolinium in gadoversetamide. Contaminants in such products can be minor components generated during chelate synthesis, other isotopes of the element, or impurities derived from inactive ingredients.

In the case of radioactive nuclides, since all of them are artificially produced either by fission process or by neutron activation, impurities can also include the parent nuclide, other nuclides produced by nuclear reaction, or the element into which the nuclide decays.

Although all diagnostic imaging agents are subject to present impurities, of those summarized in Table 5.1, only technetium, iodine, fluorine, gadolinium, and barium sulfate derived pharmaceuticals will be discussed in detail. The reason for this choice is based on either their frequent use or the amount of the product consumed by patients. While Tc-99, I-131, and F-18 are the most

Sources of Contamination in Medicinal Products and Medical Devices,
First Edition. Denise Bohrer.
© 2013 by John Wiley & Sons, Inc. Published 2013 by John Wiley & Sons, Inc.

TABLE 5.1 Selected Diagnostic Imaging Agents, Their Categories, and Dose Range

Type	Dose Range (mg/kg)					Labeling Form
X-ray attenuants	10–100					
Barium						$BaSO_4$
Iodine						Organic compounds
MR contrast agents	10^{-3}–10^{-1}					
Gadolinium						Chelates
Radioactive species	10^{-9}–10^{-5}	Species	Production	Reaction production	Half-life	
Iodine		I-123	Cyclotron	Te-124(p,2n) I-123	13.22 h	NaI and organic compounds
Iodine		I-131	Fission	U-235	8 days	NaI and organic compounds
Technetium		Tc-99m	Fission fragment	U-245 Mo-99	6 h	Chelates Protein attached formulations Red blood cells formulations
Indium		In-111	Cyclotron	Cd-112(p,2n)In-111	67.4 h	Indium chloride Chelates
Gallium		Ga-67	Cyclotron	Zn-68(p,2n)Ga-67	78 h	Gallium citrate
Thallium		Tl-201	Cyclotron	Tl-203(p,3n)Pb-201	73 h	Thallous chloride
Fluorine		F-18	Cyclotron	O-18(p,n)F-18	109.8 min	Fluorodeoxyglucose
Xenon		Xe-133	Fission	U-235	5.243 days	
Oxygen		O-15	Cyclotron	N-14(d,n)O-15	2.03 min	
Nitrogen		N-13	Cyclotron	O-16(p,a)N-13	9.98 min	

commonly used radioactive elements in nuclear medicine, large amounts of iodine and barium sulfate are consumed in procedure for X-ray attenuation, and gadolinium in magnetic resonance procedures.

5.1 RADIOPHARMACEUTICALS

Pharmacopeial compendia describe the purity of radiopharmaceutical preparations in terms of *radiochemical purity* (RCP) and *radionuclidic purity*. While RCP refers to the amount of the desired radionuclide present in the stated chemical form, radionuclidic purity is defined as the fraction of the total radioactivity in the sample that is derived from the desired radionuclide.

Radionuclidic contaminants arise in the production of radionuclides, contribute to increase the radiation dose to the patient, and may result in an incorrect counting rate or pixel intensity in images. Radiochemical impurities usually stem from parallel chemical reactions involving the radionuclide that occur in the radiolabeling process which leads to the formation of different compounds of the radionuclide. Radiochemical impurities may have different distribution patterns in the body, adding a background to the image of the desired organ. Typical RCP is higher than 95%. Also relevant are *chemical impurities*, which are related to the purity of the raw material used in the radiopharmaceutical production or may also arise from the action of the radiation on other constituents of the formulation.

Radionuclides produced by fission: The fission process takes place in a reactor containing a quantity of fissionable material, typically enriched uranium (uranium-235 [U-235]), which undergoes spontaneous nuclear fission splitting into two lighter nuclear fragments (chemical elements with lower atomic number) and emitting two or three neutrons. (U-235 enrichment means an increase in the isotopic ratio of natural uranium, which contains 99.284% U-238 and 0.711% U-235.) The U-235 fission reaction leads to more than 100 nuclides representing 20 different elements. In Figure 5.1, the mass distribution of the fission fragments is shown. It is possible to see that fragments (chemical elements) with a mass number in the ranges 85–105 and 130–150 are formed in higher proportion. The radionuclide produced by fission is called "carrier-free," which means that no stable isotope of the element of interest is formed. For example, in the production of technetium-99m (Tc-99m) by fission (see below), it is the only technetium isotope generated. This means that radionuclides produced by fission present high RCP. On the other hand, due to the lack of specificity of the fission process, the radionuclide of interest is produced together with a large amount of other radionuclides, which may lead to poor radionuclidic purity. The enrichment of the radionuclide of interest is obtained through chemical separations.

Radionuclides produced by neutron activation: When neutrons strike a target, some of them are captured by nuclei of the target atoms. Herewith, a target nucleus may be converted into a radioactive product. Two types of

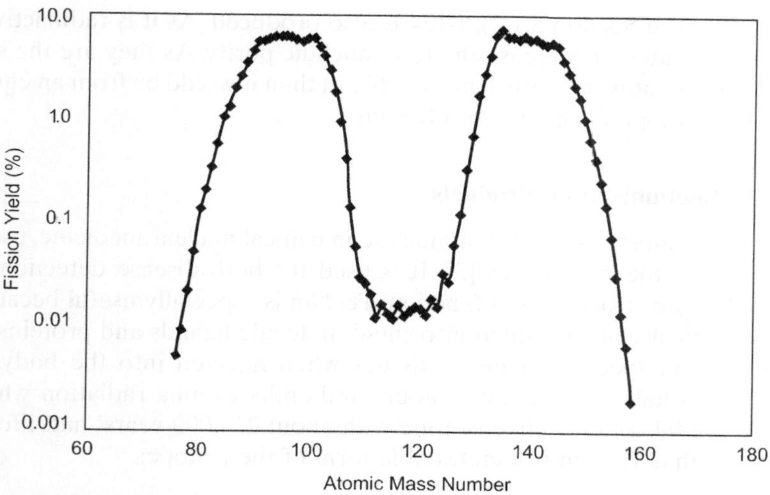

FIGURE 5.1 Mass distribution of fragments following fission of U-235 [1].

reaction can occur: After capturing a neutron, the target element reaches an excited state. The de-excitation takes place by the emission of a γ-ray (the radiation of interest). The target and product nuclei of this reaction are isotopes of the same chemical element with different masses. This reaction is represented by the *(n, γ) reaction*. In the second type of reaction, the target captures a neutron but ejects a proton, the *(n, p) reaction*. In this case, the target and product nuclei are not the same chemical element. The radionuclide produced by neutron activation through the (*n*, *γ*) reaction is not carrier-free because the product is the same element as the bombarded target material, and only a very small percentage of the target nuclei is activated; therefore, the product may present low activity and low RCP because of the presence of a large amount of unactivated stable carrier (target material).

The evaluation of RCP is generally carried out by thin-layer chromatography (TLC) since the impurities, different forms of the same element, present different chemical behaviors. In spite of TLC efficiency, other separation techniques have been proposed for this evaluation. Millar [2] investigated the purity of iodohippuric acid I-123 (I-123-IHA) by high-performance liquid chromatography (HPLC) and compared the results to TLC. The most common impurities produced in the labeling reaction of I-123-IHA, iodide, and iodobenzoic acid were determined in 10 preparations of I-123-IHA. While a mean RCP of 99.1% was found by HPLC, that of 98.3% was found by TLC.

Radionuclidic purity is more difficult to evaluate since it demands sophisticated equipment and standards to carry out the identification of radionuclides. Moreover, it is even more difficult to reduce radionuclidic impurities when they are elicited from other radioactive isotopes of the radiopharmaceutical element. I-123 serves as an example again. During the production of I-123 from Te-124 (see Section 5.1.2), I-124 is also produced. As it is radioactive as well, it contributes to decrease the radionuclidic purity. As they are the same element, the separation is much more difficult than it would be from an equally radioactive isotope from another element.

5.1.1 Technetium-Based Products

Tc-99m is the most common radionuclide in clinical nuclear medicine, used in about 80% of the procedures [3]. It is used for both disease detection and studies of organ structure and function. Tc-99m is especially useful because it can be chemically incorporated into small-molecule ligands and proteins that concentrate on specific organs or tissues when injected into the body. The isotope has a half-life of about 6 hours and emits gamma radiation when it decays to Tc-99, a radioactive isotope with about 214,000 years' half-life. The "m" means that Tc-99m is a metastable form of the isotope:

$$\text{Tc-99m}_{(t\frac{1}{2}=6\text{ h})} \rightarrow \text{Tc-99}_{(t\frac{1}{2}=214{,}000\text{ years})} + \gamma. \tag{5.1}$$

Since Tc-99m is obtained from molybdenum-99 (Mo-99), which is a fission product of U-235, the attainment of the isotope in a radiopharmaceutical form occurs in three steps: first, the production of Mo-99 and its extraction from the uranium material; second, the conversion of Mo-99 into Tc-99m; and third, the transformation of Tc-99m into a drug tolerable to the patients. The Mo-99 is called the parent isotope and Tc-99m is called the daughter isotope.

5.1.1.1 Production of Mo-99 The fission of U-235 occurs by bombarding its nucleus with neutrons. The fission fragments in higher abundance present mass numbers in the ranges of 85–105 and 130–150, and among them, Mo-99 yields approximately 6% of the fission fragments [4]. Figure 5.1 displays the mass distribution of the fission fragments:

$$\text{U-235} + \text{n} \xrightarrow{\text{Fission}} \text{Mo-99} + \text{Other radionuclides.} \tag{5.2}$$

Fission can be carried out using highly enriched uranium (HEU) or low-enriched uranium (LEU), and Mo-99 recovery can be either after alkaline or acidic dissolution of the target (irradiated material) [4]. Although almost all of the Mo-99 used worldwide is produced by only four companies, all using HEU targets, in order to reduce nuclear proliferation concerns, experimental work has been done to convert the current HEU (>93% U-235) targets to LEU (<20% U-235) [5].

Although the nature of the impurities depends on the fission material and also on the process used for Mo-99 recovery, I-125, I-131, and Xe-133, which are also used as medical isotopes, along with Ru-103, Te-132, Rh-105, and Pd/Ag-112, are the most abundant radionuclidic impurities [6] (see Table 5.2).

TABLE 5.2 Nuclides Present in Mo-99 Obtained from U-235 Fission for Different Irradiation Times [6]

Irradiation					
Test	Target Power (kW)[a]	Irradiation Time (h)	Postirradiation Decay (h)	UO_2 Mass (g)	Mo-99 Inventory[b] (TBq)[c]
1	5.0	24	24	25.6	1.6
2	3.7	8	16	25.6	0.48
3	3.8	10	16	25.2	0.56
4	3.5	24	24	25.5	1.1
5	4.1	149	6	25.4	5.6

Product Purity Specification					
	Specification	Pass/Fail Status[d]			
Nuclide	µCi/mCi of Mo-99	Test 2	Test 3	Test 4	Test 5
I-131	$\leq 5.0 \times 10^{-2}$	N	Y	Y	Y
Ru-103	$\leq 5.0 \times 10^{-2}$	Y	Y	Y	Y
Pd-112/Ag112	$<3 \times 10^{-3}$	Y	Y	N	Y
Te-132	$\leq 5.0 \times 10^{-2}$	Y	Y	Y	Y
Sr-89	$\leq 6.0 \times 10^{-4}$	N	N	Y	Y
Sr-90	$\leq 1.5 \times 10^{-5}$	Y	Y	Y	Y
Total γ	$\leq 5.0 \times 10^{-2}$	Y	Y	Y	Y
Gross α	$\leq 1.0 \times 10^{-7}$	N	N	N	Y

[a] Reactor power = 2 MW; neutron flux = 3×10^{13} neutrons/cm^2·s.
[b] Activity at processing time.
[c] ORIGEN2 calculations.
[d] No product recovered from test 1.

Alkaline dissolution is generally used for targets that contain aluminum (Al/U alloy in an Al cladding), and acidic dissolution is for targets made of uranium metal or uranium oxide. Figure 5.2 displays a flowchart of both dissolution procedures. As it is possible to see in Figure 5.2, iodine is the main contaminant of Mo-99 in both procedures since several steps are aimed at its elimination. The separate iodine includes all its isotopes because the separation is based on chemical properties of the element and not of the isotope.

Besides the radionuclides produced in uranium fission, other radioactive species may be formed during the process. Plutonium-239 (Pu-239) is a decay product of neptunium-239 (Np-239), which is formed from neutron absorption by U-238:

$$\text{U-238}\,(n, \gamma) \rightarrow \text{U-239} \tag{5.3}$$

$$\text{U-238} + n \rightarrow \text{U-239} + \gamma \rightarrow \text{Np-239} + \beta^- \rightarrow \text{Pu-239} + \beta^-. \tag{5.4}$$

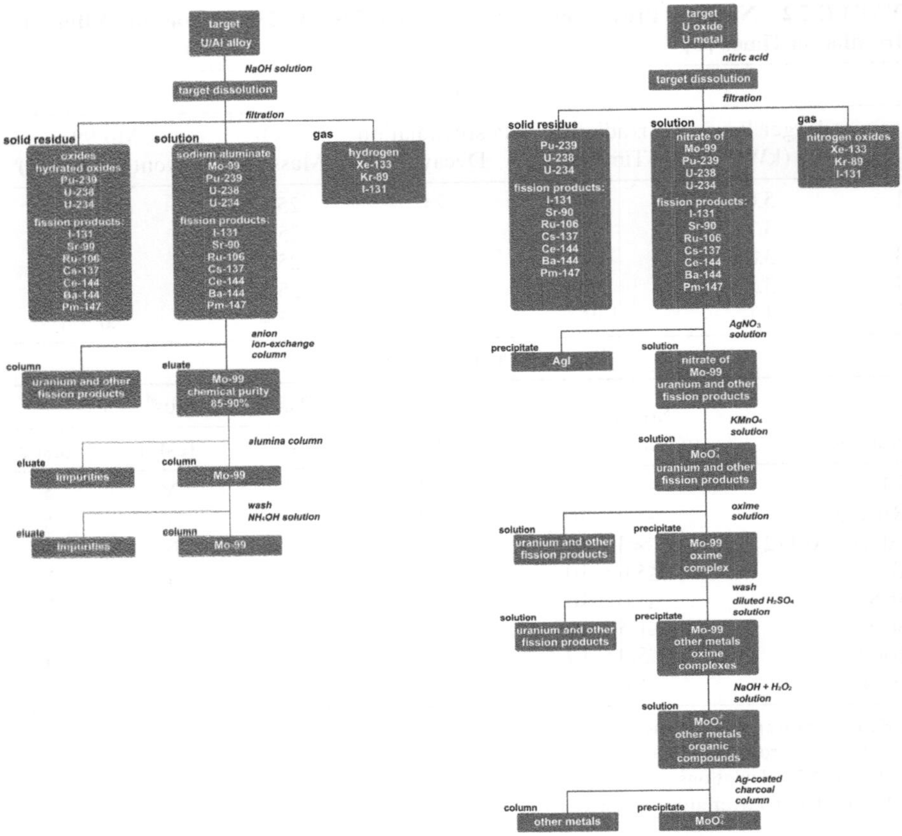

FIGURE 5.2 Flowchart of alkaline and acidic dissolution of uranium targets after bombarding.

According to Vandegrift [7, 8], 25-fold more Pu-239 is produced in a LEU target than in a HEU target. On the other hand, when uranium is enriched for HEU (93% U-235), it is even more enriched in U-234. Both Pu-239 and U-234 are α-emitter radionuclides (extremely noxious) and, considering the sum of their contributions, the α-radiation count is almost the same in the product generated by both processes.

Other subproducts that can be generated during Tc-99 production are derived from parallel reactions that occur by the interaction of radiation with other components in the system. Carson et al. [9] showed that extensive γ-irradiation of alkaline MoO_4^{2-} solutions in high-density polyethylene (HDPE) containers yielded hydrogen, which reduced the molybdate ion, leading to the formation of precipitates.

McDonald et al. [6] investigated Mo-99 purification after acidic dissolution of the target. One of their concerns was the action of irradiation on the α-benzoin oxime used to precipitate the MoO_4^{2-} from the acidic solution. They irradiated samples containing the compound for 0.5, 1.0, and 4.0 hours and collected the Fourier transform infrared spectroscopy (FTIR) spectra that are presented in Figure 5.3. Although the spectrum changed, they estimated that less than 10% of the oxime decomposed during 4 hours' irradiation, which did not affect product isolation but generated other decomposition products.

5.1.1.2 Generation of Tc-99m The separated Mo-99 (in the form of the molybdate ion, MoO_4^{2-}) is adsorbed onto an alumina (Al_2O_3) column in a cylindrical container. The column is known as a Mo-99/Tc-99m generator and is shipped in radiation-shielded cartridges. The Mo-99 in the generator decays to Tc-99m with a half-life of about 66 hours:

$$\text{Mo-99}_{(t\frac{1}{2}=66\,\text{h})} \rightarrow \text{Tc-99m}_{(t\frac{1}{2}=6\,\text{h})} + \beta^-. \quad (5.5)$$

The Tc-99m (in the form of the pertechnetate ion TcO_4^-) is typically recovered by passing a saline solution through the alumina column in the generator. The saline removes TcO_4^- but leaves MoO_4^{2-} in place. One generator can be eluted several times a day for about a week before it needs to be replaced with a fresh generator. There are two kinds of generators, wet and dry. In a dry generator, the residual saline is removed from the column after elution by means of a vacuum vial. This procedure avoids the radiolysis of the remaining water, which could lead to the formation of hydrogen peroxide or other free radicals. As a consequence, column damage, or even production of a Tc-99m species other than pertechnetate, could occur.

Although the purification processes are conceived to eliminate the other radionuclides generated by the U-235 fission, the radionuclidic purity of Tc-99m radiopharmaceuticals should be evaluated [10]. A pharmacopeial procedure [11] dictates the identification of Tc-99m and the determination of its total radioactivity. The limits are given in radioactivity units per unit of radioactivity fromTc-99m. In Table 5.3, the limits established by the United States Pharmacopeia (USP) and the British Pharmacopoeia (BP) are presented along with the radionuclides and their activity measured in raw Mo-99 after the purification process in different laboratories. All nuclides listed in Table 5.3 are gamma-emitters such as Mo-99, but there are considerable concerns regarding the presence of plutonium and other α-emitters because these elements are of high biological hazard. However, for both HEU and LEU targets, the bulk of the alpha decay in an irradiated target is from uranium isotopes, U-234, U-235, and U-238 [8]. The alpha contamination from plutonium and uranium isotopes in HEU and LEU irradiated targets, at the time that the target leaves the reactor core, is 2.47 μCi/Ci Mo-99 and 2.89 μCi/Ci Mo-99, respectively. Although the radionuclide species are not specified, USP limits the activity derived from α-emitting radionuclides to 0.01nCi/mCi Tc-99m at the time of administration.

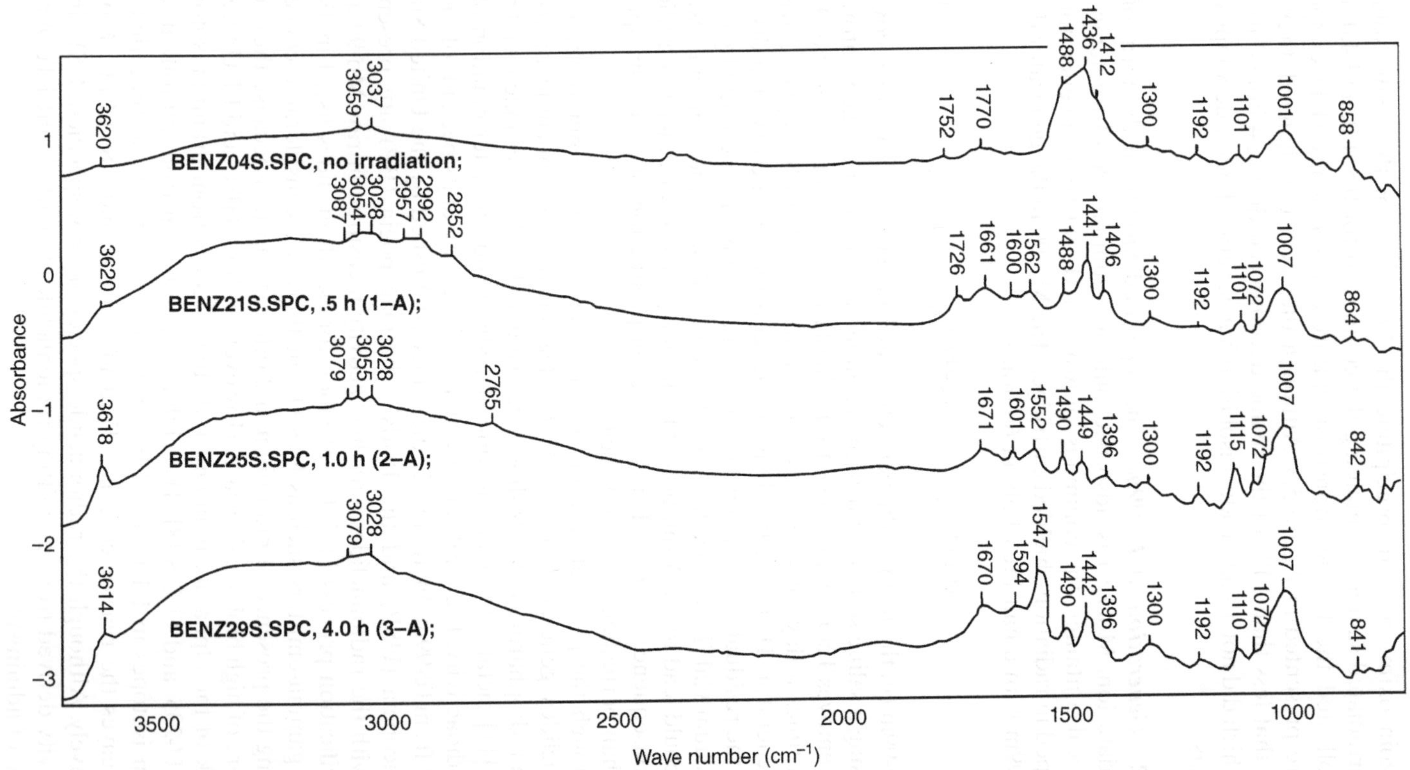

FIGURE 5.3 FTIR data for α-benzoin oxime after 0, 0.5, 1.0, and 4 hours of irradiation [6].

TABLE 5.3 Pharmacopeial Radionuclide Impurity Limits and Radionuclides Measured in Raw Mo-99 after Purification

Radionuclide	USP Limits (μCi/mCi Tc-99m)	BP Limits % of the Total Radioactivity[a]	Impurity Levels (μCi/mCi Tc-99m) [12]	Impurity Levels (μCi/mCi Tc-99m) [13]	Impurity Levels (μCi/mCi Mo-99) [5]	Impurity Levels (μCi/mCi Mo-99)[b] [7]
Ba-140					0.000036	<0.009
Ce-141					0.0000013	<0.003
Ce-143					0.0000017	<0.019
I-131	0.05	0.005	0.0015	<0.005	0.0053	0.051
I-133				<0.001	0.0063	0.479
I-135					0.00088	0.398
La-140					<0.00001	<0.070
Mo-99	0.15			<0.1		0.6
Nb-95					0.017	<0.002
Nb-97					0.00092	0.755[c]
Nd-147					<0.00026	<0.005
Np-239			0.004		<0.000019	<0.048
Pm-151					<0.0021	<0.033
Rh-105					<0.0004	0.018
Ru-103	0.05	0.005	0.001	<0.005	0.17	0.005
Sb-127					<0.043	<0.018
Sr-89	0.0006	0.00006			<0.00000023	
Sr-90	0.00006	0.000006			<0.0000000014	
Sr-91					<0.00000074	<0.017
Sr-92					<0.00000047	
Te/I-132					<0.00000071	0.022
Y-93					<0.0000008	<0.124
Zr-95					<0.0068	<0.003
Zr-97					<0.046	<0.340
Other γ-emitting impurities		0.01				
α-Emitting impurities		0.0000001				

[a] Monograph sodium pertechnetate injection obtained from Mo-99 extracted from fission products of uranium.

[b] Data collected 1 day after the end of irradiation.

[c] Nb/Zr-97.

Johansson and Mattsson [14] analyzed the first eluate of Tc-99m obtained from 31 different generators to determine if α-emitting radionuclides were present. They found significant amounts of Pu(239 + 240) in the pertechnetate eluted from the generator distributed by one manufacturer. Although generators based on fission are recognized as being potentially contaminated

TABLE 5.4 Chemical Purity of Tc-99 Eluates Measured over a Period of 2 Years Obtained from Six Different Producers [15]

Element/ Conc. (μg/L)	Producer					
	1	2	3	4	5	6
Al[a]	200–420	10–790	10–40	10–53	10–1000	10–510
B	3699	206	404	256	1441	1319
Sc	0.02	0.02	0.02	0.02	0.02	0.02
Mn	28	11	0.02	10	0.02	0.02
Ni	101	0.05	0.05	22	0.05	0.05
Zn	238	14	100	199	31	73
As	6	14	8	4	4	16
Br	573	3033	573	228	610	299
Sr	16	1	0.01	1	0.01	1
Ba	43	6	8	13	7	7
Pb, Bi, Th, U	<0.01	<0.01	<0.01	<0.01	<0.01	<0.01

[a] For Al, the authors presented the range.

with plutonium, significant amounts of plutonium were detected in the eluates derived from neutron activation produced Mo-99 generator. While fission-produced Mo-99 generators did not reveal detectable concentrations of plutonium in the eluates, the Pu(239 + 240)/Tc-99m activity ratios in the contaminated generator varied between 1.3×10^{-12} and 1.4×10^{-11}. The authors attributed the plutonium contamination to uranium impurities in the molybdenum batch.

In 1999, the Italian Association of Nuclear Medicine published the result of a multicenter survey, where 245 generators from seven different commercial sources were tested over a period of 2 years [15]. Although the authors did not relate the result to the manufacturer, these were Amersham, CSI, Mallinkrodt, Medgenix, NeN Du Pont, Sorin Biomedica, and Soreq. The parameters evaluated were, among others, chemical, radiochemical, and radionuclidic purities of the eluates. The results were presented according to suppliers and were compared to the limits established by the European Pharmacopoeia (Ph. Eur.). Table 5.4 presents the results found for chemical purity, where aluminum and other metals were determined. All samples presented Al below the level allowed by the pharmacopeia. For the other elements, for which no limits are established, boron and bromine presented the highest concentration, ranging from 206 to 3699 μg/L and from 228 to 3033 μg/L, respectively. The results of RCP reported in Table 5.5 were presented as percentage of purity and the minimum purity measured. Only one manufacturer supplied a product with an RCP below the limit of 95% established by the Ph. Eur., although the authors did not inform how many samples presented this low value. For radionuclidic purity, Mo-99, Ru-103, and I-131 were estimated. According to the results presented in Table 5.5, although these nuclides were

TABLE 5.5 Radiochemical Purity of Tc-99 Eluates Obtained from Six Different Producers [15]

Producer	% of the Total Radioactivity			Minimum Radiochemical Purity (%)
	Mo-99	Ru-103	I-131	
1	3.50×10^{-4}	1.92×10^{-6}	1.45×10^{-7}	98.8
2	5.50×10^{-4}	1.16×10^{-7}	2.22×10^{-7}	98.0
3	1.35×10^{-4}	3.36×10^{-8}	4.07×10^{-8}	99.4
4	4.60×10^{-3}	3.83×10^{-8}	1.13×10^{-7}	97.6
5	6.55×10^{-5}	1.33×10^{-8}	2.73×10^{-8}	99.5
6	7.20×10^{-3}	8.46×10^{-8}	2.16×10^{-6}	91.5
7	7.30×10^{-4}	1.17×10^{-7}	1.11×10^{-7}	99.0

detected in the eluates, their concentrations were below the limit established by the Ph. Eur.

Anderson et al. [13] evaluated the radionuclidic purity of a Tc-99m generator from a single supplier over a period of 6 months. The main contaminants found in the eluate were Mo-99, I-131, I-132, I-133, and traces of Ru-103 and Rh-105. Although they occasionally found impurities substantially larger than normal, most samples presented the levels displayed in Table 5.3. The authors observed that the largest activity ratio (I-131/Tc-99m) was found when the generator was eluted the second week after delivery and when the eluate was assayed several hours after elution. The problem with I-131 is its very long half-life (8 days) when compared with Tc-99m and even Mo-99. The ratio of I-131 activity to Tc-99m activity in the column increases with the passage of the time due to its longer half-life. Thus, the I-131 impurity becomes more significant near the end of the useful lifetime of a generator than at the beginning (assuming that only a small percentage of I-131 washes out of the column in any single elution).

5.1.1.3 Labeling Procedures The TcO_4^- obtained from Mo-99 generators is chemically treated using "kits" in order to produce the different Tc-99m complexes. These kits generally contain a reducing agent, which reduces Tc(VII) to a lower oxidation state, allowing it to bind to the complexing agent. The USP presents monographs for 27 different products containing Tc-99m. These products are shown in Table 5.6. Table 5.6 also shows the limits for radiochemical and radionuclidic purities.

The impurities associated with the T-99m generator are Mo-99 itself, which may be coeluted, Al ions either from the target material or from the alumina column of the generator where Mo-99 is adsorbed, traces of other radioisotopes, and also residues of the reducing agent and the solvent when the dissolution of the target occurs in acidic milieu (Fig. 5.2). Since Al interferes with labeling processes, pharmacopeial compendia limit its levels to 10 μg/mL in the eluate. The limits for the reducing agent, usually stannous chloride, and

TABLE 5.6 USP Tc-99m-Labeled Radiopharmaceuticals and the Respective Prescribed Radiochemical and Radionuclidic Purity

Product	Radiochemical Purity
Tc-99m albumin	>95%
Tc-99m albumin aggregated	>90.0%
Tc-99m albumin colloid	>90.0%
Tc-99m Apcitide	>90%
Tc-99m Arcitumomab	<5% free TcO_4^-
Tc-99m Bicisate	>90%
Tc-99m Depreotide	>90%
Tc-99m Disofenin	>90%
Tc-99m Etidronate	>90%
Tc-99m Exametazime	>80%
Tc-99m Fanolesomab	<10% free TcO_4^- and colloidal Tc-99m
Tc-99m Glucepate	>90%
Tc-99m Lidofenin	<10% free TcO_4^- , hydrolyzed Tc-99, and T-99 technetium tin colloid
Tc-99m Mebrofenin	>90%
Tc-99m Medronate	<10% free TcO_4^- , hydrolyzed Tc-99, and T-99 technetium tin colloid
Tc-99m Mertiatide	<6% free TcO_4^- , >90% Tc-99m Mertiatide
Tc-99m Nofetumomab merpentan	>85%
Tc-99m Oxidronate	>90%
Tc-99m Pentetate	<10% free TcO_4^- and hydrolyzed Tc-99
Tc-99m Pertechnetate	>95%
Tc-99m Ryrophosphate	>90%
Tc-99m (Pyro- and trimeta-) phosphates	>90%
Tc-99m red blood cells	>90%
Tc-99m Sestamibi	<10% free TcO_4^- and colloidal Tc-99m
Tc-99m Succimer	>85%
Tc-99m Sulfur colloid	>92%
Tc-99m Tetrofosmin	<10% free TcO_4^- and reduced hydrolyzed technetium
Radionuclidic purity is identical for all formulations	
Limits	Mo-99 < 0.15 kBq/MBq Tc-99m I-131 < 0.05 kBq/MBq Tc-99m Ru-103 < 0.05 kBq/MBq Tc-99m Sr-89 < 0.0006 kBq/MBq Tc-99m Sr-90 < 0.00006 kBq/MBq Tc-99m All other radionuclides (beta and gamma emitters) <0.01%

the solvent are 1 mg/mL Sn and 0.1%, respectively. Since the RCP for most products is limited to 90%, the 10% of impurity is related to the presence of Tc-99m in other chemical forms, which may behave differently in the body from the intended pharmaceutical form.

Technetium belongs to group VIIB of the periodic table and can exist in eight oxidation states, ranging from –1 to +7, where the most stable states are VII and IV. The stability of Tc in other oxidation states depends on the ligand and the chemical environment. In the pertechnetate ion (colorless), eluted from the Mo-99 generator, Tc is in the oxidation state VII, which does not react with any ligand, may suffer hydrolysis if the milieu is not sufficiently basic, and may be reduced to Tc (IV) to form the insoluble black oxide, TcO_2 [16]. Since Tc-99m is not administered directly to the patients and is unable to bind to any ligand, the labeling procedure, after elution, includes the reduction of Tc to a lower oxidation state. The most common oxidation state in labeled Tc-99m is V, and complexes are made up in order to reach different organs [17]. One of the events that could lead to low RCP is the incomplete conversion of Tc-99 from one oxidation state into another.

Tc-99m mertiatide (T-99m-MAG3) is an important radiopharmaceutical used in dynamic renal imaging. The method of preparation involves reconstitution of a lyophilized MAG3 with Tc-99-pertechnetate and boiling for 10 minutes. Noll et al. [18] investigated sources of radiochemical impurities in Tc-99m-MAG3, including steps for the ligand's synthesis. They observed that a by-product can be generated through the oxidation of the ligand molecule itself to form MAG3 disulfide, which also reacts with Tc-99m but is not useful for kidney imaging. The formation of Tc-99m-MAG3 disulfide can be minimized by careful preparation and storage to avoid oxidation. The authors also observed that another reason for the occurrence of impurities is the reaction time, which must be at least 10 minutes for complete labeling of Tc-99m. More recently, Seetharaman et al. [19] evaluated the importance of each step in the production of Tc99m-MAG3 for the RCP of the radiopharmaceutical. Concerns were raised regarding the information provided by the manufacturers on the preparation, analysis, and use of the product, which were alleged to be not only deficient but sometimes inadequate [20]. Moreover, the authors were wary of the practice among some radiopharmacies of reconstituting MAG3 kits with pertechnetate and of sending the product to the consumers for heating at the destination prior to administration. The authors confirmed that the boiling time should not be lower than 10 minutes and that the delay before boiling was less critical. Nevertheless, RCP purity after a 30-minute wait tended to be lower than after shorter delays (see Table 5.7).

Shattuck et al. [21] evaluated the RCP of Tc-99m-MAG3 over a 6-hour period after kit preparation using two methods: HPLC and the manufacturer's miniaturized chromatography system, Sep-Pak (Mallinckrodt Medical). The results for 36 vials reconstituted using saline obtained from plastic ($n = 16$) and glass ($n = 20$) were compared and showed no significant differences regarding RCP. However, RCP was different when determined using HPLC

TABLE 5.7 Tc-99m-MAG3 Radiochemical Purity after Different Boiling Times and Delay before Boiling [19]

	Radiochemical Purity (%)					
	1200 MBq/4 mL Preparation			1200 MBq/10 mL Preparation		
Time Delay (min)	1-min Boil	5-min Boil	10-min Boil	1-min Boil	5-min Boil	10-min Boil
1	38.1 ± 4.4	51.8 ± 11.6	92.5 ± 2.4	43.9 ± 7.2	44.4 ± 8.2	91.7 ± 2.0
15	42.3 ± 5.7	33.3 ± 0	93.0 ± 1.2	26.8 ± 1.9	39.7 ± 1.6	92.3 ± 2.0
30	38.9 ± 4.5	59.5 ± 6.8	89.6 ± 2.1	42.1 ± 0.1	42.8 ± 2.2	90.6 ± 0.6

TABLE 5.8 Radiochemical Purity of Tc-99-MAG3 at 30 Minutes and 6 Hours Postreconstitution Using 5 and 10 mL of Saline from Glass and Plastic Containers [21]

	Radiochemical Purity (%)	
Time after Reconstitution	5-mL Volume	10-mL Volume
Determination by HPLC		
30 min	91.4 (±3.5)	96.6 (±1.5)
6 h	90.8 (±1.5)	96.5 (±1.2)
Determination by Sep-Pak		
30 min	94.5 (±2.7)	98.5 (±1.3)
6 h	94.4 (±2.2)	98.0 (±1.1)

or Sep-Pak cartridge. As can be seen in Table 5.8, the RCP determined by Sep-Pak averaged at 2.5% higher than the RCP determined by HPLC ($P < 0.01$). These results show that the Sep-Pak procedure may not be as accurate as the HPLC method for the evaluation of the RCP of Tc-99m-MAG3. On the other hand, an investigation conducted by Millar and Hesselewood [22] showed that there were no statistically significant differences when comparing both techniques (HPLC and Sep-Pak) for accessing the RCP of Tc-99m-MAG3. The authors analyzed samples with RCP ranging from 43% to 98% and commented that the results of Sep-Pak are strongly dependent on operator technique, particularly the rate at which the cartridge is eluted. Moreover, unlike HPLC, no chromatogram is produced to confirm that good separation has been achieved.

In 1996, Stinger and Maltby [23] reported a year-long study on the control of RCP of Te-99m-MAG3 kits. Using 90% RCP as the acceptance limit, of approximately 300 kits, 25% failed. The failure was attributed to multiple factors such as eluate or saline solution conditions. In order to elucidate the incident, stannous chloride was added to failed kits, and it was observed

TABLE 5.9 Levels of Radiochemical Species in Tc-99m-MAG3 Resulting from Various Sodium Chloride Injection Incubation Conditions [25]

Incubation Condition	Time (min)	*n*	Hydrophilic Tc-99m Impurity (%)	Lipophilic Tc-99m Impurity (%)	Tc-99m-MAG3 (%)	Number of Fails
Plastipak	0	10	1.0 ± 0.2	2.6 ± 0.3	96.4 ± 0.5	0
Plastipak	15	10	1.1 ± 0.3	9.5 ± 5.3	89.4 ± 5.5	7
Norm-Ject	0	10	1.0 ± 0.2	2.5 ± 0.4	96.5 ± 0.5	0
Norm-Ject	15	10	1.0 ± 0.2	2.7 ± 0.4	96.3 ± 0.5	0
Monoject	15	10	0.9 ± 0.2	2.6 ± 0.2	96.6 ± 0.4	0
Lubricated rubber tips	15	5	1.4 ± 0.4	13.4 ± 6.4	85.2 ± 6.6	5
Wiped rubber tips	15	5	0.8 ± 0.3	17.0 ± 6.5	82.2 ± 6.5	5

that the RCP was recovered (>90%). The authors demonstrated that the poor RCP was caused by a disinfectant spray, containing 0.125% H_2O_2 furnished along with the kit, from which, only 3 μL was able to reduce the RCP to 62% [24]. However, by adding $SnCl_2$, the RCP was recovered, showing that the presence of the oxidizing agent was responsible for inefficient labeling of pertechnetate.

Another problem related to Tc-99m-MAG3 was observed by Waight et al. [25]. They demonstrated that the RCP of this radiopharmaceutical is affected by the residence time of the sodium chloride injection in the syringe used to reconstitute the formulation. Table 5.9 shows the RCP of Tc-99m-MAG3 formulations prepared with NaCl injection incubated for 0 and 15 minutes in different types of syringes. The results showed that for Tc-99m-MAG3 prepared in Plastipack and lubricated syringes and in syringes containing rubber plungers, the RCP decayed to unacceptably low values. Moreover, the effect was not observed in Tc-99m-sestamibi, Tc-99m-exametazime, and Tc-99m-tetrofosmin. The authors concluded that the exposure of plastic syringes to light promoted leaching of compounds into the solution, and one of these compounds destabilized the Tc-99m-tartarate, an intermediate complex in the labeling procedure [26]. This effect can be clearly seen in the chromatograms presented in Figure 5.4, where the peak of Tc-99m-MAG3 (M) is reduced to 67% after kit reconstitution with exposed sodium chloride injection and practically disappears, giving place to the pure Tc-99m-pertechnetate, after boiling in the presence of tartrate. The authors showed the influence of the fluorescent light on the leaching of substances from plastic syringes by exposing plastic and glass ampoules containing sodium chloride solution to a 50-W fluorescent light for 0, 1, 2, 4, and 7 days. These ampoules were used to prepare Tc-99m-MAG3 and the results are presented in Figure 5.5. The RCP remained

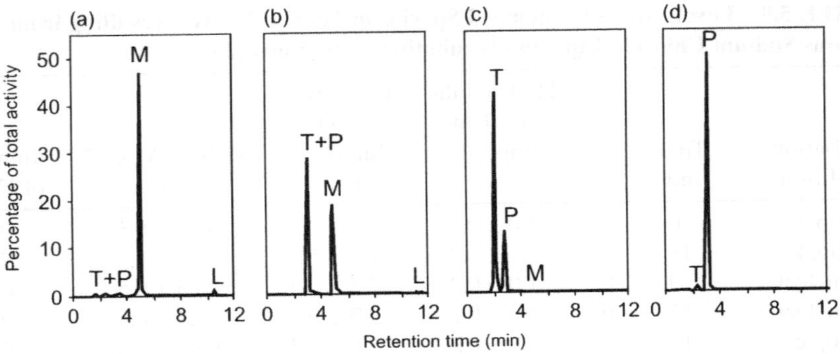

FIGURE 5.4 Chromatograms of (a) Tc-99m-MAG3 prepared with unexposed sodium chloride injection, RCP = 96%; (b) Tc-99m-MAG3 prepared with exposed sodium chloride injection, RCP = 67%; (c) MAG3 kit reconstituted with exposed sodium chloride injection, before boiling; and (d) tartrate kit reconstituted with exposed sodium chloride injection, after boiling. L, Tc-99m lipophilic impurity; M, Tc-99m-MAG3; P, Tc-99m-pertechnetate [26].

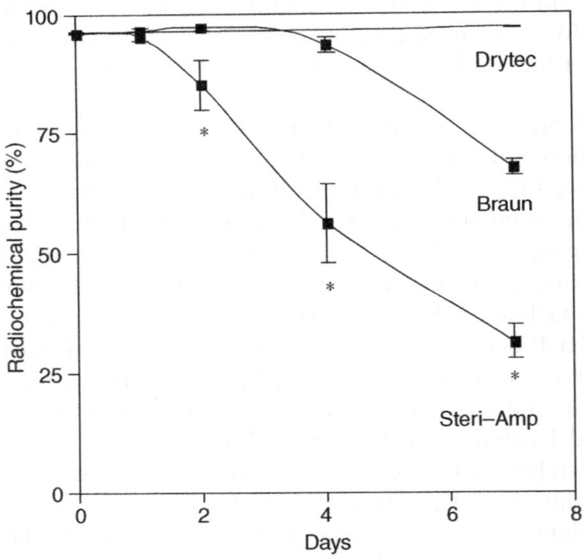

FIGURE 5.5 Radiochemical purity of Tc-99m-MAG3 of samples reconstituted using sodium chloride injection from plastic ampoules (Braun and Steri-Amp) and from glass ampoules (Drytec) [26].

acceptable for the entire period only when the glass ampoule (Drytec) was used. For the plastic ampoules, the longer the exposure, the lower the RCP of Tc-99m-MAG3. The authors attribute the oxidation of the intermediate tartrate complex to the leached substances.

The trivalent Tc-99m dimercaptosuccininc acid [99mTc(III)-DMSA], normally used for renal scintigraphy, can be converted into the pentavalent Tc-99m dimercaptosuccininc acid [99mTc(V)-DMSA] to be used as a tumor imaging agent by elevating the pH of the mixture with $NaHCO_3$. Washburn et al. [27] observed that the RCP of the product could be increased if, besides the addition of $NaHCO_3$ to the kit, oxygen were bubbled through the solution for 10 minutes. The poor RCP after $NaHCO_3$ addition was only due to the remaining excess of the reducing agent ($SnCl_2$), which was destroyed by the oxygen.

Kumar [28] carried out a careful investigation on the influence of each component on the preparation of 99mTc(V)-DMSA. It was confirmed that oxygen bubbling is an important step for maintaining the pentavalent state of technetium and therefore the RCP. An RCP of 100% was achieved within 20 minutes when compressed oxygen was bubbled for 5 minutes. When atmospheric oxygen was used instead of compressed oxygen, an RCP of 100% was attained only after 90 minutes.

Undesirable conversion of one oxidation state into another was also observed in Tc-99m hydroximethylene diphosphonate. To be used for bone scintigraphy, Tc-99m should be present as Tc (IV) in the phosphonate complex. Jacquot et al. [29] observed that the contact of this product with air led to the formation of $TcO_2.xH_2O$ and TcO_4^-, which were unable to be fixed by the bones.

Retrospective studies showed that the RCP of Tc-99m preparations increased during the 1990s [30]. The percentage of failures decreased from 6.17% in 1991 to 1.68% in 1998. Table 5.10 summarizes the RCP of each technetium compound and the number of tested samples. It is possible to see that the most critical products were Tc-99m-HIG, for which 6.7% of the samples failed, followed by Tc99m-mAb250/183, Tc-99m-HMPAO, and Tc-99m-MIBI, with 6.2%, 5.8%, and 4.8% failure, respectively. Tc-99m-MAG3, the most used Tc-99m radiopharmaceutical, presented 3.9% failure. The authors attributed the low RCP not to poor-quality kits or generators but to laboratory-specific conditions. Sample reconstitution or kit preparation involves complex chemical reactions, which may lead to other compounds with simple changes in the stoichiometry and side reactions.

5.1.2 Iodine-Based Products

Iodine isotopes, I-123 and I-131, are used in nuclear medicine for both diagnostic and therapeutic purposes. Diagnostic tests exploit the mechanism of absorption of iodine by the normal cells of the thyroid gland. The use in radiation therapy includes the treatment of thyrotoxicosis and thyroid cancer.

TABLE 5.10 Radiochemical Purity of Different Tc-99m Radiopharmaceuticals Consumed from 1991 to 1998 in Italy [30]

Radiopharmaceutical	No. of Samples Tested	Range (%)	Mean ± SD (%)	No. Failures
Tc-99m-DMSA	29	99.0–99.9	99.58 ± 0.40	0
Tc-99m-DTPA	60	95.3–99.9	99.38 ± 0.64	0
Tc-99m-MAG3	387	49.8–99.8	97.48 ± 5.13	5 (15[a])
Tc-99m-EC	273	54.5–99.7	98.58 ± 5.30	6
Tc-99m-DPD	132	92.2–99.9	99.14 ± 0.78	0
Tc-99m-MAA	85	97.5–99.9	99.55 ± 0.51	0
Tc-99m-IODIDA	9	99.0–99.9	99.64 ± 0.35	0
Tc-99m-HIDA	10	98.1–99.9	99.29 ± 0.59	0
Tc-99m-BTIDA	27	97.4–99.6	98.98 ± 0.76	0
Tc-99m-PTP	17	99.0–99.9	99.52 ± 0.40	0
Tc-99m-tin colloid	36	99.4–99.9	99.72 ± 0.15	0
Tc-99m-albumin colloid	12	96.1–99.9	98.53 ± 0.79	0
Tc-99m-mAb 431/26	35	96.9–99.8	98.76 ± 0.69	0
Tc-99m-mAb 250/183	226	67.8–99.9	96.90 ± 4.12	14
Tc-99m-IMMU-MN3 Fab'SH	11	96.7–99.3	98.18 ± 0.86	0
Tc-99m-HIG	193	41.0–99.9	97.66 ± 8.03	13
Tc-99m-HMPAO	156	66.0–97.8	88.52 ± 5.78	9
Tc-99m-ECD	134	81.6–97.1	93.72 ± 2.81	5
Tc-99m-MIBI	249	9.5–99.7	96.34 ± 7.66	1 (12[a])
Tc-99m-tetrofosmin	9	92.5–97.6	96.61 ± 2.31	0
Total	2090	9.5–99.9	96.92 ± 6.71	53 (27[a])

[a] Including preparations of fractionated kits.

Although being isotopes of the same chemical element, they are produced in different ways. While I-123 is obtained from the bombardment of enriched tellurium (Te-124) with protons, or by proton irradiation of enriched xenon (Xe-124), I-131 is produced by U-235 fission or by the bombardment of enriched tellurium (Te-124) with neutrons [31]. In both cases, the products are carrier-free and contain only minute amounts of naturally occurring I-127.

I-131 has been used to treat hyperthyroidism and thyroid cancer since the 1950s. In spite of this, I-123 has been gradually replacing I-131 as the isotope of choice in radiopharmaceuticals containing iodine. It provides a much lower radiation dose to the patient with similar specificity and sensitivity regarding the gamma ray energy used. However, I-131 is still a pharmacopeial item for consumption. The USP presents monographs for 11 radiopharmaceuticals labeled with iodine (Table 5.11), from which include six I-131's, three I-123's, and two I-125's.

The I-131 obtained from U-235 fission is produced together with Mo-99. However, while Mo-99 should be separated from the liquid phase of uranium

TABLE 5.11 USP and BP Iodinated Labeled Radiopharmaceuticals and the Respective Prescribed Radiochemical and Radionuclidic Purity

Radiopharmaceutical	Chemical Name	Radionuclidic Purity (%)	Radiochemical Purity (%)
I-123			
Sodium iodide	Sodium iodide	>90	>95.0
Iobenguane	(m-Iodobenzyl) guanidine sulfate	>97	>90.0
Iodohippurate	Sodium glycine, N-(2-iodobenzoyl)	>85	>97.0
I-131			
Sodium iodide	Sodium iodide		>95.0
Iobenguane	(m-Iodobenzyl) guanidine sulfate	>90	>99.0
Iodohippurate	Sodium glycine, N-(2-iodobenzoyl)	–	>97.0
Rose Bengal	4,5,6,7-Tetrachloro-2′,4′,5′,7′-tetraiodofluorescein	–	>90.0
Iodinated albumin	–	–	>97.0
Iodinated albumin aggregated	–	–	>94.0
I-125			
Iodinated albumin	–	–	>97.0
Sodium iothamate	Iothalamic acid	–	>98.0

target dissolution, iodine is recovered in the gas phase (see Fig. 5.2). As a consequence, contaminants for iodine are gases released in the uranium target dissolution. The yield of I-131 as a fission product of U-235 is 2.8%. It decays with a half-life of 8.02 days with beta and gamma emissions, and on decaying, I-131 transforms into Xe-131:

$$\text{I-131}_{(t\frac{1}{2}=8\text{ days})} \rightarrow \text{Xe-131}_{(t\frac{1}{2}=214{,}000\text{ years})} + \gamma + \beta^{-}. \quad (5.6)$$

Iodine-123, on the other hand, is a cyclotron (particle accelerator)-produced radioisotope. The targets used are either made of tellurium or xenon, and the choice depends on the cyclotron being used. To obtain the isotope from tellurium, either elemental tellurium or tellurium oxide has been used as a target [32].The process starts with the enrichment of Te-124, which corresponds to only 4.74% of the naturally occurring Te isotopes (see Table 5.12). The enrichment is carried out by centrifugation of a suitable gaseous compound of the element. Once Te-124 of adequate purity is available, the material

TABLE 5.12 Isotopic Composition of Natural Tellurium, after Enrichment and Iodine Isotopes Produced by Bombarding the Te Isotopes with Protons

Isotope	Natural Abundance (%)	After 91.86% Enrichment	After 99.87% Enrichment	Nuclear Reaction	Iodine Isotope	Half-Life
Te-120	0.09	<0.02%	<0.002%	(p,n)	I-120	1.35 h
Te-122	2.55	0.49%	<0.003%	(p,3n)	I-120	
				(p,2n)	I-121	2.12 h
Te-123	0.89	0.11%	0.006%	(p,4n)	I-120	
				(p,3n)	I-121	
				(p,n)	I-123	13.22 h
Te-124	4.74	91.86%	99.87%	(p,5n)	I-120	
				(p,4n)	I-121	
				(p,2n)	I-123	
				(p,n)	I-124	4.18 days
Te-125	7.07	3.17%	0.095%	(p,6n)	I-120	
				(p,5n)	I-121	
				(p,3n)	I-123	
				(p,2n)	I-124	
				(p,n)	I-125	59.408 days
Te-126	18.84	1.81%	0.020%	(p,6n)	I-121	
				(p,4n)	I-123	
				(p,3n)	I-124	
				(p,2n)	I-125	
				(p,n)	I-126	13.11 days
Te-128	31.74	1.55%	<0.010%	(p,8n)	I-121	
				(p,6n)	I-123	
				(p,5n)	I-124	
				(p,4n)	I-125	
				(p,3n)	I-126	
				(p,n)	I-128	25 min
Te-130	34.08	1.05%	0.008%	(p,8n)	I-123	
				(p,7n)	I-124	
				(p,6n)	I-125	
				(p,5n)	I-126	
				(p,3n)	I-128	
				(p,n)	I-130	12.36 h

Adapted from References 32–34.

is irradiated with protons to produce I-123, and the following reaction takes place:

$$\text{Te-124} + \text{p} \rightarrow \text{I-123}_{(t\frac{1}{2}=13.22\,\text{h})} + 2\text{n}. \tag{5.7}$$

One of the principal problems of the production of I-123 via Te-124 is the radionuclidic purity. The presence of other iodine isotopes in the irradiated product is a consequence of the enrichment of the target material. The bombardment of the other Te isotopes with protons may also lead, depending on the proton energy, to the formation of the iodine isotopes presented in Table 5.12.

Studies carried out to investigate the possibility of producing I-124 for medical purposes using natural Te targets show the relevant iodine nuclides produced by bombarding a nonenriched Te target with proton energy. El-Azony et al. [34], bombarding natural tellurium with protons of 55 MeV, produced I-123, I-124, I-125, I-126, and I-130. After 60 hours, the level of I-123 and I-130 decreased to 13% and 1%, respectively, while the levels of I-125 and I-126 increased to 3.6% and 31.5%, respectively, in relation to the percentage of I-124 produced. The production of this isotopic ratio is a consequence of the proton energy, which promotes the formation of determined nuclides, and also of the nuclides' half-lives: While I-123 and I-130 have half-lives of hours, I-125 and I-126 have half-lives of days. While the amount of the former decreases after hours in relation to the amount of I-124 (which has a half-life of 4.8 days), the amount of the latter increases, relatively speaking, since its decay occurs at a much slower rate.

Acerbi et al. [35] showed that irradiating a natural tellurium target with a proton energy of 30 MeV, in order to produce I-123, six iodine isotopes were found in the following proportion, if the yield of I-123 is considered 100%: 17% I-124, 13% I-126, 1.3% I-121, 0.6% I-128, and 0.06% I-130. By using an enriched target (91.86% Te-124), the yield of I-124 was considerably reduced and that of the other isotopes was reduced to insignificant amounts. Figure 5.6 indicates that the yield of I-124 at the end of irradiation is less that 1% of the I-123 yield. However, due to the difference in the half-life, the I-124/I-123 ratio in the product increases linearly over time (small graph in Fig. 5.6). I-124 is the main impurity in I-123 produced from Te-124. This is the reason why the Te-124 purity should be the highest possible. The sources of I-124 in the production of I-123 from tellurium targets can be seen in Table 5.12. Kondo et al. [32] observed that the utilization of 99.9% versus 91.8% Te-124 increases the radionuclidic purity of I-123 relative to I-124 by a factor of 2.

Figure 5.7 shows the production rates of iodine nuclides as a function of incident proton energy for different Te target enrichments. Although Figure 5.7a,b displays only results for I-123 and I-124 because the production of the other iodine nuclides is insignificant, it is possible to see in Figure 5.7c that at 25–30 MeV, the proton energy usually applied for I-123 production, the production of I-124 is practically 50% that of the I-123 production. Therefore, the purer the Te-124, the lower the I-124 yield.

To make the target, tellurium elemental or tellurium oxide (TeO_2) is electrodeposited onto a copper plate or mixed with aluminum powder and packed into a capsule in a He atmosphere. The capsule is mounted with the cup side facing the incident proton beam. After 3–8 hours' irradiation, the target material is mixed with 7–10 mol/L H_2SO_4 and heated under reflux at 110°C for approximately 2 hours. The complete dissolution of the material occurs after dropwise addition of 30% H_2O_2. The solution is cooled to room temperature and I-123 is recovered by adsorption [36] or by distillation [35].

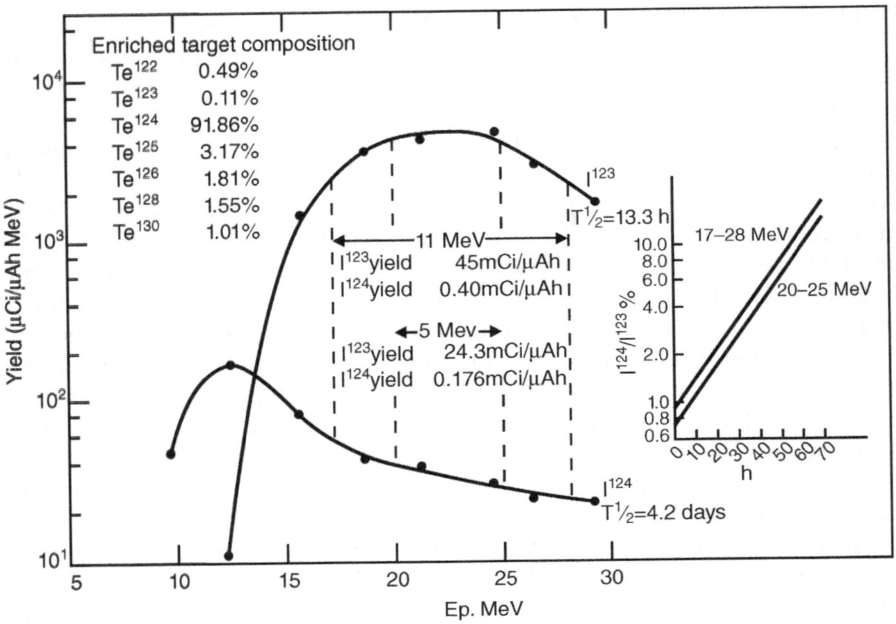

FIGURE 5.6 Excitation function for the production of I-123, I-124 in an Te-124 enriched target (91.86%) for 10- to 30-MeV protons. Percentage ratio of I-124 to I-123 as a function of the time elapsed after irradiation for two different target thicknesses of 11 and 5 MeV, respectively [35].

The adsorption takes place on a platinum felt, which is added to the solution containing the I-123. The solution is stirred for 1 hour and the liquid part decanted off. The desorption of iodine from the Pt felt is achieved by boiling the adsorbent in an alkaline solution of sodium bisulfite for 1 hour.

For the distillation, an extra portion of 30% H_2O_2 is added and the distillation is carried out at 100–120°C under a nitrogen flow. The distilled fraction is collected in 0.01 mol/L NaOH, and after 6 hours, the yield of the distillation (I-123) reached 80–90%. To recover I-123 from the distillate, iodine (I_2) is reduced to iodide (I^-) with sodium nitrite and extracted with an organic solvent (CCl_4). The extraction, however, is satisfactory only if KI is added as carrier to the solution. This is a disadvantage of the procedure since the addition of KI decreases the specific activity. The solvent can be eliminated by evaporation or the iodide can be reduced with sodium sulfite and back-extracted in physiological solution [35].

Contamination sources of I-123 production from the Te-124 target are

Te-124 purity: Depending on the level of Te-124 enrichment, other iodine isotopes with half-lives greater than that of I-123 may be produced. I-124 is the most frequent impurity.

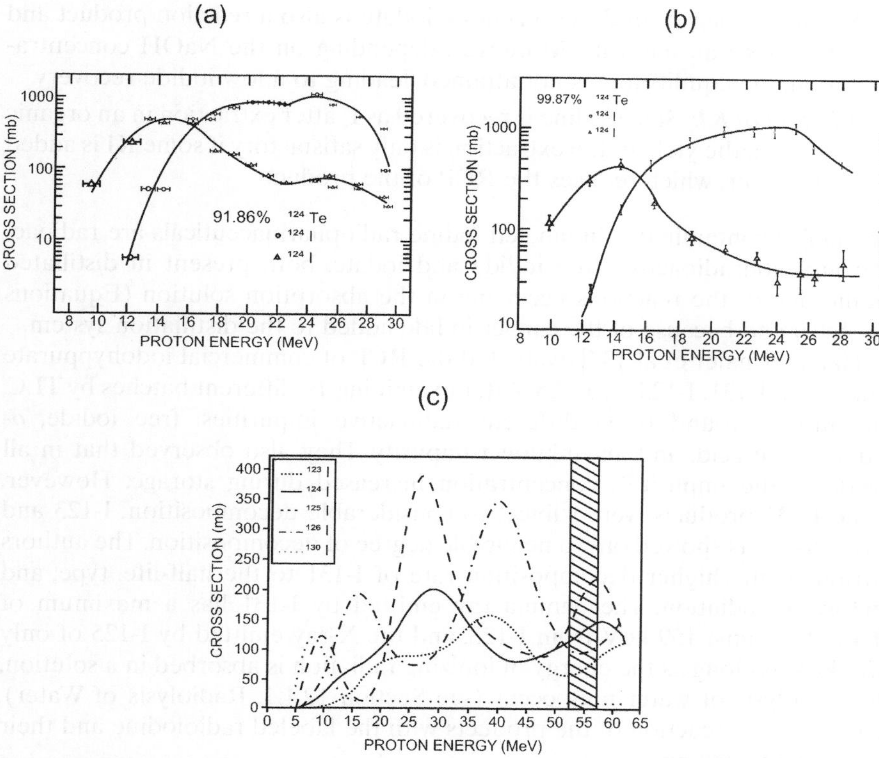

FIGURE 5.7 Excitation function of the tellurium proton irradiation: (a) excitation function for I-123 and I-124 by irradiation of 91.86% enriched Te-124, (b) excitation function for I-123 and I-124 by irradiation of 99.87% enriched Te-124, and (c) excitation function for I-123, I-124, I-125, I-126, and I-130 by irradiation of nonenriched tellurium target. Graphs (a) and (b) from Reference 32, and graph (c) from Reference 34.

Distillation temperature: Although the tellurium melting point is 450°C, if the temperature for the I-123 distillation (184°C boiling point) is too high, tellurium itself may distill, contaminating the iodine solution.

NaOH concentration in the collection solution: Since I-123 is distilled or adsorbed as I_2, the NaOH has the function of converting the volatile iodine into iodide through the following reactions:

$$I_2 + NaOH \rightarrow NaI + IOH \quad (5.8)$$

$$IOH + NaOH \rightarrow NaIO + H_2O \quad (5.9)$$

$$3NaIO \rightarrow NaIO_3 + 2NaI. \quad (5.10)$$

As can be seen from these reactions, iodate is also a reaction product and therefore an impurity. Moreover, depending on the NaOH concentration, the equilibrium is not attained, leading to a low iodide recovery.

Addition of KI: Since iodine is recovered as I_2 after extraction in an organic solvent, the yield of the extraction is only satisfactory if some KI is added as carrier, which reduces the RCP of the product.

The major contaminants in labeled iodine radiopharmaceuticals are radioactive and nonradioactive free iodide and iodate, both present in distillated iodine due to the reactions occurring in the absorption solution (Equations 5.8–5.10) and because of the carrier iodide added to the distillation system.

Hammermaier et al. [37] evaluated the RCP of commercial iodohyppurate containing I-131, I-123, or I-125. After examining 19 different batches by TLC, the authors found three different radioactive impurities: free iodide, *o*-iodobenzoic acid, and an unknown impurity. They also observed that in all products, the impurity's concentration increased during storage. However, while I-131 products were subject to considerable decomposition, I-123 and I-125 products showed only a negligible degree of decomposition. The authors attributed the higher decomposition rate of I-131 to the half-life, type, and energy of radiation. The gamma ray emitted by I-131 has a maximum of 364 keV against 159 keV from I-123, and the X-ray emitted by I-125 of only 27.5 keV. As long as the energy of ionizing radiation is absorbed in a solution, the radiolysis of water may occur (see Section 12.1.1, Radiolysis of Water), leading to the reaction of the products with the labeled radioiodine and their consequent decomposition.

Wafelman et al. [38] evaluated the stability of I-131-labeled metaiodobenzylguanidine (MIBG) regarding the formation of free I-131 as a consequence of radiolysis. Although the formulation contains benzyl alcohol, a substance known for its free radical scavenging properties, the amount of free I-131 in 23 infusion concentrate MIBG samples increased from $2.2 \pm 0.6\%$ to $3.6 \pm 0.39\%$ 1 day after production. In another study [39], the authors investigated the stability of MIBG under different storage conditions considering the drug expiration date and handling practice. In products kept on dry ice, the free iodine concentration increased from $1.9 \pm 0.34\%$ at production to $4.4 \pm 0.67\%$ at expiration (after 3 days). After thawing, at room temperature, the radiolysis effect was much higher; the percentage of free I-131 increased to 3.1% immediately after thawing, to 6.6% after 5 hours, and to 16.6% after 24 hours.

The elevated energy of the gamma rays emitted by I-131 when compared to the radiation emitted by other radionuclides used in nuclear medicine may be responsible for the accentuated radiolysis effect in I-131-labeled products. These effect lead not only to the formation of free I-131 but also to a decrease in the immunoreactivity of monoclonal antibodies (mAbs) labeled with this radioisotope.

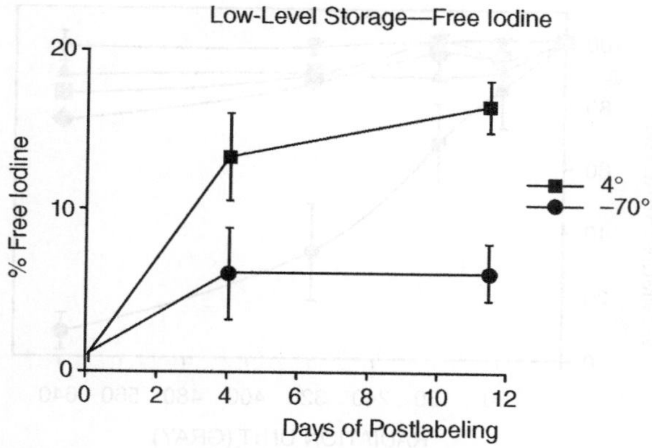

FIGURE 5.8 Increase in the free iodine concentration of a monoclonal antibody labeled with I-131, after 12 days' storage at 4 and −70°C [40].

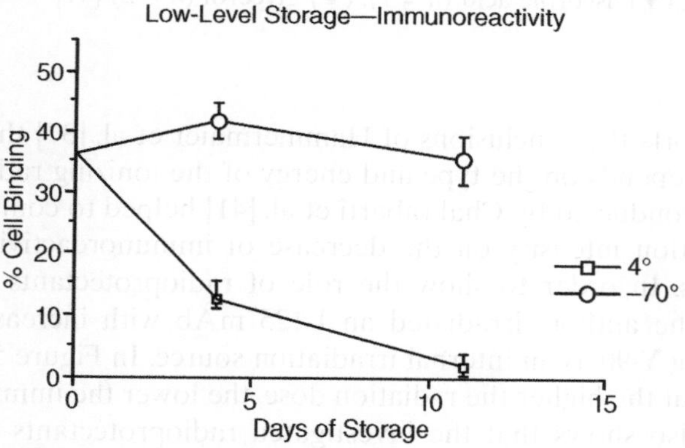

FIGURE 5.9 Decrease in the immunoreactivity of a monoclonal antibody labeled with I-131, after 12 days' storage at 4 and −70°C [40].

Wahl et al. [40] demonstrated that radiolysis promoted the release of free I-131 and protein breakdown over a relatively short time period when low therapeutic levels of I-131 were attached to mAbs. As occurred with the abovementioned I-131-labeled MIBG, freezing the samples greatly limited the ability of radiation-induced free radicals to diffuse in solution and to produce molecule damage and free iodine release (Fig. 5.8). Additionally, Figure 5.9 shows the decrease in product immunoreactivity in samples preserved at 4 and −70°C. A clear decrease in immunoreactivity was seen at 4 days' postlabeling and, particularly, after 11–12 days when stored at 4°C. This loss was essentially diminished when stored at −70°C. The authors also investigated the radiolysis provoked by I-125 under the same conditions. The more modest effect caused

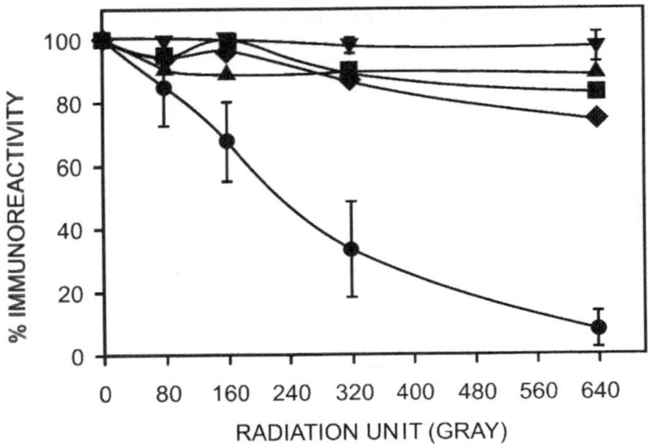

FIGURE 5.10 Decrease in the immunoreactivity of I-125-labeled monoclonal antibody evaluated after mixing with Y-90-DTPA (internal irradiation) in the presence and absence of different radioprotectants: (●) control (*n* = 6), (■) HAS (*n* = 2), (▲) cysteamine (*n* = 2), (▼) ascorbic acid (*n* = 7), (◆) glycerol (*n* = 2) [41].

by I-125 supports the conclusions of Hammermaier et al. [37] that the extent of radiolysis depends on the type and energy of the ionizing radiation.

The study conducted by Chakrabarti et al. [41] helped to confirm the influence of radiation intensity on the decrease of immunoreactivity of iodine-labeled mAbs. In order to show the role of radioprotectants (free radical scavengers), the authors irradiated an I-125 mAb with increasing doses of radiation using Y-90 as an internal irradiation source. In Figure 5.10, it is possible to see that the higher the radiation dose, the lower the immunoreactivity. Figure 5.10 also shows that the investigated radioprotectants were able to reduce the immunoreactivity loss. In another study designed to improve the quality of I-131 mAbs [42], the authors confirmed that in the absence of a free radical scavenger, the RCP of I-131-labeled mAb was greatly affected by autoradiolysis. Figure 5.11 shows the chromatograms obtained for the I-131-labeled mAb 4 days after production, before and after pH adjustment to 11 and addition of ascorbic acid. It is possible to see that without the measures taken to avoid the effect of radiolysis, iodate (IO_3^-) and other unidentified species were formed. Moreover, the authors measured the pH of the solution, which, because of radiolysis, had dropped from 11 to 7–9.

A common procedure in I-123 production is the recovery of enriched Te-124 from the dissolved target for the fabrication of new targets. The usual way to recover tellurium from the acid solution of the dissolved target has three steps: (i) Neutralize the acid with ammonium carbonate; (ii) reduce Te(VI) to Te(IV) with sodium bromide; and (iii) reduce Te(IV) to Te(0) with sodium sulfite. The subproducts of these reactions are bromine from step 2 and sodium sulfate from step 3. Zielinski et al. [43] observed that the contamination of the

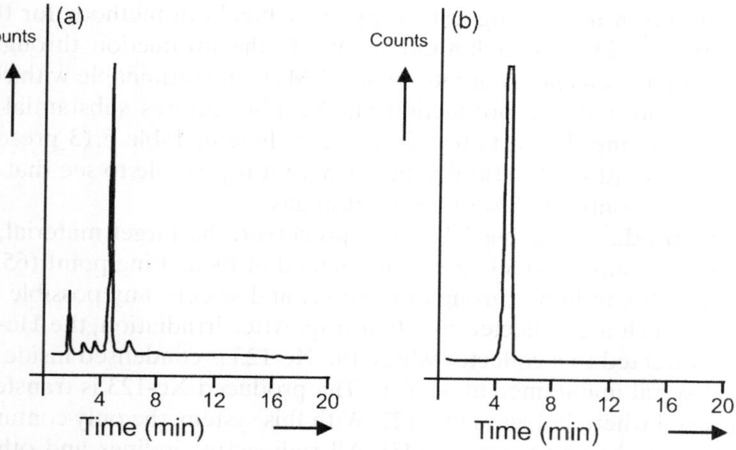

FIGURE 5.11 HPLC profile of aged I-131-labeled monoclonal antibody solution in a blackened delivery vial (3.7 GBq in 500 μL): (a) 4 days after production and (b) after adjustment of pH, transfer to reaction vial, and addition of ascorbic acid. Percentages: (a) iodate ($R_t = 1.7$ minutes) = 13%, iodide ($R_t = 5.0$ minutes) = 79%, unidentified components = 8%; (b) iodide ($R_t = 5.0$ minutes) = 100% [42].

recycled product with sodium and bromine should be avoided since the neutron flux near the target during proton irradiation is sufficient to activate sodium to Na-24 and bromine to Br-80, Br-80m, Br-82, and Br-82m. Radioactive bromine, if present, distills from sulfuric acid along with iodine during the subsequent I-123 separation.

Kudelin et al. [44] investigated the recovery of TeO_2 from a dissolved Te-124 target and the purity of the recycled target material. Although recovering TeO_2 with a purity of 95–97%, they found a series of impurities, the most abundant being Si (1.20%), Ca (0.12%), Ti (0.06%), Fe (0.05%), Mg (0.04%), Cr (0.02%), Sn (0.02%), Al (0.01%), and Zn (0.01%). The presence of Si was attributed to the quartz apparatus and the other elements to accidental melting of the target holder during irradiation.

According to Kondo et al. [32], the highest purity I-123 available is produced via Xe-123 decay since the target is in a gaseous form and is therefore easily separable from the solid I-123 produced. Xe-123 is an artificial radioactive isotope produced either by bombarding the natural isotope Xe-124 with protons or by bombarding I-127 with protons, according to the reactions represented below:

$$\text{Xe-124} + \text{p} \rightarrow \text{Xe-123}_{(t\frac{1}{2}=2.1\,\text{h})} + \text{p} + \text{n} \rightarrow \text{I-123} + \beta^+ \quad (5.11)$$

$$\text{I-127} + \text{p} \rightarrow \text{Xe-123}_{(t\frac{1}{2}=2.1\,\text{h})} + 5\,\text{n} \rightarrow \text{I-123} + \beta^+. \quad (5.12)$$

Although presenting the highest purity available, both methods for the production of Xe-123 present limitations. While the production through I-127 requires a cyclotron energy greater than 55 MeV (not attainable with compact medical cyclotrons), the production via Xe-124 requires substantial investments, considering the costs for Xe-124 enrichment. Table 5.13 presents the isotopic composition of natural xenon, where it is possible to see that Xe-124 corresponds to only 0.095% of the natural gas.

For the production using I-127 as a precursor, the target material, NaI, is placed into a stainless steel vessel and heated at its melting point (651°C). A He–Xe gas stream flows through the vessel and sweeps any possible volatile NaI or I_2, which are collected in a first trap. After irradiation, the He–Xe gas stream is directed to a collector where the Xe-123 is condensed inside a stainless steel spiral maintained at –196°C. The produced Xe-123 is transferred to a glass vessel where it decays to I-123. With this system, the only contaminants derive from radioactive xenons [45]. All radioactive iodines and other contaminants remain in the target vessel or are trapped in the first trap. The other xenon radioisotopes that can be produced and therefore can be in the distillate are Xe-121, Xe-122, and Xe-125, whose decay products are presented in Table 5.13. According to the data in Table 5.13, possible impurities originated in this process are the stable tellurium isotopes Te-121, Te-122, and Te-125, and Sb-121.

A manufacturer of NaI-123 capsules for diagnostic use states in the package insert that the radionuclidic composition of the product at calibration

TABLE 5.13 Isotopic Composition of Natural Xenon and Xenon Nuclides for Proton-Induced Reactions on I-127 Target [45]

		Xenon Products by I-127 Bombarding			
Isotope	Natural Abundance (%)	Nuclear Reaction	Xenon Isotope	Half-Life	Decay Product
Xe-124	0.095	(p,n)	Xe-127	36.406 days	I-127
Xe-126	0.089	(p,2n)	Xe-126	Stable	–
Xe-128	1.91	(p,3n)	Xe-125	17.3 h	I-125 (60.25 days) → Te-125
Xe-129	26.4	(p,4n)	Xe-124	Stable	–
Xe-130	4.07	(p,5n)	Xe-123	2.08 h	I-123 (13.22 h) → Te-123
Xe-131	21.23	(p,6n)	Xe-122	20.0 h	I-122 (3.63 min) → Te-122
Xe-132	26.91	(p,7n)	Xe-121	38.85 min	I-121 (2.12 h) → Te-121 (16.78 days) → Sb-121
Xe-134	10.44	(p,8n)	Xe-120	40 min	I-120 (1.35 h) → Te-120
Xe-136	8.86				

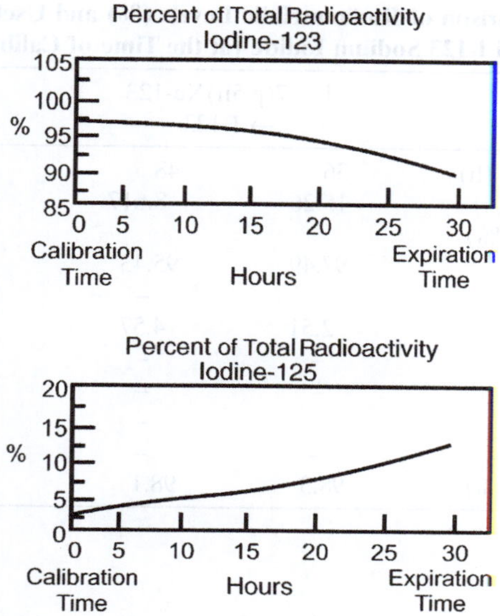

FIGURE 5.12 Percentage of the total radioactivity in a sodium iodide (I-123) capsule due to I-123 itself and due to I-125 during the shelf-life period.

(delivery time) is 97.0% I-123, 2.9% I-125, and 0.1% Te-121. Consistent with the radionuclidic impurities, this radioisotope was produced through the reaction I-127(p,5n)X-123 → I-123. Since the half-lives are different, the radionuclidic composition changes over time so that at the expiration time (30 hours), the concentration of I-123 decreased to 87.2%, while the concentration of I-125 increased to 12.4%. This change can be seen in Figure 5.12.

Jungerman and Lagunas-Solar [46] compared the radionuclidic purity of I-123 (as NaI) obtained through Te-124 and I-127 targets. While the only radionuclidic impurity found in I-123 produced from I-127 was I-125, the product obtained by Te-124 contained Na-24 and elevated contamination by I-124. Table 5.14 presents the radionuclidic levels in I-123 produced by both methods at 36 and 48 hours after the calibration time, as well as the useful proton yield. From the data presented in Table 5.14, it is possible to see that the growth of radioimpurities limits the useful photon yield of the Te-124 method particularly at the 48-hour expiration time (although the purity of the 124-Te target was not informed).

5.1.3 Fluorine-Based Products

Radionuclides used in positron emission tomography (PET) scanning are typically isotopes with short half-lives, such as carbon-11 (~20 minutes), nitrogen-13 (~10 minutes), oxygen-15 (~2 minutes), and fluorine-18 (~110 minutes). They

TABLE 5.14 Comparison of Radionuclidic Levels (%) and Useful Photon Yields for a Nominal 100 μCi I-123 Sodium Iodide (at the Time of Calibration)

I-123 Production	I-127(p,5n)Xe-123 → I-123		Te-124(p,2n)I-123	
Time after calibration (h)	36	48	36	48
Total radioactivity (μCi)	15.26	8.617	14.48	7.75
Radionuclidic levels (%)				
I-123	97.49	95.43	78.38	68.03
I-124	–	–	21.10	31.49
I-125	2.51	4.57	–	–
I-126	–	–	–	–
I-130	–	–	–	–
I-131	–	–	–	–
Na-24	–	–	0.52	0.48
Useful photon yield (%)	98.3	98.1	75.7	66.4

Source: Reference 46.

are incorporated either into compounds normally used by the body such as glucose (or glucose analogues), water, or ammonia, or into molecules that bind to receptors or other sites of drug action. These radionuclides are produced in a cyclotron by bombarding atoms able to be converted into positron-emitting isotopes.

Nuclides with a number of protons higher than the number of neutrons are unstable and decrease the number of protons through the emission of a positron (particle with a positive charge). In this reaction, a proton is converted into a neutron, and the element decays to another element with one atomic number less (see the decay of ^{18}F below). A positron is an antielectron or the antiparticle of an electron. A positron ejected by a radionuclide travels a very short distance in a tissue until combining with an electron from the medium. Because they are matter and antimatter, their collision results in annihilation, producing a pair of gamma rays moving in opposite directions. These gamma rays are then detected when they reach a scintillator in the scanning device:

$$p \rightarrow n + e^{+} + \gamma \tag{5.13}$$

$$e^{+} + e^{-} \rightarrow \gamma + \gamma. \tag{5.14}$$

Fluorine-18 ([^{18}F]) is one of the positron emitters most used for PET imaging, with a half-life of 109.8 minutes. In the production of [^{18}F], two kinds of targets may be used: water and neon gas. The water target produces fluoride ions and requires [^{18}O]-enriched water, and neon gas produces fluorine gas. Table 5.15 presents target properties and reactions involved in the production of both species.

TABLE 5.15 Nuclear Reactions for the Production of ^{18}F from Oxygen and Neon

Source	Isotopic Abundance (%)		Target	Enrichment	Nuclear Reaction	Necessary Energy (MeV)	Irradiation Time (h)
Oxygen	O-16	99.757	O-18	95–98%	O-18 + p → F-18 + n	10–19	1–2
	O-17	0.038					
	O-18	0.205					
Neon	Ne-20	90.48	Ne-20	–	Ne-20 + d → F-18 + α	8–9	1–2
	Ne-21	0.27					
	Ne-22	9.25					

By emitting a positron, [^{18}F] decays to [^{18}O] according to the reaction

$$^{18}F_{(t\frac{1}{2}=109.8\ \mathrm{min})} \rightarrow {}^{18}O + e^{+} + \nu.^{1} \tag{5.15}$$

5.1.3.1 [^{18}F]FDG Production and Labeling Although a number of [^{18}F]-labeled radiopharmaceuticals are commercially available, 2-deoxy-2-(^{18}F) fluoro-D-glucose, or fluorodeoxyglucose (^{18}F), commonly abbreviated as ^{18}F-FDG or FDG, is, to date, the most used for PET and is the PET radiochemical on which this chapter focuses. The synthesis of FDG involves the following basic steps, which are carried out in automated synthesis modules [47]: [^{18}F] Fluoride ($^{18}F^{-}$) is produced by bombarding enriched [^{18}O]water (95–98%) (0.3–3.0 mL), housed in a silver chamber at high pressure (30 bar), with protons (10–19 MeV) in a cyclotron for 1–2 hours. Once the process is finished, the irradiated water ([^{18}O]H_2O) is transferred to a $^{18}F^{-}$ separation cartridge (an anion-exchange resin in the form of HCO_3^{-}), which retains the dissolved fluoride ions, while the [^{18}O]H_2O passes through the cartridge and is collected separately. The $^{18}F^{-}$ ions are eluted into a reaction vial with a small volume (usually 0.5 mL) of a solution of potassium carbonate and Kryptofix 222 in acetonitrile (ACN), or tetramethyl ammonium carbonate or tetrabutyl ammonium bicarbonate in the same solvent. Residual water and ACN, which form an azeotropic mixture with a 16:84 ratio, respectively, are removed by distillation with the aid of a nitrogen stream. The organic precursor, mannose triflate in ACN, is added and the mixture is heated at 80–90°C for 5 minutes to form an intermediate, 2-deoxy-2-(^{18}F)fluoro-1,3,4,6-tetra-*O*-acetyl-β-D-glucopyranose. In order to generate FDG from the intermediate, the acetyl groups are hydrolyzed either with HCl by heating the solution at 130°C for 10–15 minutes or with NaOH at room temperature. The solution is then loaded into a purification column set, which encloses two ion-exchange columns (H^{+} and HCO_3^{-}), an aluminum oxide column, and a C-18 column, and allows for the elimination of synthesis reagent residuals and solvents. Finally, the [^{18}F]

[1] ν is a neutrino.

FDG is eluted with physiological saline and is sterilized by passing through a 0.2-μm membrane filter. An RCP >90% is expected [47]. Figure 5.13 shows schematically the steps of the [^{18}F]FDG production.

The following contaminants and impurities may be associated with FDG production:

- species formed during [^{18}O]H_2O irradiation
- residual solvents and components used in the labeling
- radiolysis products generated due to the elevated activity of the labeled compound.

Pharmacopeial compendia [48, 49] list the impurities that should be checked in FDG injection. Although similar, the requirements are somewhat different among compendia. Table 5.16 lists tests and limits for impurities established by two pharmacopeias, USP and Ph. Eur. Among them, radionuclidic impurities might be formed during [^{18}O]H_2O irradiation, radiochemical impurities after radiolysis of the labeled compound, and chemical impurities are synthesis residual substances.

5.1.3.2 Species Formed during [^{18}O]H_2O Irradiation In the isotopic composition of natural oxygen, only 0.205% is O-18, while the other isotopes are O-16 (99.757%) and O-17 (0.038%). For F-18 production, the O-18 percentage should increase to 95–98%. The enrichment is based on physical processes (distillation) on common water, and because ^{18}O is not itself a radioactive species, the ^{18}O[H_2O] obtained for ^{18}F production is almost free of contaminants. Table 5.17 presents the specification of enriched water supplied by different companies.

Although the target for [^{18}F]FDG is water, in a noncarrier added process, radionuclidic impurities may be formed during bombardment, arising from the water elements (formation of tritium, a radioactive hydrogen isotope) or from radio activation of the metal target housing. The target vessel is usually made of silver or titanium, covered with a thin metallic foil that allows the entrance of the proton beam. The material most commonly used for the entrance is Havar foil, a mechanically robust and chemically inert alloy, whose composition is presented in Table 5.18. These metallic components are susceptible to activation and oxidation because they are exposed to both an intense primary beam of protons and to a large flux of secondary neutrons produced by the O-18(p,n)F-18 reaction in the target.

Table 5.19 presents the concentration of metallic components found in irradiated water, not only derived from Havar foil but also from other parts of the target [50]. While Fe, Cr, Co, and Ni were originated in the Havar foil, Ag, Ti, and Cu arose from the vessel. The problem with the formation of these cationic species, at the level of concentration that they were found, is the stability of the complexes that some of them form with fluoride anions, mainly Fe. Considering that the amount of fluoride in the irradiated water [51] reached

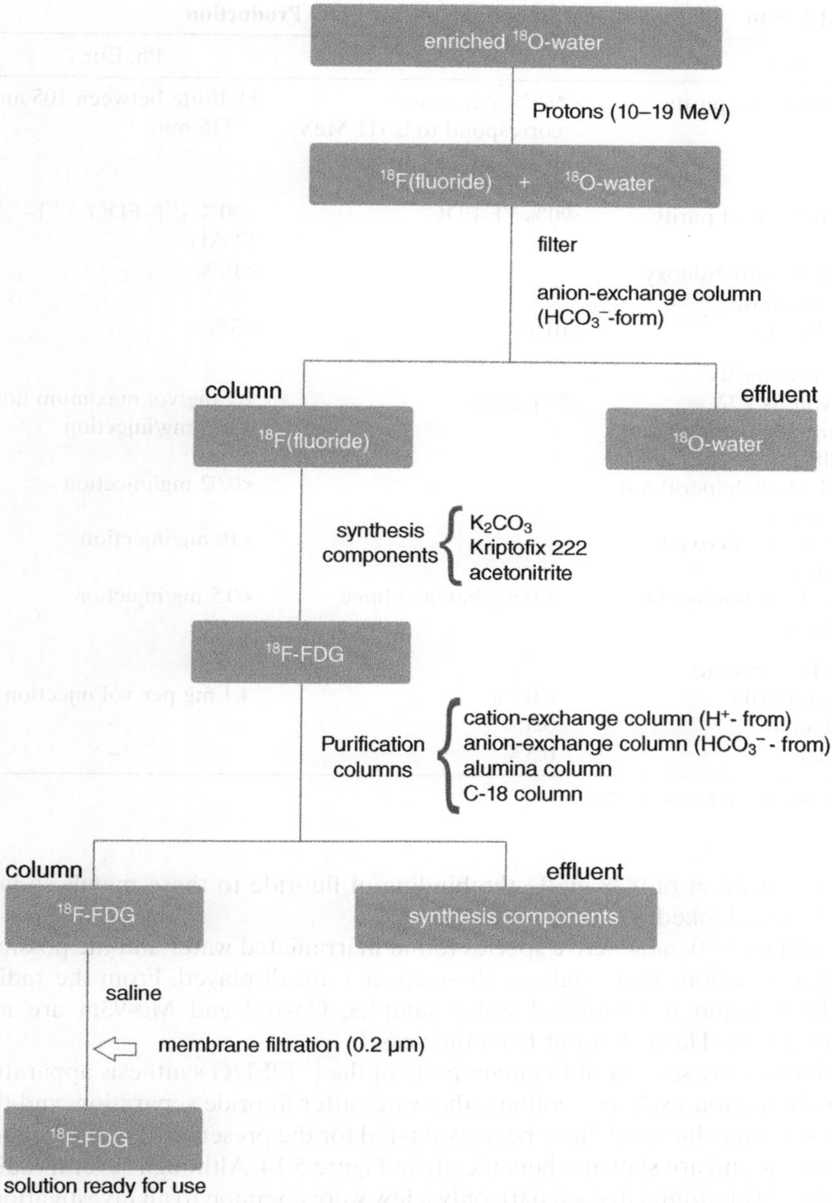

FIGURE 5.13 Schematic representation of [^{18}F]FDG production.

TABLE 5.16 Pharmacopeial Requirements for FDG Production

Parameter	USP	Ph. Eur.
Radionuclidic purity	>99.5% γ-emission correspond to 0.511 MeV, 1.022 MeV or Compton scatter peaks of ^{18}F	Half-life between 105 and 115 min
Radiochemical purity	>90% ^{18}F-FDG	>90% (^{18}F-FDG + ^{18}F-FDM)
- 2-^{18}F-Fluoro-2-deoxy-D-mannose	–	<10%
- ^{18}F-Fluoride	<10%	<5%
Chemical purity		
- Kryptofix 222	<50 μg/mL	2.2 mg/vol maximum dose
- Tetra-alkyl ammonium salt[a]	–	<2.75 mg/injection
- 4-(4-Methylpiperidino) pyridine[a]	–	<0.02 mg/injection
- 2-Fluoro-2-deoxy-D-glucose	–	<10 mg/injection
- 2-Chloro-2-deoxy-D-glucose	<1.0 mg/batch volume	<0.5 mg/injection
Residual solvents		
- Acetonitrile	0.04%	4.1 mg per vol injection
- Ethanol	0.5%	–
- Ether	0.5%	–

[a] Alternative to Kryptofix 222.

a concentration of 0.38 mg/L, the binding of fluoride to these metals should not be overlooked.

In Table 5.20, radioactive species found in irradiated water and the possible nuclear reactions that originate these species are displayed. From the radionuclides found in irradiated water samples, Cd-109 and Mo-93m are not derived from Havar foil but from the vessel.

Gamma-ray spectra of the main parts of the [^{18}F]FDG synthesis apparatus, namely, the ion-exchange column, the water after fluoride separation, and the purification column set, have been evaluated for the presence of radionuclides. Species found are shown schematically in Figure 5.14. Although several radionuclides were found in each part, only a few were common to all investigations, probably because the studies were designed differently (irradiation time and energy, reagents, and number of samples). As Co-56, Co-57, Mn-52, and Cd-109 were found in all steps of the production, they are likely the species produced at the highest concentrations. Despite being produced in considerable amounts, these impurities do not reach the final product since they are retained in the steps of labeling process and [^{18}F]FDG purification. Bowden et al. [52] drew

TABLE 5.17 Quality Parameters of O-18-Enriched Water Supplied by Different Companies

	Company								
Parameters	Huayi	Huayi	Huayi	Isoflex	Isosolutions	Global Scientific Technologies	SRICI	Isotec	Rotem
Label	10%	95%	98%	97%	97%	97%	97%	97%	98%
Isotopic purity (%)									
O-18	>10	>95.00	>98.00	≥97.00	≥97	97.3	>97	>97	≥98.0
O-17	n.i.	<1.00	<1.00	≤1.1	<2	1.0	<2	n.i.	≤1.5
O-16	n.i.	<5.00	<2.00	≤1.5	<2	1.7	<3	n.i.	≤1.8
Chemical purity (mg/L)									
Fluoride	<1.0	<0.1	<0.1	<0.3	<1	<0.3	<0.1	1	≤0.1
Chloride	<1.0	<0.5	<0.5	<0.3	<1	<0.3	<0.1	1	≤0.5
Bromide	<1.0	<0.5	<0.5	<0.11	<1	0.012–0.047	<0.1	1	≤0.5
Iodide	<1.0	<0.5	<0.5	n.i.	<1	n.i.	<0.1	1	–
Nitrate	<1.0	<0.5	<0.5	n.i.	<1	n.i.	<0.2	1	≤0.2
Sulfate	<1.0	<0.5	<0.5	n.i.	<1	n.i.	n.i.	1	≤0.5
Phosphate	<1.0	<0.5	<0.5	n.i.	<1	n.i.	<0.2	1	≤0.5
Ammonium	<1.0	<0.5	<0.5	<0.03	n.i.	<0.3	n.i.	1	≤0.2
Sodium	<1.0	<0.1	<0.1	0.15	0.02	0.18–0.95	<1	2	≤0.1
Calcium	<0.5	<0.05	<0.05	0.0011	0.005	0.004–0.15	<0.1	1	≤0.1
Potassium	<1.0	<0.1	<0.1	0.01	n.i.	0.022–0.12	<0.1	1	≤0.1
Magnesium	<0.5	<0.05	<0.05	0.0056	0.001	0.00074–0.011	<0.1	1	≤0.05
Iron	<0.5	<0.05	<0.05	0.01	0.001	0.038	<0.1	1	≤0.05
Copper	<0.5	<0.05	<0.05	0.000025	0.005	0.00027–0.0056	<0.2	1	≤0.05
Zinc	<0.5	<0.05	<0.05	0.079	0.002	0.0029–0.027	n.i.	2	≤0.05
Aluminum	n.i.	n.i.	n.i.	0.000015	0.002	0.00013–0.0026	n.i.	n.i.	–
Silicium	n.i.	n.i.	n.i.	0.17	0.001	0.11–0.69	n.i.	n.i.	–
TOC	<1.0	<1.0	<1.0	<0.5	<5	<0.5	<1.0	<5.0	≤1
pH	6.0–8.0	6.0–8.0	6.0–8.0	6.8	6–8	6.8 ± 0.2	6–7.5	5.5–8.0	6–8

n.i., not informed.

TABLE 5.18 Havar Foil Composition

Metal	Co	Cr	Fe	Ni	W	Mo	Mn	C
Concentration (%)	42.0	19.5	17.3	12.7	2.7	2.2	1.6	2.0

TABLE 5.19 Concentration of Dissolved Metallic Cations in Irradiated [^{18}O]Water

Metal	Co	Cr	Fe	Ni	Ag	Cu	Ti
Concentration (mg/L)	3.0	0.12	0.18	<2.0	2.1	<1.5	<0.2

According to References 50 and 51.

some general trends on the distribution of the impurities among the components of the synthesis apparatus. For Co-56, Co-57, Co-58, Mn-52, and Cd-109, the bulk of the activity (75–85%) was found in the used water. This distribution is to be expected as the first column is an anion-exchange resin and, consequently, should not significantly retain cations. In contrast, more than 90% of Tc-95m, Tc-96, Re-183, and Re-184 were retained in this column together with $^{18}F^-$ ions, which is an expected result since these elements exist in aqueous solutions in the form of anionic species, namely, pertechnetate (TcO_4^-) and perrhenate (ReO_4^-).

In the study conducted by Bowden, the activity of the radioactive hydrogen isotope, tritium (^3H), in irradiated water and in [^{18}F]FDG, was also investigated. Tritium may be produced from O-18 bombardment according to the reaction O-18 + p $\rightarrow$ O-16 + H-3. The results are in agreement with those found by other authors, which also measured tritium activity in [^{18}F]FDG production. Although tritium was found in the irradiated water, it was present in a very low concentration or was not detected in the final product (see Table 5.21).

Due to the high cost and limited supply of [^{18}O]water, PET centers often recycle target water. However, the use of recycled [^{18}O]water, even after purification, has been reported to give lower [^{18}F]FDG yields. Recovered water is contaminated with inorganic ions (as shown above) and with organic solvents, which should be eliminated before reuse. Inorganic ions are introduced during the bombardment of the metallic water target housing, and organic solvents arise from the column used for the separation of ^{18}F ions from the target.

Usually, the water is recycled by distillation, and different factors that could affect the [^{18}F]FDG yield when using recycled [^{18}O]water have been analyzed. Inorganic impurities, mainly chloride ions, introduced during FDG synthesis, are considered responsible for the low FDG yield, as chloride competes with fluoride in the reaction with mannose triflate (synthesis intermediate). Several authors have investigated water treatments able to correctly purify [^{18}O]water.

TABLE 5.20 Radionuclidic Impurities Found in Water after Irradiation in Different Studies and Possible Activation Reactions (According to the References)

	Possible Activation Reactions			Half-Life	Reference							
					[52]	[53][a]	[54]	[55]	[56]	[57]	[58][a]	[50]
Radionuclides from Havar foil												
Co-55	Ni-58(p,α)Co-55			17.5 h	x			x	x			x
Co-56	Fe-54(n,p2n)Co-56			5.6 days	x	x	x	x	x	x	x	x
Co-57	Ni-58(p,2p)Co-57	Ni-60(p,α)Co-57	Fe-54(n,p)Co-57	272 days	x	x	x	x	x	x	x	x
Co-58	Co-59(n,2n)Co-58	Ni-58(n,p)Co-58	Fe-56(p,n)Co-58	71 days	x	x		x	x	x	x	x
Co-60	Co-59(n,γ)Co-60	Ni-60(n,p)Co-60		5.27 years						x		
Ni-57	Ni-58(p,pn)Ni-57			35.6 h			x	x				x
Ni-59	Co-59(p,n)Ni-59			75,000 years					x			
Cr-51	Cr-52(p,pn)Cr-51	Fe-54(n, α)Cr-51		27.7 days	x		x	x	x	x	x	x
Mn-52	Cr-52(p,n)Mn-52	Fe-57(p,n)Mn-52		5.6 days	x		x	x	x	x	x	x
Mn-54	Cr-54(p,n)Mn-54	Mn-55(n,2p)Mn-54	Fe-58(p,n)Mn-54	312 days		x	x	x		x	x	x
V-48	Cr-50(p,p,2n)V-48			16 days						x	x	
Fe-55	Mn-55(p,n)Fe-55			2.7 years					x			
Tc-95 (or Tc-95m)	Mo-95(p,n)Tc-95			20 h	x			x	x	x		x
Tc-96	Mo-96(p,n)Tc-96			4.3 days	x		x	x	x	x		x
Re-181	W-182(p,2n)Re-181			19.9 h				x				
Re-183	W-183(p,n)Re-183			70 days	x	x	x			x	x	
Re-184	W-184(p,n)Re-184			38 days	x					x		
Radionuclides from metal housing												
Cd-109	Ag-109(p,n)Cd-109			453 days	x		x		x			
Mo-93m	Nb-93(p,n)			6.85 h				x				

[a] Direct Havar foil analysis.

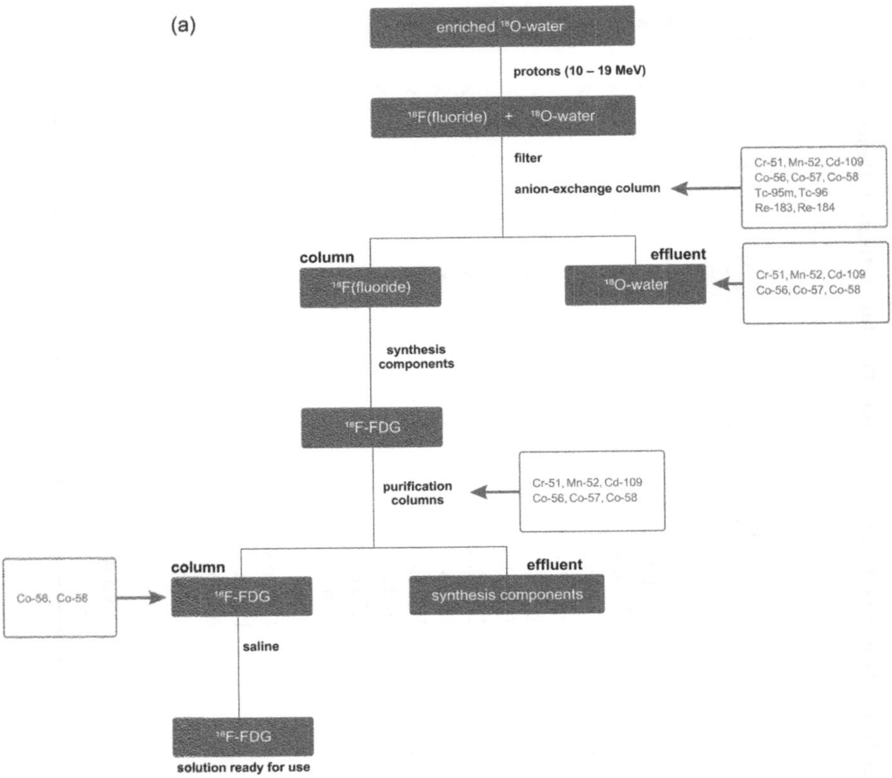

FIGURE 5.14 Radionuclides generated from the target housing during [^{18}O]water irradiation and their distribution in the steps of [^{18}F]FDG production according to different authors: (a) Bowden et al. [52]; (b) Ito et al. [56], and (c) Marengo et al. [54].

Table 5.22 presents chemical impurities found in irradiated water samples after different purification treatments. While most of the inorganic impurities are removed in the processes, organic solvents are more difficult to eliminate. It seems that the electrolysis procedure proposed by Huang et al. [61] is the most efficient in eliminating ACN and other solvents since, in the other processes [60, 62], some ACN remained in the water. Total organic carbon (TOC) and pH, parameters evaluated by Asti et al. [60], give important information regarding the purification process. The TOC value indicates the efficiency of the elimination of organics introduced during synthesis as the TOC levels in virgin [^{18}O]water are extremely low (see Table 5.17). On the other hand, the pH gives information regarding the anions in the column effluent. An elevated pH is probably related to the elution of carbonate ions when the water percolates the ion-exchange column for [^{18}F]fluoride retention. It seems that none of these impurities led to lower [^{18}F]FDG yields when recycled water was used since [^{18}F]FDG recoveries were within the expected range.

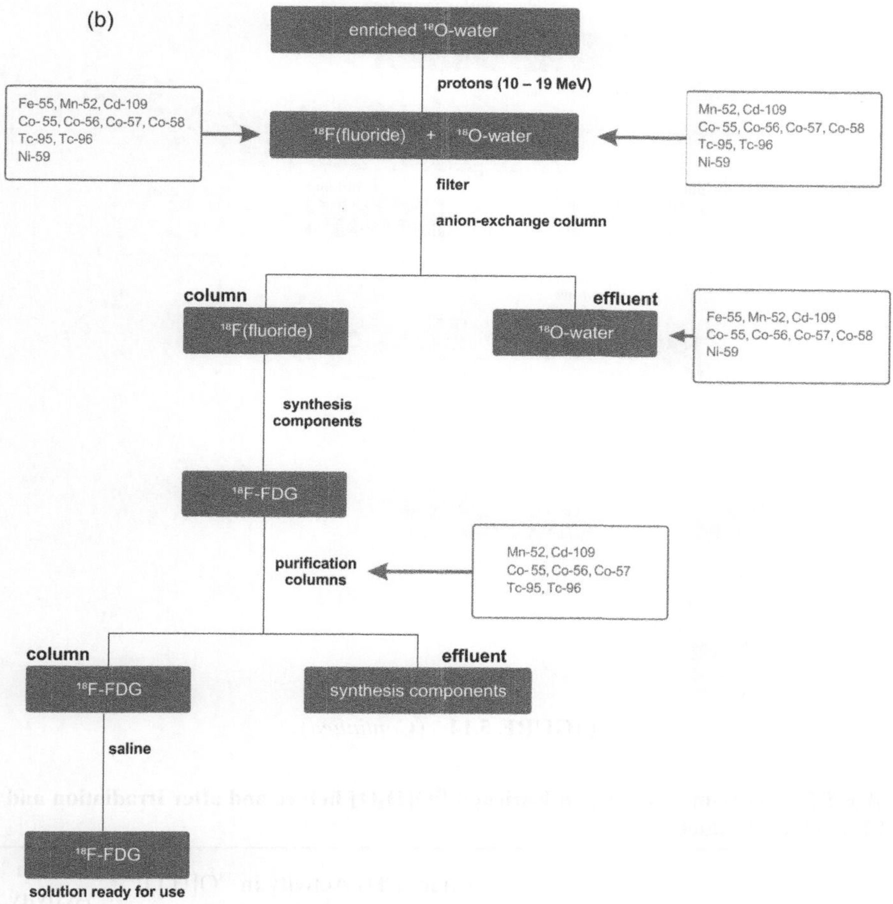

FIGURE 5.14 (*Continued*)

Huang et al. [61] included carbonate among the anions to be investigated. Carbonate is an expected contaminant in recycled [^{18}O]water since it is the counterion of the ion-exchange column for separation of [^{18}F]fluoride ions from water (see Fig. 5.13). Table 5.23 shows the levels of chloride, carbonate, and organic solvents in used and purified water samples as well as in two new and unused water samples. The results show that, although the purification was able to reduce the carbonate, which was eluted from the $^{18}F^{-}$ separation cartridge, it was still present in the purified water. In contrast, organic solvents, present in high concentrations, were practically eliminated by the purification. As carbonate was the only species present in a relatively elevated concentration in the purified water, the authors speculated on its interference in FDG production. Large amounts of carbonate in [^{18}O]water may cause the breakthrough of the anion-exchange column for [^{18}F]fluoride separation. Solutions

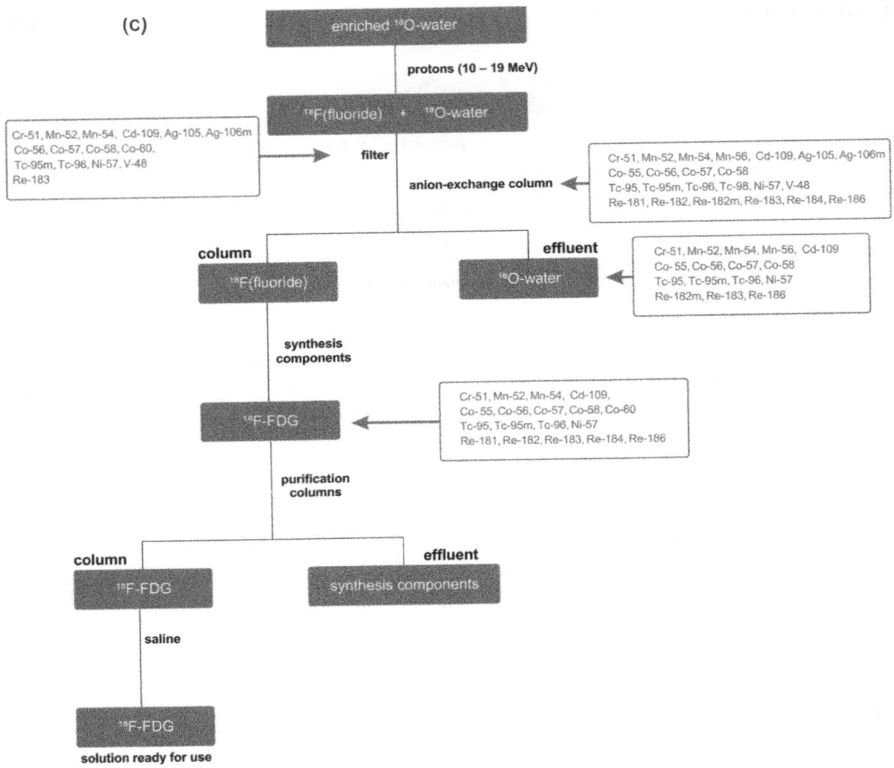

FIGURE 5.14 (*Continued*)

TABLE 5.21 Tritium Activities in Enriched $^{18}O[H_2O]$ before and after Irradiation and in the FDG Final Product

Reference	Irradiation Conditions	Time of Irradiation	Tritium (^3H) Activity in $^{18}O[H_2O]$ Bq/g — Before Irradiation	Tritium (^3H) Activity in $^{18}O[H_2O]$ Bq/g — After Irradiation	Tritium ^3H Activity in Product [^{18}F] FDG Bq/g
[52]	16-MeV protons 50-μA beam current	90 min	0.0022 ± 0.0004	146,000–183,000	0.068 ± 0.002
[59]	9.6-MeV protons 25-μA beam current	60 min	n.d.	63,500 ± 11,000	n.d.
[57]	18-MeV protons 20-μA beam current	30 min	–	134,000	–

n.d., not detected; –, not evaluated.

TABLE 5.22 Chemical Impurities in Purified Used Enriched [^{18}O]Water Using Different Treatments (All Concentrations in mg/L)

	Ozonolysis Distillation [60]		Electrolysis [61]		Photochemical Combustion Distillation [62]	
Parameters	Used	Purified	Used	Purified	Used	Purified
Fluoride	<0.05	<0.05		–	0.29	0.57
Chloride	1.8 ± 0.6	1.8 ± 0.7	0.9	–	2.54	1.23
Bromide	0.3 ± 0.02	0.2 ± 0.03	–	–	0.08	n.d.
Iodide	–	–	–	–	–	–
Nitrate	0.3 ± 0.01	0.2 ± 0.01	–	–	0.24	0.25
Sulfate	<0.1	<0.1	0.4	–	0.20	0.23
Phosphate	<1	<1	–	–	0.02	0.01
Carbonate	–	–	139	20.3	–	–
Acetate	–	–	6.7	-	–	–
Formate	–	–	1.5	-	–	–
Ammonium	<1	<1	–	–	5.44	13.57
Sodium	0.3 ± 0.1	0.15 ± 0.1	–	–	2.93	3.41
Calcium	0.15 ± 0.04	0.15 ± 0.03	–	–	1.05	0.53
Potassium	20 ± 0.3	1.3 ± 0.08	–	–	1.04	1.20
Magnesium	0.25 ± 0.05	0.07 ± 0.01	–	–	0.12	0.04
Iron	<0.05	<0.05	–	–	–	–
Zinc	<0.05	<0.05	–	–	–	–
Copper	<0.05	<0.05	–	–	–	–
Acetonitrile	33.5 ± 9.6	7.65 ± 2.2	2037	n.d.	62.7	78.1
Methanol	<0.1	<0.1	–	–	n.d.	n.d.
Ethanol	6.5 ± 1.2	2.13 ± 0.5	2430	n.d.	n.d.	n.d.
Acetone	20.5 ± 4.1	5.1 ± 1.3	210	–	n.d.	n.d.
TOC	700 ± 22	28.6 ± 3	–	–	–	–
pH	10.3	10.4	–	–	–	–

n.d., not detected; –, not informed.

containing fluoride (1 mg/L, approximately the level in the actual synthesis apparatus) and different concentrations of carbonate percolated the ion-exchange column. The results indicated that the presence of up to 200 mg/L carbonate did not affect the trapping of fluoride (>95%), but 300 mg/L carbonate decreased the trapping efficiency to 67%.

5.1.3.3 Residual Solvents and Components Used in the Labeling As mentioned above, a number of solvents are introduced during FDG synthesis: acetone, ethanol, or ethyl ether for cleaning and ACN to improve water evaporation from the ion-exchange column after elution of [^{18}F]fluoride. ACN is also used in the last step of FDG synthesis. The limits established in pharmacopeial monographs were checked in a number of procedures of FDG preparation.

TABLE 5.23 Concentration of Chloride, Carbonate, and Organic Solvents in New and Unused Water Samples, in Used Water Samples, and in Used and Purified Water Samples (All Concentrations in mg/L) [61]

New Water Samples (Unused)					Used Water Samples					Purified Water Samples				
Carbonate	Chloride	Acetone	Ethanol	ACN	Carbonate	Chloride	Acetone	Ethanol	ACN	Carbonate	Chloride	Acetone	Ethanol	ACN
0.8	<LOQ	<LOQ	<LOQ	<LOQ	79.7	0.4	204	1783	3864	9.8	<LOQ	<LOQ	<LOQ	<LOQ
1.6	<LOQ	<LOQ	<LOQ	<LOQ	299	<LOQ	685	3830	4360	6.6	<LOQ	<LOQ	<LOQ	<LOQ
					172	0.2	147	1047	1191	5.1	<LOQ	<LOQ	<LOQ	<LOQ
					156	2.3	58.0	3424	1054	62.4	<LOQ	<LOQ	<LOQ	<LOQ
					85.0	0.5	152	1613	1270	14.3	<LOQ	0.2	<LOQ	<LOQ
					39.6	2.1	12.5	2884	482.5	22.1	<LOQ	<LOQ	<LOQ	<LOQ
										21.4	<LOQ	0.8	<LOQ	<LOQ

LOQ, limit of quantification.

TABLE 5.24 Residual Solvents, Kriptofix, FDG, and ClDG in Routine [^{18}F]FDG Preparation

Impurity (mg/L)	Sample								Reference
	1	2	3	4	5	6	7	8	
Acetonitrile	41.2	67.0	79.1	68.0	57.1				[63]
Acetone	15.4	365	67.8	329	69.1				
Ethanol	10.0	37.0	198	129	80.4				
Aetonitrile	115	207	9						[64]
Ethanol	402	76	86						
Kriptofix	5.09	4.14	5.13	5.76	7.18	8.25	3.86	10.35	[65]
FDG	<0.3	<0.3	<0.3	0.85 ± 0.07	<0.3				[66]
ClDG	0.69 ± 0.05	1.23 ± 0.01	1.05 ± 0.04	52.3 ± 0.3	64.9 ± 0.6				
Kriptofix	<0.5	<0.5	<0.5						[67]
FDG	0.28 ± 0.01	0.15 ± 0.01	0.24 ± 0.01						
ClDG	0.24 ± 0.002	0.68 ± 0.003	0.48 ± 0.001						

Table 5.24 lists the levels of solvents found in these procedures. Although not exceeding the allowed limits, all solvents were found in FDG preparations. It is important to notice that while acetone, ethanol, and ethyl ether are classified as class 3 solvents (with low toxic potential), ACN is a class 2 solvent (to be limited) in the International Conference on Harmonisation of Technical Requirements for Registration of Pharmaceuticals for Human Use (ICH) [68]. Table 5.24 also shows the levels of Kriptofix, FDG, and 2-deoxy-2-chloro-D-glucose (ClDG) found in [^{18}F]FDG preparations.

The data presented in Table 5.24 show that [^{18}F]FDG formulations present impurities but within the allowed limits. It is interesting to notice that the large difference in ClDG concentrations between samples 1–3 and samples 4–5 (from Reference 66) lies in the last synthesis step, the hydrolysis process. While NaOH was used in samples 1–3, for samples 4–5, the hydrolysis was carried out with HCl, which may lead to the formation of ClDG in much higher concentrations.

Kuge et al. [69] investigated not only FDG and ClDG in [^{18}F]FDG formulations but also D-mannose, D-glucose, 2-[^{18}F]fluoro-2-deoxy-D-mannose (FDM), and radiochemical impurities. The chromatograms in Figure 5.15 show chemical impurities, detected by amperometry, and radiochemical impurities, detected with a radio detector. No radiochemical impurities, including [^{18}F]FDM, were detected (purity >99%). On the other hand, chemical impurities were detected, mainly D-glucose (90.5 ± 6.4 μg/mL), ClDG (11.8 ± 2.7 μg/mL), and D-mannose (1.7 ± 0.7 μg/mL).

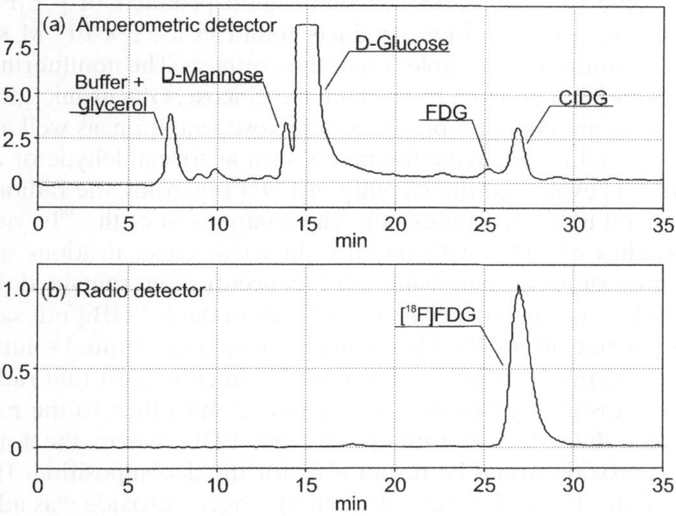

FIGURE 5.15 Chromatograms of a [^{18}F]FDG preparation showing chemical impurities detected by amperometry (a) and radiochemical impurities detected with a radio detector (b) [69].

TABLE 5.25 Content and Radiochemical Yields of the Products of Autoradiolysis of 5.5×10^{-5} M [^{18}F]FDG Solution Calculated from Mass Spectrometric Signals Obtained 30 Hours after the End of the Synthesis [70]

Product	Ion Form	Concentration (μg/mL)	Radiochemical Yield X100 eV
Glucose	$(C_6H_{12}O_6 + HCOO^-)^-$	25.4 ± 5.5	–
Arabinose	$(C_5H_{10}O_5 + HCOO^-)^-$	0.43 ± 0.05	7.3×10^{-2}
Gluconic acid	$(C_6H_{12}O_7 - H_2O + HCOO^-)^-$	0.36 ± 0.10	4.7×10^{-2}
Glucuronic acid	$(C_6H_{12}O_7 - H_2O + HCOO^-)^-$	0.10 ± 0.04	1.3×10^{-2}
2-Fluorogluconic acid	$(C_6H_{11}O_6\ {}^{19}F - H_2O + HCOO^-)^-$	<0.1	2.9×10^{-3}
2-Fluoroglucuronic acid	$(C_6H_9O_6\ {}^{19}F - H_2O + HCOO^-)^-$	<0.1	4.4×10^{-3}
Arabonic acid	$(C_5H_{10}O_6 - H_2O + HCOO^-)^-$	<0.1	$<1.3 \times 10^{-2}$
Araburonic acid	$(C_5H_8O_6 - H_2O + HCOO^-)^-$	<0.1	$<1.3 \times 10^{-2}$

5.1.3.4 Radiolysis Products Generated by Elevated Activity of the Labeled Compound PET radiopharmaceuticals are particularly susceptible to autoradiolysis due to the high activity of the labeled compounds and mainly because of their concentration in the purification column, which increases the chances of radiation-induced decomposition. Expected radiolysis products are inactive organic compounds, active ^{18}F-organic derivates and $^{18}F^-$ ions.

Buriova et al. [70], combining mass spectrometry and radiometric detection, were able to evaluate the decomposition products of [^{18}F]FDG. Table 5.25 shows the decomposition products found in a 5.5×10^{-5} M solution of [^{18}F]FDG, 30 hours after completion of the synthesis. The nonfluorinated products may be originated from both FDG and glucose. Glucuronic, gluconic, and arabonic acids are common products of hexose oxidation, as well as ring scissions, which lead to aldehydic fragments such as formaldehyde or arabinose.

Fawdry [71] evaluated the stability of FDG regarding the radioactive concentration and FDG decomposition. The comparison of the $^{18}F^-$ yields in two similar batches of FDG, with initial radioactive concentrations of 11.5 and 6.3 GBq/mL, showed that, while $^{18}F^-$ concentration remained below 5% (required pharmacopeial purity) for 14 hours in the 6.3 GBq/mL samples, this level was reached after only 4 hours in the more concentrated solution. Figure 5.16 shows the results of this stability study, which suggest that radiolysis is a cause of the instability. The authors attributed this effect to the radiolysis of water rather than the autoradiolysis of the FDG, where the formation of hydrogen peroxide would be responsible for the decomposition. The authors explained that $^{18}F^-$ yield increased when hydrogen peroxide was added to the solution, while it decreased when chemical reductants such as ascorbic acid, thiosulfate, nitrite, and iodide were added instead.

The influence of the radioactive concentration on FDG radiolysis can be seen in Figure 5.17, where the RCP (measured as $^{18}F^-$ yield) is related to

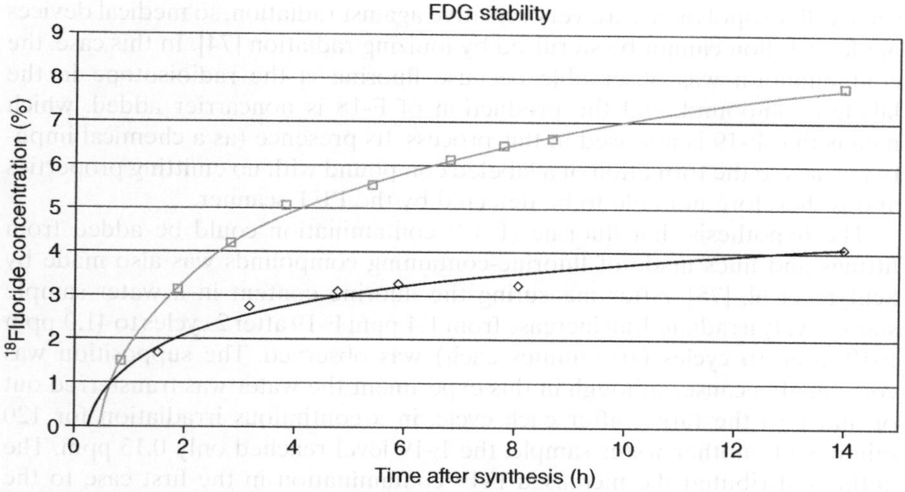

FIGURE 5.16 Free fluoride [$^{18}F^-$] buildup in [^{18}F]FDG at 22°C, pH 5.5 for initial FDG radioactive concentration of 11.5 GBq/mL (square) and 6.3 MGq/mL (diamond) [71].

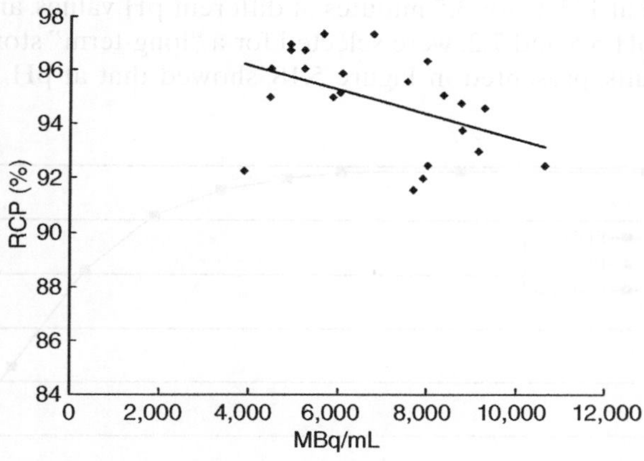

FIGURE 5.17 Radiochemical purity of [^{18}F]FDG decrease with the increase of activity concentration after storage for 4 hours at room temperature [72].

activity. The RCP was over 95% for solutions with a radioactive concentration <6 GBq/mL but declined to 92% when the concentration was over 8 GBq/mL, for 4 hours' storage [72].

Berridge et al. [73] showed that Teflon® parts of the apparatus utilized in the synthesis and purification of F-18 compounds were sources of F-19, a carrier contamination. The radiation produced by the high activity of radioactive fluorine provoked the radiolysis of Teflon, a fluorinated polymer, leading to the release of free fluoride in the milieu. Polymers with quaternary carbons, such as fluoropolymers, are very sensitive against radiation, so medical devices made of Teflon cannot be sterilized by ionizing radiation [74]. In this case, the contamination was observable because fluorine is the radioisotope in the labeled compound, and the production of F-18 is noncarrier added, which means that F-19 is not used in this process. Its presence (as a chemical impurity) leads to the formation of a labeled compound with no emitting properties and is therefore not able to be detected by the PET scanner.

The hypothesis that fluorine (F-19) contamination could be added from fittings and lines made of fluorine-containing compounds was also made by Schlyer et al. [75]. After measuring the fluorine content in a water sample successively irradiated, an increase from 1.4 ppm F-19 after 2 cycles to 41.0 ppm F-19 after 10 cycles (10 minutes each) was observed. The supposition was confirmed because, although in this experiment the water was transferred out of and into the target after each cycle, in a continuous irradiation for 120 minutes of another water sample, the F-19 level reached only 0.13 ppm. The authors attributed the increased F-19 contamination in the first case to the handling of the water with plastic materials but did not relate its release to a radiolysis process caused by F-18 itself.

The stability of FDG toward radiolysis was studied after steam sterilization by measuring free fluoride and FDM yields [76]. The sterilization procedure was conducted at 135°C for 3.5 minutes at different pH values, and two sterilized samples, pH 5.5 and 7.2, were selected for a "long-term" storage stability study. The results, presented in Figure 5.18, showed that at pH values lower

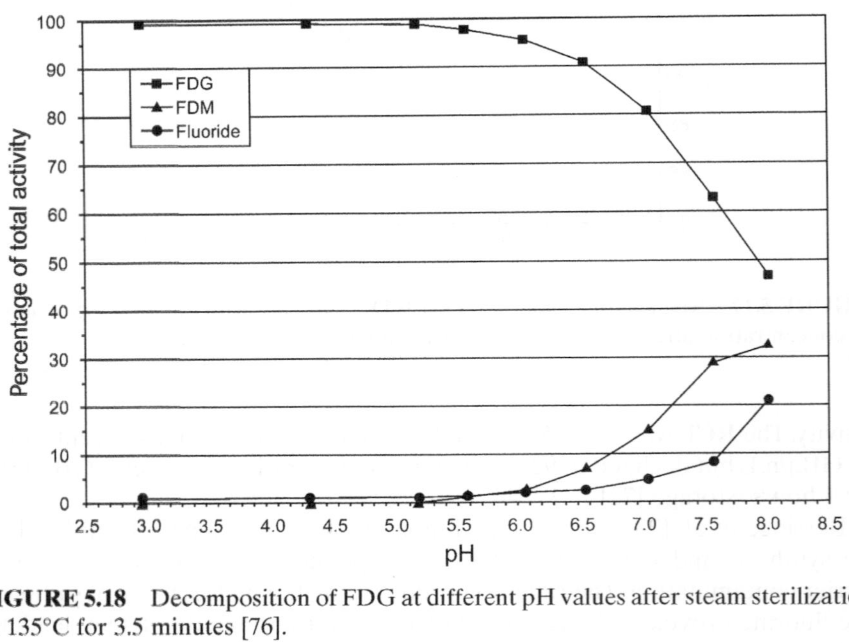

FIGURE 5.18 Decomposition of FDG at different pH values after steam sterilization at 135°C for 3.5 minutes [76].

TABLE 5.26 Radiochemical Purity of Four FDG Batches with Different Activity Concentrations Stored Either at pH 5.5 or 7.2 for up to 10 Hours [76]

Activity (MBq/mL)[a]	pH	Time				Overall Loss in Purity (%)
		45–60 min	3 h	5 h	10 h	
2800	5.5	98.6	98.0	97.5	97.0	1.6
	7.2	98.4	96.7	95.6	95.1	3.3
2003	5.5	99.4	98.7	98.4	97.8	1.6
	7.2	99.2	98.3	97.8	97.0	2.2
1614	5.5	98.4	98.1	97.8	97.6	1.2
	7.2	98.3	97.6	97.2	96.9	1.4
1368	5.5	99.1	98.5	98.2	98.0	1.1
	7.2	99.0	98.5	98.1	97.9	1.1

[a] Due to the addition of buffer to achieve a pH of 7.2, the activity concentration at this pH is 10% lower as indicated.

than 6, only very minor radiolysis occurred. The formation of free fluoride and FDM was below the allowed (Ph. Eur.) limits of 10% for FDM and 5% for free fluoride, respectively. In the long-term storage study, radiochemical purities lower than 95% were not found. However, the stability of FDG toward radiolysis at pH 5.5 was higher than at neutral pH, especially at activity concentrations exceeding 2000 MBq/mL. The highest loss of purity (3.3%) over a period of 10 hours was observed at neutral pH with the highest activity concentration of 2554 MBq/mL. In the same batch with an initial activity concentration of 2800 MBq/mL, a loss of only 1.6% occurred when the FDG was kept at a pH of 5.5. This higher stability at a lower pH could be observed for all activities (Table 5.26).

5.2 CONTRAST AGENTS

5.2.1 Gadolinium-Based Products

Gadolinium-based contrast agents or CM are widely used to enhance tissue contrast during magnetic resonance imaging (MRI) procedures. The requirements for a metal to be used as an MRI agent are high paramagnetism (magnetic moment), a stable oxidation state, and one labile coordinated water molecule. Gadolinium fulfills these conditions, making it an ideal agent for MRI. However, free gadolinium is undesirable as a drug substance due to its high toxicity. Moreover, Gd hydrolyzes under physiological conditions and precipitates in the presence of phosphate and carbonate. Consequently, a coordinating ligand is required to keep it in solution and to increase tolerance [77]. Based on this postulation, Gd-containing drug substances have been developed in the sense of enhanced stability and kinetic inertness along with tissue specificity and efficient relaxivity (a feature that characterizes an MRI agent) [78, 79].

Gadolinium occurs in nature along with 16 other elements known collectively as rare earth metals (RE). They exist in the form of oxidic compounds such as carbonates, silicates, and phosphates, and are found throughout the earth's crust in minerals such as monazite, bastnaesite, and apatite. Gadolinium oxide corresponds to 0.7–4.0% of the RE present in minerals, and the sum concentration of RE in minerals is around 4% [80]. REs are difficult to separate owing to the extraordinary similarity in chemical and physical properties between one RE and another. The process starts by heating the ore with hot caustic soda to dissolve phosphates from the mineral and to leave the RE precipitated as hydroxides, which are further dissolved in hydrochloric or nitric acid. This is followed by successive separation steps using either liquid–liquid or solid–liquid (ion-exchange) extraction procedures. Figure 5.19 displays the principle of RE separation originally proposed by the Rhône-Poulenc company, where more than 1000 separation units, using different extraction agents, are necessary for a successful separation of the RE. Although the process can produce all RE elements in a purity >99.999%, evidently at a very high cost, in the final product, each individual RE element rarely contains less than 5 ppm of the others, and the usual purity level is 99.99%. Table 5.27 shows the impurities in gadolinium oxide commercialized by five different companies. Gadolinium oxide presents not only the other RE but also other metallic oxides, such as iron oxide, as impurities.

Once obtained as a pure substance, gadolinium must be turned into a drug substance by choosing an appropriated chelating agent. Among the several chelates proposed as candidates for Gd-based contrast agents, it can be noted that (i) macrocyclic polyaminocarboxylates form thermodynamically more stable and kinetically more inert complexes than the chelates of linear polyamino carboxylates; (ii) nonionic chelates have blood compatible osmolality and viscosity, making them more tolerable than the ionic chelates; and (iii) *in vivo* biodistribution studies suggest that the macrocyclic polyaminocarboxylates chelates do not dissociate *in vivo* and have high LD_{50} in mice [81].

Nevertheless, Gd contrast agents are available not only as macrocyclic chelates but as linear chelates as well and, furthermore, can be found as ionic or nonionic preparations. As already mentioned, linear ligands are open chains that do not present as strong a binding ability as macrocyclic ligands. Due to the rigid structure of their rings, which are of almost optimal size to cage the gadolinium atom, macrocyclic ligands form more stable chelates than linear ligands. Table 5.28 presents the structure, ionicity, and formulation constituents of the nine Gd-based CM currently available for clinical use.

In order to achieve an adequate performance, a contrast agent should be administered in relatively large amounts. In Table 5.29, the molecular weight and the usual allowed dose for each commercially available Gd-based CM are shown. Table 5.29 also shows the masses of both the chelate and Gd alone, in

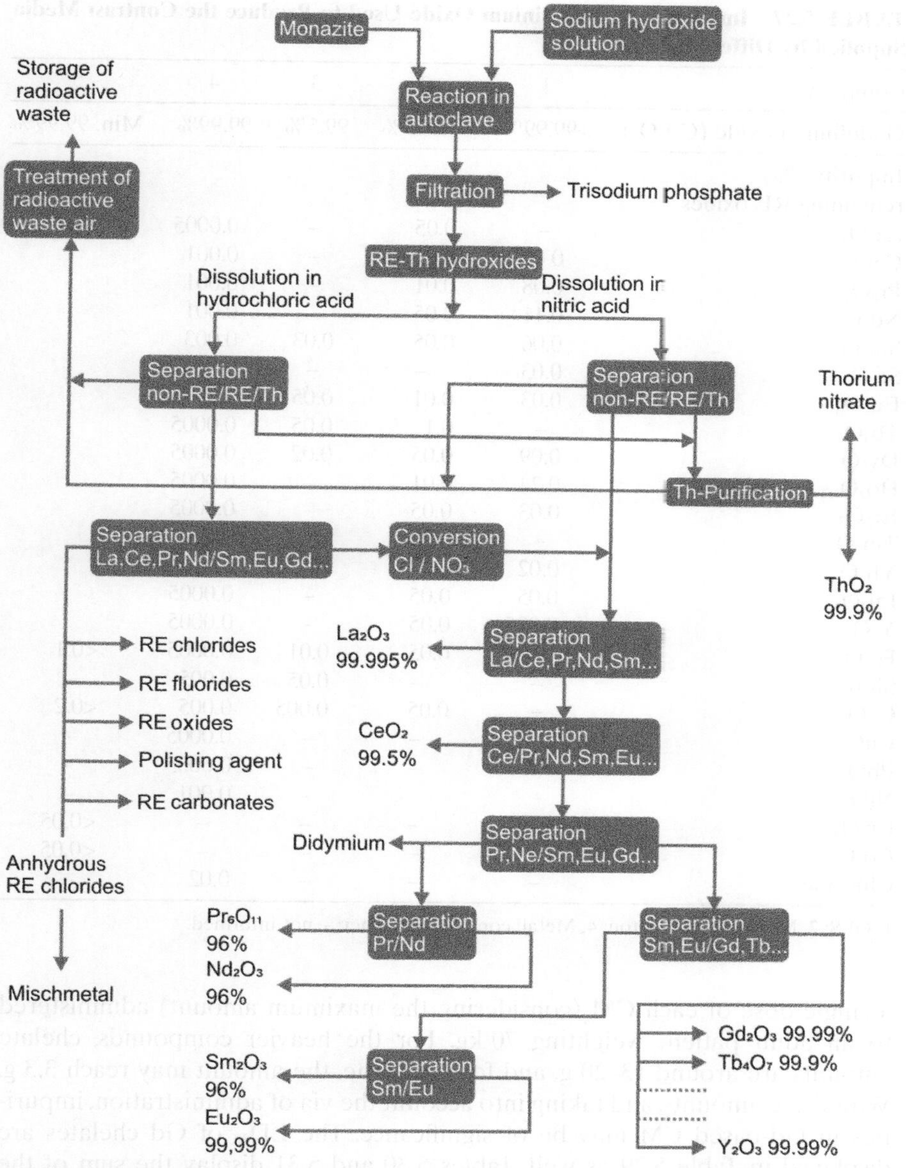

FIGURE 5.19 Rhône-Poulenc liquid–liquid extraction process for separation of the rare earth elements (reproduced with permission from McGill I. Rare earth elements. In: *Ullmann's Encyclopedia of Industrial Chemistry*. Wiley-VCH, [80].

TABLE 5.27 Impurities in Gadolinium Oxide Used to Produce the Contrast Media Supplied by Different Companies

Company	1	2	3	4	5
Gadolinium oxide (Gd_2O_3)	>99.99%	99.995%	99.5%	99.99%	Min. 99.99%
Impurity (%)					
remaining RE oxides					1
La_2O_3	–	0.05	–	0.0005	
CeO_2	0.06	0.01	–	0.001	
Pr_6O_{11}	0.08	0.01	–	0.001	
Nd_2O_3	0.11	0.05	–	0.001	
Sm_2O_3	0.06	0.05	0.03	0.003	
Sc_2O_3	0.03	–	–	–	
Eu_2O_3	0.03	0.01	0.05	0.003	
Tb_4O_7	–	0.1	0.05	0.0005	
Dy_2O_3	0.09	0.05	0.02	0.0005	
Ho_2O_3	0.24	0.01	–	0.0005	
Er_2O_3	0.03	0.05	–	0.0005	
Tm_2O_3	–	0.05	–	0.0005	
Yb_2O_3	0.02	0.1	0.02	0.0005	
Lu_2O_3	0.05	0.05	–	0.0005	
Y_2O_3	0.005	0.05	–	0.0005	
Fe_2O_3	–	0.05	0.01	0.0005	<0.1
SiO_2	–	–	0.05	0.005	–
CaO	–	0.05	0.005	0.005	<0.2
CuO	–	–	–	0.0005	–
PbO	–	–	–	0.0005	–
NiO	–	–	–	0.001	–
Cr_2O_3	–	–	–	–	<0.05
Co_3O_4	–	–	–	–	<0.05
Chloride	–	–	–	0.02	–

1, GFS; 2, Molycorp; 3, Baotou; 4, Metall.com; 5, Treibacher; -, not informed.

a single dose of each CM (considering the maximum amount) administered to an adult patient weighting 70 kg. For the heavier compounds, chelate amounts are around 13–20 g, and for Gd alone, the amount may reach 3.3 g. With these amounts, and taking into account the via of administration, impurities in Gd-based CM may be of significance. The LD_{50} of Gd chelates are displayed in Table 5.29 as well. Tables 5.30 and 5.31 display the sum of the other RE and the other metallic impurities that would be administered along with the dose of each Gd contrast agent (as calculated in Table 5.29) to an adult patient (70 kg). These values were estimated considering the certified levels of impurities in gadolinium oxides commercially available presented by the five different companies (Table 5.27).

The values presented in Table 5.30 are impressive: Depending on the origin of the raw material and the CM type, the amount of the other RE may be

TABLE 5.28 General Characteristics of Currently Commercially Available Gadolinium Chelates for Magnetic Resonance Imaging

Compound	Abbreviation	Trade Name	Company	Structure	Gd Bounding	Year Available	Structure
Gadopentate dimeglumine	Gd-DTPA	Magnevist®	Schering, Germany	Linear	Ionic	1988	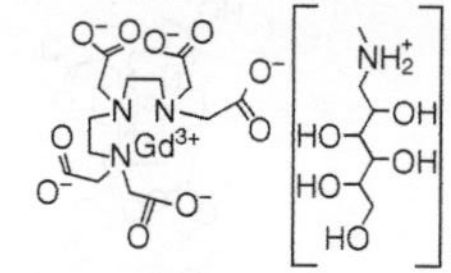
Gadodiamide	Gd-DTPA-BMA	Omniscan®	GE, United Kingdom	Linear	Nonionic	1993	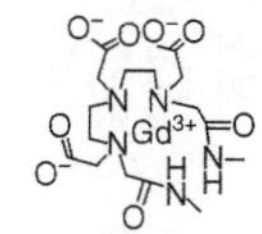
Gadoversetamide	Gd-DTPA BMEA	OptiMARK®	Tyco, United States	Linear	Nonionic	1999	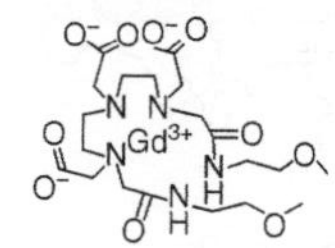
Gadobenate dimeglumine	Gd-BOPTA	MultiHance®	Bracco, Italy	Linear	Ionic	2004	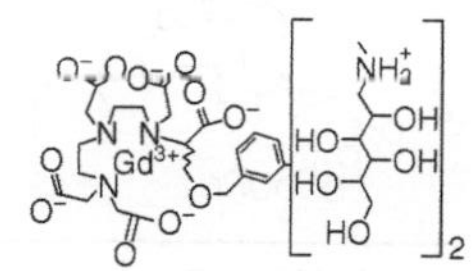

(*Continued*)

TABLE 5.28 (*Continued*)

Compound	Abbreviation	Trade Name	Company	Structure	Gd Bounding	Year Available	Structure
Gadoxetic acid disodium salt	Gd-EOB-DTPA	Primovist® Eovist®	Schering, Germany	Linear	Ionic	2004	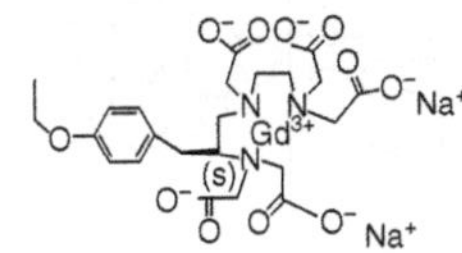
Gadofosveset trisodium	MS325	Vasovist®	Schering, Germany	linear	Ionic	2005	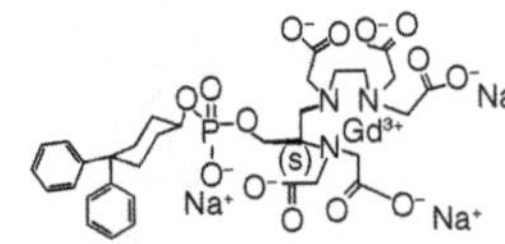
Gadoterate meglumine	Gd-DOTA	Dotarem®	Guerbet, France	Macrocyclic	Ionic	1989	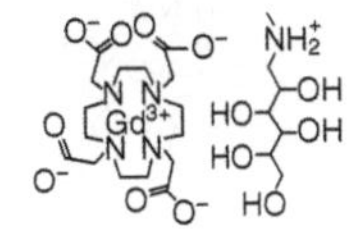
Gadoteridol	Gd-HP-DO3A	ProHance®	Bracco, Italy	Macrocyclic	Nonionic	1992	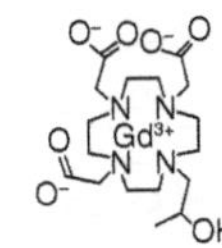
Gadobutrol	Gd-BT-DO3A	Gadovist®	Schering, Germany	Macrocyclic	Nonionic	2001	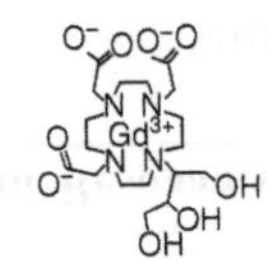

TABLE 5.29 Selected Characteristics of the Commercially Available Gadolinium Chelates, Molecular Weight, Formulation Constituents, Usual Dosage [82], LD_{50}, and Calculated Amount of Gd Chelate and Gd Alone in the Highest Dose of Each CM for an Adult Patient (70 kg)

Compound	Molecular Weight (g/mol)	Other Formulation Constituents	Approved Dosage (mmol/kg)	Amount of Chelate in a Dose (g)	Amount of Gd in a Dose (g)	LD_{50} Intravenous in Mice or Rat (mmol/kg)	LD_{50} Reference
Gadopentate dimeglumine	938	0.2% free DTPA	0.1–0.3	19.689	3.302	5.6 10	[77] [84]
Gadodiamide	573.66	25 mmol/L Ca-DTPA-BMA + Na-DTPA-BMA	0.1–0.3	12.047	3.302	14.8 34	[77] [81]
Gadoversetamide	661.77	50 mmol/LCa-DTPA-BMEA + Na-DTPA-BMEA	0.1	4.633	1.101	25–28	[85]
Gadobenate dimeglumine	1058.2	–	0.05–0.1	7.407	1.101	5.8	[86]
Gadoxetic acid disodium salt	725.72	Ca-EOB-DTPA Na-salt	0.1	5.080	1.101	7.5	[87]
Gadofosveset trisodium	976	0.325 mmol/L free ligand	0.01–0.05	3.416	0.550	[a]	[88]
Gadoterate meglumine	558.64	97.6 mg/mL meglumine	0.1–0.2	7.821	2.201	11 15	[77, 83, 87] [81]
Gadoteridol	558.7	0.1% Ca-HP-DO3A	0.1–0.2	7.822	2.201	12	[77, 81, 83, 87]
Gadobutrol	604.72	1 mmol/L Ca-HP-DO3A + Na-HP-DO3A	0.1–0.3	12.699	3.302	7–9 ~30	[77] [87]

[a] Lethality was observed in rodents after single intravenous administration at doses of at least two orders of magnitude higher than the anticipated human dose of 0.03 mmol/kg.

TABLE 5.30 Amount of Other REs Present as Impurity in One Dose of Each Gd-Based CM for an Adult Patient (70 kg), Assuming the Impurity Levels Presented in the Certificate of Analysis of Gadolinium Oxides Presented in Table 5.27

			RE Impurity per Dose (mg)				
Contrast	Usual Dose (mmol/kg)	Gd per Dose (g)	Company 1	Company 2	Company 3	Company 4	Company 5
Magnevist	0.1–0.3	3.3	26.6	21.1	5.6	0.45	33
Omniscan							
Gadovist							
ProHance	0.1–0.2	2.2	17.7	14.1	3.7	0.30	22
OptiMARK	0.1	1.1	8.9	7.0	1.9	0.15	11
Primovist/ Eovist	0.05–0.1	1.1	8.9	7.0	1.9	0.15	11
MultiHance							
Vasovist	0.01–0.05	0.55	4.4	3.5	0.9	0.07	5.5

higher than 30 mg. Nevertheless, although the amounts are very high for any impurity, mainly for these elements, which are all toxic (actually as toxic as the Gd itself), they present the same chemical behavior as gadolinium. This means that given the similarity among them, the containment of the toxic effect is achieved by chelation. Figure 5.20 shows the stability constants of selected RE chelates. For example, the stability constants of the diethylene triamine pentaacetic acid (DTPA)–RE complexes (one of the CM ligands) are as high as that of Gd, which assures the presence of these impurities in the same form as Gd, as a chelate.

In spite of the possible elevated contamination with the other RE, the main problem associated with Gd-based CM is the potential release of free Gd, and consequently the free ligand, either in the formulation or *in vivo*. The USP [89] establishes limits for free Gd in the monographs for the four Gd-based CM listed (0.3% in the highest case, for gadodiamide) and for the free ligands or related compounds (2.0% of related compounds for gadodiamide). Table 5.32 lists the impurities and their limits for the current Gd-based CM in the USP.

Gadolinium chelates are governed by a thermodynamic equilibrium among Gd, the ligand, and the complex, which defines the amount of free Gd and ligand generated by complex dissociation. Since Gd is highly toxic as are most of the ligands, the presence of these free species in the pharmaceutical formulation is an important issue. A typical dose for most CM is 0.1 mmol/kg, which encloses the same amount of Gd since the Gd/ligand ratio is 1:1. The LD_{50} for free Gd, 0.3 mmol/kg (for mice, administered as $GdCl_3$), is only threefold higher than this usual dose [90], although the LD_{50} of the pharmaceutical chelates are much higher than that of unbound Gd (see Table 5.29). Nevertheless, taking into account the great amount necessary to make the dose for a

TABLE 5.31 Amount of Impurities Other than RE in One Dose of Each Gd-Based CM for an Adult Patient (70 kg), Assuming the Impurity Levels Presented in the Certificate of Analysis of Gadolinium Oxides Presented in Table 5.27

Contrast	Usual Dose (mmol/kg)	Gd per Dose (g)	Impurity	Impurities Other than RE per Dose (mg)				
				Company 1	Company 2	Company 3	Company 4	Company 5
Magnevist	0.1–0.3	3.3	Fe_2O_3	–	1.7	0.3	0.02	3.3
Omniscan			CaO	–	1.7	0.17	0.17	6.6
Gadovist			SiO_2	–	–	1.7	0.17	–
			CuO	–	–	–	0.02	–
			PbO	–	–	–	0.02	–
			NiO	–	–	–	0.03	–
			Cr_2O_3	–	–	–	–	1.7
			Co_3O_4	–	–	–	–	1.7
			Chloride	–	–	–	0.66	–
ProHance	0.1–0.2	2.2	Fe_2O_3	–	1.1	0.2	0.01	2.2
			CaO	–	1.1	0.1	0.11	4.4
			SiO_2	–	–	1.1	0.11	–
			CuO	–	–	–	0.01	–
			PbO	–	–	–	0.01	–
			NiO	–	–	–	0.02	–
			Cr_2O_3	–	–	–	–	1.1
			Co_3O_4	–	–	–	–	1.1
			Chloride	–	–	–	0.44	–
OptiMARK	0.1	1.1	Fe_2O_3	–	0.6	0.1	0.01	1.1
Primovist/	0.05–0.1	1.1	CaO	–	0.6	0.05	0.06	2.2
Eovist			SiO_2	–	–	0.55	0.06	–
MultiHance			CuO	–	–	–	0.01	–
			PbO	–	–	–	0.01	–
			NiO	–	–	–	0.01	–
			Cr_2O_3	–	–	–	–	0.5
			Co_3O_4	–	–	–	–	0.5
			Chloride	–	–	–	0.22	–
Vasovist	0.01–0.05	0.55	Fe_2O_3	–	0.3	0.06	0.006	0.6
			CaO	–	0.3	0.3	0.03	1.1
			SiO_2	–	–	0.03	0.03	–
			CuO	–	–	–	0.006	–
			PbO	–	–	–	0.006	–
			NiO	–	–	–	0.01	–
			Cr_2O_3	–	–	–	–	0.3
			Co_3O_4	–	–	–	–	0.3
			Chloride	–	–	–	0.11	–

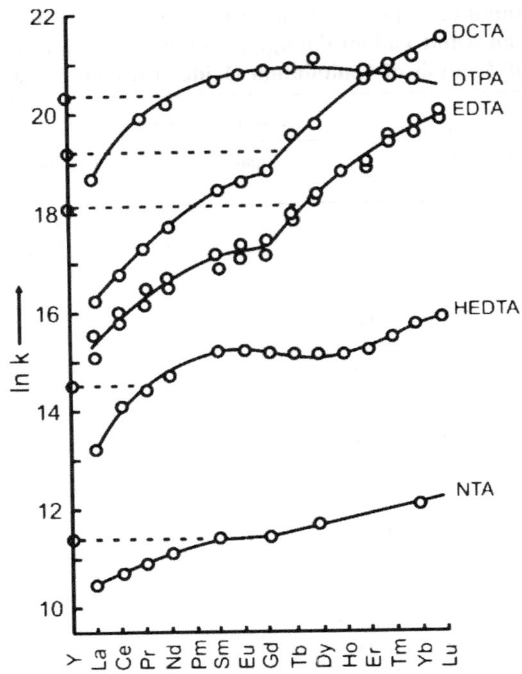

FIGURE 5.20 Stability constants of selected RE chelates (reproduced with permission from McGill I. Rare earth elements. In: *Ullmann's Encyclopedia of Industrial Chemistry*. Wiley-VCH) [80].

single procedure, the release of free Gd by any disturbance in this thermodynamic equilibrium should be taken into account. In the following paragraphs, it will be shown that the presence of cations capable of being complexed by the CM ligands is able to release free Gd, provided only that their stability constants allow for the exchange.

Gd-based contrast agents, originally introduced for MRI, were considered to be safe and non-nephrotoxic as compared to similar necessary volumes of iodinated CM [82]. However, after several indications and an extensive review of the literature, the European Society for Urogenital Radiology concluded that Gd-based CM are more nephrotoxic than iodinated CM in equivalent X-ray attenuation doses, probably due to the release of free Gd [91]. The concerns are about the accumulation of free Gd in patients with reduced renal function owing to the increase in the elimination half-life of Gd chelates. Indeed, in patients with normal renal function, the excretion of Gd chelates is rapid and the concentration of free Gd is very low. On the other hand, in patients with reduced renal function, the renal clearance is much lower than in patients with normal renal function, and this would implicate a longer

TABLE 5.32 Pharmacopeial (USP) Limits for Selected Impurities in Gadolinium Contrast Agents

Compound	Free Gadolinium	Residual Solvents	Heavy Metals	Other Impurities
Gadodiamide	0.3%	Acetone 0.2% Isopropyl alcohol 0.2% Ethylalcohol 0.2%	–	Diethylenetriamine pentaacetic acid bismethylamide: 0.7% Methylamine: 0.05% Related compounds (A and B) 2.0% Other impurities 0.5%
Gadopentetate dimeglumine	–	–	0.002%	Pentetic acid: 0.027–0.04%
Gadoteridol	0.02%	–	–	Related compounds: A: 0.02% B: 0.1% C: 0.1% Particulate matter
Gadoversetamide	0.05%	Isopropyl alcohol 0.1% Acetonitrile 0.025%	–	2-Methoxyethylamine: 0.10% Gadodiamide related compounds: A: 1.0% B: 0.5% Particulate matter

residence of the contrast in the body and the development of nephrogenic systemic fibrosis (NSF).

In this context, the stability of Gd-based CM appears to be an important factor in the pathogenesis of NSF [92, 93]. There is increasing evidence that the chemical instability of the chelates can lead to exchanges with other metal ions such as zinc and copper, which can, in turn, have biological consequences, since Gd is released unbound in the milieu by exchange with other metal cations. Typically, three different parameters are used to characterize the stability and safety of Gd chelates, the *thermodynamic stability*, the *conditional stability*, and the *kinetic stability*. All these parameters are expressed as constant values.

The thermodynamic stability governs the metal–ligand association tendency (binding constant). The equilibrium binding constant is difficult to measure accurately under physiological conditions so that the measurement is carried out *in vitro* and at a nonphysiological pH. The *thermodynamic stability constant*, K_{therm}, is measured at pH 11 or 14 by using potentiometric techniques [94]. Conversely, the *conditional stability constant*, K_{cond}, is calculated for physiological pH (7.4) from K_{therm} according to the equations

$$K_{therm} = [ML]/[M][L] \tag{5.16}$$

$$K_{cond} = [ML]/[M]\{[L]+[HL]+[H_2L]+\ldots\}, \tag{5.17}$$

where [HL] and $[H_2L]$ are the protonated forms of the free ligand species [95]. Due to different conditions and the proton competition at pH 7.4, the conditional stability (also termed apparent thermodynamic stability constant) is much lower than the thermodynamic stability. As stability constants are usually given as their log, according to the values presented in Table 5.33, Gd chelates may present more than 1000-fold difference in their stability. Since thermodynamic stability constants are obtained from experimental data, different values are found for the same complex, as can be seen in Table 5.33, depending on the investigative design.

The third useful parameter to demonstrate complex stability and safety is the kinetic stability, which is the measurement of how long it takes to release the Gd ion from its chelate. The kinetic stability may also be expressed as a *half-life* (dissociation half-life). Dissociation kinetics are often evaluated in very acidic conditions (pH = 1) to enhance the phenomenon. The complex dissociation is given by the equation

$$\text{Dissociation rate} = k_{diss}[H^+][GdL]. \tag{5.18}$$

As $[H^+]$ is considerably higher than [GdL], throughout the whole dissociation process, $[H^+]$ can be considered constant and equal to $[H^+]_0$. Therefore,

$$\text{Dissociation rate} = k_{obs}[GdL], \tag{5.19}$$

where

$$k_{obs} = k_{diss}[H^+]_0. \tag{5.20}$$

The k_{obs} has been determined for the Gd chelates, and these constants as well as the half-lives ($t_{½}$) are presented in Table 5.33. Differences in the values for the same chelate are probably related to the complexity associated to the measurement.

Figure 5.21 shows the rate at which the macrocyclic Gd contrast agents suffer dechelation at pH 1.2 and at 30°C. Of course, these are not the conditions under which Gd complexes are used as CM, but the graph demonstrates that even the most stable complexes (macrocyclic) are able to release free Gd.

The Gd ion has nine coordination sites through which it is bound to the ligands. Gd acts as a contrast agent by disturbing the relaxation of surrounding

TABLE 5.33 Stability, Thermodynamic Stability Constants, Conditional Stability Constants, Kinetic Stability Constants, and Dissociation Half-Lives of Gd Chelates

Compound	Structure	Stability [63]	Log K_{therm}	Log K_{cond}	Kinetic Stability (s^{-1})	$t_{½}$ (pH = 0.1)	Reference
Gd-DTPA	Linear/ionic	Low	22.1	17.7		<2 s (0.01 M)	[94, 96, 97]
			22.2		1.2×10^{-3}		[98]
				18.1		10 min	[94]
			22.46	14.8			[77]
					6.17×10^{-4}	18 min	[99]
Gd-DTPA-BMA	Linear/nonionic	Low	16.9	14.9		<2 s (0.01 M)	[77, 94, 96, 97]
			16.8		$>2 \times 10^{-2}$		[98]
						35 s	[94]
					20×10^{-3}	0.18 hour	[77]
Gd-DTPA BMEA	Linear/nonionic	Low	16.6	15.0			[94, 96]
			16.8		$>2.2 \times 10^{-2}$		[98]
			16.84				[77]
Gd-BOPTA	Linear/ionic	Medium	22.6				[96]
				18.4			[100]
Gd-EOB-DTPA	Linear/ionic	Medium	23.46				[96]
				18.7			[101]
MS325	Linear/ionic	Medium	22.1	18.9			[102]

(Continued)

TABLE 5.33 (*Continued*)

Compound	Structure	Stability [63]	Log K_{therm}	Log K_{cond}	Kinetic Stability (s^{-1})	$t_{½}$ (pH = 0.1)	Reference
Gd-DOTA	Macrocyclic/ionic	High	25.6	19.3			[103]
			25.8	18.8		95 days (0.01 M)	[97]
			25.6		8.4×10^{-7}		[98]
			25.8	18.8		>1 month	[94]
			23.3	18.33	0.021×10^{-3}		[77]
			24.6	18.6	0.0032×10^{-3}	60.2 h	[77]
			24.0				[77]
					2.1×10^{-5}	9 h (37°C)	[104]
					3.2×10^{-6}	60 h (37°C)	[104]
						84.2 h	[105]
					6×10^{-9}	3.6 years	[99]
Gd-HP-DO3A	Macrocyclic/nonionic	High	23.8	17.1			[96]
						1.3 days (0.01 M)	[97]
				17.21			[77]
					6.4×10^{-5}		[98]
						3 h	[94]
					6.3×10^{-5}	3 h (37°C)	[104]
					5.9×10^{-5}		[104]
						4.2 h	[105]
Gd-BT-DO3A	Macrocyclic/nonionic	High	21.8				[96]
				14.7			[106]
			21.0		2.8×10^{-6}		[98]
			21.0	14.5			[77]
				14.97			[77]
						24 h	[94]
					8.0×10^{-6}	24 h (37°C)	[104]
						18.2 h	[105]

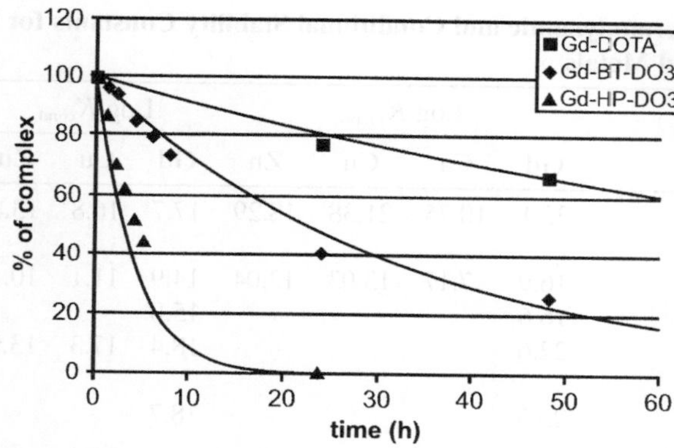

FIGURE 5.21 Dechelation kinetics at pH 1.2, 37°C, of marketed macrocyclic gadolinium contrast agents: percentage of the remaining gadolinium chelate over time [96].

water protons. In order for Gd to interact with water, one of the coordination sites must be left free. This requirement decreases the metal–ligand binding affinity and allows some dissociation [92].

In an ionic linear molecule such as Gd-DTPA, the Gd ion is coordinated with five carboxyl groups and three amino nitrogen atoms. The remaining vacant site is coordinated with a water molecule. In a nonionic linear molecule such as Gd-DTPA-BMA, the number of carboxyl groups is reduced to three and the other two carboxyl groups are replaced by a nonionic methyl amide, which is less strongly bound in comparison to carboxyl groups [107]. By contrast, macrocyclic molecules offer better protection and binding to the Gd ion because of their preorganized rigid ring of almost optimal size to cage the Gd ion. All the macrocyclic agents available for clinical use, whether ionic or nonionic, are derived from 12-membered macrocyclic polyaminocarboxylate rings.

Because of their relatively low stability, the pharmaceutical solutions of some Gd complexes include an amount of free ligand, or sodium or calcium salts of the ligand molecule (see Table 5.28). Such excipients are intended to ensure the absence of free Gd cations in the pharmaceutical product [95].

The main physicochemical parameter to evaluate *transmetallation*, the exchange of Gd in the chelate by another cation, is kinetic stability, although the available data for comparision are the thermodynamic and conditional stability constants (see Table 5.34). As the half-lives differ greatly, so does the possibility of transmetallation. It is now well established that macrocyclic ligands have much slower decomplexation reaction kinetics. On the other hand, for the less stable complexes obtained from linear ligands, important physicochemical parameters, such as as their relatively rapid decomplexation kinetics, allow a quasi-thermodynamic equilibrium *in vivo* [96].

TABLE 5.34 Thermodynamic and Conditional Stability Constants for Gadolinium and Other Selected Metals

	Log K_{therm}				Log K_{cond}			
Compound	Gd	Ca	Cu	Zn	Gd	Cu	Zn	Reference
Gadopentate dimeglumine	22.1	10.75	21.38	18.29	17.7	16.8	13.8	[96]
Gadodiamide	16.9	7.17	13.03	12.04	14.9	11.1	10.1	[96]
Gadoversetamide	16.6				15.0			[96]
Gadobenate dimeglumine	22.6				18.4	17.3	13.9	[96]
Gadoxetic acid disodium salt	23.5				18.7			[96]
Gadofosveset trisodium	22.1	10.45	21.3	17.82	18.9	15.7	13.8	[96]
Gadoterate meglumine	25.8	17.23	22.63	21.05	18.8			[96]
Gadoteridol	23.8	14.33	22.84	19.37	17.1			[96]
Gadobutrol	21.8	11.74	22.87	19.26	14.7			[77]

The stability of commercially available Gd chelates has raised some controversy due to their role in NSF [92, 104, 108, 109]. In fact, the controversy stems from the assumption that the stability (behavior) of macrocyclic chelates, measured *in vitro*, could be extended to the stability *in vivo*. However, while kinetic stability is the main criterion used to predict the fate of the Gd chelate and the amount of free Gd released either by decomplexation or by transmetallation in physiological conditions, the data show that these values (k_{obs} and $t_{½}$) differ greatly among the complexes (Table 5.33). Even for the same complex, different values are found depending on the study design. In order to be meticulous, Port et al. [105] measured the k_{obs} and $t_{½}$ of the three macrocyclic chelates, gadoterate, gadoteridol, and gadobutrol, strictly following the same conditions for all, pH = 1.2 and T = 37°C. The results showed that the order of kinetic stability is gadoterate > gadobutrol > gadoteridol, with the respective $t_{½}$ of 84.8, 18.2, and 4.2 hours. On the other hand, the authors consider that because of intrinsic biological variability and lack of sensitivity, the *in vivo* models available so far do not allow for discriminations between macrocyclic agents.

5.2.2 Iodine-Based Products

Ideally, an iodine-based contrast medium combines the properties of a high iodine concentration, osmolality close to that of blood, low viscosity at room temperature, and, of course, good tolerance with controlled physicotoxicity and chemotoxicity. Although only iodine itself is able to attenuate the X-ray beams, it is attached to a carrier molecule to reduce the toxicity that an excessive amount of free iodine in the body would present. A single dose of X-ray CM may contain more than 2000 times as much iodine as in the total physiological body content.

Because of these requirements, current iodine-based CM are organic molecules containing benzene rings with three iodine atoms replacing three hydrogen atoms. This structure allows the CM to be sterilized by hot steam with only a very small decomposition rate.

Iodine-based CM are both ionic and nonionic and contain three iodine atoms per benzene ring. Table 5.35 shows the iodine-based CM commercially available, their structure, ionicity, and amount of iodine, as well as formulation composition.

The necessity of having an elevated concentration of iodine to produce the adequate contrast in the body during the procedure requires the CM to present the drug in a concentration at the limit of its solubility. Because of this, the label of some products alerts for the risk of crystallization or evidence of a milky-cloudy appearance and/or sediment or floating crystals. This is also the reason why these products cannot be frozen. Freezing is an extreme situation and some products, depending on their concentration, cannot even be stored at temperatures below 7°C (Gastrografin), or even 15°C (Visipaque).

Dawson et al. [110] investigated the possible supersaturation of solutions of iopamidol. The authors supposed that the L and D forms of iopamidol present different solubilities and, depending on their ratio in the formulation, the solution may not be a concentrated solution of iopamidol but rather a supersaturated solution. The problem is that, when the supersaturation is broken, the substance precipitates irreversibly, and the product can become contaminated by its own solid crystals. The authors did not conduct experiments to determine either the amount of each isomer in the formulation or their solubility.

Dawson's study was criticized by Deavin [111], who claimed that Dawson's conclusions could not be reached from the procedures he adopted. In spite of this, the author did not refute the conclusion that supersaturation of iopamidol solutions depends on isomeric purity. In fact, the solubility of iopamidol is not established or at least not published, and the 370 mg I/mL formulation seems to be supersaturated.

One problem related to supersaturation or even to elevated concentrations of the active ingredient is drug incompatibility due to precipitation [112]. Iodipamide presents incompatibility (precipitate formation) with all tested drugs, and iatrizoate and ioxalate do with at least 50% of them (12–15 different drugs tested) On the other hand, iopamidol and iohexol show compatibility with almost all tested drugs (12). Drug incompatibility may be a source of impurity since events such as pH change or even precipitation lead to the appearance of unexpected species in the formulation.

Another label warning is the exposure to daylight and the direct exposure to sunlight and secondary X-rays. This warning is related to the possibility of

TABLE 5.35 Commercially Available Iodine-Based Contrast Media, Their Structure, Ionicity, and Composition

Contrast	Trade Name	Supplier	Ionicity	Chemical Name	Mol Mass	Constituents per mL	Structure
Diatrizoate	Hypaque 76 Renografin-60	GE Healthcare Bracco	Ionic	1-Deoxy-1-(methylamino)-D-glucitol 3, 5-diacetamido-2,4,6-triiodobenzoate	613.91	660 mg diatrizoate meglumine 100 mg diatrizoate sodium 370 mg iodine 520 mg diatrizoate meglumine 80 mg diatrizoate sodium 3.2 mg sodium citrate 0.4 mg edetate disodium	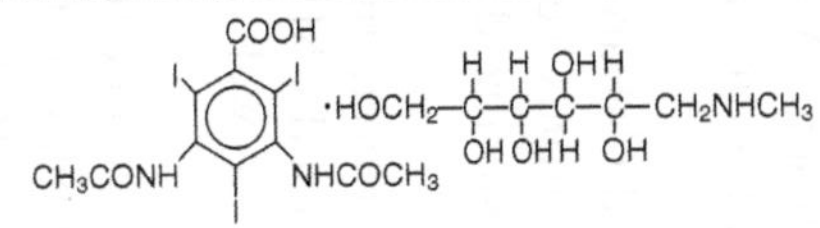
Ioxaglate	Hexabrix	Guerbet	Ionic	N-(2-Hydroxyethyl)-2,4,6-triiodo-5-[2-[2,4,6-triiodo-3-(N-methylacetamido)-5-(methylcarbamoyl) benzamido] acetamido]-isophthalamic acid	1464.10	393 mg ioxaglate meglumine 196 mg ioxaglate sodium 0.10 mg edetate calcium disodium	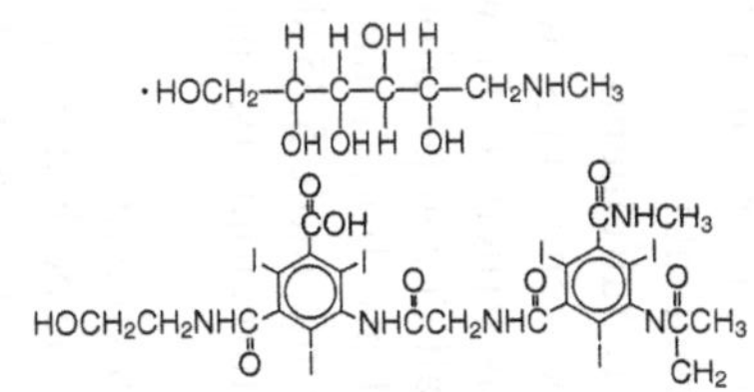
Iohexol	Omnipaque	GE Healthcare	Nonionic	1-*N*,3-*N*-*bis*(2,3-Dihydroxypropyl)-5-[*N*-(2,3-dihydroxypropyl) acetamido]-2,4,6-triiodobenzene-1,3-dicarboxamide	821.138	–	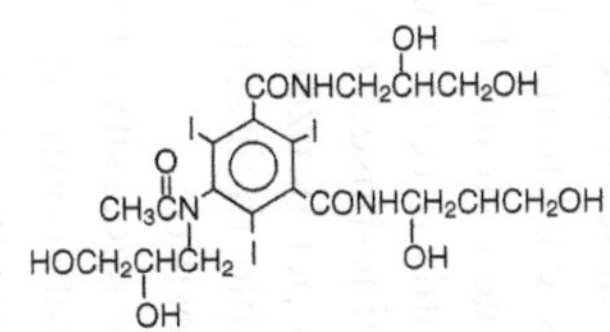

Iodixanol	Visipaque 320	GE Healthcare	Nonionic	5,5′-[(2-Hydroxy-1,3-propanediyl)bis (acetylimino)] bis[N,N′-bis(2,3-dihydroxypropyl)-2,4,6-triiodo-1,3-benzenedicarboxamide]	1550.20	320 mg iodixanol 0.044 mg calcium chloride dihydrate 1.11 mg sodium chloride 1.2 mg tromethamine 0.1 mg edetate calcium disodium	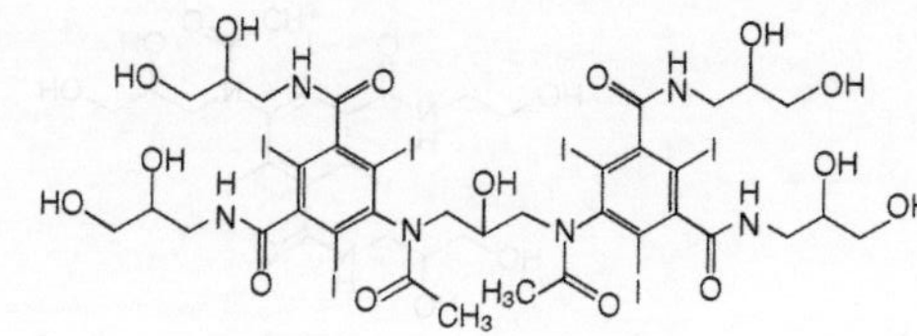
Iopromide	Ultravist	Bayer	Nonionic	1-*N*,3-*N*-*bis*(2,3-Dihydroxypropyl)-2,4,6-triiodo-5-(2-methoxyacetamido)-1-*N*-methylbenzene-1,3-dicarboxamide	791.1	498.72 mg iopromide 2.42 mg tromethamine 20.1 mg calcium edetate	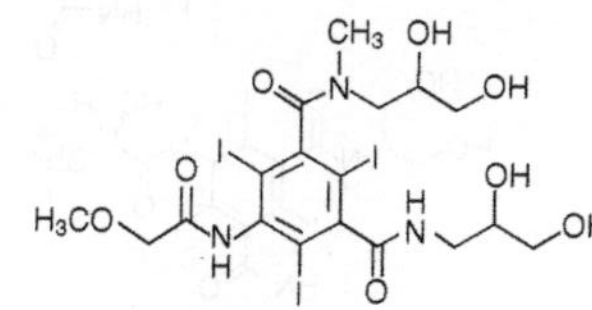
Iodipamide meglumine	Chlorgrafin	Bracco	Ionic	Benzoic acid,3,3¢-[(1,6-dioxo-1,6-hexanediyl) diimino]bis[2,4,6-triiodo-,compd.with 1-deoxy-1-(methylamino)-D-glucitol(1 : 2). 1-Deoxy-1-(methylamino)-D-glucitol 3,3¢-(adipoyldiimino) bis[2,4,6-triiodobenzoate] (2 : 1)(salt)	1530.19	520 mg iodipamide meglumine 3.2 mg sodium citrate 0.4 mg edetate disodium	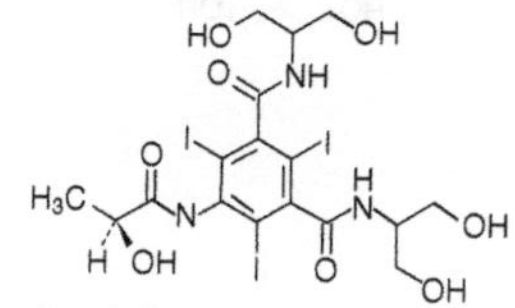

(*Continued*)

TABLE 5.35 (*Continued*)

Contrast	Trade Name	Supplier	Ionicity	Chemical Name	Mol Mass	Constituents per mL	Structure
Iothalamate meglumine	Conray	Mallinckrodt	Ionic	1-Deoxy-1-(methylamino)-D-glucitol 5-acetamido-2,4,6-triiodo-N-methylisophthalamate	809.13	172 mg of iothalamate meglumine 0.110 mg edetate calcium disodium 0.115 mg of monobasic sodium phosphate	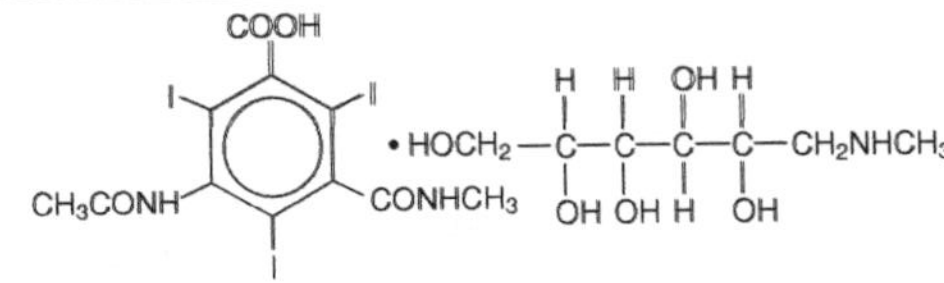
Iopamidol	Isovue	Bracco	Nonionic	(S)-N,N′-bis[2-Hydroxy-1-(hydroxymethyl)-ethyl]-2,4,6- triiodo-5-lactamidoisophthalamide.	777.09	755 mg iopamidol 1 mg tromethamine 0.48 mg edetate calcium disodium	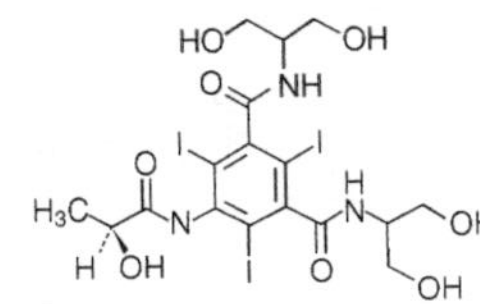
Metrizoate	Isopaque	–	Ionic	3-Acetylamino-2,4,6-triiodo-5-(N-methylacetylamino) benzoic acid sodium	649.911	–	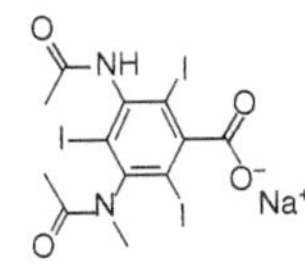
Ioxilan	Oxilan	Guerbet	Nonionic	N-(2,3-Dihydroxypropyl)-N′-(2-hydroxyethyl)-5-[N-(2,3-dihydroxypropyl) acetamido]-2,4,6-triiodoisophthal-amide	791.12	727 mg of ioxilan 350 mg of iodine/mL 0.576 mg citric acid anhydrous) (0.561 mg edetate calcium disodium)	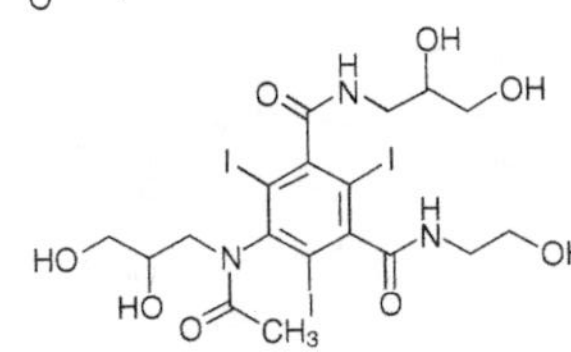

TABLE 5.36 Generation of Molecular Iodine from Potassium Iodide and Iodine Contrast Media by UV Irradiation [113]

Iodine-Containing Chemical		UV Radiation		I_2 Generated	
Chemical Name	Content of Iodine (mg/mL)	pH	Time	(μmol/L)	(nM/s)
Potassium iodide	0.635	5.0	14 min	48	57
Sodium iotalamate	200	7.2	12 h	38	0.88
Iohexol	175	7.2	12 h	n.d.	–
		6.5	48 h	26	0.15

n.d., not detected.

iodinated substances decomposing under the action of light. Shionoya et al. [113] studied the effects of UV light on the generation of molecular iodine from iodinated CM. They compared ionic and nonionic CM by exposing iotalamate and iohexol to UV light (0.013 mW/cm^2). While iodine was not detected in iohexol after 12 hours' exposure, 38 μM I_2 was produced in iotalamate. Table 5.36 presents the I_2 yield for both CM and also for a potassium iodide solution, which, even presenting a much lower iodine concentration, produced, in only 14 minutes, 48 μM I_2. These results confirm the indication that nonionic iodinated CM are more stable than ionic iodinated CM since the iodine oxidation rate increased in the following order: potassium iodide (free iodide) > iotamate (ionic) > iohexol (nonionic).The data also show that the pH affects the decomposition reaction as well. Tests were carried out only with KI and iohexol but showed that the lower the pH, the higher the amount of molecular iodine generated. These results can also be seen in Figure 5.22, which shows the amount of iodine generated in KI solutions, measured at two different pH values and for irradiation periods from 20 to 30 minutes.

Grönewäller et al. [114] also investigated the stability of iodinated CM under UV irradiation. Although the concerns in this case were related to the formation of toxic photoproducts by the action of the UV laser used during the angioplasty procedure, the results help to illustrate the effect of light on the stability of iodinated CM and the possibility of the formation of unexpected impurities during storage if ideal conditions are not maintained. The authors irradiated five different iodinated CM with up to 20,000 XeCl excimer laser pulse applications ($\lambda = 308$ nm, pulse duration 120 ns, 50 Hz) and analyzed the samples by HPLC to determine the amount of iodine released and the remaining fraction of unchanged CM. The results, presented in Figure 5.23, show an increased release of free iodide with the increase of the irradiation intensity. The graph also shows that the decomposition rates are higher for iopamidol and iopromide. The remaining amounts of CM after irradiation are presented in Table 5.37. While in iotrolan only 3% of the CM decomposed, for

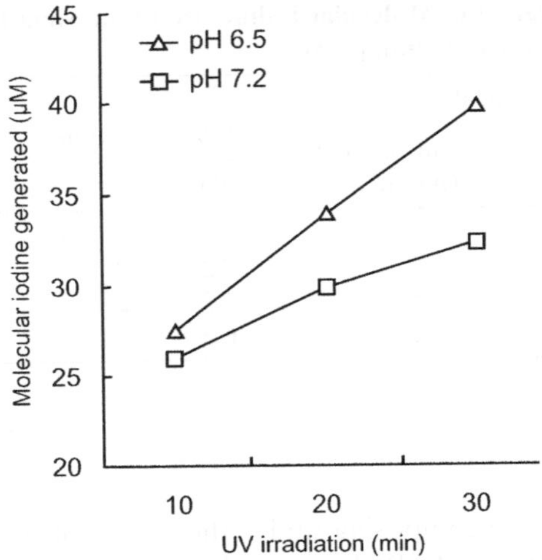

FIGURE 5.22 Generation of molecular iodine from potassium iodide solutions by UV light irradiation. A 4 mL aliquot of 1 mM potassium iodide in 0.1 M phosphate buffer (pH 6.5 or 7.2) was added to 1 mL of 1% soluble starch and was irradiated by UV light in a glass dish of 4-cm diameter with magnetic stirring [113].

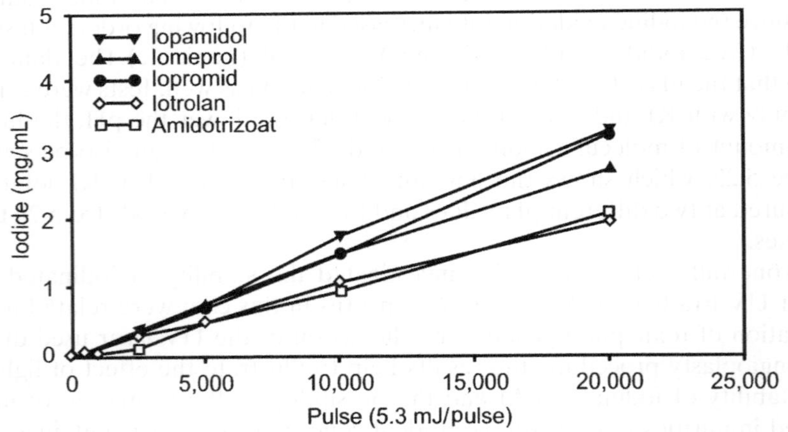

FIGURE 5.23 Iodine release after laser irradiation of a 1 mL contrast medium. Release rate for nonionic monomer compounds up to 3.3 mg iodide/mL. Reproduced from Gröneväller EF, Wahl HG, Kehlbach R, Rodemann HP, Claussen CD, Duda SH. Stability of iodinated contrast media in UV-laser irradiation and toxicity of photoproducts. Rofo-Fortschr Rontgenstr 169(5), 537–541, 1998.

TABLE 5.37 Remaining Amounts of CM after Irradiation and Percentage of the Surviving Colonies after 48 Hours' Incubation with the Irradiated CM

Contrast Medium	Ionicity	Remaining Amount after Irradiation[a] (%)	Percentage of Surviving Colonies after Treatment with CM (%)
Amidotrizoat	Ionic	96	6 ± 8
Iopromide	Nonionic	81	34 ± 9
Iopamidol	Nonionic	91	39 ± 13
Iomeprol	Nonionic	85	0
Itrolan	Nonionic, dimer	97	33 ± 22

[a] Irradiation of 1 mL CM with 20,000 pulse at 5.3 mJ (total energy = 106 J).
Adapted from Reference 114.

iopromide, 19% of the CM decomposed during exposure. The authors also evaluated the cytotoxicity of the photoproducts and the results are presented in Table 5.37, as a percentage of the surviving colonies (cell culture HSF 6) after 48 hours' incubation with the irradiated CM. Since after incubation with free iodide (10% NaI solution) no cytotoxicity was observed, the effect presented by the CM was presumed to be caused by photoproducts other than free iodide. Although presenting high iodide formation, iopamidol and iopromide showed the lowest toxicity. Conversely, iomeprol and amidotrizoate presented the highest toxicity. The authors concluded that unknown decomposition products of iomeprol and amidotrizoate (photoproducts) were probably responsible for the higher toxicity of these CM.

An interesting study to demonstrate that factors other than light can be responsible for deiodination of iodinated CM was carried out by Yu-Chang and Wang [115]. They studied the influence of pH, temperature, and Cu(II) ions in the deiodination of two CM, diatrizoic acid and iopamidol, an ionic and a nonionic CM, respectively. The temperature (autoclaving) accelerated the deiodination rate of both compounds (Figs. 5.24 and 5.25), although for iopamidol, it occurred at much slower rates. However, the decomposition was highly dependent on the presence of Cu(II) ions since in the presence of a chelating agent (EDTA), the reaction yielded very little iodide. Interestingly, the reaction was found to be independent of the Cu(II) concentration; concentrations between 0.25 and 5.08 mg/L Cu did not change the reaction rate. Another aspect related to this effect is that only Cu(II) ions catalyzed the reaction; other metallic ions such as Fe(II), Fe(III), Zn(II), Pb(II), Co(II), and Ag(I) were not able to produce detectable iodide concentrations. This result shows that if Cu(II) ions are introduced as an impurity in an iodinated CM, there is a reasonable chance of iodide formation. The authors used the *o*-, *m*-, and *p*-iodobenzoic acids as model compounds to investigate the catalyst effect of Cu(II) ions on the CM. The experiment showed that while the reaction rates

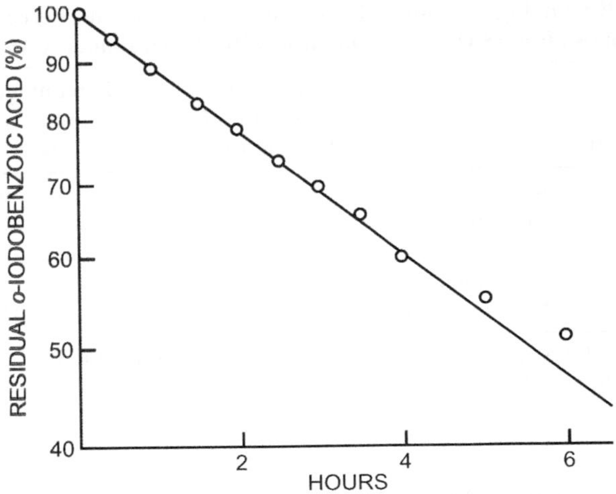

FIGURE 5.24 Production of iodide at 90°C in 0.01 M diatrizoic acid solution containing 0.0067 M, pH 7 phosphate buffer, and 4 ppm of Cu(II) [115].

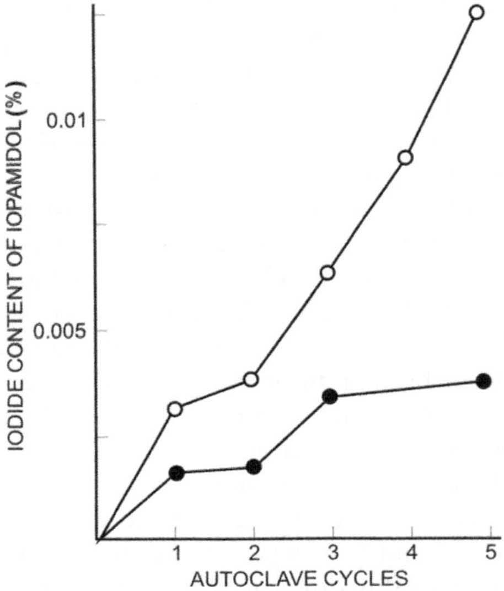

FIGURE 5.25 Amount of iodide found in 20% iopamidol solution (0.0067 M, pH 7 phosphate buffer) subjected to various numbers of autoclave cycles. Each cycle was 90 minutes at 120°C.Key: ○, without chelating agent, and ●, with 0.04% edetate disodium [115].

TABLE 5.38 Deiodination of Diatrizoic Acid and *o*-, *m*-, and *p*-Iodobenzoic Acids upon Autoclaving, 30 Minutes at 120°C [115]

Compound	Iodide Produced (M)[a]
Diatrizoic acid	0.0397
o-Iodobenzoic acid	0.0452
m-Iodobenzoic acid	0.00001
p-Iodobenzoic acid	0.00003

[a] Solutions contained 0.5 M compound, 0.04 mM $CuSO_4$, and 0.0067 M phosphate buffer to maintain the pH at 7.6.

of diatrizoic acid and *o*-benzoic acid were comparable, the *m*- and *p*-analogs of iodobenzoic acid produced almost negligible amounts of iodide. This result indicates that the two iodines adjacent to carboxylate in diatrizoic acid are more vulnerable to deiodination (see Table 5.38). The authors attributed this feature to a chelating effect of the carboxylate moiety on the Cu(II) ions, bringing them close to the reaction site. In this position. Cu(II) ions are able to polarize the carbon–iodine bond, causing iodine to leave the molecule (see Table 5.38: As the three iodines occupy intercalated positions in the benzene ring, two are *orto* and the third is *para* in relation to carboxylate or any other functional group of the molecule).

Another study on the stability of diatrizoic acid was carried out by Chellquist et al. [116]. They found that hypaque sodium solutions for oral administration contained elevated levels of inorganic iodide, ranging from 0.001% to 0.036%. They also showed that the amount of iodide was directly related to the amount of glucose found in the formulation. Moreover, they demonstrated that the basic pH (8) and the copper ions accelerated the decomposition. Different from the abovementioned study by Yu-Chang and Wang [115], they suggested that glucose reacts with copper to form an activated species, which reacts to displace iodine.

Although these studies present different mechanisms for deiodination of iodinated CM, it can be assumed that copper ions play an important role in this process. This is the reason that the label of some products warns that "iodinated contrast media, can react with metallic surfaces containing copper (eg brass), therefore the use of equipment in which the product comes into direct contact with such surfaces should be avoided."

Iodine belongs to the halogen group, whose members present seven electrons in their outer shell. This configuration is one electron less than that of the next noble gas. Thus, atoms complete their octet by either acquiring an electron, forming an ionic bond (I^-), or sharing an electron with another atom, forming in this case a covalent bond. Although halogens may exist in six oxidation states, –I, 0, +I, +III, +V, and +VII, iodine compounds present the oxidation states –I (I^-), 0 (I_2), +I (IO^-), +V (IO_3^-), and +VII ($H_3I_2O_6^-$) [117].

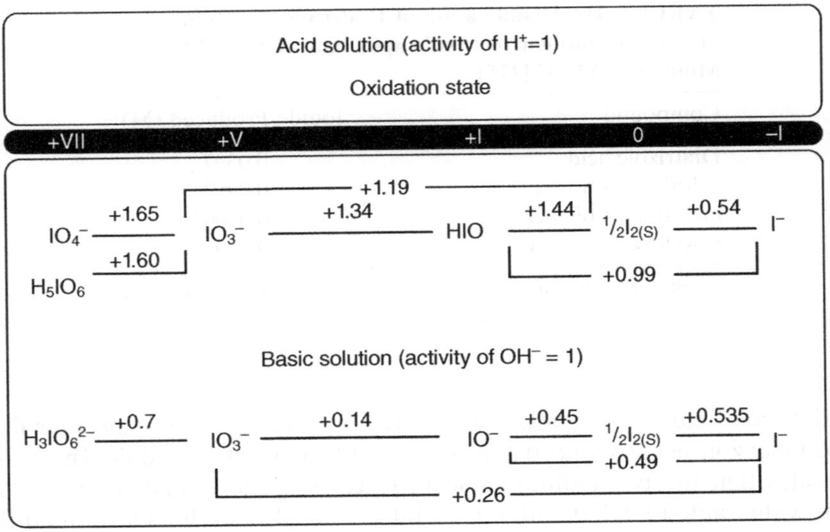

FIGURE 5.26 Standard reduction potentials for iodine species (volts).

The ionic iodinated CM do not contain iodine itself as an anion. In all compounds, iodine is covalently attached to the benzene ring, and the ionic form corresponds to a compound containing sodium or meglumine as a cation and the part containing the benzene ring as an anion.

Both iodide, I^-, and molecular iodine, I_2, are reactive species. Iodide can easily be oxidized to I_2, and I_2 can also be reduced to I^-. Schematically,

$$\text{Reducing agent} + I_2 \rightarrow I^- \tag{5.21}$$

$$\text{Oxidizing agent} + I^- \rightarrow I_2. \tag{5.22}$$

In fact, iodine (I_2) can be further oxidized to higher oxidation states, depending on the strength of the oxidizing agent. The reduction potentials to convert one species into another are presented in Figure 5.26.

Aqueous solutions of I^- are instable since it is oxidized by atmospheric oxygen:

$$4I^- + O_2 + 4H^+ \rightarrow 2I_2 + 2H_2O. \tag{5.23}$$

Moreover, an equilibrium between iodide, I^-, and molecular iodine, I_2, is established both in acidic and basic solutions:

$$I_2 + H_2O \rightarrow HIO + H^+ + I^- \quad K_{H+} = 2.0 \times 10^{-13}\ \text{mol}^2/\text{L}^2 \tag{5.24}$$

$$I_2 + 2OH^- \rightarrow IO^- + H_2O + I^- \quad K_{OH-} = 30\ \text{L}/\text{mol} \tag{5.25}$$

The stability constants show that I_2 is more stable in acidic solution than in basic solution, being disproportioned into hypoiodous and iodide in the last case [117].

According to Equations 5.23–5.25, both I^- and I_2, may be contaminants of iodinated CM since the equilibrium between both species depends on the milieu conditions. The same subject was addressed under iodine radiopharmaceuticals, where the formation of iodate (Equation 5.10) was observed as a product of hypoiodous disproportioning during the separation of the radioactive iodine from the target material.

Iodine (I_2) may react with reducing agents such as ascorbic acid, glucose, glutathione, and cysteine. If one of these or other reducing species, which are frequently used in pharmaceutical formulations, enter in contact with an iodinated CM, containing even I_2 traces, there is the risk of the formation of undesirable reaction products.

Pharmacopeial compendia do not establish a limit for I_2 in iodinated CM, but a qualitative test should be done to show that I_2 is absent from the formulation. On the other hand, the limit for iodide is usually 0.02%. Table 5.39 presents the impurity limits for eight CM, for which the USP and the BP present monographs where it is possible to see that, besides iodine, iodide and heavy metals (usual in all monographs), residual solvents, free aromatic amines, and particulate matter are also of concern in iodinated CM.

The free amine mentioned in the monographs is 5-acetamido-3-amino-2,4,6-triiodobenzoic acid. Chellquist et al. [118] studied the stability of diatrizoate sodium (DTZA) in different conditions (pH and temperature) and the formation rates of free amine and two other compounds, 2,4- and 2,6-diiodo-3,5-diacetamidobenzoic acid (DDZA). The authors prepared the drug substance in 0.1 mol/L potassium hydroxide and kept the solution at 85°C for 7 days, collecting aliquots during this time for a chromatographic assay. Although the major hydrolysis product was the free amine, a mass balance (Fig. 5.27) showed that other decomposition products were formed since the sum of the remaining diatrizoate sodium and the free amine masses was only 71% of the initial diatrizoate sodium mass. The chromatograms showed two peaks that did not corresponded to the DDZA peak. The authors suggested that DDZA is not a decomposition product of diatrizoate sodium but rather a synthetic impurity resulting from incomplete iodination of 3,5-dinitro-benzoic acid, a precursor of diatrizoic acid [119]. In another study, Farag et al. [120] found mono- and diiodo-derivatives in the analysis of a commercial DTZA, which was not submitted to any stressful conditions. Using a capillary electrophoresis system, the authors determined 2-iodo-, 4-iodo, 2,4-diiodo-, and 2,6-diiodobenzoic acid in amounts of 8.4, 5.1, 7.3, and 10.2 mg/mL, respectively. Figure 5.28 displays the electropherogram of DTZA separation and its impurities.

Particulate matter seems to be a singular problem in iodinated CM since studies have shown that different kinds of particles may be found in these formulations. As this topic is discussed under a specific chapter, the presence of particulate matter in iodinated CM is dealt with in Chapter 10.

TABLE 5.39 Pharmacopeial (USP and BP) Limits for Selected Impurities in Iodine-Based Contrast Agents

Compound	Free Iodide	Free Iodine	Residual Solvents	Heavy Metals	Others
Diatrizoate/ diatrizoic acid	0.02%	[a]	–	0.002%	Residue of ignition <0.1% Free aromatic amine: 0.05%
Ioxaglate	0.02%			0.002%	
Iohexol	0.001%	[a]	Methanol 0.005% Isopropyl alcohol 0.005% Methoxyethanol 0.002% 3-Chloro-1,2-propanediol 0.0025%	0.002%	Particulate matter[b] Free aromatic amine: 0.05% Ionic compounds 0.01% O-alkylated compounds: 0.6% Cromatographic peaks not more than 0.3%
Iodixanol	20 mg/g	[a]	Methanol 0.005% Isopropyl alcohol 0.005% Methoxyethanol 0.002%	0.002%	Calcium: 5 μg/g Free aromatic amine: 0.05% Ionic compounds 0.02% Iohexol 0.6% Total related compounds 1.5%[c]
Iopromide	0.002%	[a]	Alcohol 0.4%	0.002%	Free aromatic amine: 0.1%
Iodipamide	0.02%	[a]	–	0.002%	Free aromatic amine: 0.05%
Iothalamate	0.02%		–	0.002%	Free aromatic amine: 0.05%
Iopamidol	0.04 mg iodide/mL	[a]	–	0.001%	Residue of ignition <0.1% Particulate matter[b] Free aromatic amine: 0.02%
Ioxilan	200 μg/mL	[a]	Methanol 0.005%	0.002%	Serinol: <0.5% Cromatographic peaks not more than 1.5%
Ioversol	0.02%			20 μg/g	Cromatographic peaks not more than 1.5%

[a] Toluene layer shows no red color.
[b] Meets the requirements for small-volume injections.
[c] Iodixanol related compounds (see USP monograph).

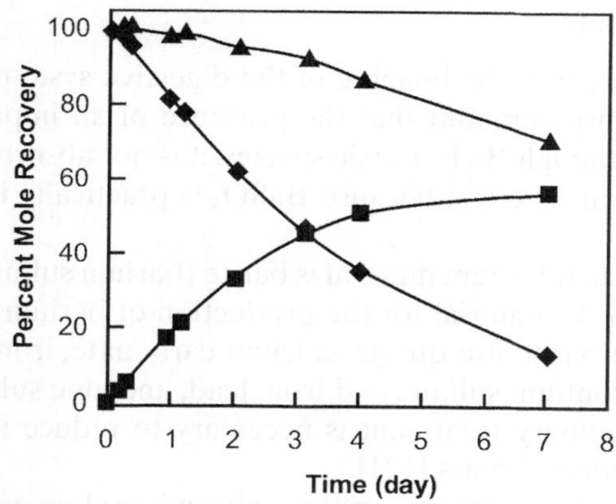

FIGURE 5.27 Decomposition of DTZA Na in 0.1 M potassium hydroxide at 85°C where ▲ is mass balance, ◆ is DTZA Na, and ■ is the free amine [118].

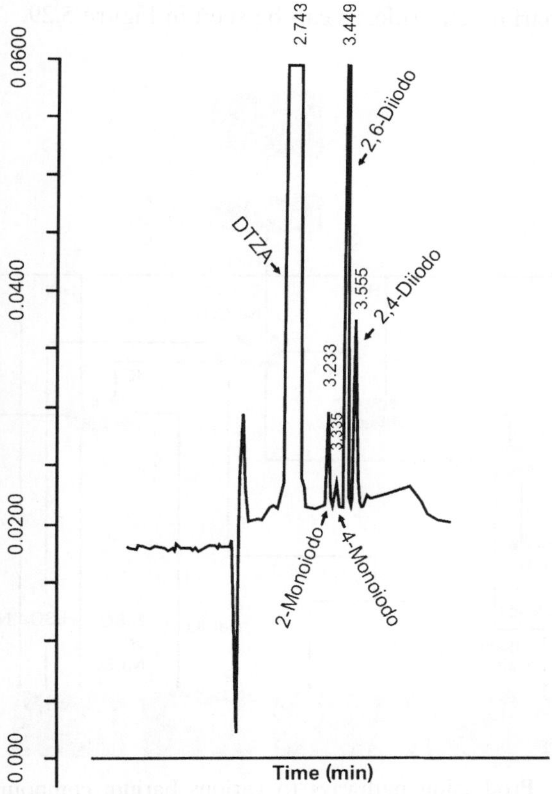

FIGURE 5.28 Separation of DTZA (2.743 minutes), 2-iodo- (3.233 minutes), 4-iodo (3.335 minutes), 2,4-diiodo- (3.555 minutes), and 2,6-diiodo- (3.449 minutes) compounds by capillary electrophoresis (CE) [120].

5.2.3 Barium Sulfate

Barium sulfate is used in the imaging of the digestive system. Barium sulfate is used in such a large amount that the presence of an impurity may play a significant role. Although Ba is a toxic species, it is not absorbed in the gastro-intestinal tract during the test because $BaSO_4$ is practically insoluble even in the gastric juice.

The most important barium mineral is barite (barium sulfate), almost exclusively used as the raw material for the production of barium compounds. The main impurities in barite are quartz, calcium carbonate, iron and manganese oxides, fluorite, strontium sulfate, and iron, lead, and zinc sulfides. Depending on the level, preliminary treatment is necessary to reduce impurities to less than 10% of the mineral mass [121].

Since barite rarely occurs as a pure white mineral, generally, its color is yellowish, reddish brown, dark gray, or black, and the pure form of the compound is obtained after the reduction of barite with carbon to give a water-soluble barium sulfide. Barium sulfate can be recovered either via barium carbonate or barium chloride, as can be seen in Figure 5.29.

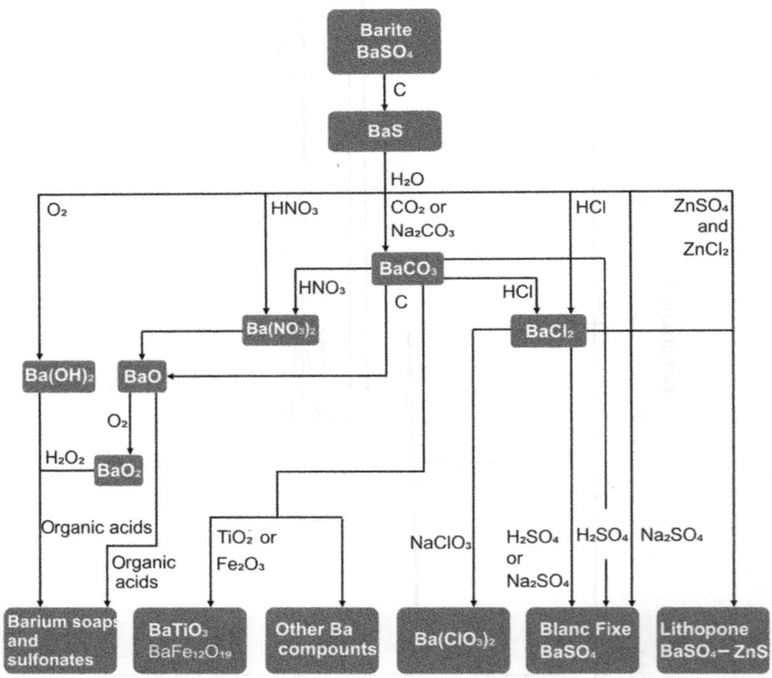

FIGURE 5.29 Production pathways to various barium compounds starting from barite (reproduced with permission from Kresse R, Baudis U, Jäger P, Riechers HH, Wagner H, Winkler J, Wolf HU. Barium and barium compounds. In: *Ullmann's Encyclopedia of Industrial Chemistry*. Wiley-VCH) [121].

TABLE 5.40 Solubility of Some Barium Salts and Their LD_{50} for Experimental Animals and Humans When Available

Salt	Solubility in Water (20°C) (g/L)	Solubility in HCl	LD_{50} (mg Ba/kg)	Test for LD_{50}
$BaSO_4$	0.00137	Insoluble	>2000[a]	Rat
$BaCO_3$	0.0002	Soluble	800	Rat, oral
			29	Human, oral
$BaCl_2$	263	Soluble	19.2	Mouse, IV
			11.4	Human, oral
$Ba(NO_3)_2$	84	Soluble	8.5	Mouse, IV
			355	Rat, oral
Ba acetate	666.7	Soluble	23.31	Mouse, IV

[a] Some sources consider barium sulfate nontoxic since it is hardly absorbed, and therefore present the LD_{50} as not available.

Residual impurities in barium sulfate may arise from the partial conversion of barium carbonate into barium sulfate through its reaction with sulfuric acid and from the incomplete washing out of sodium sulfide when preparing barium sulfate directly by reacting barium sulfide with sodium sulfate. The presence of such impurities is foreseen by pharmacopeial compendia so that USP, for example [122], limits the presence of soluble barium salts to 0.001%, and sulfide to 0.5 μg/g barium sulfate.

The problem with the presence of barium-soluble salts is their elevated toxicity. Table 5.40 presents the LD_{50} of some barium salts [123]. The effect of the presence of soluble barium compounds in the barium sulfate used as a contrast agent was tragically demonstrated in 2003 in Brazil. A pharmaceutical industry, instead of buying barium sulfate to formulate the product "Celobar," bought the cheaper intermediate barium carbonate. The industry planned to convert barium carbonate into barium sulfate by reacting it with sulfuric acid. Although the reaction occurred, approximately 15% of the barium carbonate was not converted into barium sulfate. This was enough to kill 21 patients who received the formulation for exams of the digestive system. Barium carbonate is also insoluble in water but not in hydrochloric acid, the acid present in the gastric juice. After ingesting the contrast, a lethal dose of barium was solubilized in the stomach and absorbed by the gastrointestinal tract.

Studies on barium sulfate impurity are rare in the literature, although formulations contain not only barium sulfate. Various compounds are added to lengthen shelf life, enhance coating ability, and improve taste [124]. Table 5.41 displays the most important additives used in oral barium sulfate contrast preparations, their function in the formulation, and the possible contaminants they can present [122].

TABLE 5.41 Components of Barium Sulfate Formulations, Their Functions, Main Additives for Each Function with Their Respective Limits of Impurities[a]

Component	Action	Additive	Limits of Impurities						
			Heavy Metals	As	Pb	Sulfate	Sulfated Ash	Others	
Deflocculation agents	Avoid formation of aggregates	Ammonium chloride	≤10 ppm	–	–	≤150 ppm	≤0.1%	Ca Fe	≤200 ppm ≤20 ppm
		Sodium citrate	≤10 ppm	≤2 ppm	–	≤150 ppm	–	Oxalate Chloride	≤10 ppm ≤10 ppm
		Sodium tartarate							
		EDTA	≤50 ppm	≤2 ppm	–	–	–	Fe NTA	≤80 ppm ≤0.1%
Suspending agents	Suspension stabilization	Carboxymethylcellulose	≤20 ppm	≤10 ppm	–	≤0.96%	20–33%	Chloride Silicate Naglycolate	≤0.64% ≤0.5% ≤0.4%
		Pectin	–	≤3 ppm	≤5 ppm	–	–		
		Magnesium/aluminum/ silicate	–	≤3 ppm	≤15 ppm	–	–		
		Gelatin	≤50 ppm	≤1 ppm	–	–	–	SO_2 Fe Cr Zn Hg Peroxides	0.15% ≤30 ppm ≤10 ppm ≤30 ppm ≤0.1 ppm ≤10 ppm
		Tragacanth	≤20 ppm	–	≤10 ppm	–	4% (ash)	Foreign matter	≤1.0%

Dispersing agents	Mixing the suspension with other fluids	Sodium citrate	≤10 ppm	≤2 ppm	–	≤150 ppm	–	Oxalate Chloride	≤300 ppm ≤50 ppm
		Sodium polyphosphate							
Wetting agents	To form a thin and uniform surface film	Natural gums Artificial gums							
Antidrying agents	To maintain humidity	Sorbitol	≤5 ppm	≤1.3 ppm	≤0.5 ppm	≤0.006%	–	Chloride Ni Red. sugars	≤0.005% ≤1 ppm ≤0.3%
		Mannitol	≤5 ppm	≤1.3 ppm	≤0.5 ppm	≤0.001%	–	Chloride Ni Red. sugars	≤0.007% ≤1 ppm ≤0.2%
Antifoaming agents	To prevent bubble formation	Simethicone Siloxane	≤5 ppm	–	–	–	–	Silica	≤0.7%
Flavoring agents	To improve the taste	Sugar	≤5 ppm	–	–	≤0.006%	–	Chloride	≤0.014%
		Saccharin	≤20 ppm	–	–	–	≤0.2%		

[a] Limits of impurities according to USP, BP, Japanese Pharmacopeia (JP) (when the values presented by the pharmacopeias did not correspond, the highest level was chosen).

CONCLUDING REMARKS

Expected impurities in pharmaceutical products include by-products generated during synthesis, species derived from inactive ingredients, residual solvents, and, of course, the omnipresent "heavy metals."

Dealing with diagnostic imaging agents, impurities and contaminants are particularly different due to the unusual way they are produced. Although the process involves synthesis and the use of solvents, all of them have a metallic base, making them subject to a variety of contaminants and impurities that are not typical in regular pharmaceutical products.

For most radiopharmaceuticals, drug manufacture comprises thee steps: (i) production of the radionuclide, either by fission or by bombardment, which demands a fissionable material (uranium) or a target element to be converted into the radionuclide product by bombardment; (ii) isolation of the radionuclide of interest; and (iii) labeling procedure. In all of these steps, impurities may be already present or may be generated in the process.

Technetium radiopharmaceuticals may contain I-131, Ru-103, and Mo-99 from the radionuclide production, and free pertechnetate or technetium in an oxidation state other than that which is adequate for the labeling procedure.

Common to all radiopharmaceuticals are the products generated by self-decomposition or radiolysis due to the inherent ionizing effect of radiation. In this case, besides the decomposition of the drug itself, the decomposition of water with free radical formation can cause molecules to break down, mainly in products containing proteins or other large molecules. Due to its higher energy radiation, I-131-labeled products are more susceptible to low RCP due to radiolysis than the products labeled with other iodine radioisotopes. The main contaminant in this case is the free iodine, which is not able to reach the target organ as is the labeled compound.

Also common to all radiopharmaceuticals is the presence of the decay nuclide, which may be or may be not a dangerous contaminant. In the case of I-124, the decay product is Se-124, but for Tc-99, for example, the decay nuclide is Mo-90.

Contaminants in iodine radiopharmaceuticals also arise from the radionuclide production. Although I-131 is obtained in the same way as Tc-99m, from U-235 fission, its possible contaminants are not the same as those of Tc-99m. As I-131 is collected in gaseous form, the contaminants are nuclides that are also separated from the target as a gas, such as Te-124. On the other hand, I-123 is a cyclotron radioisotope produced either from tellurium (Te-124) or xenon (Xe-123) bombardment. In I-123, produced from Te-124, the most probably contaminants are other iodine isotopes, mostly I-124, which, due to its longer half-life, will over time present an increased concentration as compared to I-123.

Gadolinium-based CM are mostly contaminated by other RE metals due to the extraordinarily demanding task of separating these metals. Iodinated formulations have long been established CM for X-ray imaging. However, due to their high iodine concentration, events like light exposure or mixing with other drugs may lead to drug decomposition or drug precipitation.

REFERENCES

1. Cherry SR, Sorenson JA, Phelps ME. *Physics in Nuclear Medicine*, 3rd ed. Saunders, Philadelphia, 2003, p. 47.
2. Millar AM. High performance liquid chromatographic method for determination of the radiochemical purity of iodohippuric acid (I-123) injection. J Pharm Pharmacol 34(1), 14–17, 1982.
3. Massey CD, Miller DL, Carson SD, Wheeler TA, Longley SW, Coats RL, Parma EJ, Mcdonald M, Vernon ME, Bourcier SC, Mills SG, Trennel AJ, Boldt KR. Feasibility Study of Medical Isotope Production at Sandia National Laboratories, SAND95-2703 Rev. O, Sandia National Laboratories, 1995.
4. Committee on Medical Isotope Production without Highly Enriched Uranium, National Research Council. *Medical Isotope Production without Highly Enriched Uranium*. National Academies Press, Washington, DC, 2009.
5. Committee on Radiation Source Use and Replacement, National Research Council. *Radiation Source—Use and Replacement: Abbreviated Version*. National Academies Press, Washington, DC, 2008.
6. McDonald MJ, Carson SD, Naranjo GE, Wemple JA. Challenges of extracting and purifying fission-produced molybdenum-99. Ind Eng Chem Res 39, 3146–3150, 2000.
7. Vandegrift GF, Conner C, Hofman GL, Leonard RA, Mutalib A, Sedlet J, Walker DE, Wiencek TC, Snelgrove JL. Modification of targets and processes for conversion of 99Mo production from high- to low-enriched uranium. Ind Eng Chem Res 39, 3140–3145, 2000.
8. Vandegrift G. Facts and Myths Concerning 99Mo Production with HEU and LEU Targets. International RERTR Meeting, Boston, Massachusetts, USA, November 6–10, 2005. Available at http://www.rertr.anl.gov/RERTR27/PDF/S8-1_VandeGrift.pdf.
9. Carson SD, Tallant DR, McDonald MJ, Garcia MJ, Simpson RL. Radiation chemistry of simulated Mo-99 product. Ind Eng Chem Res 39(9), 3151–3156, 2000.
10. Bond AH, Rogers RD, Dietz ML. Nuclear separations for radiopharmacy: the need for improved separations to meet future research and clinical demands. Ind Eng Chem Res 39, 3130–3134, 2000.
11. United States Pharmacopeia. USP 30, 2007, p. 3284.
12. Billinghurst MW, Hreczuch FW. Contamination from 1-131, Ru-103, and Np239 in the eluate of Mo-99-Tc-99m generators loaded with (n,γ)-produced M-99. J Nucl Med 17, 840–843, 1976.
13. Anderson DW, Raeside DE, Ficken VJ. Determination of impurity activities in fission-product generator eluate. J Nucl Med 15, 889–891, 1974.
14. Johansson L, Mattsson S. Plutonium in Tc-99m pertechnetate for clinical use: concise communication. J Nucl Med 21, 1091–1094, 1980.

15. Marengo M, Aprile C, Bagnara C, Bolzati C, Bonada C, Candini G, Casati R, Civollani S, Colombo FR, Compagnone G, Del Dottore F, Di Guglielmo E, Ferretti PP, Lazzari S, Minoia C, Pancaldi D, Ronchi A, Sanità Di Toppi G, Saponaro R, Torregiani T, Uccelli L, Vecchi F, Piffanelli A. Quality control of M-99/Tc99m generators: results of a survey of the radiopharmacy working group of the Italian association of nuclear medicine (AIMN). Nucl Med Commun 20, 1077–1084, 1999.

16. Lee JD. *Concise Inorganic Chemistry*, 4th ed. Chapman & Hall, London, 1993, p. 747.

17. Banerjee S, Pllai MRA, Ramamoorthy N. Evolution of Tc-99m in diagnostic radiopharmaceuticals. Semin Nucl Med XXXI(4), 260–277, 2001.

18. Noll B, Johannsen B, Spies H. Sources of radiochemical impurities in the Tc-99m/S unprotected MAG3 system. Nucl Med Biol 22, 1057–1062, 1995.

19. Seetharaman S, Sosabowski MH, Ballinger JR. Validation of alternative methods of preparing Tc-99m-MAG3. Appl Radiat Isot 65, 1240–1243, 2007.

20. Hung JC, Ponto JA, Gradient KA, Frie JA, Aksamit CM, Enquist CL, Carrels KE. Deficiencies of package insert directions for the preparation of radiopharmaceuticals. J Am Pharm Assoc 44(1), 30–35, 2004.

21. Shattuck LA, Eshima D, Taylor AT, Jr., Anderson TL, Graham DL, Latino FA, Payne SE. Evaluation of the hepatobiliary excretion of technetium-99m-MAG3 and reconstitution factors affecting radiochemical purity. J Nucl Med 35, 349–355, 1994.

22. Millar AM, Hesselewood SR. A comparison of SeoPak and high-performance liquid chromatography as technique for measuring the radiochemical purity of Tc-99m-MAG3. Nucl Med Commun 25, 1049–1051, 2004.

23. Stinger RE, Maltby PJ. A one year study of factors affecting Tc-99m-MAG3 kit failure. Nucl Med Commun 17, 993, 1996.

24. Stinger RE, Schroeder RE, Maltby PJ. MAG3 failure is due to inadvertent oxidant contamination. Nucl Med Commun 18, 294, 1997.

25. Waight CC, Beattie LA, O'Brien LM, Millar AM. Preparation of Tc-99m-MAG3: radiochemical purity is affected by the residence time of sodium chloride injection in a three-part syringe. Nucl Med Commun 27, 197–200, 2006.

26. Beattie LA, O'Brien LM, Waight CC, Homer NZ, Andrew R, Millar AM. Preparation of Tc-99m-MAG3: the effect on radiochemical purity of using sodium chloride injection from plastic ampoules that have been exposed to light. Nucl Med Commun 29, 649–653, 2008.

27. Washburn LC, Biniakiewick DS, Maxon IIIHR. Reliable preparation of Tc-99m(V) DMSA by a simple modified method using a commercial kit for Tc-99m(III) DMSA. Nucl Med Biol 22, 689–691, 1995.

28. Kumar V. Evaluation of stannous oxidation in the preparation of ultrahigh-purity Tc-99m(V)-DMSA. Nucl Med Commun 22, 1261–1266, 2001.

29. Jacquot Y, Zoubir B, Xicluna A, Bidet AC. Research of $99mTcO_4^-$ and $99mTcO_2^-$ in injectable solutions of 99mTc-HMDP by inverse phase HPTLC. Ann Pharm Fr 64(1), 63–67, 2006.

30. Decristoforo C, Siller R, Chen F, Riccabona G. Radiochemical purity of routinely prepared Tc-99m radiopharmaceuticals: a retrospective study. Nucl Med Commun 21, 349–354, 2000.

31. Schlyer DJ. Production of radioactive iodine. Technical meeting of project counterparts on cyclotron production of iodine 123, São Paulo, Brazil, 2001, p 1–14.
32. Kondo K, Lambrecht RM, Wolf AP. Iodine-123 production for radiopharmaceuticals—XX excitation functions of the Te-124(n,2p)I-123 and Te-124(n,p)I-124 reactions and the effect of target enrichment on radionuclidic purity. Int J Appl Radiat Isot 28, 395–401, 1977.
33. Kim D, Chang J. Calculation of proton-induced reactions on tellurium isotopes below 60 MeV for medical radioisotope production. J Korean Nucl Soc 32, 361–371, 2000.
34. El-Azony KM, Suzuki K, Fukumura T, Szelécsényi F, Kovacs Z. Proton induced reactions on natural tellurium up to 63 MeV: data validation and investigation of possibility of I-124 production. Radiochim Acta 96, 763–769, 2008.
35. Acerbi E, Birattari C, Castiglioni M, Resmini F. Production of I-123 for medical purposes at the Milan AVF cyclotron. Int J Appl Radiat Isot 26, 741–747, 1975.
36. Kondo K, Lambrecht RM, Norton EF, Wolf AP. Cyclotron isotopes and radiopharmaceuticals—XXII Improved targetry and radiochemistry for production of I-123 and I-124. Int J Appl Radiat Isot 28, 765–771, 1977.
37. Hammermaier A, Reich E, Bögl W. Radiochemical purity and in vitro stability of commercial hippurans. J Nucl Med 27, 850–854, 1986.
38. Wafelman AR, Auchi R, Hoefnagel CA, Beijnen JH. Radiochemical purity of I-131-labeled metaiodobenzylguanidine infusion fluids—a report from clinical-practice. Eur J Nucl Med 20(7), 614–616, 1993.
39. Wafelman AR, Hoefnagel CA, Maes RAA, Beijnen JH. Radiochemical purity, at expiry, and radiochemical stability of iodine-131 labelled meta-iodobenzylguanidine concentrates for intravenous infusion. Nuklearmedizin 35(4), 122–125, 1996.
40. Wahl RL, Wissing J, del Rosario R, Zasadny KR. Inhibition of autoradiolysis of radiolabeled monoclonal antibodies by cryopreservation. J Nucl Med 31(1), 84–89, 1990.
41. Chakrabarti MC, Le N, Paik CH, DeGraff WG, Carrasquillo JA. Prevention of radiolysis of monoclonal antibody during labeling. J Nucl Med 37(8), 1384–1388, 1996.
42. Visser GW, Klok RP, Gebbinck JWK, ter Linden T, van Dongen GA, Molthoff CF. Optimal quality I-131-monoclonal antibodies on high-dose labeling in a large reaction volume and temporarily coating the antibody with IODO-GEN. J Nucl Med 42(3), 509–519, 2001.
43. Zielinski FW, MacDonald NS, Robinson GD, Jr. Production by compact cyclotron of radiochemically pure iodine-123 as iodide for synthesis of radiodiagnostic agents. J Nucl Med 18, 67–69, 1977.
44. Kudelin BK, Gromova EA, Gavrilina LV, Solin LM. Purification of recovered tellurium dioxide for re-use in iodine radioisotope production. Appl Radiat Isot 54, 383–386, 2001.
45. Lagunas-Solar MC, Carvacho OF, Liu B-L, Jin Y, Sun ZX. Cyclotron production of high purity I-123 1. A revision of excitation functions, thin-target and cumulative yields for I-127(p,xn) reactions. Appl Radiat Isot 37, 823–833, 1986.
46. Jungerman JA, Lagunas-Solar MC. Cyclotron production of high purity iodine-123 for medical applications. J Radioanal Chem 65, 31–45, 1981.

47. Yu S. Review of 18F-FDG synthesis and quality control. Biomed Imaging Interv J 2(4), 1–11, 2006.
48. United States Pharmacopeia. USP-30—Fludeoxyglucose F 18 injection, pp. 2158–2160, 2007.
49. Pharmacopeia Europeia. Ph Eur—Monograph 1325, Fludeoxyglucose [^{18}F] Injection, 2009.
50. Schueller MJ, Alexoff DL, Schlyer DJ. Separating long-lived metal ions from F-18 during (H_2O)-O-18 recovery. Nucl Instrum Methods Phys Res B 261(1-2), 795–799, 2007.
51. Kilbourn MR, Jerabek PA, Welch MJ. An improved [^{18}O]water target for [^{18}F]fluoride production. Int J Appl Radiat Isot 36(4), 327–328, 1985.
52. Bowden L, Vintro LL, Mitchell PI, O'Donnell RG, Seymour AM, Duffy GJ. Radionuclide impurities in proton-irradiated [O-18]H_2O for the production of F-18(-): activities and distribution in the [F-18]FDG synthesis process. Appl Radiat Isot 67(2), 248–255, 2009.
53. Manickam V, Brey RR, Jenkins PA, Christian PE. Measurements of activation products associated with Havar foils from a GE PET trace medical cyclotron using high resolution gamma spectroscopy. Health Phys 96(2 Suppl), S37–S42, 2009.
54. Marengo M, Lodi F, Magi S, Cicoria G, Pancaldi D, Boschi S. Assessment of radionuclidic impurities in 2-[^{18}F]fluoro-2-deoxy-d-glucose ([^{18}F]FDG) routine production. Appl Radiat Isot 66(3), 295–302, 2008.
55. Avila-Rodriguez MA, Wilson JS, McQuarrie SA. A quantitative and comparative study of radionuclidic and chemical impurities in water samples irradiated in a niobium target with Havar vs. niobium-sputtered Havar as entrance foils. Appl Radiat Isot 66(12), 1775–1780, 2008.
56. Ito S, Sakane H, Deji S, Saze T, Nishizawa K. Radioactive by-products in [^{18}O]H_2O used to produce 18F for [^{18}F]FDG synthesis. Appl Radiat Isot 64(3), 298–305, 2006.
57. Mochizuki S, Ogata Y, Hatano K, Abe J, Ito K, Ito Y, Nishino M, Miyahara H, Ishigure N. Measurement of the induced radionuclides in production of radiopharmaceuticals for positron emission tomography (PET). J Nucl Sci Technol 43(4), 348–353, 2006.
58. O'Donnell RG, Vintro LL, Duffy GJ, Mitchell PI. Measurement of the residual radioactivity induced in the front foil of a target assembly in a modern medical cyclotron. Appl Radiat Isot 60(2-4), 539–542, 2004.
59. Ito S, Saze T, Sakane H, Ito S, Ito S, Nishizawa K. Tritium in [^{18}O]water containing [^{18}F]fluoride for [^{18}F]FDG synthesis. Appl Radiat Isot 61(6), 1179–1183, 2004.
60. Asti M, Grassi E, Sghedoni R, De Pietri G, Fioroni F, Versari A, Borasi G, Salvo D. Purification by ozonolysis of O-18 enriched water after cyclotron irradiation and the utilization of the purified water for the production of [F-18]-FDG (2-deoxy-2-[F-18]-fluoro-D-glucose). Appl Radiat Isot 65(7), 831–835, 2007.
61. Huang BX, Channing MA, Plascjak PS, Kiesewetter DO, Der M, Ma Y, Eckelman WC. Routine quality control of recycled target [O-18]water by capillary electrophoresis and gas chromatography. Nucl Med Biol 30(7), 785–790, 2003.
62. Nishijima K, Kuge Y, Tsukamoto E, Seki K, Ohkura K, Magata Y, Tanaka A, Nagatsu K, Tamaki N. Increased [F-18]2-fluoro-2-deoxy-D-glucose ([F-18]FDG)

yield with recycled target [O-18]water: factors affecting the [F-18]FDG yield. Appl Radiat Isot 57(1), 43–49, 2002.

63. Channing MA, Huang BX, Eckelman WC. Analysis of residual solvents in 2-[F-18] FDG by GC. Nucl Med Biol 28(4), 469–471, 2001.
64. Kanu AB, Dixon M, Yin Zu Y, Bailey J, Gillies JM, Zweit J, Thomas CLP. Rapid screening of 2-[^{18}F]-fluoro-2-deoxy-D-glucose infusions for volatile organic compound contaminants by solid phase microextraction with gas chromatography—selective ion monitoring mass spectrometry (SPME-GC-SIMMS). Appl Radiat Isot 58(2), 193–200, 2003.
65. Ma Y, Huang BX, Channing MA, Eckelman WC. Quantification of Kryptofix 2.2.2 in 2-[F-18]FDG and other radiopharmaceuticals by LC/MS/MS. Nucl Med Biol 29(1), 125–129, 2002.
66. Nakao R, Ito T, Yamaguchi M, Suzuki K. Improved quality control of [^{18}F]FDG by HPLC with UV detection. Nucl Med Biol 32(8), 907–912, 2005.
67. Nakao R, Ito T, Yamaguchi M, Suzuki K. Simultaneous analysis of FDG, ClDG and Kryptofix 2.2.2 in [^{18}F]FDG preparation by high-performance liquid chromatography with UV detection. Nucl Med Biol 35(2), 239–244, 2008.
68. International Conference on Harmonization—ICH Q3C(R5)—Impurities: Guidelines for Residual Solvents, 2010.
69. Kuge Y, Nishijima K, Nagatsu K, Seki K, Ohkura K, Tanaka A, Sasaki M, Tsukamoto E, Tamaki N. Chemical impurities in [^{18}F]FDG preparations produced by solid-phase 18F-fluorination. Nucl Med Biol 29(2), 275–279, 2002.
70. Buriova E, Macasek F, Melichar F, Kropacek M, Prochazka L. Autoradiolysis of the 2-deoxy-2-[F-18]fluoro-D-glucose radiopharmaceutical. J Radioanal Nucl Chem 264(3), 595–602, 2005.
71. Fawdry RM. Radiolysis of 2-[^{18}F]fluoro-2-deoxy-D-glucose (FDG) and the role of reductant stabilisers. Appl Radiat Isot 65(11), 1193–1201, 2007.
72. Zhang JM, Li YG, Liu J, Zhang XJ, Tian JH. The analysis of radiolysis impurities in F-18-FDG and methods of repurification. J Radioanal Nucl Chem 283(3), 565–569, 2010.
73. Berridge MS, Apana SM, Hersh JM. Teflon radiolysis as the major source of carrier in fluorine-18. J Label Compd Radiopharm 59, 543–548, 2009.
74. Kowalski JB, Morrissey RF. Sterilization of implants. In: *Biomaterials Science*, Ratner BD, Hoffman AS, Schoen FJ, Lemons JE (eds.). Academic Press, San Diego, CA, 1996, pp. 415–420.
75. Schlyer DJ, Firouzbakht ML, Wolf AP. Impurities in the [^{18}O] water target and their effect on the yield of an aromatic displacement reaction with [^{18}F] fluoride. Appl Radiat Isot 44(12), 1459–1465, 1993.
76. Karwath P, Sartorb J, Griesb W, Wodarskib C, Dittmarb C, Biersacka HJ, Guhlkea S. Steam sterilization and automatic dispensing of [^{18}F]fludeoxyglucose (FDG) for injection. Appl Radiat Isot 62(4), 577–586, 2005.
77. Caravan P, Ellison JJ, McMurry TJ, Lauffer RB. Gadolinium(III) chelates as MRI contrast agents: structure, dynamics, and applications. Chem Rev 99, 2293–2353, 1999.
78. Pulukkody KP, Norman TJ, Parker D, Royale L, Broan CJ. Synthesis of charged and uncharged complexes of gadolinium and yttrium with cyclic

polyazaphosphinic acid ligands for in vivo applications. J Chem Soc Perkin Trans 2, 605, 1993.

79. Kumar K, Tweedle MF. Macrocyclic polyaminocarboxilate complexes of lanthanides as magnetic resonance imaging agents. Pure Appl Chem 65, 515–520, 1993.
80. McGill I. Rare earth elements. In: *Ullmann's Encyclopedia of Industrial Chemistry*, Vol. 31. Wiley-VCH, 2012, pp. 183–228.
81. Kumar K. Macrocyclic polyamino carboxylate complexed of Gd(III) as magnetic resonance imaging contrast agents. J Alloys Compd 249, 163–172, 1997.
82. Ledneva E, Karie S, Launay-Vacher V, Janus N, Deray G. Renal safety of gadolinium contrast media in patients with chronic renal insufficiency. Radiology 250, 618–628, 2009.
83. Watson AD. The use of gadolinium and dysprosium chelate complexes as contrast agents for magnetic-resonance-imaging. J Alloys Compd 207, 14–19, 1994.
84. Weinmann HJ, Brasch RC, Press WR, Wesbey GE. Characteristics of gadolinium-DTPA complex - A potential NMR contrast agent. Am J Roentgenol 142(3), 619–624, 1984.
85. Wible JH, Jr., Troup CM, Hynes MR, Galen KP, MacDonald JR, Barco SJ, Wojdyla JK, Periasamy MP, Adams MD. Toxicological assessment of gadoversetamide injection (OptiMARK), a new contrast-enhancement agent for use in magnetic resonance imaging. Invest Radiol 36(7), 401–412, 2001.
86. Vittadini G, Felder E, Tirone P, Lorusso V. B-19036, A potential new hepatobiliary contrast agent for MR proton imaging. Invest Radiol 23, 246–248, 1988.
87. Weinmann HJ, Schuhmann-Giampieri G, Schmitt-Willich H, Vogler H, Frenzel T, Gries H. A new lipophilic gadolinium chelate as a tissue-specific contrast medium for MRI. Magn Reson Med 22(2), 233–237, 1991.
88. Steger-Hartmann T, Graham PB, Müller S, Schweinfurth H. Preclinical safety assessment of Vasovist (Gadofosveset trisodium), a new magnetic resonance imaging contrast agent for angiography. Invest Radiol 41(5), 449–459, 2006.
89. United States Pharmacopeia. USP 30, 2007.
90. Bousquet JC, Saini S, Stark DD, Hahn PF, Nigam M, Wittenberg J, Ferrucci JT, Jr. Gd-DOTA: characterization of a new paramagnetic complex. Radiology 166, 693–698, 1988.
91. Thomsen HS, Almèn T, Morocos SK. Contrast media safety committee of the European Society for Urogenital Radiology (ESUR). Gadolinium-containing contrast media for radiographic examinations: a position paper. Eur Radiol 12(10), 2600–2605, 2002.
92. Morcos SK. Nephrogenic systemic fibrosis following the administration of extacellular gadolinium based contrast agents: is the stability of the contrast agent molecule an important factor in the pathogenesis of this condition? Br J Radiol 80, 73–76, 2007.
93. Idée JM, Port M, Medina C, Lancelot E, Fayoux E, Ballet S, Corot C. Possible involvement of gadolinium chelates in the pathophysiology of nephrogenic systemic fibrosis: a critical review. Toxicology 248(2-3), 77–88, 2008.
94. Morcos SK. Extracellular gadolinium contrast agents: differences in stability. Eur J Radiol 66, 175–179, 2008.

95. Idée JM, Port M, Raynal I, Schaefer M, Le Greneur S, Corot C. Clinical and biological consequences of transmetallation induced by contrast agents for magnetic resonance imaging: a review. Fundam Clin Pharmacol 20, 563–576, 2006.
96. Port M, Idée JM, Medina C, Robic C, Sabatou M, Corot C. Efficiency, thermodynamic and kinetic stability of marketed gadolinium chelates and their possible clinical consequences: a critical review. Biometals 21, 469–490, 2008.
97. Tweedle MF. The ProHence story: the making of a novel MRI contrast agent. Eur Radiol 7(5), S225–S230, 1997.
98. Rofsky NM, Sherry AD, Lenkinski RE. Nephrogenic systemic fibrosis. A chemical perspective. Radiology 247, 608–612, 2008.
99. Brugieres P, Gaston A, Degryse HR, Parizel PM, Deschepper AM, Berry I, Manelfe C, Lebras F, Marsault C, Wichmann W, Valavanis A. Randomized double-blind trial of the safety and efficacy of 2 gadolinium complexes (Gd-DTPA and Gd-DOTA). Neuroradiology 36(1), 27–30, 1994.
100. Uggeri IF, Aime S, Anelli PL, Botta M, Brocchetta M, Dehaen C, Ermondi G, Grandi M, Paoli P. Novel contrast agents for magnetic-resonance-imaging—synthesis and characterization of the ligand BOPTA and its Ln(III) complexes (Ln=Gd, La, Lu)—X-ray structure of disodium (tps-9-145337286-c-s)-[4-carboxy-5,8,11-tris(carboxymethyl)-1-phenyl-2-oxa-5,8,11-triazatridecan-13-oato(5-)]gadolinate(2-) in a mixture with its enantiomer. Inorg Chem 34, 633, 1995.
101. Schmitt-Willich H, Brehm M, Ewers CLJ, Michl G, Muller-Fahrnow A, Petrov O, Platzek J, Raduchel B, Sulzle D. Synthesis and physicochemical characterization of a new gadolinium chelate: the liver-specific magnetic resonance imaging contrast agent Gd-EOB-DTPA. Inorg Chem 38, 1134, 1999.
102. Carvan P, Comuzzi C, Crooks W, McMurry TJ, Choppin GR, Woulfe SR. Thermodynamic stability and kinetic inertness of MS-325, a new blood pool agent for magnetic resonance imaging. Inorg Chem 40, 2170, 2001.
103. Moreau J, Guillon E, Pierrard JC, Rimbault J, Port M, Aplincourt M. Complexing mechanism of the lanthanide cations Eu^{3+}, Gd^{3+}, and Tb^{3+} with 1,4,7,10-tetrakis(carboxymethyl)-1,4,7,10-tetraazacyclododecane (DOTA)—characterization of three successive complexing phases: study of the thermodynamic and structural properties of the complexes by potentiometry, luminescence spectroscopy, and EXAFS. Chem Eur J 10, 5218–5232, 2004.
104. Schmitt-Willich H. Stability of linear and macrocyclic gadolinium based contrast agents. Br J Radiol 80, 581–583, 2007.
105. Port M, Idée JM, Medina C, Dencausse A, Corot C. Stability of gadolinium chelates and their biological consequences: new data and some comments. Br J Radiol 81, 258–259, 2008.
106. Bellini MF, Vasile M, Morel-Precetti S. Currently used non-specific extracellular MR contrast media. Eur Radiol 13(12), 2688–2698, 2003.
107. Dawson P. Gadolinium chelates: chemistry. In: *Textbook of Contrast Media*, Dawson P, Cosgrove DO, Grainer RG (eds.). Isis Medical Media, Oxford, England, 1999, pp. 291–296.
108. Tweedle MF. "Stability" of gadolinium chelates. Br J Radiol 80, 583–584, 2007.
109. Morcos SK. Author's reply. Br J Radiol 80, 584–585, 2007.

110. Dawson P, Pitfield J, Skinnemoen K. Isomeric purity and super saturation of iopamidol. Br J Radiol 56(670), 711–713, 1983.
111. Deavin JC. Solubility of iopamidol solution. Br J Radiol 57(673), 107, 1984.
112. Trissel L. *Handbook on Injectable Drugs*, 15th ed. American Society of Health-System Pharmacists, Betheseda, MD, 2009.
113. Shionoya H, Sugihara Y, Okano K, Sagami F, Mikami T, Katayama K. Studies on experimental iodine allergy: 2 iodinate protein antigens and their generation from inorganic and organic iodine-containing chemicals. J Toxicol Sci 29, 137–145, 2004.
114. Grönewäller EF, Wahl HG, Kehlbach R, Rodemann HP, Claussen CD, Duda SH. Stability of iodinated contrast media in UV-laser irradiation and toxicity of photoproducts. Rofo-Fortschr Rontgenstr 169(5), 537–541, 1998.
115. Yu-Chang JW. Deiodination kinetics of water-soluble radiopaques. J Pharm Sci 69(6), 671–675, 1980.
116. Chellquist E, Wheeler S, Straster J, Wlofe H, Storflor H. The effect of glucose and copper on the stability of diatrizoic acid. PDA J Pharm Sci Technol 53(5), 247–251, 1999.
117. Greenwood NN, Earnshaw A. *Chemistry of the Elements*. Pergamon Press, Oxford, 1984.
118. Chellquist EM, Nelson WO, Storflor HL. Assay and purity analysis of diatriazoate sodium in drug product by LC. J Pharm Biomed Anal 16, 39–45, 1997.
119. Lerner HH. Diatrizoic acid. In: *Analytical Profile of Drug Substances*, Vol. 4, Florey K (ed.). Academic Press, New York, 1975, pp. 137–167.
120. Farag SA, Wells CE. Capillary electrophoresis determination of diatrizoic acid and its impurities in diatrizoate radiopaque solutions. Mikrochim Acta 126, 141–145, 1997.
121. Kresse R, Baudis U, Jäger P, Riechers HH, Wagner H, Winkler J, Wolf HU. Barium and barium compounds. In: *Ullmann's Encyclopedia of Industrial Chemistry*, Vol. 4. Wiley-VCH, 2011, pp. 621–640.
122. United States Pharmacopoeia. USP 30, 2007. Barium sulfate—official monograph, p. 1485.
123. Merck Index, p. 974–1010.
124. O'Connor SD, Summers RM. Revisiting oral barium sulfate contrast agents. Acad Radiol 14, 72–80, 2007.

6

CONTAINERS

6.1 GLASS CONTAINERS

Glass containers for packaging pharmaceutical preparations should meet stability specifications, which are related to the presence of certain components in the structure of the glass and to the type and use of the formulation. According to pharmacopeial prescriptions, four different types of glass may be used for storing pharmaceutical preparations. Table 6.1 presents the classification of glass containers in accordance with the product's nature and pharmaceutical usage.

Though mostly made of silica, the different glass types are obtained by adding or subtracting certain glass components. Table 6.2 lists the principal components of pharmaceutical glasses based on the criteria presented in Table 6.1. Type I glass must have at least 10% boron oxide and a higher concentration of aluminum oxide than the ordinary soda-lime glass to improve its resistance. The reduced amount of sodium oxide lessens the solubility of Type I glass in water. Type II glass is made of common soda-lime glass after a dealkalization process on the inside surface at the time of manufacture. In this process, known as sulfur treatment, sulfur dioxide is introduced as newly formed bottles pass from the forming machine, and it reacts on the surface of the glass to form sodium sulfate, according to the equation

$$SO_2 + \tfrac{1}{2}O_2 + Na_2O_{(glass)} \rightarrow Na_2SO_4. \qquad (6.1)$$

Sources of Contamination in Medicinal Products and Medical Devices, First Edition. Denise Bohrer.
© 2013 by John Wiley & Sons, Inc. Published 2013 by John Wiley & Sons, Inc.

TABLE 6.1 Glass Classification According to Pharmacopeial Prescription and Their Pharmaceutical Usages

Glass Type	General Description	Uses
I	Highly resistant borosilicate glass	Parenteral preparations
II	Treated soda-lime glass	Acidic or neutral parenteral preparations
III	Soda-lime glass	Not for parenteral preparations
NP	Soda-lime glass	Oral or topic use

TABLE 6.2 The Typical Composition of Glass Containers According to the Pharmacopeial Classification

Component	Type I Borosilicate (%)	Type II, III, IV Soda Lime (%)
SiO_2	70–73	69–73
B_2O_3	10	0–1
Na_2O	2–9	13–14
Al_2O_3	6–7	2–4
BaO	0.1–2.0	0–2
K_2O	1–2	0–3
CaO	0.7–1.0	5–7
MgO	0–0.5	3–4
ZnO	0–0.5	–

Sources: Corning Life Sciences, Schott and Wheaton Science International. Technical Data Bulletins.

The container is delivered with a haze of sodium sulfate, which should be rinsed away before filling.

Stability tests for glass containers prescribed by pharmacopeial compendia are limited to the action of water on the glass surface to measure the alkalinity released (Water Attack). After autoclaving the container filled with water at 121°C for 60 minutes, 100 mL of the resulting extract is titrated with 0.01 M HCl. The volume of the acidic titrant should not be greater than a specific value based on the container's capacity. Briefly, Type I and II glass containers with a capacity of less than 10 mL should consume less than 2.0 mL titrant, whereas for Type III glass containers, the limit is 20 mL. For containers with a capacity of between 20 and 500 mL, the titrant consumption should be from 0.8 to 0.2 mL for Type I and II glass and from 8.1 to 2.2 mL for Type III glass. For NP glass, no limit of titrant is established.

The released alkalinity is, however, not enough to show the actual stability of a glass container. Glass is not as inert as it appears, and water is not the only substance that attacks the glass surface. In spite of this, the only test

TABLE 6.3 Metal Profile of a Type I Clear Glass Vial

Metal	Amount (%)	Metal	Amount (%)
Si	>30	K	0.1
Al	1.6	Ca	0.036
Na	2.4	Fe	0.024
B	5.5	Zr	0.021
	(μg/g)		(μg/g)
Mg	61	Pd	1.6
Ba	21	Ga,	1.2
Ce	8.8	Pb	1.0
Ti	6.7	Pr, Cd, Ge, Sb	0.5
Hf	6.0	Nd, Nb	0.4
Mo	4.8	As, U	0.3
Y	2.8	Mn, Cr, Th, Ir, Tb	0.2
La	2.5	Cu, Bi, W, Yb	0.1
Sr	1.7		

Adapted from Reference 1.

prescribed by the United States Pharmacopeia (USP) for toxic impurities in glass is for arsenic (<660> Containers, USP 31), which specifies that Type I glass containers should be assayed according to the Water Attack procedure and present not more than 0.1 ppm As. Though not listed as a glass constituent, arsenic may be used as a fining agent in the glass industry (see below) and, therefore, may be present in the glass structure.

Several other metallic impurities may be present in glass, even in high-quality Type I glasses. Zuccarello et al. [1] evaluated the metal profile of a Type I glass vial after promoting its microwave digestion in a mixture of hydrofluoric acid and nitric acid. The measured elements and their concentrations are presented in Table 6.3. Besides the usual constituents, unusual elements such as Zr and Ce were found in relatively high amounts.

Studies have shown that even Type I glass is not completely inert. Bohrer et al. [2] showed that the hydrolytic resistance of glass containers is diminished in the presence of some substances commonly present in infusion solutions. Table 6.4 presents the released alkalinity and also the concentration of some glass constituents (silicate, sodium, boron, aluminum) in the aqueous extract obtained after the hydrolytic resistance test.[1] These data show that, despite complying with pharmacopeial prescriptions, glass can release its constituents by the action of species in solutions and even in pure water. Basic solutions of bicarbonate and gluconate presented the highest level of glass constituents

[1] The hydrolytic resistance test described in this paper was the "powdered glass test" and not the "water attack test" as described above; see USP Chapter <660>.

TABLE 6.4 Volume H_2SO_4 Solution Consumed in the Titration of Solutions before and after Heating in Contact with Glass Mass for 30 Minutes at 121°C and Species Extracted from Glass during the USP Powdered Glass Test Carried Out with Clear Ampoules in the Presence of Different Solutions. Concentration of the Solutions: 0.01 mol/L [2]

	Volume 0.02 N H_2SO_4 (mL)		Species Concentration (mg/L) ± SD (n = 3)					
Sample	Before Heating	After Heating	Na	Silicate	Borate	Al	Cu	Pb
NaCl	[a]	0.56 ± 0.01	–	8.30 ± 2.01	2.22 ± 0.11	0.15 ± 0.01	0.09 ± 0.02	0.03 ± 0.01
KCl	[a]	0.69 ± 0.01	14.42 ± 1.51	9.96 ± 1.71	1.58 ± 0.03	0.19 ± 0.01	0.14 ± 0.08	0.04 ± 0.02
$CaCl_2$	[a]	0.45 ± 0.02	13.41 ± 1.33	7.31 ± 2.08	2.00 ± 0.11	0.18 ± 0.01	0.06 ± 0.01	0.02 ± 0.01
$MgCl_2$	[a]	0.13 ± 0.01	15.25 ± 1.64	6.54 ± 0.53	3.25 ± 0.21	0.16 ± 0.01	0.09 ± 0.03	0.02 ± 0.01
Sodium gluconate	2.68 ± 0.22	1.83 ± 0.05	–	246.0 ± 18.7	1.91 ± 0.09	2.04 ± 0.02	0.49 ± 0.07	0.05 ± 0.01
Sodium hydrogenophosphate	[a]	10.22 ± 0.91	–	10.21 ± 3.41	2.66 ± 0.09	4.93 ± 0.12	0.05 ± 0.01	0.05 ± 0.02
Potassium hydrogenophosphate	[a]	8.78 ± 0.95	18.38 ± 2.34	45.32 ± 3.76	1.73 ± 0.06	4.30 ± 0.07	0.13 ± 0.04	0.04 ± 0.02
Sodium bicarbonate	24.3 ± 0.41	12.68 ± 1.03	–	271.7 ± 23.2	5.98 ± 0.14	5.22 ± 0.24	0.88 ± 0.09	0.06 ± 0.03
Citric acid	[a]	1.92 ± 0.04	19.82 ± 1.90	16.46 ± 0.87	1.44 ± 0.10	6.36 ± 0.53	0.49 ± 0.03	0.10 ± 0.05
Glucose	[a]	1.08 ± 0.05	13.86 ± 1.22	11.10 ± 0.88	3.18 ± 0.17	6.34 ± 0.33	0.28 ± 0.01	0.15 ± 0.03

[a] pH ≤ 7.

TABLE 6.5 Metals Leached from Three Types of Glass into Four Extracting Media Kept for 23 Weeks at 60°C [3]

Metal	Type I Clear Glass				Type I Amber Glass				Type I SiO_2 Coated Glass			
(mg/L)	Water	Buffer	Tween 80	EDTA	Water	Buffer	Tween 80	EDTA	Water	Buffer	Tween 80	EDTA
Mg	–	–	–	0.14	–	–	–	0.05	–	–	–	–
Ca	–	0.13	–	0.16	0.01	0.04	–	0.19	–	0.01	–	–
Mn	–	–	–	–	–	–	–	0.05	–		–	–
Al	0.53	0.08	0.18	4.71	0.56	0.16	–	0.39	–	0.18	–	0.16
Fe	–	–	–	–	–	–	–	0.56	–	–	–	–
Pb	0.02	0.04	–	0.06	–	0.03	–	0.04	0.01	0.01	–	0.02

–, not detected.

extracted, confirming the potential of basic solutions to attack and dissolve the glass network. Glucose and citric acid interacted with the glass surface, selectively extracting not only Al but also Cu and Pb (traces also present in the glass). This specific action of citrate and glucose could be related to their metal-complexing ability.

To diminish the interaction between drug and container surface, the glass industry has developed special inner-coated glass containers, which are more resistant to acidic or alkaline solutions. Coated glass containers were compared to conventional Type I clear and amber glasses for metal leaching [3]. Four different extracting media were used: USP water, 10 mM ammonium acetate pH 6.8, 0.2% polysorbate 80 in water, and 0.1% EDTA in water. The most prevalent leachable was Al from all glasses, reaching 4.71 mg/L in EDTA solution stored in clear Type I glass (Table 6.5). This is an expected result since, among the investigated metals, Al is present in higher concentrations in Type I glass and because EDTA is a complexing agent, and, therefore, the leaching ability of its solution is higher than in the other media. The SiO_2 coated glass proved to be more resistant than the conventional Type I glasses, and only Al and Pb were extracted into EDTA solution and in low amounts.

The presence of arsenic as a contaminant in pharmaceutical products is a pharmacopeial concern. Practically all monographs present limits for arsenic, ranging from 0.1 to 3.0 ppm. Bohrer et al. [4] have shown that arsenic is a ubiquitous contaminant in raw materials, but its presence in final products occurs in concentrations much higher than in the corresponding raw substances [5]. According to the authors, the main source of arsenic in injectable formulations is glass containers [4]. As mentioned above, glass can contain arsenic because arsenic oxide (III) is a fining agent to improve glass transparency [6], a feature especially important for solutions for intravenous administration that must be subjected to visual inspection of the content before use [7]. Arsenic oxide (III) is supposed to react with potassium nitrate in the glass melt to release oxygen and nitrogen oxides. These gases form large bubbles that rapidly rise to the surface, stirring the bath and sweeping small bubbles,

TABLE 6.6 Arsenic Species Present as Contaminant in Commercial Parenteral Solutions [5]

Product	Total As (μg/L)	As(V) (μg/L ± RSD)	As(III) (μg/L ± RSD)
KCl 19.1%[10]	41.3	41.3 ± 1.6	n.d.
KCl 10%[4]	23.6	23.6 ± 0.8	n.d.
NaCl 20%[10]	43.8	40.9 ± 5.7	2.9 ± 0.2
NaCl 20%[4]	15.9	15.9 ± 3.2	n.d.
Sodium acetate 2 meq/mL[1]	46.1	41.8 ± 2.0	4.3 ± 0.6
Sodium phosphate 0.5 mol/L[10]	36.7	36.7 ± 2.5	n.d.
Sodium bicarbonate 8.4%[5]	248.6	227.8 ± 0.5	20.8 ± 2.0
Sodium bicarbonate 8.4%[9]	198.3	147.2 ± 1.2	51.1 ± 0.8
Calcium gluconate 10%[11]	73.1	46.8 ± 4.1	26.3 ± 2.3
Calcium gluconate 10%[9]	92.7	72.8 ± 0.5	19.9 ± 1.9
Calcium gluconate 10%[4]	239.6	176 ± 7.2	63.6 ± 2.0
Magnesium sulfate 50%[11]	15.7	15.7 ± 0.2	5.7 ± 0.2
Magnesium sulfate 50%[10]	33.5	12.2 ± 1.2	21.3 ± 1.8
Magnesium sulfate 50%[9]	53.8	39.9 ± 5.6	13.9 ± 3.0
Glucose 25%[6]	21.8	16.5 ± 2.6	5.3 ± 0.6
Glucose 25%[6]	18.6	14.0 ± 2.2	4.6 ± 0.7
Vitamins (Tiaminose[a])	103.7	86.1 ± 0.9	17.6 ± 3.2
Vitamins (Dextrovitase[b])	10.2	10.2 ± 0.2	3.8 ± 0.1
Vitamins (Frutovena[c])	61.7	21.5 ± 1.7	40.2 ± 6.7
Amino acids 10%[3]	2.5	2.5 ± 0.1	n.d.
Amino acids 10%[1]	15.4	15.4 ± 2.3	n.d.
Amino acids 10%[2]	41.0	41.0 ± 0.9	n.d.
Amino acids 8%[2]	21.1	16.7 ± 1.2	4.4 ± 0.4
Amino acids 8%[2]	94.7	87.9 ± 5.7	6.8 ± 1.0
Lipid emulsion 10%[2]	0.9	0.9 ± 0.3	n.d.
Lipid emulsion 20%[2]	1.7	1.7 ± 0.6	n.d.
Heparin 5000 UI/mL[7]	56.7	17.2 ± 2.8	39.5 ± 2.3
Heparin 5000 UI/mL[8]	79.4	79.4 ± 4.9	23.3 ± 1.2

[a] (Glucose 3 g, ascorbic acid 0.25 g, thiamin chloridrate 0.015 g in 10 mL).
[b] (Glucose 2 g, ascorbic acid 2 g, pyridoxine chloridrate 20 mg, nicotinamide 30 mg, riboflavin in 20 mL).
[c] (Frutose 5 g, ascorbic acid 1 g, pyridoxine chloridrate 20 mg, sodium pantothenate 10 mg, riboflavin 4 mg in 20 mL).
Brands: [1]B. Braun, [2]Fresenius, [3]Baxter, [4]Halex Istar, [5]JP, [6]Merck, [7]EMS, [8]Elkins Sinn, [9]Ariston, [10]Geyer, [11]Hipolabor.
n.d., not detected (below limit of detection [LOD]); RSD, relative standard deviation.

formed by the decomposition of batch materials. Table 6.6 shows the arsenic levels measured by these authors in solutions for parenteral nutrition. Among these products, sodium bicarbonate and calcium gluconate are the most contaminated, presenting levels even higher than the limit of 0.1 mg/L for arsenic in infusion solutions.

Other elements such as chromium, barium, and zinc have also been found in solutions for parenteral nutrition, and, though not stated in the studies, the origin of these elements is probably the glass packaging. Since glass is an inorganic material, inorganic species may arise as contaminants from glass containers. Exceptions could be Zn and Ba, which, besides being present in Type I glasses, are components of plastic additives as well. Tin is also a plastic additive component and organotin compounds are stabilizers for PVC. Table 6.7 summarizes inorganic elements as contaminants in different infusion solutions and their respective concentrations.

Aluminum is a special case in this chapter because of its special behavior and toxicity. Since the discovery in 1976 by Alfrey et al. [14] that Al can accumulate in patients with reduced renal function causing toxic manifestations including neurological diseases, much has been done to reveal and reduce sources of aluminum contamination. Nowadays, Al is also considered toxic for patients with normal renal function who receive parenteral nutrition, mainly preterm infants [15]. Recognized sources of Al include water used for dialysis, solutions for total parenteral nutrition (TPN), and oral aluminum-containing compounds. Though additives and raw materials are sources as well, the most important source of Al in solutions for parenteral nutrition is the storage of parenteral preparations in glass containers. Figure 6.1 shows the Al level in different kinds of solutions stored in plastic and glass containers for a period of 3 months. While not more than 50 μg/L Al was leached from PE containers, over 3000 μg/L Al was found in solutions stored in glass containers, confirming that glass is a valid and continual source of Al.

The same authors also showed that, while glass is the common source, Al leaching depends on the nature of the substance in contact with its surface [17, 18]. In an experiment with glass and an ion exchanger containing Al, complexing agents and also amino acids were able to extract Al from both sources, but following an extraction yield related to the affinity of the ligand for the metal. The results of this interaction, measured during 2 months, can be seen in Figure 6.2. Based on this conclusion, the notable contamination by Al in gluconate and phosphate solutions can be attributed to the affinity of these species for Al. They are able to selectively withdraw Al from the glass network, when in contact with the glass containers (see Table 6.8).

Figure 6.3 shows the result of 1 year of storage of different amino acids, components of parenteral solutions, in Type I glass containers. Not all the solutions were contaminated by Al, but rather only those solutions whose amino acids presented affinity with this element showed Al contamination.

Table 6.9 presents the Al level in parenteral formulations, collected by different authors in different countries. Not all of them mention glass packaging as for the origin of Al contamination, but, overall, it seems to have been the major source. These data also confirm that the nature of the formulation components plays a key role in selective Al leaching.

TABLE 6.7 Metallic Contaminants Found in Commercial Solutions for Parenteral Nutrition

	Element Concentration (μg/L)												
Sample	Zn	Cr	Fe	Pb	As	Mn	Ba	Sn	Ge	Cu	Cd	V	Reference
TPN bag	9.1	–	–	–	–	–	–	–	–	–	–	–	[8]
Amino acids	60–4,970	n.d.	–	–	–	–	–	–	–	–	–	–	[9]
L-Cysteine HCl	32,000–86,000	110–230	–	–	–	–	–	–	–	–	–	–	[9]
NaCl, KCl, Na acetate	350–560	20–230	–	–	–	–	–	–	–	–	–	–	[9]
Ca gluconate	280–2,380	–	–	–	–	–	–	–	–	–	–	–	[9]
Phosphate	910–2,330	390–440	–	–	–	–	–	–	–	–	–	–	[9]
Ca gluconate	47–244	–	237–655	–	–	–	–	–	–	–	–	–	[10]
Albumin	–	–	–	–	–	–	–	–	–	–	–	1.1–677	[11]
TPN	233–703	–	84	–	–	–	–	–	–	–	–	–	[10]
Standard adult TPN formula, Baxter	–	86.8	–	0.6	288.0	309.9	11.0	0.5	5.5	–	–	–	[12]
Standard adult TPN formula, Abbott	–	25.8	–	1.1	65.0	109.9	16.0	0.4	5.5	–	–	–	[12]
Standard adult TPN renal formula, Abbott	–	99.8	–	0.6	298.0	259.9	28.0	2.2	28.9	–	–	–	[12]
Standard adult TPN formula	–	139.8	–	0.4	65.9	299.9	22.0	1.4	15.9	–	–	–	[12]
Standard pediatric TPN formula, Baxter	–	15.8	–	0.7	61.0	47.9	73.0	0.6	9.2	–	–	–	[12]
Standard adult amino acid solution, Fresenius	131.2	–	–	4.99	–	–	–	–	–	2.96	0.35	–	[13]
Standard adult amino acid solution, B. Braun	88.92	–	–	16.80	–	–	–	–	–	6.66	4.37	–	[13]
Standard adult amino acid solution, Baxter	1.39	–	–	4.39	–	–	–	–	–	40.81	0.71	–	[13]

n.d., not detected; –, not measured.

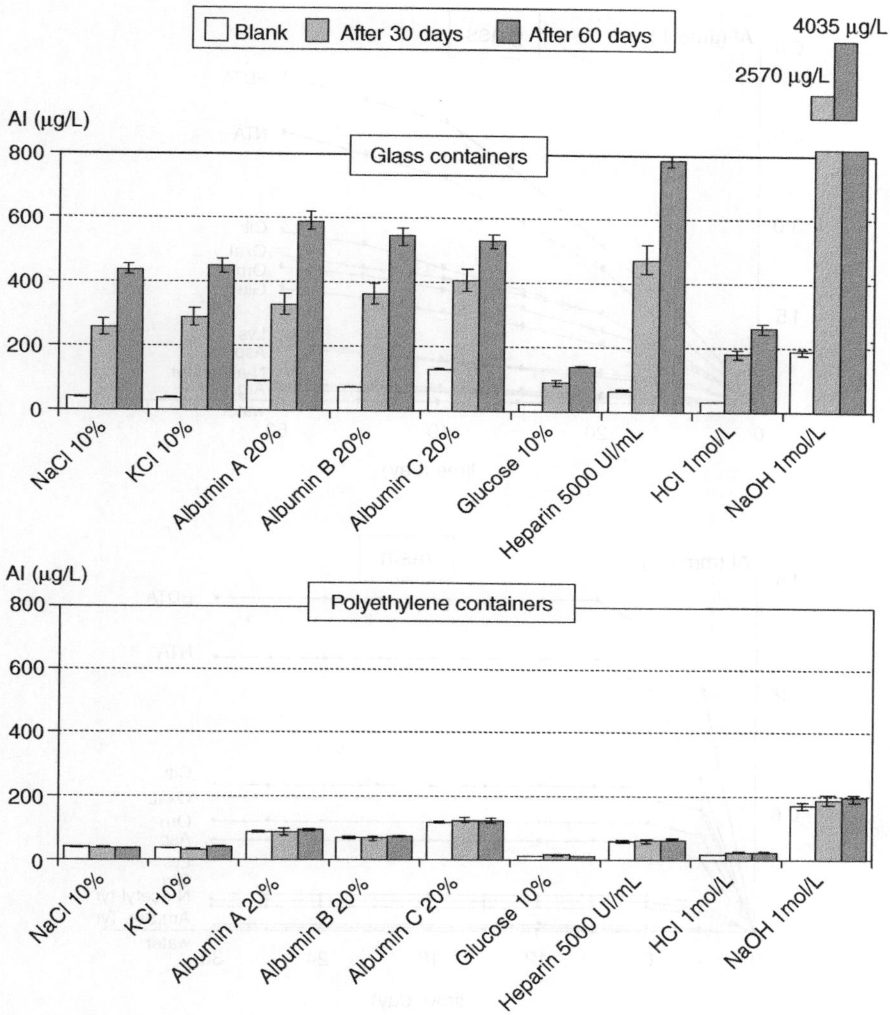

FIGURE 6.1 Aluminum extracted from Type II glass containers and from polyethylene containers by action of NaCl, KCl, albumin, glucose, heparin, HCl, and NaOH solutions after 30 and 60 days' storage at room temperature. The three different albumins are A, bovine (Merck); B, bovine (Reagen), and C, egg (Sigma) [16].

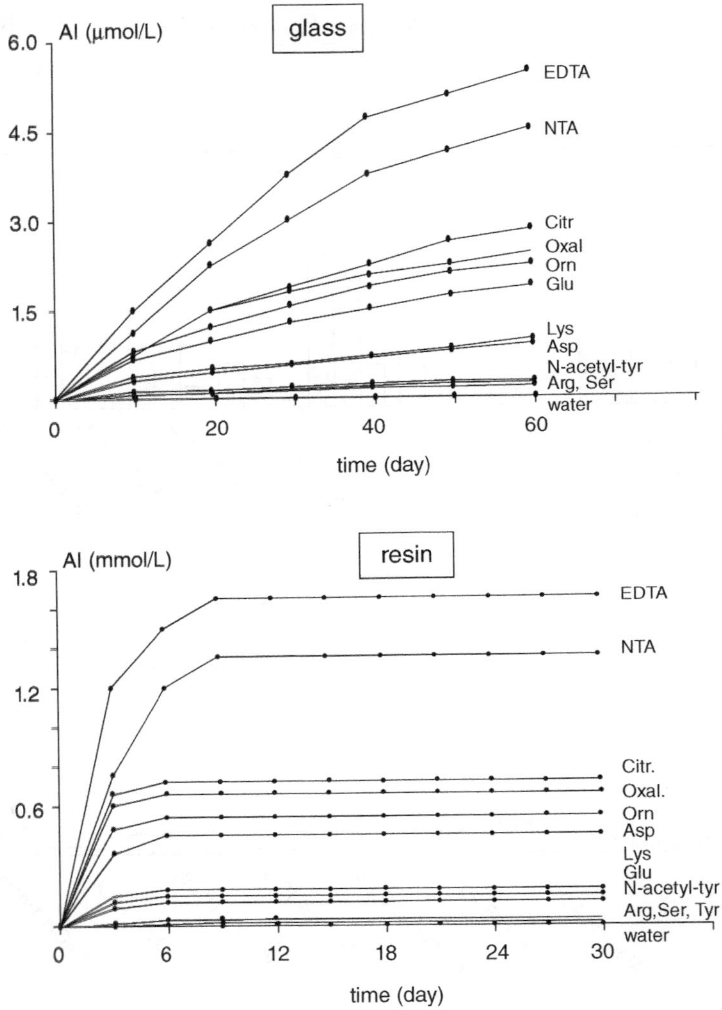

FIGURE 6.2 Aluminum extracted from glass particles (18 mesh) and Al-form exchanger (18 mesh) as a function of time by the action of some amino acids and complexing agents. Concentration of ligands: 0.05 mol/L [17].

TABLE 6.8 Aluminum Present as Contaminant in Commercial Parenteral Solutions, and Al Present in Container Materials

Product	Al Solution ±SD (μg/L)	Container	Al Container (%)
NaCl 20%	149 ± 10	Glass ampoule	1.43
	13 ± 4	Polyethylene	0.04
KCl 10%	68 ± 6	Glass ampoule	1.25
	23 ± 5	Polyethylene	0.06
Magnesium sulfate 50%	560 ± 85	Glass ampoule	1.25
	380 ± 288	Glass ampoule	1.43
Sodium acetate 2 meq/mL	45 ± 7	Glass ampoule	2.14
	17 ± 8	Polyethylene	0.05
Potassium phosphate 2 meq/mL	988 ± 76	Glass ampoule	1.98
	1325 ± 142	Glass ampoule	2.45
Sodium phosphate 0.5 M	933 ± 88	Glass ampoule	1.65
	879 ± 203	Glass ampoule	2.05
Calcium gluconate 10%	5621 ± 1165	Glass ampoule	1.51
	5960 ± 62	Glass ampoule	2.21
Sodium bicarbonate 8.4%	833 ± 141	Glass bottle	0.99
	922 ± 102	Glass bottle	1.03
Oligoelements[a]	1129 ± 33	Glass ampoule	2.14
Oligoelements[b]	1854 ± 744	Glass ampoule	2.21
Amino acids 10%	164 ± 6	Glass bottle	0.82
		Rubber closure	3.91
Amino acids 10%	116 ± 30	Glass bottle	0.76
		Rubber closure	4.23
Amino acids 10%	93 ± 23	Glass bottle	0.84
		Rubber closure	5.34
Amino acids 10%	65 ± 13	Glass bottle	0.89
		Rubber closure	5.70
Amino acids 10%	23 ± 8	Plastic bag	0.01
Glucose 50%	13 ± 1	Polyethylene	0.04
	293 ± 14	Glass ampoule	1.15
Glucose 25%	9 ± 3	Polyethylene	0.08
	370 ± 23	Glass ampoule	1.87
Albumin 20%	644 ± 58	Glass flask	0.67
		Rubber closure	4.06
	149 ± 23	Glass flask	0.66
		Rubber closure	3.99
Heparin 5000 UI/mL	732 ± 23	Glass ampoule	2.88
	738 ± 54	Glass ampoule	3.03

All solutions were within their guaranteed period of shelf-life [18].

[a] Composition: 22.0 mg $ZnSO_4$, 6.3 mg $CuSO_4$, 2.46 mg $MnSO_4$, 102.5 mcg $CrCl_3$ per ampoule.

[b] Composition: 8.8 mg $ZnSO_4$, 1.60 mg $CuSO_4$, 123.04 mcg $MnSO_4$ 20.50 mcg $CrCl_3$ per ampoule.

Suppliers: Abbott, Ariston, Aster, Baxter, Behring, B. Braun, Darrow, Fresenius, Fujisawa, Gayer, Halex Istar, Hypofarma, Santisa, Roche, Zenalb.

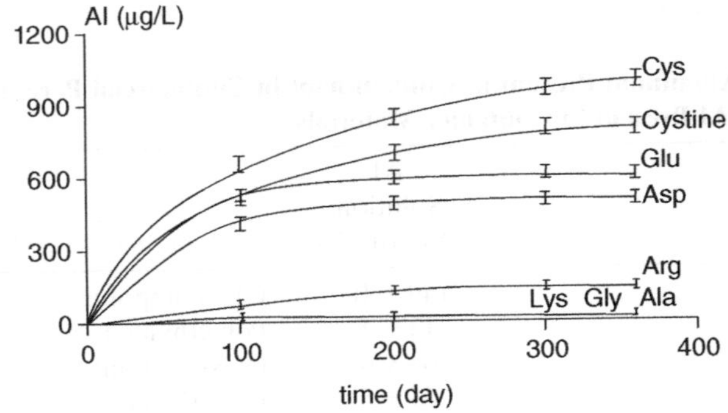

FIGURE 6.3 Aluminum leached from glass containers by amino acids as a function of the time at room temperature. Amino acid concentration: 0.028 mol/L [19].

TABLE 6.9 Aluminum Present in Commercial Solutions for Parenteral Administration

Product	Specification	Brand	Al Concentration (μg/L)	Reference
Electrolytes				
NaCl	10%	Abbott	3	[20]
	10%	Commercial Polfa	14	[21]
	5.85%	Braun	1	[22]
	1 M	Kabi	22	[22]
	10%	Abbott	43 ± 8[a]	[23]
	10%	Elkins Sinn	78 ± 7[a]	[23]
KCl	15.0%	Commercial Polfa	18	[21]
	–	Abbott	4 (5–11)	[20]
	–	Elkins Sinn	3	[20]
	7.45%	Braun	<0.6	[22]
	7.45%	Kabi	12	[22]
	7.45%	Braun	33 ± 5[a]	[23]
	10%	Abbott	97 ± 8[a]	[23]
	10%	Aster	23 ± 5[a]	[23]
Ca chloride	–	Elkins Sinn	15 (12–19)	[20]
	–	Abbott	5	[20]
	0.5 M	Boehringer	27	[22]
	1 N	Kabi	224	[22]
Na acetate	2 meq/mL	Braun	17 ± 8[a]	[23]
K acetate	–	Abbott	<5	[20]
Mg sulfate	50%	Abbott	5	[20]
	50%	Braun	606	[22]
	–	Geyer	560 ± 85[a]	[23]
	–	Ariston	380 ± 288[a]	[23]
Na phosphate	–	Invenex	2236 (2026–2370)	[20]
	0.5 M	Abbott	933 ± 88[a]	[23]
	0.5 M	Geyer	430 ± 177[a]	[23]

TABLE 6.9 (*Continued*)

Product	Specification	Brand	Al Concentration (μg/L)	Reference
K phosphate	–	Invenex	92	[20]
	–	Abbott	2189 (2069–2301)	[20]
	–	Braun	188	[22]
	–	Kabi	2826	[22]
	2 meq/L	Fresenius	1021 ± 188[a]	[23]
	2 meq/L	Braun	988 ± 76[a]	[23]
	2 meq/L	Aster	332 ± 27[a]	[23]
Ca gluconate	10%	Elkins Sinn	3973 (1095–5565)	[20]
	10%	Lyphomed	2245 (2000–2586)	[20]
	10%	Braun	4734	[22]
	10%	Fresenius	6549	[22]
	10%	Pharma Hameln	4421	[22]
	10%	Commercial Polfa	4400	[21]
	10%	Elkins Sinn	3987 ± 993[a]	[23]
	10%	Braun	4530 ± 1072[a]	[23]
	10%	Ariston	5960 ± 62[a]	[23]
	10%	Halex Istar	6781 ± 1842[a]	[23]
Trace elements				
Zinc chloride	–	Abbott	99 (81–123)	[20]
TE	Tracitrans	Fresenius	994	[22]
Pediatric TE	Ped-el	Pharmacia	3000	[21]
Pediatric TE	Peditrace	Pharmacia	130	[21]
Pediatric TE	Ped-element	Darrow	1423 ± 68[a]	[23]
TE	Ad-element	Darrow	3574 ± 237[a]	[23]
TE	Tracitrans	Fresenius	5712 ± 988[a]	[23]
Amino acids				
	Freamine 8.5%	McGaw	12 (5–24)	[20]
	Travasol 10%	Travenol	7 (6–8)	[20]
	HepatAmine 8%	McGaw	22	[20]
	Aminoplasmal 10%	Braun	55	[22]
	Aminopaed 10%	Kabi	38	[22]
	Aminosteril 8%	Fresenius	17	[22]
	Primene 10%	Clintec	120	[21]
	Aminomel 12.5%	Clintec	121	[21]
	Vaminolact 6.5%	Pharmacia	30	[21]
	Aminoplasmal L10	Braun	160 ± 48[a]	[23]
	Pediamino 10%	Braun	116 ± 30[a]	[23]

(*Continued*)

TABLE 6.9 (*Continued*)

Product	Specification	Brand	Al Concentration (μg/L)	Reference
	Aminoped 10%	Fresenius	195 ± 27[a]	[23]
	Nefroamino AEH	Braun	272 ± 66[a]	[23]
	Primène 10%	Baxter	65 ± 13[a]	[23]
Carbohydrates				
	Glucosteril 70%	Fresenius	9	[22]
	Glucose 40%	Fresenius	20	[22]
	Glucose 50%	Schiwa/ Hormonchemie	18	[22]
	Glucose 20%	Schiwa/ Hormonchemie	3	[22]
	Glucose 10%	Schiwa/ Hormonchemie	<0.6	[22]
	Dextrose (10, 20, 50%)	Abbott	<5	[20]
	Dextrose (10, 50%)	McGaw	<5	[20]
	Dextrose (5, 10, 50%)	Travenol	<5	[20]
	Glucose 20%	Commercial Polfa	16	[21]
	Glucose 25%	Darrow	9 ± 3[a]	[23]
		Equiplex	9 ± 2[a]	[23]
	Glucose 50%	Fresenius	15 ± 3[a]	[23]
		Darrow	17 ± 3[a]	[23]
		J.P. Ind. Farm	8 ± 2[a]	[23]
		B. Braum Lab.	13 ± 1[a]	[23]
		Ariston	11 ± 4[a]	[23]
Lipids				
	Intralipid 10%	Kabi Vitrum	<5	[20]
	Intralipid 20%	Kabi Vitrum	<5	[20]
	Intralipid 10%	Pfrimmer Kabi	5	[22]
	Intralipid 20%	Pfrimmer Kabi	7	[22]
	Lipofundin 20%	Braun	35	[22]
	Lipofundin 20%	Braun	14	[22]
	Intralipid 20%	Pharmacia	30	[21]
	Lipofundin 20%	Braun	180	[21]
	Lipofundin 10%	Braun	29 ± 6[a]	[23]
	Lipofundin 20%	Braun	34 ± 7[a]	[23]

[a] Mean value ± standard deviation of three samples of the same lot.
–, not informed.

6.2 PLASTIC CONTAINERS

6.2.1 Polymer Formation

Plastics are organic and polymeric in nature. A polymer is a large molecule built up by the repetition of small and simple chemical units. The repeated unit of the polymer is usually equivalent or nearly equivalent to the monomer, or the starting material from which the polymer is formed. The structural units of the polymers most used to manufacture plastic containers, along with their uses for pharmaceutical purposes, are shown in Table A.1 in the Appendix.

Modern polymer technology is founded on catalysis, and catalytic methods are extensively used in the production of plastics. Catalysts, since they only catalyze reactions, do not count as polymer constituents but may be present as impurities in the polymeric material. Table A.2 in the Appendix lists the usual catalysts used for the polymerization of the polymers in Table A.1, Appendix, which can be potential contaminants for formulations stored in plastic containers.

Although the basic structure of a polymer is its monomer or monomers (in the case of copolymers), additives are necessary to turn the raw polymer into adequate containers:

$$\text{Monomer} + \text{cross linker}^{2} \rightarrow \text{polymer}$$

$$\text{Polymer} + \text{additives} \rightarrow \text{useful material.}$$

While polyethylene (PE), for example, may be used without any additive, the other plastics are virtually useless alone but are converted into highly serviceable products by combining them with other substances or materials. The additives most commonly found in plastics used for pharmaceutical products are antioxidants, heat stabilizers, lubricants, plasticizers, fillers, and colorants. These additives can be in liquid, solid, or fine particle forms and are used in amounts varying from less than 1% to more than 50% of the plastic mass. The additives necessary for each of the selected types of polymers are described in Table A.2, Appendix.

Additives authorized for use in plastics for pharmaceuticals, along with their limits in the polymer mass (according to the British Pharmacopoeia [BP] and European Pharmacopoeia (Ph. Eur.), are summarized in Table 6.10. These restrictions are not posed by the USP. Since additives, together with the polymers, provide a large variety of substances, and the leachability of such components cannot be predicted a priori, the pharmacopeial compendium has determined procedures to investigate the biological and physicochemical properties of plastics. In general, tests do not include the measurement of the polymeric material or the additives themselves, but the biological reactivity

[2] Agent to link polymer chains.

TABLE 6.10 Acceptable Additives Allowed in Plastics for Pharmaceuticals According to the BP and Ph. Eur. and Their Limits

No.	Additive	Polymer	Limit
	Plasticizer		
1	Di(2-ethylhexyl)phthalate	PVC	40%
2	Zinc octanoate	PVC	1%
3	N′N′-diaylethylenediamines	PVC	1%
4	Epoxidized soya oil	PVC	10%
5	Epoxidized linseed oil	PVC	10%
	Antioxidants		
7	Butylhydroxytoluene	PE, PP, EVA	0.125%
8	Ethylene bis[3,3-bis[3-(1,1-dimethylethyl)-4-hydroxyphenyl]butanoate]	PE, PP	0.3%
9	Pentaerythrityl tetrakis[3-(3,5-di-*tert*-butyl-4-hydroxyphenyl)propionate]	PE, PP EVA	0.3% 0.2%
10	2,2′,2″,6,6′,6′-Hexa-*tert*-butyl-4,4′,4″ [(2,4,6-trimethyl-1,3,5-benzenetriyl)-trismethylene]triphenol	PE, PP EVA	0.3% 0.2%
11	Octadecyl 3-(3,5-di-*tert*-butyl-4-hydroxyphenyl) propionate	PE, PP EVA	0.3% 0.2%
12	Tri(2,4-di-*tert*-butylphenyl) phosphite	PE, PP EVA	0.3% 0.2%
13	1,3,5-Tris(3,5-di-*tert*-butyl-4-hydroxybenzyl)-*s*-triazine-2,4,6(1*H*,3*H*,5*H*)-trione	PE, PP	0.3%
14	2,2′-Bis(octadecyloxy)-5,5′-spirobi[1,3,2-dioxaphosphinane]	PE, PP	0.3%
15	Dioctadecyl disulfide	PE, PP	0.3%
16	Didodecyl 3,3′-thiodipropionate	PE, PP	0.3%
17	Dioctadecyl 3,3′-thiodipropionate	PE, PP	0.3%
18[a]	2,4-Bis(1,1-dimethylethyl)phenyl biphenyl-4,4′-diyldiphosphonite	PE, PP	0.1%
	2,4-Bis(1,1-dimethylethyl)phenyl biphenyl-3,4′-diyldiphosphonite		
	2,4-bis(1,1-Dimethylethyl)phenyl biphenyl-3,3'-diyldiphosphonite		
	2,4-Bis(1,1-dimethylethyl)phenyl biphenyl-4-ylphosphonite		
	2,4-Bis(1,1-dimethylethyl)phenyl phosphite		
	2,4-Bis(1,1-dimethylethyl)phenyl 4′-[bis[2,4-bis(1,1-dimethylethyl)phenoxy]phosphanyl]biphenyl-4-ylphosphonate		
	2,4-Bis(1,1-dimethylethyl)phenol		
	Lubricants and fillers		
19	Stearic acid		
20	Oleamide	EVA	0.5%
21	Erucamide	EVA	0.5%

TABLE 6.10 (*Continued*)

No.	Additive	Polymer	Limit
22[a]	Copolymer of dimethyl butanedioate and 1-(2-hydroxyethyl)-2,2,6,6-tetramethylpiperidin-4-ol	PE, PP	0.3%
23	Hydrotalcite	PE, PP	0.5%
24	Alkanamides	PE, PP	0.5%
25	Alkenamides	PE, PP	0.5%
26	Sodium silico-aluminate	PE, PP	0.5%
27	Sílica	PE, PP	0.5%
28	Sodium benzoate	PE, PP	0.5%
29	Fatty acid esters	PE, PP	0.5%
30	Trisodium phaphate	PE, PP	0.5%
31	Liquid paraffin	PE, PP	0.5%
32	Zinc oxide	PE, PP	0.5%
33	Magnesium oxide	PE, PP	0.5%
34	Calcium stearate	PE, PP	0.5%
35	Zinc stearate	PE, PP	0.5%
36	Talc	PP	0.5%
	Colorants		
37	Titanium dioxide	PE, PP	4%
38	Ultramarine blue	PVC	n.i.

[a] Only for nonparenteral preparations.
PE for parenteral and ophthalmic preparations may not contain additives.
PP for parenteral and ophthalmic preparations may contain only three antioxidants.
n.i., not informed.

(toxicity screening) of their extracts (<87> *Biological reactivity tests, in vitro* and <88> *Biological reactivity tests, in vivo*). Extracts are generally obtained by autoclaving the plastic container filled with an extraction medium at one of the three standard temperatures: 50, 70, or 121°C. Only containers for parenteral and ophthalmic preparations have a more controlled regulation since their risk of toxicity is more significant, and therefore assays for plastic additives are foreseen in the pharmacopeial compendia.

On the other hand, for plastics for general use as pharmaceutical containers, the pharmacopeial compendia set limits for impurities other than additives. These are inorganic species (metallic and nonmetallic) that can also be present in plastic containers, and their determination serves as a quality criterion for the plastic material. They may or may not be related to any of the additives listed in Table 6.10. For example, the investigation of aluminum in PE and polypropylene (PP) is related to the use of aluminum-based catalysts for the attainment of the polymers; barium as a contaminant in a plastic container is indicative of the presence of a barium soap stabilizer, possibly used in plastic compounding. The tests, including chemicals such as heavy metals and sulfated

TABLE 6.11 Inorganic Impurities and Their Limits in Extracts of Plastic Containers for Pharmaceuticals

No.	Impurity	Polymer	Use	Limit
1	Heavy metals	PE, PP	Injectable, ophthalmic	2.5 ppm
		PVC	Aqueous parenteral, blood, tubing	50 ppm
2	Sulfated ash	PE, PP	Injectable, ophthalmic	0.2%, 1.0%
		EVA	Parenteral container and tubing	1.2%
		PET	Oral forms	0.5%
3	Aluminum	PE, PP	Injectable, ophthalmic	1 ppm
		PET	Oral forms	1 ppm
4	Chromium	PE, PP	Injectable, ophthalmic	0.05 ppm
5	Titanium	PE, PP	Injectable, ophthalmic	1 ppm
		PET	Oral forms	1 ppm
6	Vanadium	PE, PP	Injectable, ophthalmic	0.1 ppm
	Zinc	PE, PP	Injectable, ophthalmic	1 ppm
		PVC	Aqueous parenteral, blood	0.2%
		PET	Oral forms	1 ppm
7	Zirconium	PE	Injectable, ophthalmic	0.1 ppm
8	Barium	PVC	Aqueous parenteral, blood, tubing	5 ppm
		PET	Oral forms	1 ppm
9	Cadmium	PVC	Aqueous parenteral blood, tubing	0.6 ppm
10	Calcium	PVC	Aqueous parenteral, blood	0.07%
11	Tin	PVC	Aqueous parenteral, blood, tubing	20 ppm
12	Ammonium	PVC	Blood	2 ppm
13	Chloride	PVC	Blood	0.4 ppm
14	Antimonium	PET	Oral forms	1 ppm
15	Cobalt	PET	Oral forms	1 ppm
16	Germanium	PET	Oral forms	1 ppm
17	Manganese	PET	Oral forms	1 ppm

Adapted from Ph. Eur.

ash residue, are listed in Table 6.11, with the respective limits for each species in the different plastic materials.

While not mandatory from regulatory guidelines, much research has been carried out to investigate the extractability of plastic additives in contact with a variety of pharmaceutical formulations, mainly those for parenteral use. The research initially concentrated on the extractability of plasticizer phthalates, mainly di-2-ethylhexylphthalate (DEHP) from polyvinyl chloride (PVC), into the blood, blood components, and infusion solutions. In recent years, this research has been extended to other polymeric materials and additives. Potentially extractable species (characterized by evaluating the polymeric material itself) or species leached from plastic containers due to interactions between the formulation and the polymeric material are currently being investigated. The expression "extractables and leachables" (E&L) has been coined to characterize chemicals released from packaging materials into pharmaceutical formulations. Additives to plastic and elastomeric container/closure system components, as well organic and inorganic residues released from container/closure systems made of metal and glass comprise E&L [24].

Table A.3 in the Appendix presents the extractables from usual plastic materials, where tests were carried out using conventional solvents such as water or ethanol, or following tests prescribed by pharmacopeial monographs for plastic containers. The leaching of plastic container constituents by the action of specific formulation components is presented below.

Plastics for storing pharmaceutical formulations are made of PE, PP, PVC, ethylene vinyl acetate (EVA), polyethylene terephthalate (PET), and multi-layer materials. Because PVC is the most used and also the most contentious material, this chapter discusses PVC and the other plastics separately.

6.2.2 PVC Containers

PVC's intrinsic properties make it suitable for a wide variety of applications. It is cheap, durable, and easy to assemble. PVC is the third most widely produced plastic, after PE and PP. It is biologically and chemically resistant, making it the plastic of choice for most household applications where corrosion would limit the use of metal (rigid PVC). By adding plasticizers, it can become flexible enough to be used to make fabrics and wire insulator (flexible PVC). In the medical area, devices and containers for preparing and delivering infusions and blood are mainly made of flexible PVC. The amount of additive necessary to turn rigid PVC into a flexible material may reach up to 40% of the polymer mass. Moreover, since there are no chemical bonds between the polymer and the plasticizer, it is a potentially extractable species. Although a group of compound classes is able to perform this function (see Table A.2, Appendix), DEHP is the leading plasticizer for PVC.

Limits for DEHP in formulations stored in PVC bags are not established in pharmacopeial compendia. The BP prescribes a standard test whose results should show that the amount of DEHP extracted by the action of ether on PVC bags does not exceed 40% of the polymer mass. This test is valid for PVC for all uses: dry dosage forms, noninjectable aqueous solutions, intravenous aqueous infusions, blood and blood components, and tubing used in sets for blood and blood components. On the other hand, the Food and Drug Administration (FDA) launched a nonregulatory publication entitled "Safety Assessment of Di-2-ethylhexylphthalate (DEHP) Released from PVC Medical Devices" [25], where concerns about its toxicity are discussed in detail. This publication intends to provide risk managers with the information necessary to make decisions about the safety of DEHP exposure from medical devices. Conclusions of the study were reached by calculating the dose of DEHP received by patients undergoing different medical procedures. For the pharmaceutical industry, what is truly relevant is the concern about the presence of DEHP in IV infusion fluids and drugs and in parenteral nutrition, because of their storage in plastic bags and the delivery of these preparations through

PVC tubing sets. The study concluded that there is little to no risk posed by patient exposure to the amount of DEHP released from PVC IV bags following the infusion of crystalloid fluids (e.g., normal saline, Ringer's lactate). There is a small risk, however, posed by the exposure to the amount of DEHP released from PVC bags used to store and administer drugs that require a pharmaceutical vehicle for solubilization. The dose of DEHP received by adult patients receiving TPN is estimated to be less than the tolerable intake (TI),[3] suggesting that there is little need for concern about DEHP-mediated effects in these patients. The dose of DEHP received by neonates undergoing TPN is uncertain. Depending on the data used to derive the TI/dose ratio, neonates may be at an increased risk of DEHP-mediated adverse effects.

In fact, studies have shown that, depending on the constituents in the formulation, the extractable DEHP may be much higher than expected. Our knowledge of the nature of the polymer and its additives, as well as of standard migration, is not great enough to predict container–formulation interactions. Migration is a unique phenomenon, and predictions based only on the structure and physical or chemical properties may not be effective for solving interaction problems. Table 6.12 shows the level of DEHP found in different kinds of parenteral formulations stored in PVC bags. It is clear that the amount of DEHP leached into the formulation depends on the nature of its constituents. While in saline solutions the amount did not exceed 24 μg/L in 14 months, in oleaginous vehicles, it reached over 300 μg/L in 2 hours, which corroborates with FDA studies. Due to the lipophilic character of DEHP, its migration process from the packaging into hydrophilic preparations is limited to a random dissolution on the contacting surface. On the other hand, in lipophilic preparations such as lipid emulsions, DEHP is encouraged to leave the polymer surface through a dissolution process into the lipophilic milieu.

There are drug products whose interaction with PVC bags and infusion sets are so high that they must include labeling precautions for use with PVC containers. These drugs include antineoplastics such as paclitaxel, docetaxel, tacrolimus, and teniposide, and others such as ciprofloxacin, cefoperazone sodium, fluconazole, metronidazole HCl, cimetidine, and propofol. However, apparently DEHP leaching is much more related to the presence of surfactants in the formulation than to the drug itself. Because of their poor solubility in water, formulations of cyclosporine, miconazole, and paclitaxel, among others, contain large amount of the surfactants or solvents (ethanol, polyethylene glycol, and propylene glycol).

The extent of leaching of DEHP from PVC bags due to the influence of surfactants and solvents was evaluated in at least two studies. In one [36], test solutions (surfactants, solvents, and drugs) were prepared in 5% dextrose (except for phenytoin sodium solution, which was prepared in 0.9% NaCl), stored at room temperature, and agitated by bending and shaking before each

[3] The tolerable intake values for DEHP are outlined in the AAMI/ISO 10993-17 standard, Method for the Establishment of Allowable Limits for Leachable Substances.

TABLE 6.12 Concentration of DEHP in Infusion Solutions Stored in PVC Containers

Sample	Vehicle	Diluent	Volume (mL)	Time of Storage	Temperature Storage (°C)	DEHP (μg/L)	Reference
0.9% Saline	–	–	1,000	5 months	–	7	[26]
				14 months	–	24	
0.9% Saline	–	–	500	0.5 months	–	13	
				3 months	–	<4	
0.9% Saline	–	–	100	6 months	–	8	
5% Glucose	–	–	1,000	5 months	–	4	
				12 months	–	4	
5% Glucose	–	–	500	0.5 months	–	34	
				3 months	–	7	
ACO bag	–	–	100	12 months	–	7	[27]
Travenol bag	–	–	100	12 months	–	5	
0.9% Saline	–	–	n.i.	1 month	22	0.22	[28]
				2 months		4.93	
				7 months		3.59	
				9 months		6.45	
TPN—3.85% lipid	–	–	2,200	24 h	4	330	[29]
	–	–		1 week	4	450	
TPN—2.50% lipid	–	–	2,200	24 h	4	230	
				1 week	4	260	
TPN—1.85% lipid	–	–	650	24 h	4	300	
	–	–		1 week	4	360	

(*Continued*)

TABLE 6.12 (*Continued*)

Sample	Vehicle	Diluent	Volume (mL)	Time of Storage	Temperature Storage (°C)	DEHP (μg/L)	Reference
TPN—1.00% lipid	–	–	800	24 h	4	240	
	–	–		1 week	4	270	
Paclitaxel 300 mg	50% Cremophor EL and 50% ethanol	5% Glucose	500	3 h	Room T	26,000	[30]
Paclitaxel 600 mg	50% Cremophor EL and 50% ethanol	5% Glucose	500	3 h	Room T	61,000	
Paclitaxel 0.3 mg/mL	2.5% Cremophor EL and 2.5% ethanol	5% Glucose	100	9 h	20–23	150,000	[31]
				25 h	20–23	340,000	
Paclitaxel 0.3 mg/mL	5% Cremophor EL and 5% ethanol	5% Glucose	100	9 h	20–23	290,000	
				25 h	20–23	600,000	
Miconazole 1%	11.5% Cremophor EL	5% Glucose	250	24 h	4	16,900	[32]
				24 h	22–26	57,900	
Cyclosporine	4% Cremophor EL and 2% ethanol	–	50	8 h	–	116,000 ± 24,000	[33]
				24 h	–	373,000 ± 47,000	
Cyclosporine 50 mg/mL	4% Cremophor EL and 2% ethanol	0.9% NaCl	Dil 1:20	12 h	25	3,370 ± 350	[34]
			Dil 1:100	12 h	25	4,300 ± 580	
Quiniforme 500 mg	–	5% Dextrose	500	48 h	4	<4.5	[35]
Quinimax 400 mg	–	5% Dextrose	500	72 h	4	13.3	[35]

n.i., not informed.

sample collection. In the other [37], docetaxel and paclitaxel and solutions containing only the vehicles of these formulations were also diluted in 5% dextrose infusion solution. The results from both studies showed that the surfactants polyethoxylated castor oil (Cremophor EL) and polysorbate 80 (Tween 80) were responsible for the leaching of DEHP from PVC bags (Table 6.13). Moreover, the data in Table 6.13 allowed one to conclude that solvents and the drugs themselves did not promote DEHP to leach. While DEHP was not detected in digoxin, lorazepam, pentobarbital, phenobarbital, phenytoin, or trimethoprim sulfamethoxazole, which do not contain any surfactants, in all the other drug products, its concentration varied from 3 to 238 μg/mL up to 24 hours.

In a study conducted by Waugh et al. [31], it was demonstrated that the extent of DEHP leaching into Taxol was directly dependent on the concentrations of Cremophor EL and ethanol, and not on the paclitaxel itself. The graphics in Figure 6.4 clearly demonstrate that increasing the concentration of these vehicles in the formulation leads to greater amounts of DEHP leached, regardless of the presence of the drug. Comparing both extraction performances, Cremophor EL is more proficient in releasing DEHP than polysorbate 80. While 10% polysorbate 80 leached approximately 230 mg/L DEHP in 24 hours, Cremophor EL, at the same concentration, extracted more than 800 mg/L in the same time interval.

The data in Table 6.12 confirm the role of surfactants in DEHP leaching. Pure glucose and saline solutions do not promote its release. Lipid emulsions, due to their lipophilic character, promote its release in some extent. By contrast, all formulations containing Cremophor EL promoted the release of extremely elevated values of DEHP in less than 24 hours.

Another problem related to the leachability of phthalates from PVC bags is the contamination of blood products. Out of 78 samples of blood products analyzed, including red blood cell concentrate, whole blood, blood platelet, and frozen plasma, all were contaminated by DEHP and mono-2-ethylhexyl phthalate (MEHP) [38]. The most contaminated samples were whole blood and the least contaminated were blood platelets (Table 6.14). The authors of the study attributed this difference to the higher lipid content of the whole blood and, therefore, higher solubility of the nonpolar phthalates in these samples. Among the blood platelet samples, 6 were stored in PVC bags and 12 in polyolefin bags,[4] and both phthalates leached from the polyolefin bags, although in smaller amounts (Fig. 6.5). These results help to demonstrate how challenging the assessment of "E&L" is in pharmaceutical products since substitution of PVC by polyolefin bags for blood products aimed to avoid DEHP contamination, and failed in this case (for sources of MEHP, see Chapter 9, Section 9.2, Extracorporeal Circuits).

Although differences in the methods of the studies shown in Table 6.12 (time of contact, volume of solution, temperature, etc.) do not allow for direct

[4] Not informed, but probably PE or PP.

TABLE 6.13 Concentration of DEHP Leached from PVC Bags by Action of Active Ingredients or Formulation Components (Vehicles) Dissolved in 5% Dextrose Injection and after 4, 8, and 24 Hours' Storage at 24°C

Solution	Vehicle	Vehicle Concentration	DEHP Concentration (μg/mL)		
			4 h	8 h	24 h
5% Dextrose	Ethanol	25%	n.d.	n.d.	n.d.
5% Dextrose	Polyethylene glycol	25%	n.d.	n.d.	n.d.
5% Dextrose	Propylene glycol	25%	n.d.	n.d.	n.d.
5% Dextrose	Polyolyethylated castor oil + ethanol	5% + 5%	46.6–58.8	–	197.8–193.4
5% Dextrose	Polysorbate 80	1%	7.0–7.2	–	36.6–37.2
5% Dextrose	Polysorbate 80	5%	27.3–30.1	–	155.8–162.5
5% Dextrose	Polysorbate 80	10%	42.4–45.0	–	224.6–231.1
5% Dextrose	Polysorbate 80	25%	76.1–73.3	–	237.7 – 226.1
5% Dextrose	Polysorbate 80 + ethanol	1% + 1%	16.2–18.2	–	82.6–86.2
5% Dextrose	Polysorbate 80 + ethanol	5% + 5%	41.5–42.7	–	178.5–179.2
Chlodiazepoxide hydrochloride	Polysorbate 80	0.16%	n.d.	<1	3.2–3.1
Cyclosporine	Polyolyethylated castor oil	3.9%	10.2–13.1	27.2 ± 0.4	90.3–104.4
Digoxin	–	–	n.d.	n.d.	n.d.
Etoposide	Polysorbate 80	0.16%	n.d.	<1	2.6–2.3
Lorazepan	–	–	n.d.	n.d.	n.d.
Miconazole	Polyolyethylated castor oil	3.0%	6.5–8.4	21.5 ± 0.5	63.3–67.6
Multivitamins	Polysorbate 80	0.032%	n.d.	n.d.	< 1
Pentobarbital sodium	–	–	n.d.	n.d.	n.d.
Phenobarbital sodium	–	–	n.d.	n.d.	n.d.
Phenytoin sodium	–	–	n.d.	n.d.	n.d.
Teniposide	Polyolyethylated castor oil	1.0%	2.9–3.2	7.5 ± 1.7	20.9–22.2
Trimethoprim sulfamethoxazole	–	–	n.d.	n.d.	n.d.
Docetaxel	Polysorbate 80 + ethanol	2.4% + 1%	20 ± 0.6	38 ± 0.9	–
5% Dextrose	Polysorbate 80 + ethanol	2.4% + 1%	22 ± 1.3	43 ± 5.5	–
Paclitaxel	Polyolyethylated castor oil	10.5% + 1%	138 ± 4.0	198 ± 7.5	–
5% Dextrose	Polyolyethylated castor oil	10.5% + 1%	139 ± 4.7	217 ± 11.4	–

Adapted from References 36 and 37.
–, not measured; n.d., not detected.

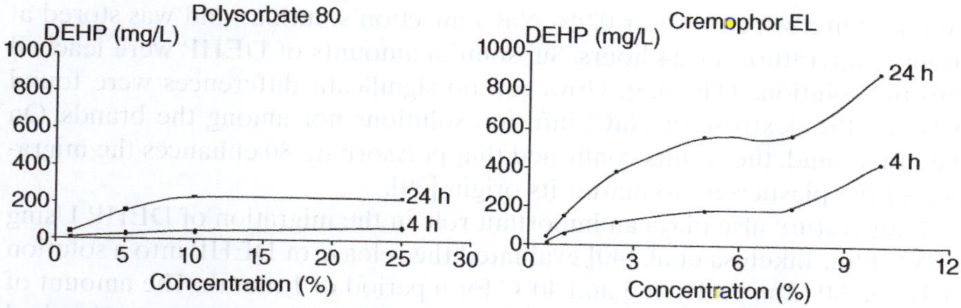

FIGURE 6.4 DEHP leached from PVC bags as a function of the concentrations of polysorbate 80 and Cremophor EL, after 4 and 24 hours' contact at 24°C (adapted from Reference 31).

TABLE 6.14 Concentration of DEHP and MEHP in Blood Products Stored in PVC Bags

Product	DEHP (mg/L)	MEHP (mg/L)
Whole blood	15.0–83.2	0.5–9.7
Red blood cell concentrate	6.8–36.5	0.3–4.3
Irradiated blood cell concentrate	7.4–36.1	0.3–3.3
Blood platelet	1.8–15.0	0.1–4.6
Fresh frozen plasma	11.6–18.5	1.1–2.5

Adapted from Reference 38.

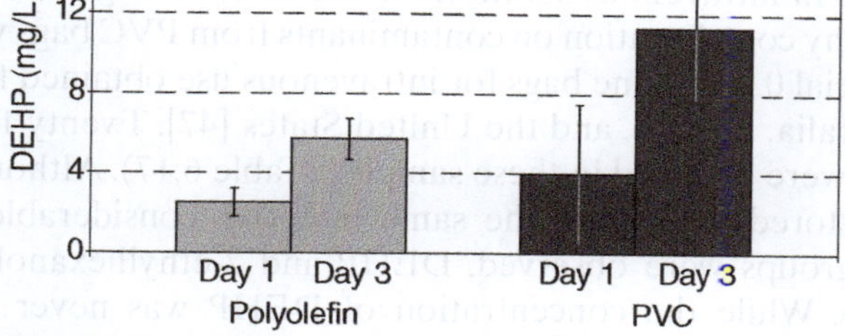

FIGURE 6.5 DEHP extracted into blood platelet samples stored in polyolefin ($n = 6$) and PVC ($n = 3$) bags, 1 and 3 days after storage (adapted from Reference 38).

comparisons, the data presented suggest that the leachability of DEHP is highly dependent on the nature of the pharmaceutical formulation. The question of migration of DEHP into solutions is relevant not only to lipophilic vehicles but also to drugs such as paclitaxel and etoposide, in which the amount of DEHP is much higher than in simple lipophilic lipidic emulsions.

In order to compare the release of DEHP from PVC bags from different manufacturers, the same drug (etoposide 0.4 mg/mL in polysorbate 80) was prepared in 5% dextrose or 0.9% NaCl injection solutions and was stored at room temperature for 24 hours. Substantial amounts of DEHP were leached into the solutions (Fig. 6.6). However, no significant differences were found between the dextrose or NaCl infusion solutions nor among the brands. On the other hand, the results confirmed that polysorbate 80 enhances the migration of the plasticizer, no matter its origin [39].

Temperature also plays an important role in the migration of DEHP. Using a PVC film, Takehisa et al. [40] evaluated the release of DEHP into a solution of Tween 80 (2 mg/mL) at 5 and 40°C for a period of 1 month. The amount of DEHP leached at 40°C was almost four times higher than the amount leached at 5°C (Fig. 6.7). The problem here is that many formulations are delivered at the body's temperature or to neonates in warmed incubators, which increase the leaching rate.

Although most studies focus on DEHP, it is not the only phthalate used as a plasticizer in PVC. Other parent phthalates or their decomposition products are also found among E&L from PVC containers. Phthalic acid, 2-ethylhexyl hydrogenphthalate, and 2-ethylhexanol were found in infusion solutions stored in PVC bags, in concentrations even higher than that of DEHP (Table 6.15).

Cyclohexanone has been widely utilized as an industrial solvent for cellulose acetate and vinyl resins, rubber, waxes, and fats. It is also used as an adhesive solvent in the manufacture of a variety of medical devices made of PVC. Due to its low volatility, cyclohexanone may remain in the PVC products and therefore may be released into solutions stored in PVC bags or into the internal airspace of medical devices. Though it is classified as a class 2 solvent, with a concentration limit of 3880 ppm, when present in intravenous formulations, harmful effects can be intensified [42]. Table 6.16 presents the levels of cyclohexanone in intravenous solutions stored in PVC bags.

A noteworthy corroboration on contaminants from PVC bags was an analysis of commercial 0.9% saline bags for intravenous use obtained from hospital stores in Australia, Europe, and the United States [47]. Twenty-four different contaminants were detected in these samples (Table 6.17). Although the same product was stored in bags of the same material, considerable differences between the groups were observed. DEHP and 2-ethylhexanol were found in all samples. While the concentration of DEHP was never greater than 25 μg/L, the concentration of 2-ethylhexanol reached 1200 μg/L. Cyclohexanone, present only in samples from Australia and the United States, was in a very high concentration. On the other hand, *t*-butyl cyclohexanone and *t*-butyl cyclohexanol were detected in samples from Australia and Europe and in high concentrations as well. To evaluate the sources of such compounds, a qualitative thermodesorption analysis of different parts of the bags was carried out. The analysis of the PVC plastic itself, the joints at the base of the bag, and the printing ink on the outside of the bag, revealed that *t*-butyl cyclohexanol and *t*-butyl cyclohexanone originated from the printing ink. Cyclohexanone was detected mainly in segments from the joints at the bases of the bags because it is an adhesive solvent for PVC, and 2-ethylhexanol was present in all

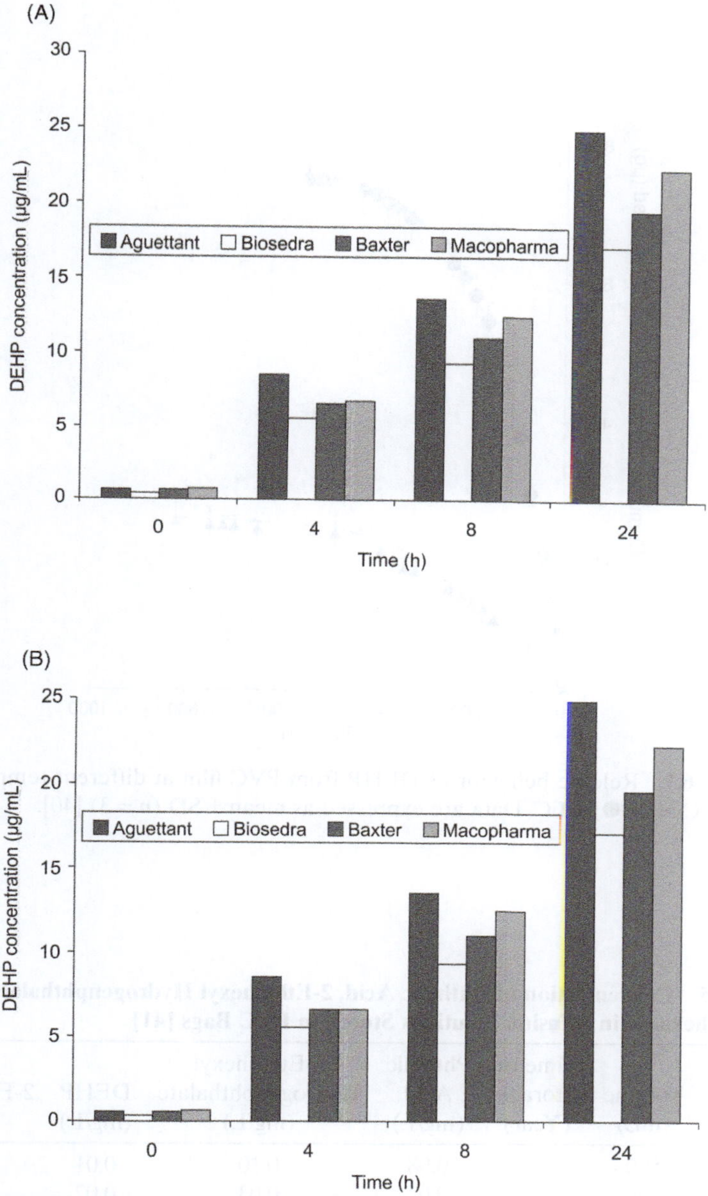

FIGURE 6.6 Concentration of DEHP leached from (A) 0.9% sodium chloride injection of etoposide 0.4 mg/mL and (B) from 5% dextrose injection of etoposide 0.4 mg/mL into PVC bags from different manufacturers after storage at room temperature during 24 hours [39].

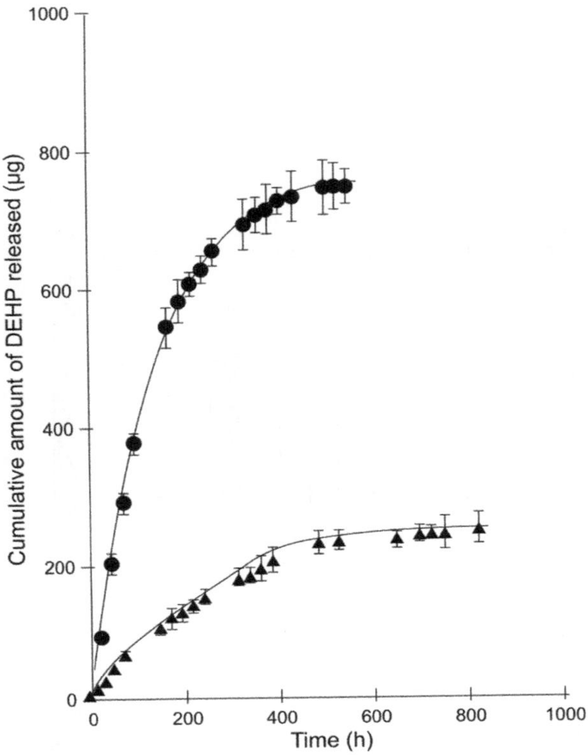

FIGURE 6.7 Release behavior of DEHP from PVC film at different temperatures of: (▲) 5°C and (●) 40°C. Data are expressed as mean ± SD ($n = 3$) [40].

TABLE 6.15 Concentration of Phthalic Acid, 2-Ethylhexyl Hydrogenphthalate, DEHP, and 2-Ethylhexanol in Infusion Solutions Stored in PVC Bags [41]

Sample	Volume (mL)	Time of Storage (Year)	Phthalic Acid (mg/L)	2-Ethylhexyl Hydrogenphthalate (mg/L)	DEHP (mg/L)	2-Ethylhexanol (mg/L)
0.9% NaCl	500	2.5	0.08	0.10	0.01	0.04
5% Glucose	500	3.5	0.07	0.03	0.02	0.02
0.9% NaCl/5% glucose	500	2.5	0.08	0.03	0.01	0.02
Ringer's lactate	500	4.5	0.03	0.15	0.01	n.d.

TABLE 6.16 Concentration of Cyclohexanone in IV Solutions Stored in PVC Bags

Sample	Volume (mL)	Time of Storage	Cyclohexanone (mg/L)	Reference
0.9% NaCl (lot 1)	500	2.5 years	2.9	[43]
0.9% NaCl (lot 1)	500	2.5 years	8.4	
0.9% NaCl (lot 2)	500	3.5 years	4.8	
0.9% NaCl (lot 2)	500	3.5 years	7.0	
5% Glucose	500	3.5 years	6.5	
Ringer's lactate	500	2 years	15.7	
0.9% NaCl	50	n.i.	44.96	[44]
0.9% NaCl	100	n.i.	9.31	
0.9% NaCl	500	n.i.	3.50	
0.9% NaCl	1000	n.i.	2.05	
5% Glucose	50	n.i.	24.78	
5% Glucose	100	n.i.	9.15	
5% Glucose	500	n.i.	11.32	
Dialysis fluid (CAPD)	2000	n.i.	1.83–2.75	[4]
0.9% NaCl	500	n.i.	6.05	
0.9% NaCl	50	n.i.	9.24	
0.9% NaCl	1000	n.i.	0.024–1.08	[42]
Ringer's lactate	1000	n.i.	0.010–0.068	
Immunoglobulin 5% or 10%	200 (after dil.)	–	1.34 ± 0.6	[46]

n.i., not informed; CAPD, continuous ambulatory peritoneal dialysis.

segments. The results are summarized in Table 6.18, where the contaminants are related to the segment in which they were identified. Phthalic anhydride, 2-ethylhexanol, and acetic anhydride, the latter of which was not detected in the saline samples themselves, are thermal breakdown products of DEHP, usually occurring together with DEHP. These results are similar to those presented in Table 6.15, where the amount of degradation products of DEHP leached is even higher than the amount of DEHP itself. Moreover, this study showed that, though DEHP is the most prominent leachable PVC, it is not the only one nor is it found in the greatest amounts.

Tin, in the form of organotin compounds, is used as a stabilizer for PVC, mostly for unplasticized rigid PVC, whose use is restricted in pharmaceutical packaging or in medical devices. In spite of this, the BP establishes limits for tin in PVC based materials. A limit of 20 ppm Sn is set for containers for blood and blood components and for intravenous infusions (made of plasticized flexible PVC). For containers of noninjectable aqueous solutions, limits are established for organotin compounds. Table 6.19 lists the allowed organotin stabilizers and their limits for PVC for noninjectable drugs.

Despite being limited by pharmacopeia, research on the migration of organotin compounds into pharmaceuticals is not frequent. Based on records of the

TABLE 6.17 Contaminants Leached into 0.9% Saline Stored in PVC Bags for Intravenous Use Obtained from Hospital Stores in Australia, Europe, and the United States [47]

	Sample Origin					
	Australia[a]		Europe[b]		United States[a]	
Contaminant/Concentration (μg/L)	1	2	1	2	1	2
2-Ethylhexanol	160	150	1200	1200	70	60
3-Hydroxy-4-methoxy benzaldehyde	–	–	–	–	40	30
Benzyl alcohol	–	–	–	–	50	50
Diethylhexyl phthalate	14	11	24	25	2	3
Butoxyethyl ethanol	–	–	–	–	40	50
Caproalctam	–	–	20	20	–	–
Cyclohexanone	1100	900	–	–	1200	1300
Dibutyl phthalate	15	10	–	–	–	–
Diethyl phthalate	–	–	–	–	0.5	0.6
Dimethyl gluterate	–	–	–	–	20	30
Isobenzofuranone	30	40	–	–	–	–
Methyl phenols	20	18	–	–	–	–
Phenoxy propanol	–	–	–	–	45	40
Phthalic anhydride	15	12	–	–	4	4
t-Butyl cyclohexanol	650	600	1100	1100	–	–
t-Butyl cyclohexanone	70	65	280	300	–	–
Toluenesulfonamides	–	–	–	–	40	50

[a] 100 mL, Baxter Healthcare.
[b] 500 mL, SIFRA (Italy).

TABLE 6.18 Overview of the Contaminants and the Sample Segment in Which They Were Identified [47]

	Sample Segment		
Contaminant	Printed	Unprinted	Base Joint
2-Ethylhexanol	x	x	x
2-Ethyl hexanone	x	x	x
Acetic anhydride	x	x	x
Cyclohexanone	–	–	x
Phthalic anhydride	x	x	x
t-Butyl cyclohexanol	x	–	–
t-Butyl cyclohexanone	x	–	–

presence of these compounds in PVC for the food industry and on their migration into food simulants, it can be presumed that if similar PVC polymer materials are used in pharmaceutical packaging, organotin compounds would migrate into the products. There has been a great deal of attention focused on the toxicological aspects of organotins. There is special concern over the use

TABLE 6.19 Acceptable Organotin Stabilizers in PVC for Noninjectable Drugs According to the British Pharmacopoeia and Their Limits [48]

Compounds	Limit Amount
73% Di(iso-octyl)2,2′-[dioctylstannylene)bis(thio)]diacetate and 27% tri(iso-octyl)2,2′,2″-[(mono-octylstannylidyne) tris(thio)] triacetate	0.25% as tin
76% Di(iso-octyl)2,2′-[dimethylstannylene)bis(thio)]diacetate and 24% tri(iso-octyl)2,2′,2″-[(mono-methylstannylidyne) tris(thio)] triacetate	0.25% as tin

of tributyltin because of its biocidal properties. However, the organotin compounds used as stabilizers and catalysts, approved for use in PVC and other polymers, are mono- and diorganotin with no biocidal properties. The compounds, listed in Table 6.20, are mono- and diorganotin compounds, which are approved as PVC stabilizers for food contact.

6.2.3 Other Plastic Containers

Infusion bags made from EVA are offered as an alternative to PVC bags and are considered DEHP free. Perez-Feas et al. [52] demonstrated that not only DEHP but also other phthalates may be leached into solutions for parenteral nutrition for neonates stored in EVA bags. The authors divided the samples into those containing and not containing vitamins, and observed that the number and amount of phthalates was higher in samples containing vitamins. They attributed this to the lipophilic character of some vitamins, which would favor the release. Table 6.21 shows the results for dimethyl phthalate (DMP), diethyl phthalate (DEP), butyl benzyl phthalate (BBP), dibutyl phthtalate (DBP), dioctyl phthalate (DOP), and DEHP in nine samples prepared at the same place and under the same conditions.

Marcato et al. [53] investigated the migration of antioxidants from polyolefinic plastics into oleaginous vehicles. The study concentrated on two antioxidants, Irganox 1010 and Irgafos 168, used in PE and PP, and their copolymers (additives 9 and 12 in Table 6.10). The study was carried out by dipping plastic sheets containing 0.15% of antioxidants into a mixture of five oils (caprylic, capric triglyceride, cyclomethicone, dicaprylyl ether, isohexadecane, and $C_{12\text{-}15}$ alkyl benzoate) commonly used as pharmaceutical vehicles. Table 6.22 shows the percentage of antioxidants left by the polyolefin in the mixture stored at 25°C for 1 year. The results showed that the amount of antioxidant leached into the oleaginous mixture varied remarkably in relation to the polymeric material and decreased in the order EP > RACO > PP > high-density polyethylene (HDPE) (RACO and EP are ethylene–propylene copolymers). The authors concluded that the migration of both antioxidants is related to polymer crystallinity and structure: The higher the crystallinity, the lower the additive

TABLE 6.20 Organotin Compounds Leached from Different PVC Packaging Products

Packaging Item	Extraction Medium	Concentration: Monoctyltin	Concentration: Dioctyltin	Concentration: Mono or Dibutyltin	Reference
Food industry PVC	Methylene chloride	–	0.8–8751 μg/g	0.3–4.7 μg/g	[49]
Water pipe	Methylene chloride	–	n.d.	0.9–5985 μg/g	
Water pipe	Tetrahydrofuran	0.97 ± 0.10 mg Sn/g	0.24 ± 0.03 mg Sn/g	–	[50]
Cooking oil bottle	Tetrahydrofuran	0.98 ± 0.13 mg Sn/g	1.08 ± 0.13 mg Sn/g	–	
Film A	Water	n.d.	n.d.	n.d.	[51]
Film A	3% Acetic acid	n.d.	n.d.	n.d.	
Film B	Water	n.d.	n.d.	n.d.	
Film B	3% Acetic acid	11.5 μg/L	n.d.	n.d.	
Film C	Water	n.d.	n.d.	n.d.	
Film C	3% Acetic acid	n.d.	n.d.	6.83 μg/L	
Container A	Water	n.d.	n.d.	n.d.	
Container A	3% Acetic acid	14.3 μg/L	n.d.	110 μg/L	
Container B	Water	10.6 μg/L	n.d.	n.d.	
Container B	3% Acetic acid	18.6 μg/L	n.d.	73.0 μg/L	
Water pipe	Water	n.d.	n.d.	n.d.	
Water pipe	3% Acetic acid	n.d.	n.d.	n.d.	

n.d., not detected; –, not measured.

TABLE 6.21 Concentration of Phthalates in Different Samples for Parenteral Nutrition Containing and Not Containing Vitamins (n = 3) [52]

Sample	DMP (μg/L)	DEP (μg/L)	BBP (μg/L)	DBP (μg/L)	DEHP (μg/L)	DOP (μg/L)
Without vitamins						
1	–	–	1.3 ± 0.1	–	–	–
2	–	–	–	13.3 ± 3.7	7.0 ± 0.5	–
3	2.4 ± 0.9	–	–	–	9.3 ± 3.2	–
4	–	–	11.6 ± 0.2	12.9 ± 5.4	22.2 ± 1.8	–
5	–	–	–	–	19.5 ± 7.0	–
6	–	–	–	–	17.0 ± 0.5	–
7	–	–	–	21.6 ± 4.4	36.5 ± 0.8	–
With vitamins						
8	6.8 ± 0.1	–	6.1 ± 0.3	35.3 ± 1.5	215.1 ± 3.9	–
9	7.1 ± 0.1	–	5.9 ± 0.1	42.4 ± 4.7	993.7 ± 4.2	–

DMP, dimethyl phthalate; DEP, diethyl phthalate; BBP, butyl benzyl phthalate; DBP, dibutyl phthtalate; DOP, dioctyl phthalate. Stored at 4°C; storage time not informed.
–, not detected.

TABLE 6.22 Percentage of Antioxidants Left in Polyolefin Bottles Filled with a Mixture of Oils Used as Pharmaceutical Vehicles and Stored at 25°C for 1 Year [53]

Polyolefin	Antioxidant Left (%) Irganox 1010	Irgafos 168
High-density polyethylene (HDPE)	92.7 ± 5.7	47.8 ± 5.2
Polypropylene (PP)	76.3 ± 7.1	47.0 ± 4.1
Ethylene–propylene copolymer (RACO)	57.9 ± 6.3	36.4 ± 4.7
Ethylene–propylene copolymer (EP)	0	0

release. They suggested that PP and HDPE are satisfactory for the manufacture of oleaginous formulations.

Akapo and McCrea [54] investigated residual solvents in extracts of pharmaceutical packaging components of overwrapped low-density polyethylene (LDPE) vials. Extracts were obtained by filling the vial with purified water or by sealing foil pouches in contact with purified water and were kept at 70°C for 24 hours. Although solvents were found in the extracts (Table 6.23), they were in concentrations lower than the allowed limits for residual solvents in pharmaceuticals.

The risk of leaching of polymer monomers or degradation products of a multilayer plastic material (PP, polyamide, and PE) into 0.5% metronidazole solution was evaluated by Sarbach et al. [55]. Among several unidentified compounds, caprolactam, a phthalic derivative, and butylhydroxytoluene

TABLE 6.23 Residual Solvents Extracted from Packaging Components of Pharmaceutical-Grade Overwrapped LDPE Vials Obtained by Filling the Vials with Pure Water and Keeping Them at 70°C for 24 Hours [54]

	Amount Solvent (µg/L)			
Solvent	Unpouched LDPE Vial	Pouched LDPE Vial (n = 12)	Plain Foil Pouch	Preprinted Foil Pouch (n = 5)
Acetone	100	40–150	n.d.	n.d.
Isopropyl alcohol	2300	n.d. – 3200	n.d.	n.d. – 260
Ethyl acetate	n.d.	n.d. – 290	n.d.	n.d.
2-Butanone methyl ethyl ketone	n.d.	n.d. – 260	n.d.	n.d.
n-Heptane	1.4	1.0–2.0	2.1	1.0–1.6
n-Propyl acetate	2.2	1.1–3.8	1.7	n.d. – 1.1
Toluene	0.1	0.1–0.8	0.6	2.1–3.2
Diacetone alcohol 4-hydroxy-4-methyl-2-pentanone	4600	1100–3800	1200	800–1100
1-Propoxy-2-propanol	40	n.d. – 60	10	8.5–18

n.d., not detected.

(BHT) were disclosed. Caprolactam, a polyamide monomer, migrated through the PE inner layer into the solution, and BHT was present due to the decomposition of Irganox 1010, which is constituted by at least one entity of BHT, and is an antioxidant used either for PP and PE.

Not only container constituents can be leached into formulation products. Pan et al. [56] demonstrated that an unknown compound found in an ophthalmic solution was, in fact, originated from the varnish applied to the packaging label. After analyzing by liquid chromatography–mass spectrometry (LC-MS) and gas chromatography–mass spectrometry (GC-MS), the water/methanol extracts obtained from the LDPE bottle, plug, and cap, the label (Avery GEXPBE), the varnish (EC001245 Film III 16Y5), and the label ink (QY001645 Film III 16Y5), a monomethyl derivative of mephenesin was identified. The compound itself was not identified, but the ion fragmentation patterns in the sample and the varnish extracts matched well. Although the sample was stored under accelerated aging conditions (6 months at 40°C and 20% relative humidity), the results showed that even the product used to coat the labels migrated through the LDPE bottle and reached the product.

Although much less common, metallic contaminants are also leached from polymeric materials. While there are limits for their presence in plastic containers for pharmaceuticals, little has been investigated on their migration capability.

Zuccarello et al. [1] compared the actual metallic content of three plastic container materials with their extractability in 2% citric acid aqueous solution.

TABLE 6.24 Actual Metallic Content (Measurement after Sample Digestion) of Three Plastic Container Materials and Their Extractability in 2% Citric Acid Aqueous Solution [1]

Element	HDPE		Polyester		Polycarbonate	
(μg/g)	Digestion	Extraction	Digestion	Extraction	Digestion	Extraction
Al	89	24	–	–	–	–
Ca	8	7	21	14	12	5
Fe	5	<1	2	<1	<1	<1
Sb	–	–	151	7	n.d.	n.d.
Sn	<5	n.d.	n.d.	n.d.	n.d.	n.d.
Ti	3860	<1	16	<1	<1	<1
Zn	241	55	10	4	13	2

–, not measured; n.d., not detected.

TABLE 6.25 Metals Leached from Three Types of Polymeric Materials into Four Extracting Media Kept for 23 Weeks at 60°C [3]

Metal	LDPE				HDPE				PET[a]			
(mg/L)	Water	Buffer	Tween 80	EDTA	Water	Buffer	Tween 80	EDTA	Water	Buffer	Tween 80	EDTA
Mg	0.28	0.25	0.25	0.27	0.03	0.01	0.01	0.05	–	–	–	–
Ca	0.26	0.13	0.13	0.16	0.05	–	–	0.03	–	0.14	–	–
Al	–	–	1.50	0.50	–	–	–	–	–	0.16	–	0.16
Fe	–	–	–	0.02	–	–	–	0.02	–	–	–	–
Pb	0.01	0.01	0.01	–	0.01	0.03	0.03	–	–	–	–	0.03
Zn	0.23	0.03	0.03	0.05	–	–	–	–	–	–	–	–

[a] Temperature = 40°C.
–, not detected.

Samples were digested in a mixture of nitric, sulfuric, and perchloric acids, and the analysis was carried out by inductively coupled plasma mass spectrometry (ICP-MS). The elements analyzed and their concentrations related to polymer mass, are presented in Table 6.24. Significant amounts of Al, Ti, and Zn were present in HDPE, yet low levels were extracted into the citrate solution. The same occurred with Sb in polyester; although present in elevated concentrations in the polymer, very little was released into the extraction solution. Polycarbonate presented very low concentrations of the investigated metals and, therefore, low extraction yields.

In another study, LDPE, HDPE, and polyethylene terephthalate (PET) were evaluated for metal leaching [3]. Four different extracting media were used: USP water, 10 mM ammonium acetate, pH 6.8, 0.2% polysorbate 80 in water, and 0.1% EDTA in water. The main leachables were Mg, Ca, Al, Fe, Pb, and Zn (Table 6.25). HDPE and PET showed very low levels of extractable metals. By contrast, LDPE leached practically all metals investigated, and high levels of Al were found in Tween 80 and EDTA solutions.

6.3 METAL CONTAINERS

The main metals currently in use for pharmaceutical packaging are tinplate, stainless steel, and aluminum. They are used in shallow drawn containers, rigid and collapsible tubes, rigid containers, especially aerosols, and also in closure systems, where they serve as components such as lids, aerosol valves, and diaphragm seals. Blisters, strips, and sachet packaging are also made of a metallic component, usually aluminum foil.

In order to avoid direct contact, metallic containers are internally lacquered. The choice of an internal lacquer is based on satisfying product compatibility requirements such as resistance to acid or alkali, and freedom of flavor or taint. The resin must adhere well and must have sufficient flexibility to withstand the container fabrication process and the stress of handling.

Contamination may occur when an intimate contact between product and pack takes place, which leads to chemical interactions or exchange of ingredients. Therefore, contact between liquid or semisolid dosage forms and the container material is of significance.

Aluminum tubes are the most commonly used container/closure system for semisolid dosage forms. Although Al can be produced with up to 99.99% purity, at this purity level, it is mechanically weaker than aluminum of lower purity. Therefore, a small percentage of iron, (0.4%), silicon (0.2%), and traces of manganese and copper are often present.

Usually, the protective coating for aluminum tubes consists of epoxy resins. Epoxy resins are polymers that contain, on average, two or more epoxide groups per molecule. Although a vast number of epoxy resins have been developed, more than 70% of those commercialized worldwide are derived from bisphenol A (BPA) and epichlorohydrin. Epoxy resins produced by coupling BPA and epichlorohydrin are termed glycidyl-based resins. Only after curing will the resin attain the desirable strength, chemical resistance, and adhesive properties. Monomers, catalysts, and curing agents used to manufacture glycidyl-based resins are presented in Table 6.26.

Aluminum tubes are used to store semisolid dosage forms (creams, gels, ointments), which can act on the resin coating, mainly because of the in-use conditions (mechanical stress of the tube), and release resin constituents or its derivatives. When these coatings are made with epoxy resins, bisphenol and its derivatives are extractable candidates.

Diglycidyl ethers of bisphenol A (BADGE) and bisphenol F (BFDGE) and their hydrolyzed and chlorinated derivatives have received much attention lately due to the suspected mutagenic, genotoxic, and antiandrogenic effects of these compounds. Due to these concerns, regulatory agencies established migration limits for BADGE (and its derivatives) and prohibited the use of BFDGE and novolac glycidyl ethers (NOGE) [58].

In 2006, Soeborg et al. [59] conducted migration experiments of bisphenol compounds from lacquered aluminum tubes using a model cream (cetostearyl alcohol, macrogol cetostearyl ether, paraffin, and water) under accelerated

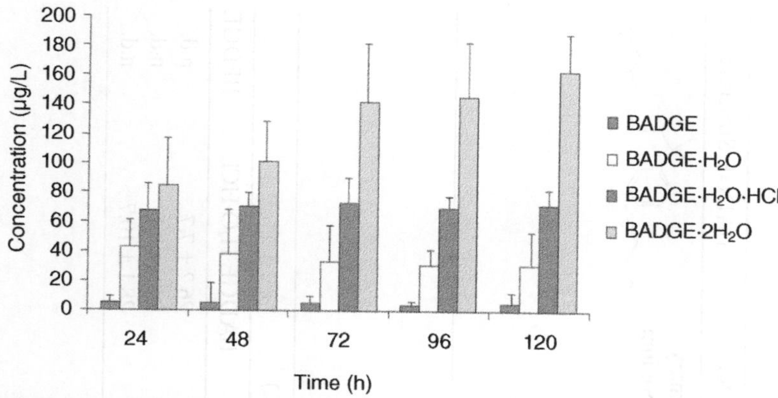

FIGURE 6.8 Concentrations of BADGE, BADGE·H_2O, BADGE·H_2O·HCl and BADGE·$2H_2O$ extracted from aqueous cream from 5-mL lacquered aluminum tubes for 120 hours at 70°C [59].

extraction conditions (70°C for up to 120 hours). BFDGE, banned in 2005 (EC regulation [58]), was not found in the extraction medium, but all the BADGE derivatives were (Fig. 6.8). The accelerated extraction conditions resembled an extreme worse-case situation, but, even so, the concentration of BADGE compounds was lower than the specific limit set by the European Commission of 1 mg/kg for the sum of all derivatives. They also evaluated migration into commercial creams obtained from a local community pharmacy among used drugs delivered back for safe destruction. Although they were commercialized before the new rule, BFDGE was not found in these samples (Table 6.27).

In another study, Haverkamp et al. [60] evaluated the migration not only of BADGE and its derivatives but also of the BPA itself from lacquered aluminum tubes into different kinds of semisolid dosage forms and under in-use conditions (mechanical stress by squeezing the tubes once a week). The samples, a lipophilic eye ointment, a water-based carbomer gel, and a nonionic hydrophilic cream were stored in aluminum lacquered tubes for up to 1 year at 30 and 40°C. While the amount of BADGE extracted by the eye ointment was not higher than 20 μg/kg (50 weeks, 40°C), in the nonionic hydrophilic cream, it reached more than 120 μg/kg. Additionally, while no BADGE·H_2O was observed in the eye ointment, in the nonionic hydrophilic cream, its amount was similar to the amount of BADGE, in the same time interval. This allowed the authors to conclude that BADGE primarily leaches from the coating into the matrix and BADGE·H_2O is mainly the hydrolysis product of BADGE after leaching. In the mechanically stressed tubes containing the nonionic hydrophilic cream, the extent of migration was increased 10-fold, as can be seen in Table 6.28, and even products not found in the nonstressed sample were measured. The data presented in Table 6.28 also confirm that the primarily leached compounds are BPA and BADGE, and the BADGE

TABLE 6.26 Monomers, Catalysts, and Curing Agents Used to Manufacture Epoxy Resins

Epoxy Group	Bisphenol A	Epichlorohydrin	Curing Agents	Catalyst	BADGE Structure
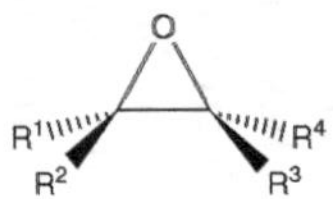	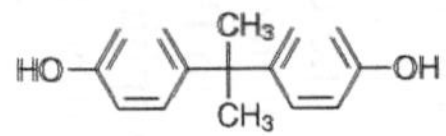	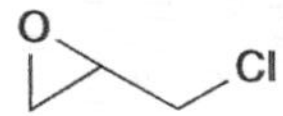	Triethylenetetramine 4,4′-Methylenedianiline Methylbicyclo(2,2,1) heptene- -2,3-Dicarboxylic anhydride- benzyldimethylamine BF3-methyethylamine	Quaternary ammonium salts	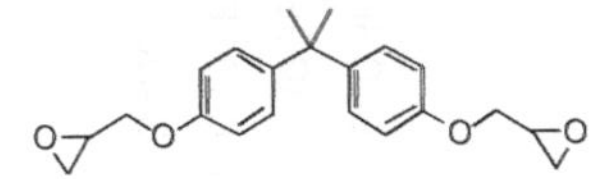

From Reference 57.

TABLE 6.27 Concentration of BADGE and BADGE Derivatives in Used Commercial Creams

			Compound (μg/L)				
Sample	Size (g)	Expiring	BADGE	BADGE·H_2O	BADGE·$2H_2O$	BADGE·H_2O·HCl	BFDGE
Cream 1	15	09/2004	2.4 ± 0.8	25.7 ± 5.9	13.3 ± 0.9	26.2 ± 7.7	n.d.
Cream 2	15	01/2003	n.d.	24.4 ± 9.4	6.4 ± 0.6	24.9 ± 8.7	n.d.
Cream 3	10	10/2003	n.d.	26.2 ± 15.2	12.4 ± 4.9	28.1 ± 10.7	n.d.

Analysis realized after 9/2004 and before 4/2005; $n = 3$.
Adapted from Reference 59.
n.d., not detected.

TABLE 6.28 BPA and BADGE Derivatives Leached into a Nonionic Hydrophilic Cream from Mechanically Stressed and Nonstressed Lacquered Aluminum Tubes [60]

	Nonstressed Samples Conc. (μg/kg)			Stressed Samples Conc. (μg/kg)		
Compound	6 Weeks	3 Months	6 Months	6 Weeks	3 Months	6 Months
BPA	12.9	12.9	18.8	47.9	117	142
BADGE	120	91.5	57.3	956	1100	725
BADGE·H_2O	7.5	<LOQ	n.d.	110	93.6	43.7
BADGE·$2H_2O$	54.1	69.4	136	392	1010	1790
BADGE·H_2O·HCl	<LOQ	4.3	7.7	19.3	55.2	92.1
BADGE·HCl	9.1	5.1	<LOQ	62.1	85.2	63.8
BADGE·2HCl	n.d.	n.d.	n.d.	n.d.	n.d.	n.d.

Limit of quantification (LOQ): BADGE·H_2O = 3.4 μg/kg; BADGE·H_2O·HCl = 4.0 μg/kg; BADGE·HCl = 3.9 μg/kg; $n = 3$.
n.d., not detected.

derivatives are formed after their hydrolysis since BADGE concentration decreased in 6 months, while the amount of BADGE·$2H_2O$ increased in the same period.

CONCLUDING REMARKS

Although pharmacopeial tests for containers are good indicators of the toxicity of extractables, there are species that do not cause an acute toxic reaction, but rather a chronic reaction, as is the case of phthalates (DEHP) and metallic species such as aluminum. Moreover, there are species that are extractable from packaging materials only by the action of formulation constituents and therefore are not present in the extracts obtained by conventional pharmacopeial compendium tests. In effect, significant variations in the contaminant and level of contamination depend on the formulation constituents, brand, and, mostly, on the nature of the packaging material.

REFERENCES

1. Zuccarello DJ, Murphy MP, Meyer RF, Winslow PA. A Comprehensive approach for the determination of extractable and leachable metals in pharmaceutical products by inductively-coupled plasma. PDA J Pharm Sci Technol 63, 339–352, 2009.
2. Bohrer D, do Nascimento PC, Becker E, Bortoluzzi F, Depoi F, de Carvalho LM. Critical evaluation of the standard hydrolytic resistance test for glasses used for containers for blood and parenteral formulations. PDA J Pharm Sci Technol 58, 96–105, 2004.

3. Fliszar KA, Walker D, Allain L. Profiling of metal ions leached from pharmaceutical packaging materials. PDA J Pharm Sci Technol 60, 337–342, 2006.
4. Bohrer D, Becker E, do Nascimento PC, de Carvalho LM, Marques MS. Arsenic release from glass containers by action of intravenous nutrition formulation constituents. Int J Pharm 315, 24–29, 2006.
5. Bohrer D, do Nascimento PC, Becker E, de Carvalho LM, Dessuy M. Arsenic species in solutions for parenteral nutrition. J Parenter Enteral Nutr 29, 1–7, 2005.
6. Scholze H. *Glas, Natur, Struktur und Eigenschaften*, 3rd ed. Springer-Verlag, Heidelberg, 1988, pp. 128–324.
7. Bacon FR. Glass containers for parenterals. In: *Pharmaceutical Dosages Forms: Parenteral Medications*, Vol. 2, Avis KE, Lachman L, Lieberman HA (eds.). Marcel Dekker, New York, 1986, pp. 55–110.
8. Pluhator Murton MM, Fedorak RN, Audette PJ, Marriage B, Yatscoff RW. Extent of trace-element contamination from simulated compounding of total parenteral nutrient solutions. Am J Health Syst Pharm 53, 2299–2303, 1996.
9. Hak EB, Storm MC, Helms RA. Chromium and zinc contamination of parenteral nutrient solution components commonly used in infants and children. Am J Health Syst Pharm 55, 150–154, 1998.
10. Borg C, Constant H, Fusselier M, Aulagner G. Zinc, copper and iron in total parenteral nutrition mixtures: a contamination study. Nutrition 13, 325–326, 1994.
11. Heinemann G, Braun S, Overbeck M, Page M, Michel J, Vogt W. The effect of vanadium-contaminated commercially available albumin solutions on renal tubular function. Clin Nephrol 53(6), 473–478, 2000.
12. Buchman AL, Neely M, Grossie JB, Truong L, Lykissa E, Ahn C. Organ heavy-metal accumulation during parenteral nutrition is associated with pathologic abnormalities in rats. Nutrition 17, 600–606, 2001.
13. do Nascimento PC, Marques MS, Hilgemann M, de Carvalho LM, Bohrer D, Pomblum SG, Schirmer S. Simultaneous determination of cadmium, copper, lead, and zinc in amino acid parenteral nutrition solutions by anodic stripping voltammetry and sample digestion by UV irradiation. Anal Lett 39, 1–14, 2006.
14. Alfrey AC, Le Gendre GR, Kaenhy WD. The dialysis encephalopathy syndrome, possible aluminum intoxication. N Engl J Med 294, 184–188, 1976.
15. Driscoll WR, Cummings JJ, Zorn W. Aluminum toxicity in preterm infants. N Engl J Med 337, 1090–1091, 1997.
16. Bohrer D, do Nascimento PC, Binotto R, Pomblum SG. Influence of the glass packing on the contamination of pharmaceutical products by aluminium. Part I: salts, glucose, heparin and albumin. J Trace Elem Med Biol 15, 95–101, 2001.
17. Bohrer D, do Nascimento PC, Martins P, Binotto R. Availability of aluminum from glass and an Al-form exchanger in presence of complexing agents and amino acids. Anal Chim Acta 459, 267–276, 2002.
18. Bohrer D, do Nascimento PC, Binotto R, Becker E. Influence of the glass packing on the contamination of pharmaceutical products by aluminium. Part III: interaction container-chemicals during heating for sterilisation. J Trace Elem Med Biol 17, 107–115, 2003.

19. Bohrer D, do Nascimento PC, Binotto R, Carlesso R. Influence of the glass packing on the contamination of pharmaceutical products by aluminium. Part II: amino acids for parenteral nutrition. J Trace Elem Med Biol 15, 103–108, 2001.
20. Koo WW, Kaplan LA, Horn J, Tsang R, Steichen J. Aluminum in parenteral nutrition solutions—sources and possible alternatives. J Parenter Enteral Nutr 10, 591–595, 1986.
21. Popinska K, Kierkus J, Lyszkowska M, Socha J, Pietraszek E, Kmiotek W, Ksiazky J. Aluminum contamination of parenteral nutrition additives, amino acid solutions, and lipid emulsions. Nutrition 15, 683–686, 1999.
22. Recknagel S, Brätter P, Chrissafidou A, Gramm HJ, Kotwas J, Rösick U. Parenteral aluminum loading in critical care medicine. Part I: aluminum content of infusion solutions and solutions for parenteral nutrition. Infusionsther Transfusionsmed 21, 266–273, 1994.
23. Bohrer D, do Nascimento PC, Binotto R, Becker E, Pomblum SG. Contribution of the raw material to the aluminum contamination in parenterals. J Parenter Enteral Nutr 26, 382–388, 2002.
24. Norwood DL, Jenke D, Manolescu C, Pennino S, Grinberg N. HPLC and LC/MS analysis of pharmaceutical container closure system leachables and extractables. J Liq Chromatogr Relat Technol 32(11-12), 1768–1827, 2009.
25. *Safety Assessment of Di-2-ethylhexylphthalate (DEHP) Released from PVC Medical Devices*. FDA, Center for Devices and Radiological Health, Rockville, MD, 1999.
26. Arabin A, Östelius J. Determination by electron-capture gas chromatography of mono (2-ethylhexyl) phthalate and di(2-ethylhexyl) phthalate in intravenous solutions stored in poly(vinyl chloride) bags. J Chromatogr B 193, 405–412, 1980.
27. Arabin A, Jacobsson S, Hagman A, Östelius J. Studies on contamination of intravenous solutions from poly(vinyl chloride) bags with dynamic headspace gas chromatography-mass spectrometry and gradient liquid chromatography diode array techniques. Int J Pharm 28, 211–218, 1986.
28. Štrac IV, Pušić M, Gajski G, Garaj-Vrhovac V. Presence of phthalate esters in intravenous solution evaluated using gas chromatography-mass spectrometry method. J Appl Toxicol 2011, doi: 10.1002/jat.1741, in press
29. Kambia K, Dine T, Gressier B, Bah S, Germe AF, Luyckx M, Brunet C, Michaud L, Gottrand F. Evaluation of childhood exposure to di(2-ethylhexyl) phthalate from perfusion kits during long-term parenteral nutrition. Int J Pharm 262, 83–91, 2003.
30. Allwood MC, Martin H. The extraction of diethylhexylphthalate (DEHP) from polyvinyl chloride components of intravenous infusion containers and administration sets by paclitaxel injection. Int J Pharm 127, 65–71, 1996.
31. Waugh WN, Trissel LA, Stella VJ. Stability, compatibility, and plasticizer extraction of Taxol (nsc-125973) injection diluted in infusion solutions and stored in various containers. Am J Hosp Pharm 48(7), 1520–1524, 1991.
32. Faouzi ME, Dine T, Luyckx M, Brunet C, Mallevais ML, Goudaliez F, Gressier B, Cazin M, Kablan J, Cazin JC. Stability, compatibility and plasticizer extraction of miconazole injection added to infusion solutions and stored in PVC containers. J Pharm Biomed Anal 14(6), 695–705, 1996.
33. Venkataramanan R, Burckart GJ, Ptachcinski RJ, Blaha R, Logue LW, Bahnson A, Giam CS, Brady JE. Leaching of diethylhexyl phthalate from polyvinyl-chloride

bags into intravenous cyclosporine solution. Am J Hosp Pharm 43(11), 2800–2802, 1986.

34. Gotardo MA, Monteiro M. Migration of diethylhexyl phthalate from PVC bags into intravenous cyclosporine solutions. J Pharm Biomed Anal 38(4), 709–713, 2005.
35. Faouzi MA, Khalfi F, Dine T, Luyckx M, Brunet C, Gressier B, Goudaliez F, Cazin M, Kablan J, Belabed A, Cazin JC. Stability, compatibility and plasticizer extraction of quinine injection added to infusion solutions and stored in polyvinyl chloride (PVC) containers. J Pharm Biomed Anal 21(5), 923–930, 1999.
36. Pearson SD, Trissel LA. Leaching of diethylhexyl phthalate from polyvinyl-chloride containers by selected drugs and formulation components. Am J Hosp Pharm 50(7), 1405–1409, 1993.
37. Mazzo DJ, NguyenHuu JJ, Pagniez S, Denis P. Compatibility of docetaxel and paclitaxel in intravenous solutions with polyvinyl chloride infusion materials. Am J Health Syst Pharm 54(5), 566–569, 1997.
38. Inoue K, Kawaguchi M, Yamanaka R, Higuchi T, Ito R, Saito K, Nakazawa H. Evaluation and analysis of exposure levels of di(2-ethylhexyl) phthalate from blood bags. Clin Chim Acta 358(2), 159–166, 2005.
39. Demore B, Vigneron J, Perrin A, Hoffman MA, Hoffman M. Leaching of diethylhexyl phthalate from polyvinyl chloride bags into intravenous etoposide solution. J Clin Pharm Ther 27(2), 139–142, 2002.
40. Takehisa H, Naoko E, Masahiko S, Katsuhide T, Moriyuki O, Keizoh S, Mutsuko T, Kenji K, Shin'ichiro N, Toshio O. Release behavior of diethylhexyl phthalate from the polyvinyl-chloride tubing used for intravenous administration and the plasticized PVC membrane. Int J Pharm 297(1-2), 30–37, 2005.
41. Ulsaker GA, Hoem RM. Determination by gas-chromatography—single-ion monitoring mass-spectrometry of phthalate contaminants in intravenous solutions stored in PVC bags. Analyst 103(1231), 1080–1083, 1978.
42. Thompson-Torgerson CS, Champion HC, Santhanam L, Harris ZL, Shoukas AA. Cyclohexanone contamination from extracorporeal circuits impairs cardiovascular function. Am J Physiol Heart Circ Physiol 296(6), H1926–H1932, 2009.
43. Ulsaker GA, Korsnes RM. Determination of cyclohexanone in intravenous solutions stored in PVC bags by gas-chromatography. Analyst 102(1220), 882–883, 1977.
44. Khalfi F, Dine T, Luyckx M, Gressier B, Brunet C, Ballester L, Cazin M, Cazin JC. Determination of cyclohexanone after derivatization with 2,4-dinitrophenyl hydrazine in intravenous solutions stored in PVC bags by high performance liquid chromatography. Biomed Chromatogr 12(2), 69–72, 1998.
45. Falk O, Jacobsson S. Determination of cyclohexanone in aqueous-solutions stored in PVC bags by isotope-dilution gas-chromatography mass-spectrometry. J Pharm Biomed Anal 7(10), 1217–1220, 1989.
46. Parti R, Mankarious S. Stability assessment of lyophilized intravenous immunoglobulin after reconstitution in glass containers and poly(vinyl chloride) bags. Biotechnol Appl Biochem 25(1), 13–18, 1997.
47. Story DA, Leeder J, Cullis P, Bellomo R. Biologically active contaminants of intravenous saline in PVC packaging: Australasian, European, and North American samples. Anaesth Intensive Care 33(1), 78–81, 2005.

48. British Pharmacopoeia. Appendix XX A. Materials based on poly(vinyl chloride), 2003.
49. Forsyth DS, Dabeka R, Sun WF, Dalglish K. Speciation of organotins in poly (vinyl chloride) products. Food Addit Contam 10(5), 531–540, 1993.
50. Ou QR, Whang CW. Determination of butyltin and octyltin stabilizers in poly(vinyl chloride) products by headspace solid-phase microextraction and gas chromatography with flame-photometric detection. Anal Bioanal Chem 386(2), 376–381, 2006.
51. Papaspyrou SD, Thomaidis NS, Lampi EN, Lioupis A. Determination of migration of n-butyltins and n-octyltins to food simulants by gas chromatography–mass spectrometry. Appl Organomet Chem 21(6), 412–424, 2007.
52. Perez-Feas C, Barciela-Alonso MC, Sedes-Diaz A, Bermejo-Barrera P. Phthalates determination in pharmaceutical formulae used in parenteral nutrition by LC-ES-MS: importance in public health. Anal Bioanal Chem 397(2), 529–535, 2010.
53. Marcato B, Guerra S, Vianello M, Scalia S. Migration of antioxidants additives from various polyolefinic plastics into oleaginous vehicles. Int J Pharm 257, 217–225, 2003.
54. Akapo SO, McCrea CM. SPME-GC determination of potential volatile organic leachables in aqueous-based pharmaceutical formulations packaged in overwrapped LDPE vials. J Pharm Biomed Anal 47(3), 526–534, 2008.
55. Sarbach C, Yagoubi N, Sauzieres J, Renaux C, Ferrier D, Postaire E. Migration of impurities from a multilayer plastics container into a parenteral infusion fluid. Int J Pharm 140(2), 169–174, 1996.
56. Pan C, Harmon F, Toscano K, Liu F, Vivilecchia R. Strategy for identification of leachables in packaged pharmaceutical liquid formulations. J Pharm Biomed Anal 46(3), 520–527, 2008.
57. Feldman D, Barbalata A. *Synthetic Polymers: Technology, Properties, Application*. Chapman & Hall, London, 1996, pp. 318–337.
58. Commission Regulation (EC) No. 1895/2005. On the restriction of use of certain epoxy derivatives in materials and articles intended to come into contact with food. Off J Eur Union 48, 28–32, 2005.
59. Soeborg T, Hansen SH, Halling-Sorensen B. Determination of bisphenol diglycidyl ethers in topical dosage forms. J Pharm Biomed Anal 40(2), 322–330, 2006.
60. Haverkamp JB, Lipke U, Zapf T, Galensa R, Lipperheide C. Contamination of semi-solid dosage forms by leachables from aluminium tubes. Eur J Pharm Biopharm 70(3), 921–928, 2008.

7

CLOSURES

Closures for pharmaceutical products are generally made of polymeric materials, which may be of either synthetic or natural origin. While brittle closures, such as screw caps, are made of conventional thermoplastics with a single composition, elastomeric closures are made of complex mixtures of many ingredients. Because elastomers or rubbers can be molded into an almost limitless variety of permanent shapes and are easily penetrated by needles and reseal after needle withdrawal, they are primary closures for parenteral vials and for preparations intended for repeated use.

The polymeric materials usually used to manufacture rigid closures are practically the same as those seen under plastic containers (Table A.1, Appendix). The same impurities are therefore to be expected in these packaging components. On the other hand, though made of polymeric materials, elastomeric closures present a different structure. In the manufacture of rubber, the chief component, elastomer, is combined with other chemicals to produce a material with specific properties that meet target needs, such as its abovementioned ability to reseal after repeated use. Table 7.1 lists the common elastomers used in the pharmaceutical industry and their monomeric structures.

The substances listed in Table 7.1 correspond to the basic structure of elastomeric closures. The other components in rubber formulations are curing or vulcanizing agents, accelerators, activators, antidegradants, plasticizers, fillers, and pigments. The most common additives used to compound rubber for the

Sources of Contamination in Medicinal Products and Medical Devices, First Edition. Denise Bohrer.
© 2013 by John Wiley & Sons, Inc. Published 2013 by John Wiley & Sons, Inc.

TABLE 7.1 Common Elastomers Used to Compound Rubbers for the Pharmaceutical Industry

Common Name	Chemical Name	Structure
Butyl rubber	Poly(isobutylene-isoprene)	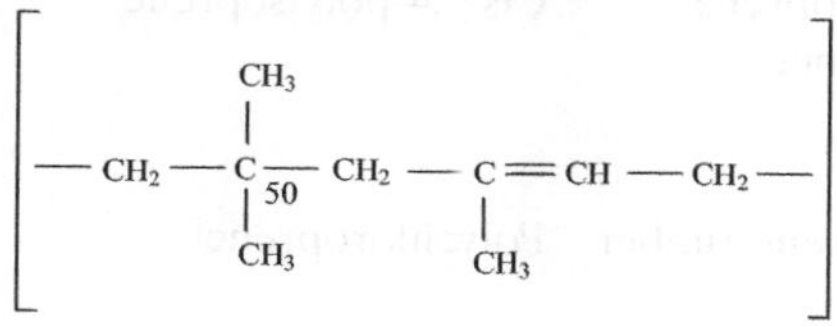
Halobutyl rubber	Halogenated poly(isobutylene-isoprene)	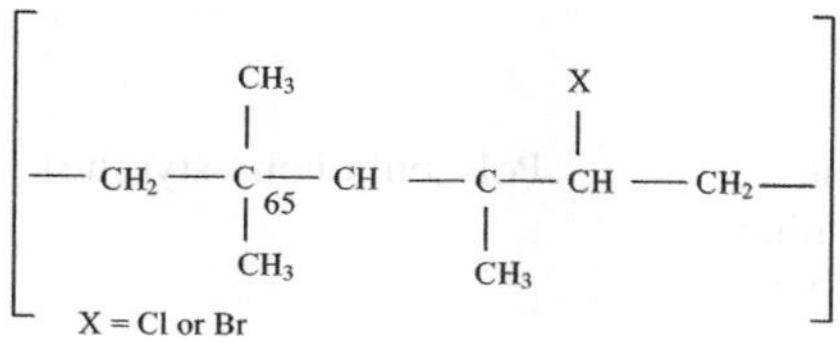
Ethylene-propylene rubber	Poly(ethylene-propylene)	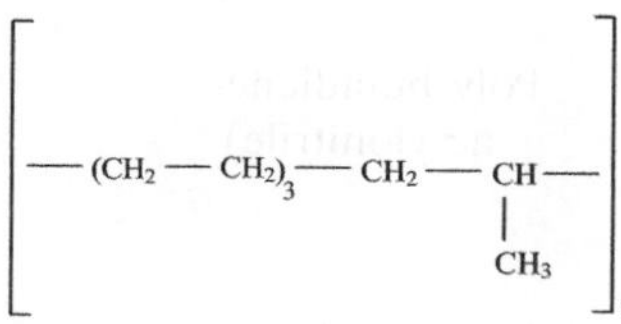
Ethylene propylene-diene rubber	Poly(ethylene-propylene-diene)	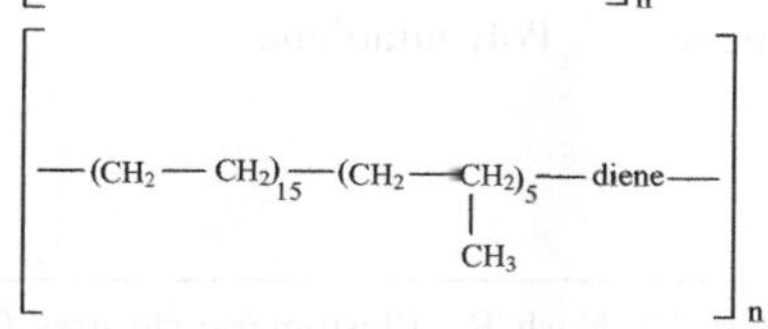
Silicone rubber	Polydimethilsiloxane	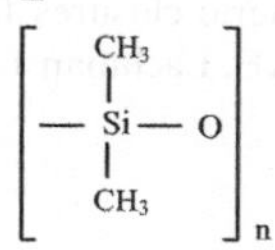
Urethane rubber	Adipic acid-ethylene glycol polyester	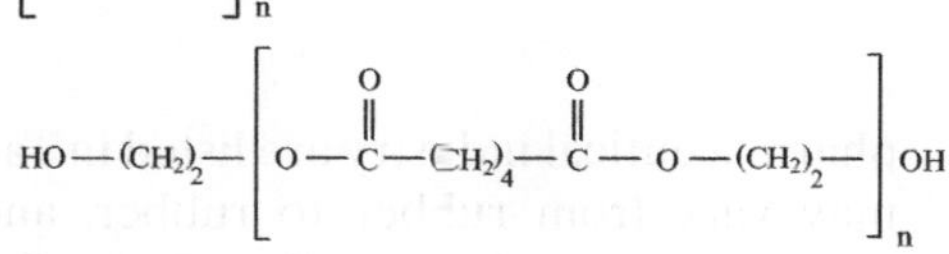
Fluoroelastomers	Poly(tetrafluoroethylene)	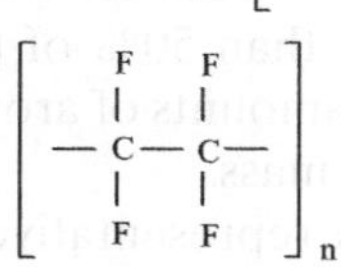
Natural rubber	Cis-1,4-polyisoprene	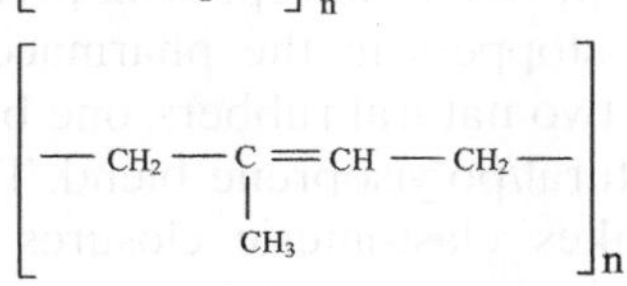

(*Continued*)

TABLE 7.1 (*Continued*)

Common Name	Chemical Name	Structure
Polyisoprene rubber	Cis-1,4-polyisoprene	$\left[-CH_2-C(CH_3)=CH-CH_2-\right]_n$
Neoprene rubber	Polychloroprene	$\left[-CH_2-C(Cl)=CH-CH_2-\right]_n$
Styrene butadiene rubber	Poly(butadiene-styrene)	$\left[-(CH_2-CH=CH-CH_2)_4-CH_2-CH(C_6H_5)-\right]_n$
Nitrile rubber	Poly(butadiene-acrylonitrile)	$\left[-(CH_2-CH=CH-CH_2)_5-(CH_2-CH(CN))_2-\right]_n$
Polybutadiene	Polybutadiene	$\left[-CH_2-CH=CH-CH_2-\right]_n$

Source: Smith EJ, Nash RJ. Elastomeric closures for parenterals. In: *Pharmaceutical Dosages Forms: Parenteral Medications*, Vol. 2, Avis KE, Lachman L, Lieberman HA (eds.). Marcel Dekker, New York, 1986 [1].

pharmaceutical industry are listed in Table 7.2. The amount of each component may vary from rubber to rubber, and, depending on the component, the amount can reach more than 50% of the total mass of a formulation. While accelerators are used in amounts of around 1%, fillers may make up more than 50% of the formulation mass.

Table 7.3 presents six representative rubber formulations commonly used for molded stoppers in the pharmaceutical industry. The formulations are made up of two natural rubbers, one butyl, one bromobutyl, one chlorobutyl, and one natural/polyisoprene blend. The presence of such a variety of components makes elastomeric closures a potential source of contaminants. Although polymeric materials present reasonable inertness, the possibility of other rubber components reaching the drug product should be considered. Moreover, these ingredients, though intended for pharmaceutical purposes and, therefore, meeting pharmacopeial requirements, may still present their own impurities. The rubbers in Table 7.3 differ from each other in the type

TABLE 7.2 Ingredients Other Than Elastomers Used to Compound Rubbers for the Pharmaceutical Industry

Curing Agents	Accelerators	Activators	Antidegradants	Plasticizers	Fillers	Pigments
Sulfur	Hexamethylene tetramine	Zinc oxide	Amines	Paraffinic waxes	Carbon black	Carbon black
Sulfur-containing chemicals	Ditiocarbamates	Stearic acid	Hindered phenols	Silicon oil	Aluminum silicates (clays)	Titanium dioxide
Peroxides	Sulfenamides		Waxes	Paraffinic oils	Magnesium silicate (talc)	Iron oxide
Cadmium oxide	Thiuram			Naphthenic oils	Barium sulfate	Chromium oxide
Magnesium oxides	Thiazole			Organic phosphates	Zinc oxide	Organic dyes
Zinc oxide					Silica	

Source: Smith EJ, Nash RJ. Elastomeric closures for parenterals. In: *Pharmaceutical Dosages Forms: Parenteral Medications*, Vol. 2, Avis KE, Lachman L, Lieberman HA (eds.). Marcel Dekker, New York, 1986 [1].

TABLE 7.3 Composition of Six Representative Rubber Formulations Commonly Used for Molded Stoppers in the Pharmaceutical Industry

	Rubber Composition (%)					
Ingredient	A	B	C	D	F	G
Natural rubber	60.6	24.7	93.5	–	–	–
Butyl	–	–	–	47.4	–	–
Bromobutyl	–	–	–	–	46.9	–
Chlorobutyl	–	–	–	–	–	47.9
Polyisoprene	–	37.0	–	–	–	–
Calcined clay	30.3	30.8	–	47.4	46.9	47.9
Titanium dioxide	–	1.2	–	–	1.0	–
Carbon black	3.0	0.2	–	–	0.1	–
Zinc oxide	3.0	3.0	1.4	1.4	1.4	1.4
Sulfur	1.2	1.2	1.8	0.8	–	0.2
Stearic acid	1.0	0.9	1.4	–	–	–
Tetramethyl thiuram monosulfide (TMTM)	0.2	0.3	0.2	–	–	–
Tetramethyl thiuram disulfide (TMTD)	–	–	–	0.5	–	–
2,2′-Dibenzthiazyl disulfide (MBTS)	–	–	–	0.2	–	–
Zinc 2-mercaptobenzothiazole (ZMBT)	–	–	0.4	–	–	–
Zinc dimethyl dithiocarbamate (ZMDC)	–	–	–	–	–	0.4
2-Mercaptobenzothiazole (MBT)	–	–	0.4	–	–	–
Paraffin	–	–	–	0.9	0.9	0.9
Iron oxide	–	–	–	0.2	–	0.2
Polyethylene 617A	–	–	–	0.9	0.9	0.9
Phenolic resin SP-1045	–	–	–	–	1.8	–
Antioxidant 425	0.6	0.6	0.9	–	–	–

Adapted from Reference 2.

of elastomer; nevertheless, there are ingredients common to almost all formulations. These include calcined clay in amounts over 30%, zinc oxide, and sulfur.

The purity of zinc oxide for pharmaceutical rubbers is presented in Table 7.4. The main contaminant is lead, which may be present in amounts of up to 0.2%.

Jenke [3] evaluated the extraction of zinc from an elastomer (synthetic polyisoprene) containing zinc oxide. The extraction was carried out in buffered solutions, with pH from 2.5 to 11.0, at room temperature for 13 days. Other rubber components were measured as well: calcium, stearic acid, and palmitic acid. The results show that pH had a great influence on the extractability of these species. At high pH, stearic acid and zinc oxide were extracted in amounts high enough to precipitate zinc stearate. Zinc and calcium were extracted in the greatest amounts at low and high pH, whereas the extraction of stearate and palmitate occurred above pH 8. For materials containing both stearic acid

TABLE 7.4 Purity of Zinc Oxide for Use in the Rubber Industry from Different Manufacturers

	Manufacturer			
Component (%)	A	B	C	D
ZnO	≥99.0	99.8	99.0	≥99.0
Lead	≤0.002	0.04	0.2	≤0.002
Cadmium	≤0.0015	0.04	0.005	≤0.0005
Iron	≤0.002	0.01	0.04	≤0.0005
Arsenic	≤0.003	–	–	≤0.0001
Manganese	–	–	0.003	≤0.0005
Copper	–	–	0.005	≤0.0005
Ca, Na	–	–	–	≤0.001
Co, Cr, Mg, Ni	–	–	–	≤0.0005
Sulfur/sulfate	–	–	0.08	≤0.005
Chlorine	–	–	0.1	≤0.001
Moisture	<0.3	–	0.2	–
H_2O soluble salts	–	0.02	–	–

A, Votorantin; B, Horsehead; C, Grillo; D, Fluka (Puriss); –, not informed.

TABLE 7.5 Purity of Calcined Clay for Use in the Rubber Industry from Different Manufacturers

	Manufacturer			
Component (%)	A	B	C	D
SiO_2	51.0–52.4	58.30	57.0	57.0
Al_2O_3	42.1–44.3	23.40	40.0	39.0
Fe_2O_3	0.5	1.70	0.93	1.00
TiO_2	1.56–2.50	–	0.54	0.70
K_2O	–	5.10	1.2	1.70
Na_2O	–	10.30	–	[a]
MgO	–	0.02	–	[b]
CaO	–	0.51	–	0.35
Loss of ignition	0.1	0.67	0.23	0.3

[a] Included in K_2O amount.
[b] Included in CaO amount.
A, Burgess; B, Unimin; C, CB Chrystal; D, U.S. Silica; –, not informed.

and zinc oxide, the potential for zinc stearate to precipitate is such that the formation of particulate matter should be a concern.

Table 7.5 presents the components of calcined clays for the rubber industry. Clays do not have a defined composition; however, their main constituents are silica and alumina. All other oxides are present in concentrations lower than 10%. However, due to the elevated amount of calcined clay in any rubber,

TABLE 7.6 Major and Trace Elements in Clay Samples from Different Origins in Africa [4]

Element	Clay Sample 1	2	3	4	5	6	7
Major (%)							
SiO_2	54.85	59.23	47.56	46.09	48.17	50.49	64.61
Al_2O_3	26.98	24.48	33.11	33.09	33.06	25.68	21.27
Fe_2O_3	2.09	2.64	3.38	3.35	2.86	1.98	1.83
MgO	0.14	0.12	0.22	0.49	0.27	0.89	0.93
CaO	0.09	0.07	0.07	0.08	0.09	0.05	0.06
Na_2O	0.19	0.13	0.27	0.14	0.11	0.19	0.17
K_2O	0.85	0.86	1.09	1.19	1.04	5.07	5.07
TiO_2	1.62	1.49	1.63	1.41	1.33	0.93	0.91
P_2O_5	0.04	0.06	0.07	0.07	0.06	0.03	0.03
H_2O	0.50	0.50	0.90	0.80	0.80	0.70	0.06
Total	98.05	98.58	99.30	99.72	99.29	98.02	99.81
Trace (μg/g)							
Ba	299.25	288.76	419.60	390.45	191.97	568.38	558.78
Co	12.29	11.67	12.59	18.20	12.53	6.76	6.89
Cr	155.18	139.81	112.16	46.35	63.76	57.23	45.58
Cu	24.04	15.05	11.64	22.94	22.59	14.46	35.22
Mn	2.75	2.47	1.63	20.73	8.15	0.44	0.57
Ni	42.88	37.07	34.02	47.25	42.84	12.69	12.42
Pb	26.42	21.02	27.89	30.74	13.79	12.15	13.22
Sr	50.97	56.74	82.68	26.93	16.80	28.05	37.47
V	82.75	117.51	146.92	57.70	78.77	59.94	79.43
Zn	70.06	65.20	75.86	98.59	97.57	43.29	46.97
Zr	270.37	230.69	236.36	161.72	151.06	105.89	108.74

minor trace elements and their respective concentrations may be of relevance. Since the major constituents make up almost 100% of the mineral, little attention is paid to such traces. A study conducted with edible clays [4], however, revealed that clays contain several trace elements. Table 7.6 presents the composition of these clays. The major components are qualitatively and quantitatively similar to those of the clays listed in Table 7.5, corresponding to more than 98% of the minerals. Nevertheless, some elements found in these minerals, such as barium, reached amounts greater than 0.5 mg/g. Zirconium is the second most abundant element, and lead, nickel, and chromium are present in all samples. As the concentrations of traces seem to be equivalent among the samples, it is reasonable to suppose that there is a pattern for trace elements in clays. Therefore, these would be minor components also to be expected in calcined clays for rubber compounding.

Impurities in sulfur (Table 7.7) vary from one supplier to another, either qualitatively or quantitatively. While several metallic impurities were

TABLE 7.7 Purity of Sulfur for Use in the Rubber Industry from Different Manufacturers

	Manufacturer			
Component (%)	A	B	C	D
Sulfur content	99.9	≥99.9	≥99.5	99.5–100.5
Ash	<0.01	≤0.03	≤0.2	≤0.3
Carbon	<0.1	–	–	–
Arsenic	<0.00001	≤0.0001	–	≤0.004
Moisture	<0.5	≤0.1	–	≤0.5
Acidity (H_2SO_4)	<0.02	≤0.003	–	≤0.02
Organic matter	–	≤0.03	–	–
Chloride, sulfate	–	–	≤0.005	≤0.01
Sulfide	–	–	–	≤0.001
Fe	0.0026	–	≤0.005	–
Si	0.0018	–	–	–
Ca	0.0015	–	≤0.005	–
Al	0.0012	–	–	–
P	0.0007	–	–	–
K	0.0005	–	≤0.01	–
Na	n.d.	–	≤0.01	–
Sn	0.0003	–	–	–
Cr, Mg, Mn, Ba, Cu, Sb, Pb, Ni, Cd	<0.0001	–	–	–
Mo, Tl, Be, Se, Te, Co, V, Ag, Hg	<0.00001	–	–	–
Cd, Co, Cu, Ni, Pb, Zn	–	–	≤0.005	–

A, Montana sulfur; B, Be-Long Int'l Group; C, Fluka (Purum); D, Sigma (USP); n.d., not detected; –, not informed.

evaluated in product A, only arsenic was determined in product B, which presented the same nominal purity of 99.9% as product A. In fact, only arsenic is specified in pharmacopeial compendia as an impurity to be determined in sulfur.

Titanium dioxide is also a regular rubber component. In fact, titanium dioxide is also used to opacify plastic materials, and its purity is, therefore, prescribed by pharmacopeias. Table 7.8 presents impurities of commercial titanium dioxide from different suppliers.

Carbon black, whose amount in rubber may reach 50%, is usually an impure material and can contain polynuclear aromatic hydrocarbons (PAHs) [5]. In spite of that, the presence of PAHs in pharmaceutical products stored in contact with rubber closures has not been effectively evaluated. Polycyclic aromatic hydrocarbons are chemical compounds that consist of fused aromatic rings and do not contain heteroatoms or carry substituents. The U.S. Environmental Protection Agency (EPA) has identified 16 priority PAHs, based on concerns that they do or might cause cancer in animals and humans. One of the best characterized and most toxic PAH is benzo[a]pyrene, which

TABLE 7.8 Purity of Titanium Dioxide for Use in the Rubber Industry from Different Manufacturers

	Supplier			
Component (%)	A	B	C	D
TiO_2	99.0–100.5	99.0–100.5	≥99	>99.00
As	0.0001	≤0.0001	≤0.0005	–
Sb	0.01	≤0.005	–	–
Ba	–	in accordance	–	–
Fe	0.02	≤0.005	≤0.005	<0.005[a]
Pb	0.002	≤0.002	≤0.005	–
Heavy metals	<0.002	≤0.002	–	–
Water soluble	<0.25	≤0.25	–	–
HCl (0.5 M) soluble	<0.5	≤0.5	–	–
Moisture	<0.5	≤0.5	–	<0.2
Loss of ignition	<0.5	≤0.5	≤0.5	–
Sulfur	–	–	–	–
Nb_2O_5	–	–	–	<0.02
Cr_2O_3	–	–	–	<0.0006
V_2O_5	–	–	–	<0.004

[a] As Fe_2O_3.
A, Merck puriss (USP); B, Sigma puriss (USP); C, Sigma Technical; D, GMS; -, not informed.

is generally used as the indicator of PAH exposure. For this reason, the Food and Drug Administration (FDA) imposes carbon black purity specifications based on limits for total PAHs (0.5 mg/kg) and for benzo[a]pyrene (5 µg/kg) [6].

Besides PAHs, suppliers of carbon black should also analyze the other parameters displayed in Table 7.9. Table 7.10 displays the levels of PAHs and inorganic impurities in two carbon blacks from different manufacturers. The major inorganic impurities were chloride, silicon, sodium, and sulfur, present in amounts higher than 100 mg/kg. All 16 of the U.S. EPA priority PAHs, excluding 2-bromonaphthalene and benzo[*ghi*]perylene, were present (the first 16 are listed in Table 7.10), although not surpassing the limits for benzo[a] pyrene and for total PAHs.

Table 7.11 presents PAHs extracted from a carbon black sample and also the PAHs extracted from three different commercial rubber closures [7]. Almost all PAHs evaluated were present in the carbon black sample. Benzo[a] antracene was found in the highest concentration, and it was also the only PAH extracted from gray rubber (the carbon black analyzed was not the one used to compound the rubbers). Red and black rubbers released several PAHs, but amounts were substantially higher from black rubber. This is an expected result, considering that carbon black is the pigment of black rubber.

Metered dose inhalers (MDIs) are devices designed to deliver a drug in the form of an aerosol. As the system is pressurized, a metering valve is part of

TABLE 7.9 Purity Parameters and Their Limits[a] for Carbon Black and Purity Levels of Two Commercial Samples

Parameter	Limits	Cancarb Powder N991	Timcal ENSACO® 150G
Ash content (%)	–	<0.02	–
Toluene extracts (%)	≤0.1	<0.5	<0.1
Heat loss (%)	–	<0.1	–
Total sulfur (%)	–	<0.006	–
UV absorption at 386 nm (cyclohexane extract)	<0.02 AU (1-cm cell) <0.1 AU (5-cm cell)	–	<0.02 (1-cm cell)
Benzo[*a*]pyrene (μg/kg)	<250	–	<2
Total PAHs (μg/kg)	–	–	<20
Heavy metal (mg/kg)	–	–	<100

[a] Commission European Community: Directive 2008/39/EC of 6 March 2008.
–, not informed.

TABLE 7.10 Organic (PAHs) and Inorganic Impurities in Two Selected Samples of Commercial Carbon Black

Organic Impurities	Cabot Black Pearls 4350 (mg/kg)	Inorganic Impurities	Cancarb N-990 Standard (mg/kg)
PAHs			
Acenaphthene	0.001	Ash	1200
Acenaphthylene	0.023	Chloride	100
Anthracene	0.002	Iron	20
Benz[a]anthracene	<0.001	Silicon	135
Benzo[*a*]pyrene	0.001	Sodium	425
Benzo[*b*]fluoranthene	0.001	Sulfur	150
Benzo[*ghi*]perylene	–	Aluminum	2
2-Bromonaphthalene	–	Antimony	<0.5
Chrysene	0.001	Arsenic	<0.1
Dibenz[a,h]anthracene	<0.001	Bismuth	<0.5
Fluoranthene	0.016	Boron	<0.5
Fluorene	0.002	Cadmium	<0.5
Indeno[1,2,3-cd]pyrene	<0.001	Calcium	1.0
Naphthalene	0.086	Chromium	<0.5
Phenanthrene	0.020	Copper	1.0
Pyrene	0.082	Gallium	<1.0
Cyclopenta[cd]pyrene	0.006	Lead	<0.5
Benzo[e]pyrene	0.002	Lithium	<0.5
Benzo[*k*]fluoranthene	0.001	Magnesium	1.0
1,12-Benzperylene	0.001	Manganese	<0.5
Anthanthrene	<0.001	Mercury	<1.0
Coronene	<0.001	Nickel	<0.5
Benzo[ghi] fluoranthene	0.006	Phosphorous	<1.0
Perylene	<0.001	Potassium	2.0
Total	0.257	Tin	<1.0
		Titanium	<1.0
		Vanadium	<1.0
		Zinc	<1.0

TABLE 7.11 PAHs Extracted from a Sample of Carbon Black and from Different Types of Rubber Closures [7]

Hydrocarbon	Carbon Black (μg/g)	Gray Rubber (μg/g)	Red Rubber (μg/g)	Black Rubber (μg/g)
Acenaphthene	3.5 ± 3.0	n.d.	n.d.	0.9 ± 1.6
Acenaphthylene	n.d.	n.d.	n.d.	n.d.
Anthracene	n.d.	n.d.	n.d.	n.d.
Benz[a]anthracene	7.4 ± 4.5	0.2 ± 2.3	0.2 ± 3.6	0.3 ± 2.2
Benzo[*a*]pyrene	2.4 ± 1.9	n.d.	n.d.	n.d.
Benzo[*b*] fluoranthene	3.7 ± 4.0	n.d.	0.3 ± 5.0	0.5 ± 3.2
Benzo[*ghi*]perylene	6.8 ± 3.2	n.d.	0.2 ± 2.0	5.6 ± 2.3
2-Bromonaphthalene	2.0 ± 2.1	n.d.	n.d.	n.d.
Chrysene	n.d.	n.d.	n.d.	n.d.
Dibenz[a,h] anthracene	n.d.	n.d.	n.d.	n.d.
Fluoranthene	0.9 ± 5.0	n.d.	0.1 ± 1.9	0.1 ± 2.0
Fluorene	3.2 ± 2.8	n.d.	n.d.	0.6 ± 2.9
Indeno[1,2,3-cd] pyrene	3.6 ± 2.5	n.d.	0.4 ± 3.8	2.7 ± 3.3
Naphthalene	4.0 ± 2.1	n.d.	n.d.	2.9 ± 4.2
Phenanthrene	2.0 ± 2.6	n.d.	n.d.	1.0 ± 4.5
Pyrene	6.5 ± 3.0	n.d.	0.2 ± 4.1	0.2 ± 2.9
Total	46.0	0.2	1.4	14.8

n.d., not detected.

the device. This valve, incorporating rubber sealing components, may leach contaminants into the product, mainly because the propellant is kept under pressure and, therefore, its extracting properties are enhanced. Norwood et al. [8] evaluated the leaching of PAHs into MDI formulations. Table 7.12 presents the PAHs and their levels found in three different MDI products. Benzo[a] pyrene was present in two of the inhalers, though in low concentrations, and phenanthrene, fluoranthene, and pyrene presented the highest concentrations in all the samples.

Three kinds of infusions stored in containers with rubber stoppers were analyzed for their PAH content: amino acid solutions, lipid emulsions, and propofol emulsions [7]. Five out of 16 PAHs were found in the samples (Table 7.13). With the exception of 8% Aminosteril Hepa, all products had at least one PAH, and propofol was the most contaminated sample. Despite the differences in origin, product type, and container system, the PAHs found in highest concentrations in these samples were practically the same as those in MDIs.

Pharmacopeial monographs do not set limits for additives for rubber closures as they do for plastic polymers. Tests are limited to showing the content of sulfur, sulfated ash, volatile sulfides and extractable zinc, ammonium, and heavy metals (Table 7.14). Extractability tests prescribed by regulatory

TABLE 7.12 PAHs Found in MDI Samples [8]

Product	PAH (μg/inhaler)										Total (μg/inhaler)
	Benzo[*a*]pyrene	Benzo[*b*]fluoranthene	Benzo[*ghi*]perylene	Fluoranthene	Pyrene	Naphtalene	Acenaphthylene	Fluorene	Phenantrene	Anthracene	
A	<0.05	n.d.	0.08	1.20	1.26	0.29	0.43	<0.05	1.96	0.10	5.50
B	n.d.	n.d.	0.03	0.53	0.61	0.15	0.22	n.d.	0.88	n.d.	2.45
C	<0.05	n.d.	0.06	1.37	2.13	0.57	0.45	<0.05	2.14	0.12	7.02

n.d., not detected (below the limit of detection).

TABLE 7.13 PAHs Found in Infusion Solutions Stored in Container with Rubber Closures [7]

Product	PAH (μg/L) ± RSD ($n = 3$)					Total (μg/L)
	Benzo[*a*]pyrene	Benzo[*b*]fluoranthene	Benzo[*ghi*]perylene	Fluoranthene	Pyrene	
Primène (lot 1)	0.02 ± 5.10	0.03 ± 2.96	0.13 ± 3.51	0.13 ± 2.30	n.d.	0.31
Primène (lot 2)	0.02 ± 0.83	0.03 ± 3.77	0.17 ± 14.80	0.13 ± 3.84	n.d.	0.35
Primène (lot 3)	0.03 ± 2.14	0.03 ± 2.35	0.17 ± 6.15	0.14 ± 1.53	n.d.	0.37
Aminoped	0.11 ± 2.54	0.04 ± 0.81	n.d.	n.d.	0.16 ± 11.20	0.31
Aminosteril	n.d.	n.d.	n.d.	n.d.	0.11 ± 7.89	0.11
Aminosteril Hepa 8%	n.d.	n.d.	n.d.	n.d.	n.d.	0
Aminoplasmal	0.16 ± 1.01	n.d.	n.d.	n.d.	0.21 ± 2.97	0.37
Nefroamino	0.07 ± 1.07	n.d.	n.d.	0.17 ± 3.83	0.18 ± 7.90	0.42
Lipofundin 10%	0.14 ± 3.30	n.d.	n.d.	n.d.	0.07 ± 3.37	0.21
Lipofundin 20%	0.15 ± 0.88	n.d.	n.d.	n.d.	0.07 ± 7.30	0.22
Lipovenos 10%	0.14 ± 4.10	n.d.	n.d.	0.15 ± 4.44	0.08 ± 9.30	0.37
Lipovenos 20%	0.19 ± 4.45	n.d.	n.d.	0.13 ± 2.99	0.08 ± 2.30	0.40
Fresofol 1%	0.31 ± 4.54	0.24 ± 4.42	n.d.	0.19 ± 3.11	0.18 ±3.61	0.92
Fresofol 2%	0.24 ± 3.88	0.19 ± 2.54	n.d.	0.16 ± 1.20	0.15 ± 2.90	0.74

n.d., not detected (below the limit of detection); RSD, relative standard deviation.

TABLE 7.14 Pharmacopeial Tests for Elastomeric Closures

	USP (<381>)			Ph. Eur. (3.2.9)	JP (59)
	Water	Isopropyl Alcohol	Drug Vehicle	Water	Water
	121°C for 30 min	Reflux for 30 min	Reflux for 30 min	121°C for 30 min	121°C for 60 min
Turbidity	Difference between blank and sample	Difference between blank and sample	Difference between blank and sample	Less opalescent than reference suspension	–
Appearance of solution (transparency at 430 and 650 nm)	–	–	–	–	>99.0%
Absorbance (from 220–360 nm)	–	–	–	<0.2	<0.2
Foam (shake for 3 min)	–	–	–	–	Foam disappears within 3 min
pH change	Difference between blank and sample	Difference between blank and sample	–		<1.0 difference against blank
Acidity or alkalinity (titration with 0.01 M HCl or NaOH)	–	–	–	<0.3 mL	–

Extractables (residue of evaporation)	Difference between blank and sample	Difference between blank and sample	Difference between blank and sample	<2 mg/50 mL type I closure <4 mg/50 mL type II closure	<0.2 mg/100 mL
Reducing agents (titration with $KMnO_4$)	–	–	–	Difference between blank and sample <3.0-mL type I closure <7.0-mL type II closure	<2.0-mL difference against blank
Reducing agents (titration with Iodine)	Compare with the blank	–	–	–	–
Ammonium	–	–	–	<2 ppm	–
Heavy metals	Compare with the blank	Compare with the blank	–	<2 ppm as Pb	–
Zinc	–	–	–	<5 μg/mL	<0.01 mg/mL
Volatile sulfides (reaction with lead acetate paper)	–	–	–	Not more intense than the standard: 0.154 mg sulfide/50 mL	–
Cadmium (carried out with the residue of incineration)	–	–	–		<0.001 mg/2 g closure
Lead (carried out with the residue of incineration)	–	–	–		<0.01 mg/2 g closure

agencies for closures for drug packaging [9–11] are also limited to the amount of extractable residues or evaluating the *in vivo* reaction of the extractable residue when the material fails in the *in vitro* tests. These tests, however, are neither qualitative (in the sense of showing which substances can be extracted) nor specifically quantitative since they show only the total amount of extractable as a residue. The United States Pharmacopeia (USP) <381> "Elastomeric Closures for Injection," for example, recommends the calculation of the weight of the residue after evaporating the solvent (purified water, drug vehicle, or isoprpyl alcohol) used for extraction. Tests *in vivo* are recommended only when the material does not meet the requirements of the *in vitro* tests. However, since extractability also depends on the interaction between the container and the formulation constituents, they are unable to show further contamination in the final product.

Milano and Bailey [12] showed that pharmacopeial tests are inadequate to evaluate the potential for extraction of contaminants from rubber stoppers. Extracts obtained from 10 different commercial rubber closures were analyzed following USP prescriptions and by gas chromatography–mass spectrometry (GC-MS) with identification of the main peaks. The results, presented in Table 7.15, showed that, despite meeting pharmacopeial criteria, 5 out of 10 rubber stoppers presented identified compounds in their extracts (butylhydroxytoluene [BHT] among them). Compendia tests for "extractables," consisting of a series of nonspecific wet chemical analyses with water extracts, were unable to indicate any further interaction that could occur between the active ingredients or excipients and the rubber closures.

In spite of possible interactions, very little is known about the leachability of rubber constituents through the direct contact of closures with formulations. Delaunay-Bertoncini et al. [13] analyzed two representative commercial rubbers by GC-MS and detected more than 100 different compounds. The rubbers, mixtures of isobutylene and isoprene, were analyzed after being cryogenically ground and submitted to two different extraction procedures: a Sohxlet extraction with a series of solvents and a static headspace extraction, which entailed placing the sample in a 20-mL sealed vial in an oven at 110°C for 5, 20, or 50 minutes. Although these are not the conditions to which pharmaceutical products are submitted, the results may give an idea of which compounds could be expected from these materials. Residual monomers, isobulylene in the dimeric or tetrameric form, and compounds derived from the scission of the polymeric chain were found in the extracts. Table 7.16 presents an overview of the nature of the compounds identified in the headspace and Soxhlet extracts of the polymers. While the liquid-phase extraction yielded less-volatile compounds, the headspace technique was able to show the presence of compounds with low molecular mass in the extract. A complete screening of the extracted substances can be seen in Figure 7.1, where the graphics depict the abundance of the different classes of compounds detected in the extracts of different solvents for both rubbers. Not only oligomers, but also phthalates, phenols, and acidic compounds were found in the extracts, mainly from rubber P1.

TABLE 7.15 Results of the Evaluation of 10 Different Rubber Closures Following Pharmacopeial Requirements

Closure	Color	Turbidity		Total Extractables		Reducing Agents	Heavy Metals	pH Change	UV Absorption[a]	GC-MS Analysis
		Water	Isopropanol	Water	Isopropanol				220 nm	
Halobutyl	Red	0.3	8.6	0.00	0.01	n.d.	n.d.	–0.03	4.0 > A > 0.2	2,2′Methylenebis[6-(1,1 dimethylethyl)-4-ethyl] phenol 4-(2,2,3,3-Tetramethyl butyl) phenol
Chlorobutyl	Gray	0.3	13.8	0.00	0.01	n.d.	n.d.	–0.03	<0.2	–
Natural rubber	Black	6.6	7.0	0.00	0.05	n.d.	n.d.	0.04	4.0 > A > 0.2	Butylhydroxytoluene Diphenylamine
Natural rubber	Pink	2.1	3.5	0.00	0.03	n.d.	n.d.	–0.04	4.0 > A > 0.2	Butylhydroxytoluene Diphenylamine
Chlorobutyl	Gray	0.3	11.9	0.00	0.00	n.d.	n.d.	–0.01	4.0 > A > 0.2	–
Bromobutyl	Gray	0.3	0.7	0.00	0.00	n.d.	n.d.	–0.05	<0.2	–
Bromobutyl	Gray	0.4	1.0	0.00	0.00	n.d.	n.d.	–0.02	<0.2	–
Bromobutyl	Gray	0.4	2.2	0.00	0.01	n.d.	n.d.	–0.02	<0.2	–
Chlorobutyl	Gray	0.6	1.8	0.00	0.01	n.d.	n.d.	–0.03	<0.2	–
Butyl	Gray	0.6	6.0	0.00	0.01	n.d.	n.d.	–0.05	4.0 > A > 0.2	–

Extracts for the tests were obtained according to the USP [12].
[a] After 2 hours' extraction.
n.d.: not detected.

TABLE 7.16 Nature of the Compounds Identified by GC-MS in the Headspace and Soxhlet Extracts of the Two Rubbers under Study [13]

	Compound Identified in the Extracts	
	Headspace	Soxhlet
Alkanes	C_5 and higher	C_9–C_{30}
Aromatics	92 < MW < 132	168 < MW < 182
Fatty acids	No	228 < MW < 352
Esters	No	Yes
2,6-Di-tert-butyl-p-cresol	Yes	Yes
2,6-Di-tert-butyl-p-benzoquinone	Yes	Yes
Ketone	MW = 198	MW = 198

MW, molecular weight.

Jaehnke et al. [14] evaluated the release of volatile compounds from butyl rubbers. Butyl, chlorobutyl, and bromobutyl stoppers were heated at 80°C for 18 hours and the volatiles released were analyzed by GC. In the chromatograms shown in Figure 7.2, it is possible to see that several compounds were released. Most of the peaks were identified, and they are collated in Table 7.17. Chlorobutyl rubber released the largest number of volatiles. Peaks 1–5 in Figure 7.2B are residual solvents from the chlorination of butyl rubber in hexane. The presence of cyclic polydimethylsiloxanes in the headspace of chlorobutyl and bromobutyl rubbers suggests the use of silicone oil either as a mold release or as a coating agent. Bromobutyl closures showed a small release of hydrocarbons since brominated oligomers are not volatile. Because the closures were analyzed as received (cleaned by the manufacturer and ready to use), the authors attributed the low concentration of residual solvents (peaks 1–5) to an extensive stripping of bromobutyl rubber. Moreover, because phenolic antioxidants are sensitive to bromination, BHT, found in butyl and chlorobutyl rubbers (peak 33), might be replaced by epoxidized soy bean oil in the bromobutyl rubber. A group of peaks (retention time 34–36 minutes) only found in the headspace of bromobutyl rubber might be originated from the epoxidized oil.

Similar results were found in another study where coated chloro- and bromobutyl rubbers were compared [15]. Using thermal desorption GC-MS (200°C), total extractables were measured, but only one phenolic antioxidant was characterized. The data in Table 7.18 show that in this case as well, bromobutyl rubbers released the lowest amount of extractables. Moreover, the amount of the phenolic antioxidant released, about 100 times less than the amount released by the chlorobutyl rubbers, supports the premise that phenolic antioxidants, sensitive to bromination, are less used in bromobutyl rubbers.

A potential source of contaminants in lyophilized powders is the release of volatile compounds into the headspace of vials sealed with rubber closures.

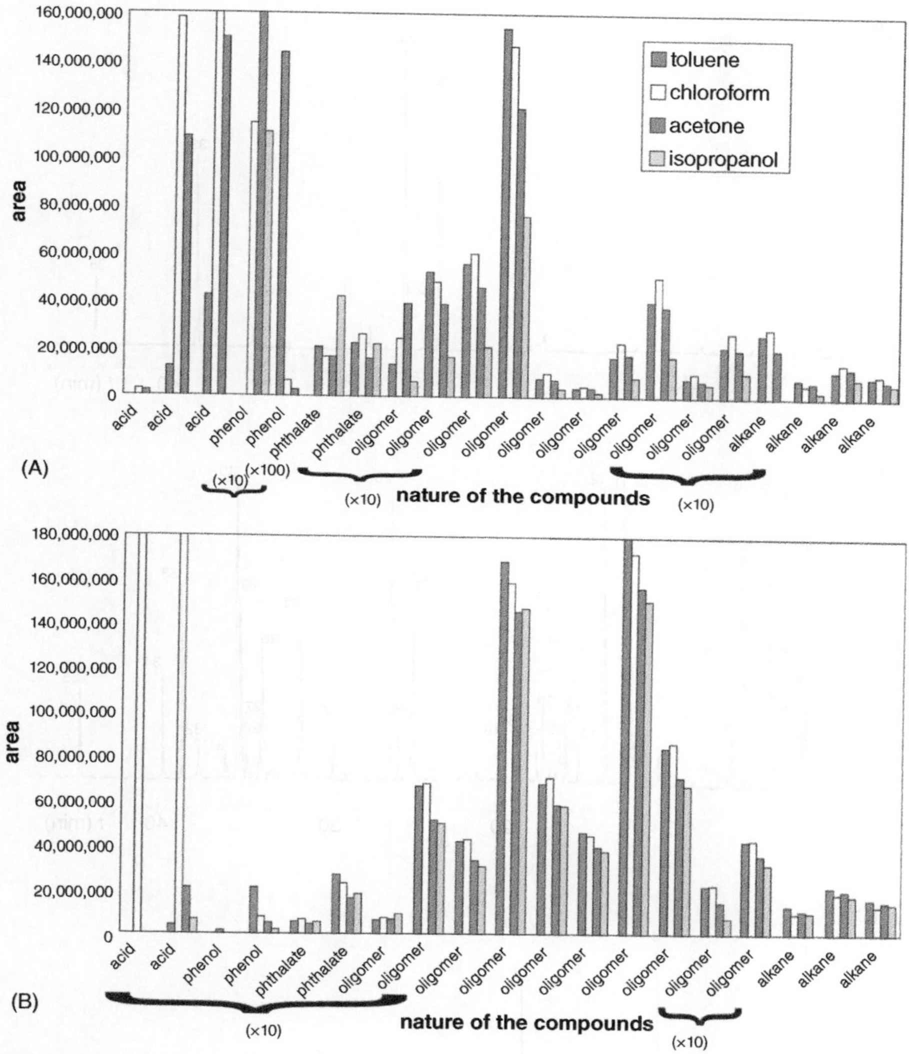

FIGURE 7.1 Influence of the solvent used for Soxhlet extractions of (A) rubber P1 and (B) rubber P2 on the area of the chromatographic peaks of different classes of compounds detected in GC-MS [13].

This phenomenon is frequent and can be identified when the reconstituted solution shows a haze formation due to adsorbed volatiles onto the powder. To investigate the role of the drug product in the adsorption of volatile compounds released from rubber stoppers, Jaehnke et al. [16] employed four powdered drugs and five lactose powder samples with different particle sizes. Volatiles released from a butyl rubber (80°C, 18 hours) were quantified before

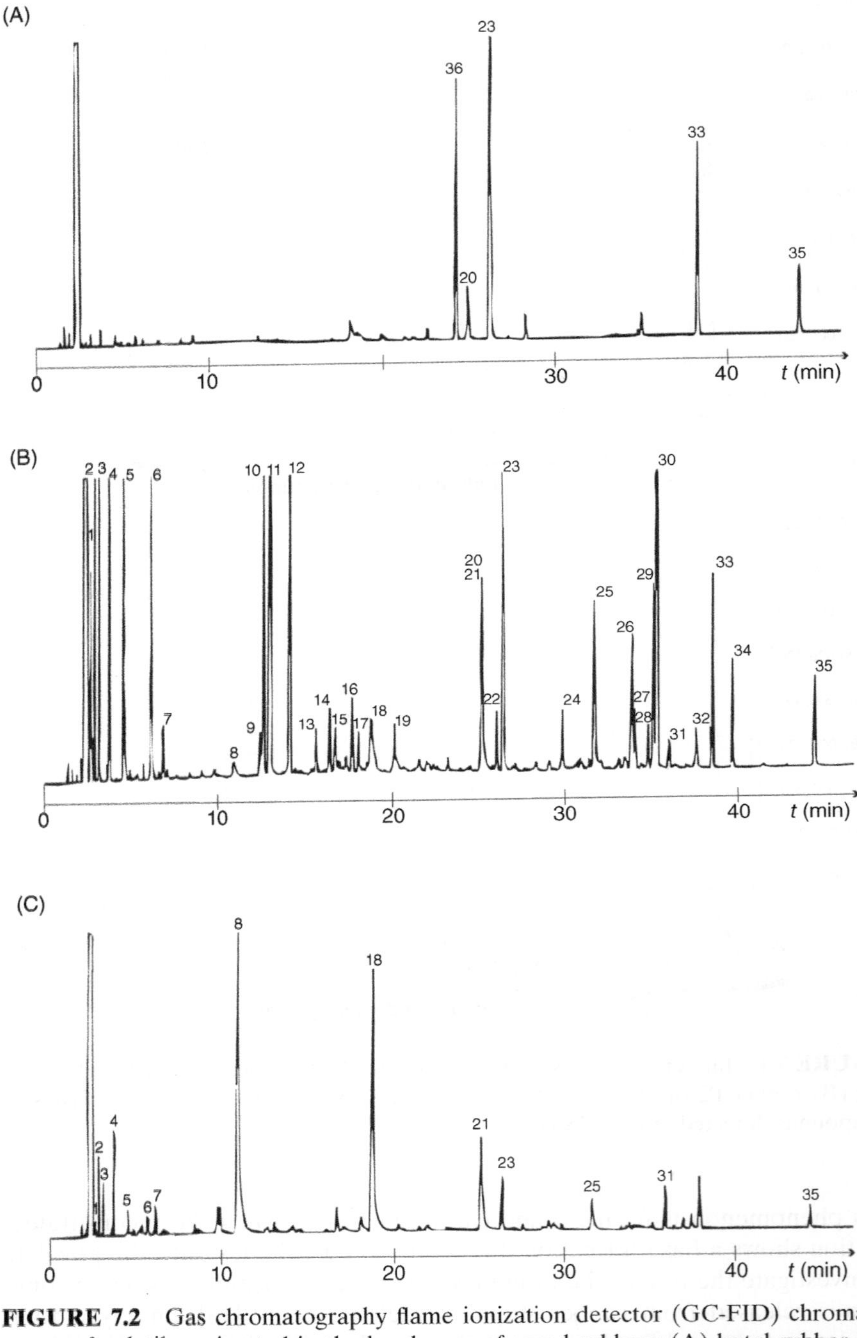

FIGURE 7.2 Gas chromatography flame ionization detector (GC-FID) chromatograms of volatiles migrated in the headspace of cured rubbers: (A) butyl rubber, (B) chlorobutyl rubber, and (C) bromobutyl rubber. Explanation of peaks are seen in Table 7.17 [14].

TABLE 7.17 Identification of Volatile Compounds Released from Butyl, Chlorobutyl, and Bromobutyl Rubber Closures Displayed in Chromatograms in Figure 7.3 [14]

Peak	Compound	Peak	Compound
1	2-Methylpentane	19	Unidentified
2	3-Methylpentane	20	IIR-trimmer
3	N-Hexane	21	Pentasiloxane
4	Methylcyclopentane	22	Unidentified
5	Cyclohexane	23	1-Esopropenyl-2,2,4,4-tetramethyl-cyclohexane
6	α-Diisobutene	24	Unidentified
7	β-Diisobutene	25	Hexasiloxane
8	Hexamethylcyclotrisiloxane	26	CIIR-trimmer
9	Monochlorinated isoprene	27	Unidentified
10	Ethylbenzene	28	CIIR-trimmer
11	*m*- and *p*-xylene	29	1-(1-Chloromethylethenyl)-2,2,4,4-tetramethyl-cyclohexane
12	*o*-Xylene	30	1-(1-Chloromethylethenyl)-2,2,4,4-tetramethyl-cyclohexane
13	C-9-Alkylbenzene	31	Heptasiloxane
14	CIIR dimmer	32	Heptasiloxane
15	Unidentified	33	Butylhydroxytoluene
16	CIIR dimmer	34	Unidentified
17	Unidentified	35	IIR-tetramer
18	Octamethylcyclotetrasiloxane		

TABLE 7.18 Total Extractables and Phenolic Antioxidant Amounts Released from Rubber Stoppers after Thermal Desorption GC-MS

Sample	Rubber	Coating	Supplier	Total Extractables (μg/Stopper)	Antioxidant (μg/Stopper)
1	Bromobutyl	No	A	337	4
2	Bromobutyl	Yes	A	554	6
3	Chlorobutyl	Yes	B	2488	756
4	Chlorobutyl	Yes[a]	B	2088	490
5	Bromobutyl	Yes	B	1534	54
6	Butyl	Yes	C	1136	0
7	Bromobutyl	Yes	D	1186	3

[a] Different coating from sample 3.

Adapted from Reference 15.

TABLE 7.19 Relationship between Surface Area, Level of Adsorbed Volatiles, and Haze Formation in Reconstituted Solutions of Solids for Parenteral Use and Different Samples of Lactose [16]

Product	Form	Specific Surface Area (BET) (m^2/g) ± SD	Amount of Volatiles Adsorbed to the Solid in the Dry State (ng) ± SD	Intensity of Haze in Solutions of Reconstituted Solid (NTU) ± SD
Apalcillin	Lyophilizate	0.17 ± 0.01	Traces	16 ± 1
Pirenzepine	Lyophilizate	0.66 ± 0.01	98 ± 38	33 ± 8
Pirenzepine	Powder	1.06 ± 0.03	103 ± 41	42 ± 3
Mezlocillin	Powder	5.15 ± 0.51	160 ± 30	39 ± 4
Apalcillin	Powder	9.62 ± 0.37	379 ± 25	95 ± 8
Cefodizim	Powder	11.53 ± 1.58	625 ± 0	154 ± 3
Lactose	Powder	0.08 ± 0.00	–	13.3 ± 0.9
Lactose	Powder	0.12	–	18.5 ± 3.5
Lactose	Powder	0.17 ± 0.03	–	22.0 ± 2.2
Lactose	Powder	0.63	–	45.0 ± 11.3
Microlactose	Powder	1.32 ± 0.03	–	65.7 ± 5.3

Lactose amount = 11.5% (m/v).
NTU, nephelometric turbidity units; –, not evaluated.

and after contact with the powdered samples. Turbidity of the aqueous solutions of reconstituted samples after volatile adsorption as well as specific surface areas of lactose and drug samples was evaluated. The data, presented in Table 7.19, allow the conclusion that after dissolution of the powdered sample in water, the volatiles remain insoluble and form an emulsion that appears as a haze. The level of turbidity depends on the amount of volatiles present in the solution and their amounts on the powder. The higher the powder surface area, the higher the amount of volatiles adsorbed.

Sendo et al. [17] observed not only haze but also particulate formation after lyophilized amphotericin B reconstitution with water for injection (WFI), which was probably caused by the same phenomenon observed by Jaehnke et al. [16]. Amphotericin B exists as a colloid suspension by building a complex with sodium deoxycholate, which, when in lyophilized form, certainly presents a large surface. The authors observed higher numbers of particles when the vial was vigorously shaken before or after solvent addition than when it was gently turned upside down. The release of adsorbed rubber volatiles, and consequently particle formation, is probably much more efficient by vigorous shaking.

2-Mercaptobenzothiazole (MBT), although an important contact allergenic agent, is commonly used as an accelerator in the vulcanization of natural and synthetic rubber. Rubber C in Table 7.3 contains MBT as an ingredient.

TABLE 7.20 Amount of Mercaptobenzothiazole Found in Different Drug Products Stored in Prefilled Syringes or in Syringe Cartridges Containing a Rubber Part

Drug	Drug Concentration	Mercaptobenzothiazole (μg/mL)	
		Sample 1	Sample 2
Digoxin	0.25 mg/mL	8.2	6.6
Dexamethasone sodium phosphate	4 mg/mL	1.1	7.9
Epinephrine	1:1000	9.7	11.1
Lidocaine hydrochloride	2%	n.d.	2.8
Mepivacaine hydrochloride	2%	n.d.	1.5
Pilocarpine hydrochloride	4%	1.0	0.7
Sodium phenobarbital	60 mg/mL	3.2	–
Sodium phenobarbital	130 mg/mL	3.4	5.1

Adapted from Reference 19.
n.d., not detected; –, not analyzed.

Because of its wide use, the European Committee for Standardization (CEN) [18] set a migration limit of 8-mg/kg rubber for MBT in rubber articles that come into contact with the skin to minimize sensitization risk. However, as for many other plastic and rubber additives, limits for leaching into pharmaceutical products have not yet been set. MBT was detected in injectable drugs stored in prefilled syringes or syringe cartridges containing a rubber part [19]. MBT was not found in only 2 out of 15 samples. These 15 samples comprise eight different formulations from different suppliers (Table 7.20). With amounts varying from 0.7 to 11.1 mg/mL, the highest concentrations were found in epinephrine formulations.

Thiurams are also accelerators for the vulcanization of rubber. Two thiurams, mono- and disulfide thiuram, were used to compound the rubbers presented in Table 7.3. As thiuram disulfides may remain unreacted in a finished rubber product, they could react with each other or with thiol group to give mixed disulfide products, according to the reactions in Figure 7.3. Because drugs may have thiol groups in their structure, it was speculated that unreacted thiuram could react with drugs containing thiol groups [20]. To investigate this possibility, captopril, a thiol group-containing drug, and two different thiurams, tetramethylthiuram disulfide (TMTD) and dibutylthiuram disulfide (TBTD), were mixed and allowed to react. The mixtures, prepared separately at different pH values, were sealed in glass vials, kept at 25°C, and analyzed at different time intervals for 24 hours. In all cases, adducts were formed, reaching the highest level at pH 6. Thirty-two percent of the captopril present was converted into captopril-TMTD and 8% into captopril-TBTD in 5 hours, and in both cases, the products subsequently degraded. However, at pH 2.2, the conversion increased linearly with the time reaching about 14% in 24 hours for

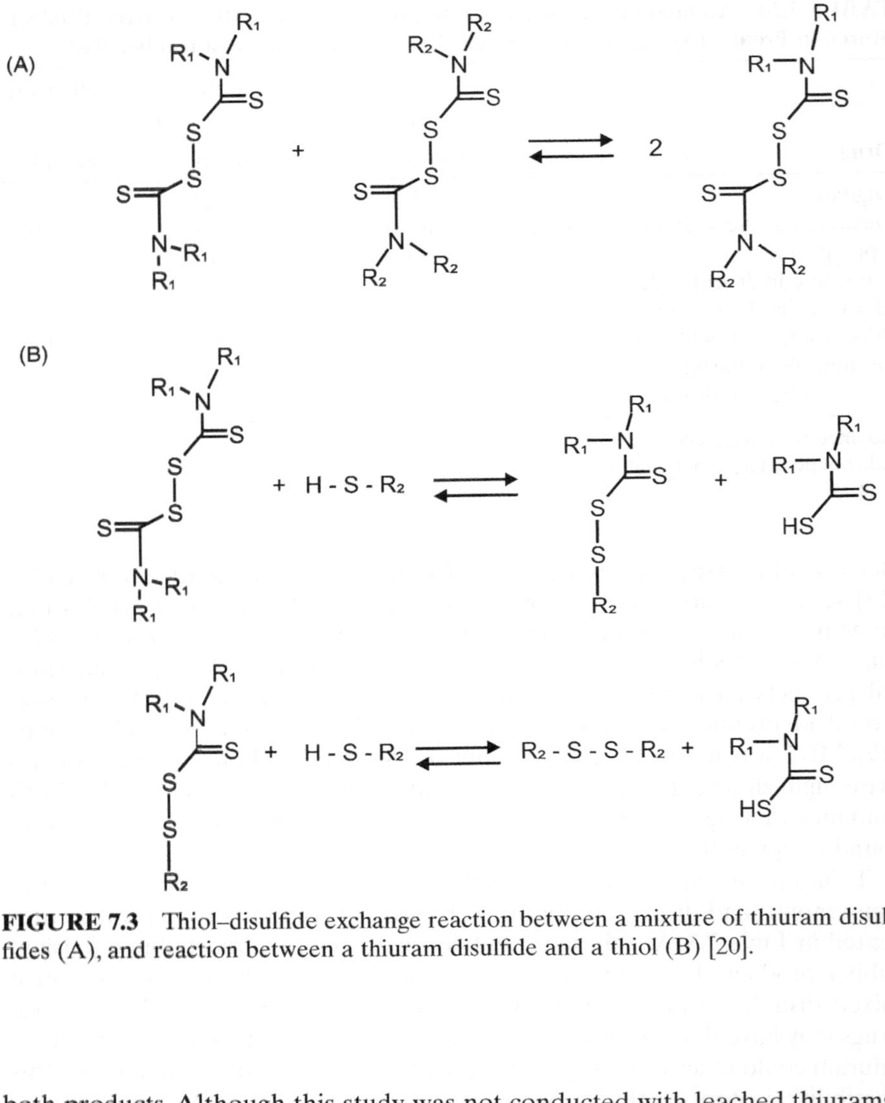

FIGURE 7.3 Thiol–disulfide exchange reaction between a mixture of thiuram disulfides (A), and reaction between a thiuram disulfide and a thiol (B) [20].

both products. Although this study was not conducted with leached thiurams, the warning is valid for drugs containing thiol groups stored in containers with rubber closures.

Although the number of additives necessary to make an elastomeric material is limited, the number of extractables and leachables may be much greater due to reactions occurring during rubber processing. Table 7.21 lists extractable compounds from rubber closures. Although only identified peaks are presented, there were, in some cases, several unidentified peaks in the respective chromatograms. The chromatogram B in Figure 7.2 is an example: At least eight peaks were not identified.

TABLE 7.21 Organic Extractables from Rubber for Pharmaceutical Use

Rubber	Sample	Extraction Conditions	Compounds Extracted	Reference
Not specified	Sealing gasket of aerosol valve	Soxhlet Methylene chloride	2,6-Di-t-butyl-4-bromomethyl phenol	[21]
			Dimethyl terephthalate	
			Dibutyl fumarate	
			3-Ethyl-3-methylpentanedioic acid	
			Diphenyl amine	
			4,6-Dimethyl-2-(2-methyl cyclohexyl) phenol	
			Palmitic acid	
			Phenanthrene	
			Dibutyl phthalate	
			Bis-(3-t-butyl-2-hydroxy-5-methyl-phenyl)methane	
			Stearic acid	
			N,N-Dimethyl-9-octadecenoamide	
			2-Propen-1-one,1-(methylphenyl)-3-phenyl oxime	
			Fluoranthene	
			Hexadecanoamide	
			Bis-(2-ethylhexyl)phthalate	
			Pyrene	
			Triphenylphosphine oxide	
Carbon black nitrile-based elastomer	Closure	2-Propanol refluxed 7 h	Acenaphtylene	[22]
			Phenanthrene	
			Anthracene	
			Fluoranthene	
			Pyrene	

(Continued)

TABLE 7.21 (*Continued*)

Rubber	Sample	Extraction Conditions	Compounds Extracted	Reference
Not specified	Closure	Acetonitrile, refluxed 8 h	4-(1,1-Dimethyl-propyl)-phenol Sulfur 2,6-Di-tert-butyl-[1,4]benzoquinone Furan-2-yl-(5-hydroxymethyl-furan-2-yl)-methanol 2-Bromo-4-(1,1-dimethylpropyl)-phenol	[23]
Not specified	Lyophilization stopper	Acetonitrile 48-h room temperature	Butylhydroxytoluene (BHT)	[24]
Synthetic polyisoprene	Injection site septum	Water pH 3 (KCl/HCl) 1 h at 121°C	Tinuvin 770 Tinuvin RS Tinuvin-related substances Decane	[25]
Synthetic polyisoprene	Injection site septum	Ethanol/water 3 days at 55°C	Bis(2,2,6,6-tetramethyl-4-piperidyl) sebacate (Tinuvin 770) Tinuvin RS Tinuvin-related substances Hexadecanoic acid Stearic acid Decanedioic acid,1-(2,2,6,6-tetramethyl-4-piperindyl) mono ester Triisopropanolamine Tetradecanoic acid Irgafos 168 degradation product 2,6-Di-t-butyl phenol	[25]
Synthetic polyisoprene		Water or NaCl 0.9% solution for 1 h at 121°C	Aniline Diphenylguanidine Dedenzylamine Triisopropanolamine	[26]

Not only organic components are released from elastomeric materials. Zuccarello et al. [27] evaluated the inorganic composition (metals) of a gray butyl rubber stopper (Lyo/Serum) and analyzed the components leached into a vehicle solution containing 0.9% NaCl, 0.8% polysorbate 80, and 0.1% sodium phosphate, at pH 4, exposed to rubber for 1 week at 50°C. Table 7.22 shows the metals present in the analyzed rubber and Table 7.23 shows those released

TABLE 7.22 Metal Profile of a Butyl Rubber (Gray) Lyo/Serum Stopper

Metal	Amount (%)	Metal	Amount (%)
Al	0.13	Ca	0.064
Mg	0.93	Fe	0.033
Ti	0.81	K	0.012
	(μg/g)		(μg/g)
Na	87	La	2.0
Zn	81	B, Nd	1.7
P	43	V	1.3
Sr	25	Pb	1.1
Si	14	Ni, Y	1.0
Mn	7.8	Li, Sn, Zr	0.7
Ba	3.6	Dy, Pr, Gd	0.4
As	3.0	Sm	0.3
Cr	2.6	Co, Er, Th	0.2
Cu	2.3	Ga, U	0.1

Adapted from Reference 27.

TABLE 7.23 Rubber Components Leached into a Vehicle Solution Containing 0.9% NaCl, 0.8% Polysorbate 80, and 0.1% Sodium Phosphate, pH 4, after 1-Week Storage at 50°C

Element	Concentration (mg/L)	Element	Concentration (mg/L)
Al	3800	Cr	3
Zn	1400	Mg	3
Ti	180	Si	3
Na	49	V	3
Ca	2	Cu	1
Fe	31	Sr	1
K	23	Zr	1
P	10		

Adapted from Reference 27.

TABLE 7.24 Elemental Composition of Rubber Stoppers Based on the Element Concentration in the Aqueous Extract, According to Mannermaa et al. [28]

	Amount (μg/g)		
	Sample 1	Sample 2	Sample 3
Element	Old Formulation of a Bromobutyl Siliconized Rubber	New Formulation of a Bromobutyl Siliconized Rubber	New Formulation of a Chlorobutyl Siliconized Rubber
Ti	500	1,600	200
Fe	7,500	8,000	4,500
Cu	10	10	–
Zn	50	5	20,000
Br	100	7500	50
Pb	2	–	–

into the vehicle. Although containing several metal elements as constituents and as impurities, the extractability test revealed that only a few are really leached, and only Al and Zn were of significance.

Mannermaa et al. [28] determined the inorganic composition of extracts obtained from three different types of rubber stoppers used to seal parenteral solutions. After immersing the stoppers in 0.9% NaCl solution and autoclaving at 121°C for 15 minutes, the analysis of the aqueous extract was carried out by proton-induced X-ray emission. Table 7.24 presents the elemental composition of the rubber stoppers investigated in microgram per gram dry weight. The concentration of zinc varied from sample to sample by a factor of 4000 (5–20,000 ppm). Ti, Fe, and Br were present in all stoppers and even Pb (2 ppm) was found in one extract.

Because primary packaging for sterile products may consist of a container/closure system, different configurations of glass containers and elastomeric closures were evaluated for aluminum extractability and total organic carbon (TOC) release in WFI during the sterilization procedure [29]. Combinations of four glass vials (Type I glass, SiO_2 coated and not coated) and four rubber closures (coated and not coated) were heated at 121°C for 40 minutes. The results in Table 7.25 show that the only possible source of aluminum is the glass vial. While aluminum was not detected in extracts from vials A and D, regardless of the closure type, aluminum was found in the extracts of glass vials B and C. Vials A and D are SiO_2 coated, and vials B and C are conventional Type I glass vials. On the other hand, as TOC cannot be originated from the glass vial, it seems that the coating does not impede the release of organic carbon from rubber closures since, in some extracts of coated closures, TOC was even higher than in the extracts of uncoated closures.

In another study conducted with the entire container/closure system, five different rubber stoppers commonly used for injectable products combined with a Type I flint glass vial were tested [30]. In this case, three vehicles,

TABLE 7.25 Aluminum and Total Organic Carbon (TOC) Released into Water for Injection from Four Different Types of Container/Closure Systems Submitted to a Sterilization Procedure at 121°C for 40 Minutes

	Glass Vial			
Rubber Closure	A	B	C	D
		Al (μg/L)		
West	<1	12.5	130	<1
West coated	<1	4.9	93	<1
Helvoet	<1	5.0	58	<1
Helvoet coated	<1	51.5	6	<1
		TOC (μg/L)		
West	506	551	363	509
West coated	404	457	426	356
Helvoet	1383	888	686	740
Helvoet coated	–	1140	976	1023

Adapted from Reference 29.

TABLE 7.26 Total Extractables and Phenolic Antioxidant Amounts Released Rubber Stoppers after Thermal Desorption GC-MS Stored at 25°C and 60% RH for up to 24 Months

	Vehicle					
	Citrate Buffer pH 3		Phosphate Buffer pH 8		Castor Oil/Ethanol	
Stopper	Extractable	Amount (mg/mL)	Extractable	Amount (mg/mL)	Extractable	Amount (mg/mL)
A	–		Silicon	16–24	–	
B	–		Silicon	15–23	–	
C	–		Silicon	13–23	–	
D			Silicon	15–23		
	Isopropanol	19–21	Isopropanol	18–35		
	t-Butyl-alcohol	13–22	t-butyl-alcohol	16–28	t-Butyl-alcohol	9–19
E	–		Silicon	15–24	–	

Adapted from Reference 30.

representative of a wide range of formulations, were chosen. The stoppers were kept in contact with the vehicles, citrate buffer pH 3 with 20% w/v sulfobutylether-β-cyclodextrin, phosphate buffer pH 8, and polyoxyethylated castor oil/dehydrated alcohol 50/50 v/v, for up to 24 months at 25°C and 60% relative humidity. Several potential extractables were evaluated, but, as a threshold of 10 mg/L was established, only silicon, isopropanol, and t-butyl alcohol were measured above this limit (Table 7.26). Silicon, found in all vials filled with phosphate buffer pH 8, was a result of contact between this buffer and glass. Isopropanol and t-butyl alcohol, measured in extracts of stopper D, were probably solvents used in rubber compounding.

CONCLUDING REMARKS

Elastomeric closures are made of a complex mixture of many ingredients to give them the necessary feature of resealing after needle perforation and withdrawal. The chief component is the elastomer, and the others correspond to vulcanizing agents, accelerators, activators, antidegradants, plasticizers, fillers, and pigments. They comprise organic and inorganic substances, which can be leached into a packaged product. The leaching depends mainly on the nature of the formulation and the time or contact. It is possible, however, that even when there is no contact, the release of rubber constituents will occur. Lyophilized powders adsorb volatile compounds released into the headspace of vials sealed with rubber closures, which are noticeable due to a turbidity of the reconstituted solution.

REFERENCES

1. Smith EJ, Nash RJ. Elastomeric closures for parenterals. In: *Pharmaceutical Dosages Forms: Parenteral Medications*, Vol. 2, Avis KE, Lachman L, Lieberman HA (eds.). Marcel Dekker, New York, 1986, pp. 155–215.
2. A Parenteral Drug Association Task Force Report. Effects of gamma irradiation on elastomeric closures. J Parenter Sci Technol 46(2), 1–13, 1992.
3. Jenke D. Extraction of stearate salts from plastic materials used in pharmaceutical applications. PDA J Pharm Sci Technol 64, 200–210, 2010.
4. Kikouama JRO, Konan KL, Katty A, Bonnet JP, Balde L, Yagoubi N. Physicochemical characterization of edible clays and release of trace elements. Appl Clay Sci 43(1), 135–141, 2009.
5. Accardi-Dey A, Gschwend PM. Reinterpreting literature sorption data considering both absorption into organic carbon and adsorption onto black carbon. Environ Sci Technol 37, 99–106, 2003.
6. Food and Drug Administration. CRF 21- Sec. 178.3297, Colorants for polymers. Available at http://www.accessdata.fda.gov/scripts/cdrh/cfdocs/cfcfr/CFRSearch.cfm?fr=178.3297.
7. Moura JF. Assessment on the presence of polycyclic aromatic hydrocarbons (PAHs) in injectable pharmaceutical formulations, MSc thesis, Federal University of Santa Maria, Brazil, 2009.
8. Norwood DL, Prime D, Downey BP, Creasey J, Sethi SK, Haywood P. Analysis of polycyclic aromatic hydrocarbons in metered dose inhaler drug formulations by isotope dilution gas chromatography/mass spectrometry. J Pharm Biomed Anal 13(3), 293–304, 1995.

9. Guidance for Industry, Container Closure Systems for Packaging Human Drugs and Biologics, Federal Drug Administration, 1999.
10. Pharmaceutical Package Integrity, Technical Report 27, PDA-Parenteral Drug Association, 1998, USA.
11. ISO 8871-1. Elastomeric parts for parenterals and for devices for pharmaceutical use—part 1: extractables in aqueous autoclavates. 2003.
12. Milano CJ, Bailey LC. Evaluation of current compendial physiochemical test procedures for pharmaceutical elastomeric closures and development of an improved HPLC procedure. PDA J Pharm Sci Technol 53(4), 202–210, 1999.
13. Delaunay-Bertoncini N, van der Wielen FWM, de Voogt P, Erlandsson B, Schoenmakers PJ. Analysis of low-molar-mass materials in commercial rubber samples by soxhlet and headspace extractions followed by GC-MS analysis. J Pharm Biomed Anal 35, 1059–1073, 2004.
14. Jaehnke RW, Kreuter J, Ross G. Interaction of rubber closures with powders for parenteral administration. J Parenter Sci Technol 44(5), 282–288, 1990.
15. Zweiben C, Shaw JA. Use of thermal desorption GC-MS to characterize packaging materials for potential extractables. PDA J Pharm Sci Technol 63, 353–359, 2009.
16. Jaehnke RWO, Kreuter J, Ross G. Content container interactions—the phenomenon of haze formation on reconstitution of solids for parenteral use. Int J Pharm 77(1), 47–55, 1991.
17. Sendo T, Hirakawa M, Makino K, Nakashima K, Kataoka Y, Oishi R. Particulate contamination of lyophilized amphotericin B preparation during reconstitution process. J Clin Pharm Ther 26(2), 87–91, 2001.
18. European Commission Scientific Committee on Consumer Products. Opinion on 2-Mercaptobenzothiazole (SCCP/0883/05), 2005.
19. Reepmeyer JC, Juhl YH. Contamination of injectable solutions with 2-mercaptobenzothiazole leached from rubber closures. J Pharm Sci 72(11), 1302–1305, 1983.
20. Corredor C, Tomasella FP, Young J. Drug interactions with potential rubber closure extractables: identification of thiol–disulfide exchange reaction products of captopril and thiurams. J Chromatogr A 1216(1), 43–48, 2009.
21. Kim-Kang H, Gilbert SG, Malick AW, Johnson JB. Methods for predicting migration to packaged pharmaceuticals. J Pharm Sci. 79(2), 120–123, 1990.
22. Paskiet DM. Strategy for determining extractables from rubber packaging materials in drug products. PDA J Pharm Sci Technol 51(6), 248–251, 1997.
23. Zhang F, Chang A, Karaisz K, Feng R, Cai J. Structural identification of extractables from rubber closures used for pre-filled semisolid drug applicator by chromatography, mass spectrometry, and organic synthesis. J Pharm Biomed Anal 34(5), 841–849, 2004.
24. Wu SL, Wang YJ, Hu J, Leung DM. The detection of the organic extractables in a biotech product by liquid chromatography on-line with electrospray mass spectrometry. PDA J Pharm Sci Technol 51(6), 229–237, 1997.
25. Story J, Gill M, Liu N, Hua Y, Jenke D. Generation of the extractables profile for an elastomeric material and investigation of the accumulation behavior of targeted leachables including bis(2,2,6,6-tetramethyl-4-piperidyl) sebacate (tinuvin 770) and a related substance. PDA J Pharm Sci Technol 64, 101–111, 2010.

26. Jenke DR. Utilization of extraction profiles to estimate the accumulation of extractables from polymeric materials. J Appl Polym Sci 63, 843–848, 1997.
27. Zuccarello DJ, Murphy MP, Meyer RF, Winslow PA. A comprehensive approach for the determination of extractable and leachable metals in pharmaceutical products by inductively-coupled plasma. PDA J Pharm Sci Technol 63, 339–352, 2009.
28. Mannermaa JP, Räisänen J, Hyvönen-Dabek M, Spring E, Yliruusi J. Use of proton-induced X-ray emission (PIXE) analysis in the evaluation of large volume parenteral rubber stoppers. Int J Pharm 103, 125–129, 1994.
29. White F, Koberda M, Chilamkurti R. A systematic approach for screening glass containers and elastomeric closures for use with parenteral solutions. PDA J Pharm Sci Technol 62, 157–176, 2008.
30. Corredor CC, Haby TA, Young JD, Shah PA, Varia SA. Comprehensive determination of extractables from five different brands of stoppers used for injectable products. PDA J Pharm Sci Technol 63, 527–536, 2009.

8

DELIVERY SYSTEMS AND FILTERS

Pharmaceutical formulations have direct contact with plastic materials through delivery sets, transfer tubing, and devices, as well as during the phases of their production, via gaskets, filters, and transportation.

Although there is no specific pharmacopeial prescription for testing these materials, they should present the same profile stipulated for containers and closures. Although several materials can be used, flexibility is a required feature; therefore, they are made mostly of polyvinyl chloride (PVC), silicone, or elastomeric materials.

It is more difficult to generalize about the leachability of plastic tubing materials than it is for plastic containers or closures themselves since it depends not only on the nature of the plastic material but also on the flow rate, temperature, solvent, and time of contact with the tubing.

Studies conducted by different authors on the release of chemical substances from medical devices, mainly those used for infusing solutions, show that these are potential sources of contamination for pharmaceutical formulations. One of the most studied is di-2-ethylhexylphthalate (DEHP), the same plasticizer found in PVC infusion bags to give flexibility. The same concerns about the use of PVC bags for the storage of lipids or lipophilic formulations are valid for tubing.

Sources of Contamination in Medicinal Products and Medical Devices,
First Edition. Denise Bohrer.
© 2013 by John Wiley & Sons, Inc. Published 2013 by John Wiley & Sons, Inc.

TABLE 8.1 DEHP Loaded in Infusion Solutions after Exposure to PVC Infusion Lines

Sample	Volume (mL)	No. of Samples	DEHP (μg/mL)	Range DEHP (μg/mL)
Amino acid–glucose	140	06	0.31 ± 0.56	0.0–1.05
Lipid emulsion	24	10	422.78 ± 47.39	329.15–490.40
Midazolam	24	03	0.90 ± 0.38	0.55–1.30
Propofol infusion	10	10	654.87 ± 96.49	423.85–736.10
Fentanyl	28.8	10	3.63 ± 0.92	1.95–5.05
Imipenem	8	03	0	−0.10–0.05

Adapted from Reference 1.

8.1 DELIVERY SYSTEMS MADE OF PVC

Table 8.1 presents the amount of DEHP leached from typical PVC infusion lines from a delivery set of 2.25 m kept at 27°C. A sample volume from 8 to 140 mL was perfused through the lines over 24 hours and collected for analysis. It was possible to see that while amino acids did not promote the migration of DEHP from PVC tubing, the lipophilic lipid emulsion and propofol promoted the extraction of large amounts of DEHP into the preparations.

An investigation of DEHP extractability from PVC tubing into a large variety of pharmaceutical formulations used for different purposes was carried out by Haishima et al. [2]. The authors divided the IV drugs into five groups according to the properties of their active ingredients and additives. Group 1 included drugs that were practically insoluble or insoluble in water and contained additives such as surfactants, oils, glycerine, ethanol, and benzyl alcohol. Group 2 included drugs equally insoluble in water but soluble in acidic or basic solutions. Formulations slightly soluble in water were placed in group 3, and formulations very soluble or freely soluble in water were classified in groups 4 and 5. The difference between groups 4 and 5 was the presence of additives suspected to induce DEHP migration in group 4. Tests were carried out by placing the drug formulations into a PVC tubing of 10 cm in length with a 2.13 mm inner diameter for 1 hour at room temperature under shaking. As shown in Table 8.2, the products included in group 1 promoted the release of large amounts of DEHP, with the exception of insulin and dinoprost, probably due to the absence of oleaginous excipients. No significant DEHP release was observed by most of the other drugs assigned to groups 2–5, with the exception of human serum albumin and antithrombin III (both group 4), and phenytoin solution (group 2), probably due to the presence of propylene glycol and ethanol in the formulation. These results confirm the affinity of DEHP to lipophilic media.

Due to their utilization modus, many studies conducted with tubing materials are essentially kinetic, where tubing length, time of contact, flow rate, and temperature are important parameters. Kambia et al. [3] studied the kinetics of DEHP migration into total parenteral nutrition (TPN) solutions from PVC

TABLE 8.2 DEHP Released from PVC Tubing into Intravenous Drug Preparations after 1-Hour Contact at Room Temperature

Active Ingredient	Concentration	Additives	DEHP (μg/L)	SD (μg/L)
Group 1				
Cyclosporin	500 μg/mL	Polyoxythene castor oil, ethanol	27,363.9	384.8
Tacrolymus hydrate	10 μg/mL	Absolute ethanol, HCO-60	4,091.9	31.9
Propofol	10 mg/mL	Soybean oil, concentrated glycerin, egg yolk lecithin, edetate	19,451.2	852.5
Flurbiprofen axetil	10 mg/mL	Soybean oil, egg yolk lecithin, concentrated glycerin	17,838.5	821.6
Vitamins—fat soluble	–	Sodium citrate, sodium pyrosulfite, sodium thioglycollate, HCO-60, benzyl alcohol, polysorbate 80	1,157.1	5.1
Menatetrenone	5 mg/mL	Aminoethylsulfonic acid, sesame oil, soybean lecithin, D-sorbitol, concentrated glycerin	8,457.5	62.9
Insulin human	40 units/mL	Concentrated glycerin, m-cresol	281.6	6.0
Dinoprost	2 mg/L	–	185.8	17.3
Miconazole	1 mg/L	HCO-60	30,098.3	423.3
Diazepam	5 mg/L	Propylene glycol, ethanol, benzyl alcohol, sodium benzoate, benzoic acid	2,008.8	257.6
Prednisolone sodium succinate	10 mg/L	Sodium carbonate, sodium hydrogenphosphate, sodium dihydrogenphosphate	915.6	182.3
Prednisolone sodium succinate	1 mg/L	Sodium carbonate, sodium hydrogenphosphate, sodium dihydrogenphosphate	407.1	2.4
Group 2				
Famotidine	20 mg/L	L-Aspartic acid, D-manitol	166.0	0.9
Droperidol	2.5 mg/mL	p-Oxymethil benzoate, p-oxypropyl benzoate	171.0	0.6
Droperidol	50 μg/mL	p-Oxymethil benzoate, p-oxypropyl benzoate	167.4	24.6
Sivelestat sodium	1 mg/L	D-Mannitol, sodium hydroxide, propylene glycol, ethanol	885.7	10.6
hydrate Phenytoin	50 mg/L			

(*Continued*)

TABLE 8.2 (*Continued*)

Active Ingredient	Concentration	Additives	DEHP (µg/L)	SD (µg/L)
Methotrexate	0.2 mg/mL	Sodium chloride, sodium hydroxide	372.8	6.8
Haloperidol	5 mg/mL	Glucose, lacit acid, sodium hydroxide	50.6	2.5
Epinephrine	0.25 mg/mL	Chorobutanol, sodium hydrogen sulfite, hydrochloric acid, sodium chloride	290.3	24.6
Group 3				
Methilergometrine maleate	0.2 mg/mL	–	462.7	4.2
Vecuronium bromide	2 mg/mL	D-Mannitol	192.7	1.5
Panipenem betamipron	5 mg/mL	–	237.0	1.2
Mynocycline hydrochloride	1 mg/mL	–	150.0	8.9
Nicardipine hydrochloride	0.1 mg/mL	D-Sorbitol	211.6	24.0
Bromhexine hydrochloride	2 mg/mL	Glucose	174.9	23.7
Ceftazidime	10 mg/mL	Sodium carbonate	301.0	0.5
Fluconazole	1 mg/mL	–	210.5	0.2
Aspoxicillin	50 mg/mL	Sodium chloride	296.7	2.6
Carbazochrome sodium sulfonate	0.05 mg/mL	Sodium hydrogensulfite, D-sorbitol, propylene glycol	246.1	3.0
Group 4				
Oxytocin	0.01 units/mL	Chlorobutanol	423.1	0.8
Hydroxyzine hydrochloride	0.05 mg/mL	Benzyl alcohol	430.8	144.4
Ranitidine hydrochloride	0.1 mg/mL	Phenol	197.9	29.5
Human immunoglobulin G	50 mg/mL	D-Sorbitol	243.9	14.3

Panthenol	250 mg/mL	Benzyl alcohol	412.1	18.2
Human serum albumin	250 mg/mL	Sodium N-acetyl tryptophan, sodium caprylate, sodium hydrogen carbonate	10,080.8	84.1
Human antithrombin III	25 units/mL	Sodium chloride, sodium citrate, D-mannitol	2,008.2	21.8
Nitroglycerin	0.5 mg/mL	D-Mannitol	267.6	8.9
Sulpyrine	2.5 mg/mL	Benzyl alcohol	302.8	3.8
Erythromycin lactobionate	2.5 mg/mL	Benzyl alcohol	92.2	0.7
Clindamycin phosphate	3 mg/mL	Benzyl alcohol	274.9	4.0
Group 5				
Imipenem cilastatin sodium	5 mg/mL	Sodium hydrogencarbonate	205.1	1.6
5% Glucose	50 mg/mL	–	284.6	4.8
Ferric oxide, saccharated	0.4 mg/mL	–	244.5	5.5
Maltose	–	Sodium chloride, magnesium chloride, potassium dihydrogenphosphate, sodium acetate	262.8	5.0
Atropine sulfate	0.5 mg/mL	–	200.7	5.1
Ampicillin sodium	10 mg/mL	–	262.3	6.8
Aminophyline	0.5 mg/mL	Ethylenediamine	301.1	4.0
Fosfomycin sodium	20 mg/mL	Glucose solution	289.6	6.7
Calcium gluconate	85 mg/mL	–	179.4	4.3
Cefazolun sodium hydrate	10 mg/mL	–	215.1	0.9
Amino acids, electrolytes	–	Sodium hydrogen sulfite	328.5	5.0
Suxamethonium chloride	2 mg/mL	–	228.6	2.1
Ioversol	320 mg/mL	–	404.0	79.5
l-Isoprenaline hydrochloride	1 μg/mL	Sodium hydrogen sulfite, L-cysteine hydrochloride	326.3	8.6

Adapted from Reference 2.

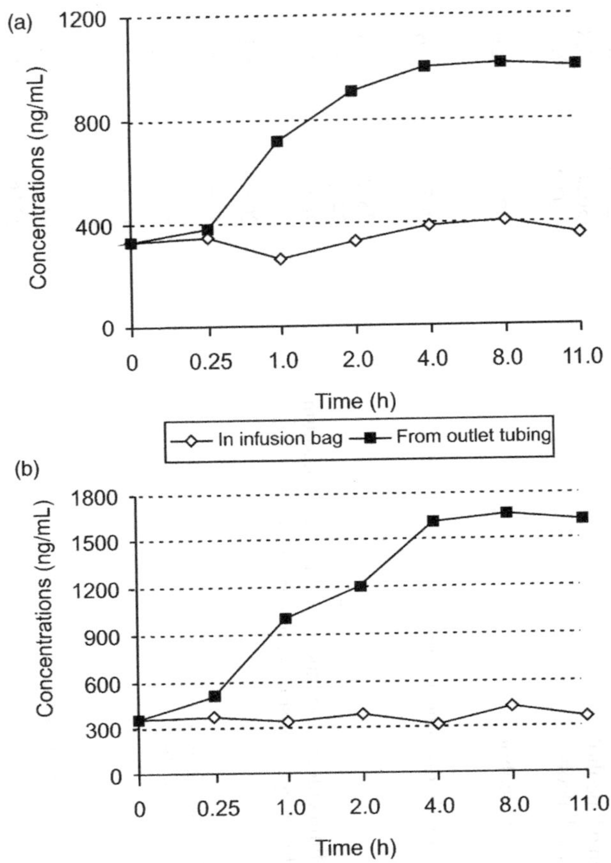

FIGURE 8.1 Comparative kinetics of DEHP leachability during simulated infusion of TPN 24 hours after reconstitution of the preparation ($n = 2$). (a) Infusion volume of 2200 mL, flow rate 177 mL/h, lipid concentration 3.85%. (b) Infusion volume of 650 mL, flow rate 46 mL/h, lipid concentration 1.85%. Reproduced from Kambia K, Dine T, Gressier B, Bah S, Germe AF, Luyckx M, Brunet C, Michaud L, Gottrand F. Evaluation of childhood exposure to di(2-ethylhexyl) phthalate from perfusion kits during long-term parenteral nutrition. Int J Pharm 262(1-2), 83–91, 2003.

tubing. They quantified the amount of DEHP leached into two types of emulsions 24 hours after preparation and storage at 4°C. The authors concluded that the extraction depended on the lipid content of the formulations and the flow rates. Figure 8.1 shows the DEHP concentration time course obtained from bags and outlet tubing during 10–11 hours' simulated infusion of TPN after 24 hours' storage at 4°C.

The influence of the temperature on the leachability of DEHP from PVC infusion lines is an important issue because in newborn intensive care units, the temperature is 31°C and may reach up to 37°C in the incubator. Typical newborn

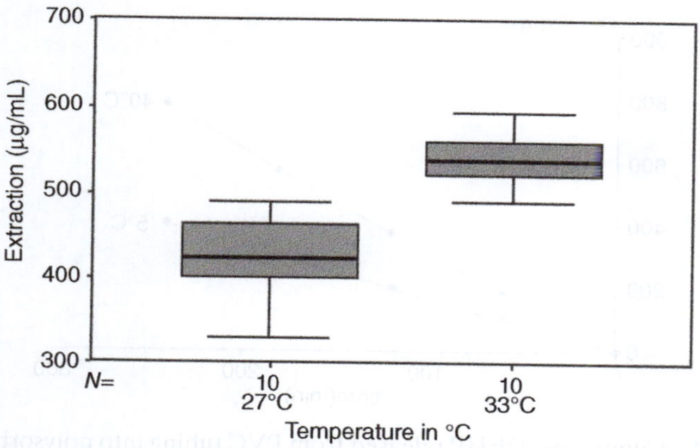

FIGURE 8.2 Extraction of di-2-ethylhexylphthalate (DEHP) dependent on temperature [4].

24 hours' fat infusion was perfused in PVC lines kept at 27 and 33°C [4]. The infusion was collected every 4 hours and analyzed for DEHP. The results, presented in Figure 8.2, showed that the increase of the temperature from 27 to 33°C was responsible for an increase of 28% in the release of DEHP. Calculations on the amount of DEHP administered to a typical 2-kg newborn through the nutrition solution gave values reaching 13 mg/day. Moreover, depending on the use of additional medications or the administration of blood products, this value could rise to >20 mg, which corresponds to >10 mg/kg of body weight.

Similar testing was carried out with a solution of polysorbate 80, a formulation vehicle known for its ability to promote the release of DEHP from PVC [5]. The cumulative amount of DEHP leached from 1 m PVC tubing into a 2 mg/mL polysorbate 80 solution reached 400 μg in 4 hours' circulation, 66-mL/h flow rate, at 5°C and practically doubled when the temperature was kept at 40°C (Fig. 8.3).

The leachability of DEHP by the action of the formulation vehicles, polysorbate 80 and Cremophor EL, from PVC lines shows the same behavior as from PVC bags, as already discussed under plastic containers. Although not tested under exactly the same conditions, the graph in Figure 8.4 shows that Cremophor EL is more able to promote the release of DEHP from PVC infusion lines than polysorbate 80.

Because Cremophor EL displays this great capacity to leach DEHP from PVC, Hanawa et al. [6] investigated the role of sodium chloride and glucose in the interaction Cremophor EL–DEHP. The experiments comprised the dissolution of Cremophor EL in water, sodium chloride, and glucose solutions and the dissolution of Cremophor EL in solutions with different concentrations of glucose. The release seemed to be affected by the presence and concentration of glucose. Compared with pure water and saline solution, the release of DEHP in 50% glucose solution was much lower (Fig. 8.5). Even

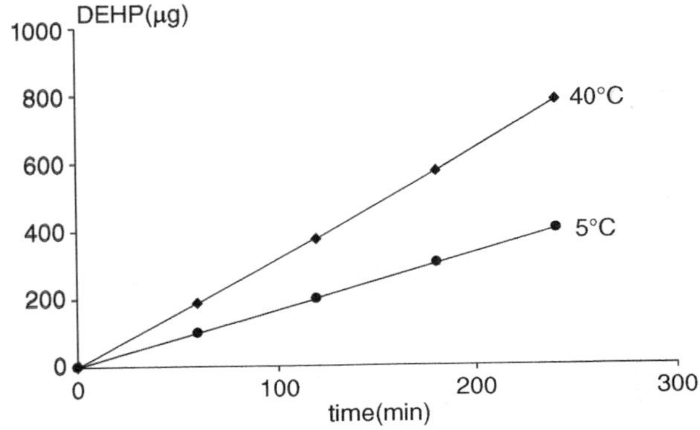

FIGURE 8.3 Cumulative DEHP released from PVC tubing into polysorbate 80 solution (2 mg/mL) at 5 and 40°C. Tubing length: 1 m. Flow rate 66 mL/h (adapted from Reference 5).

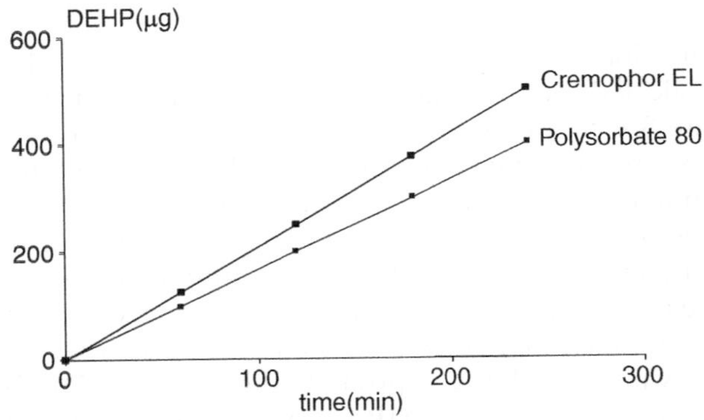

FIGURE 8.4 Cumulative DEHP released from PVC tubing into Cremophor EL and polysorbate 80 solutions at room temperature. Tubing length: 1 m. Cremophor EL: concentration: 3 mg/mL and flow rate 60 mL/h. Polysorbate 80: concentration: 2 mg/mL and flow rate 66 mL/h (adapted from References 5 and 6).

more diluted solutions of glucose were able to reduce the amount released (Fig. 8.6). While a 10 mg/mL solution of Cremophor EL in pure water leached approximately 1000 μg DEHP in 4 hours' perfusion, in 10% glucose, the amount was reduced to 800 μg DEHP. The authors attributed the reduction to a molecular interaction between glucose and Cremophor EL, which would occur through the restriction of Cremophor EL mobility due to an increase in microscopic viscosity.

To overcome the problem of DEHP leaching into parenteral infusions containing lipophilic components, a triple-layered tubing material has been used. This type of tubing is made of a PVC outer layer, while its innermost layer, which comes into contact with the drug solution, is made of inert

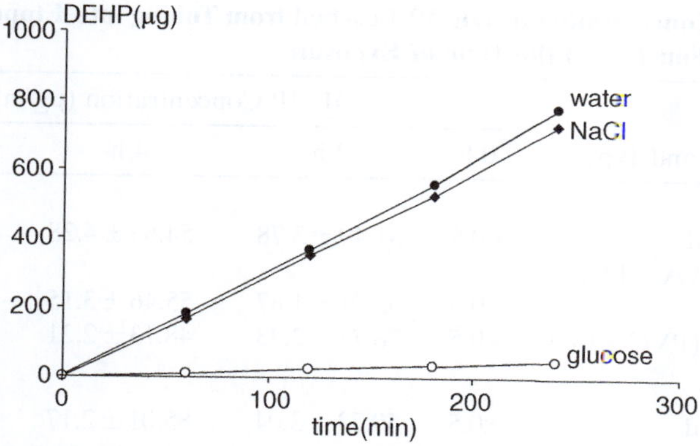

FIGURE 8.5 Cumulative DEHP released from PVC tubing into Cremophor EL solution in water (●), in 0.9% NaCl (◆), and 50% glucose (○). Cremophor EL concentration: 5 mg/L in water and glucose, and 6 mg/L in NaCl (adapted from Reference 6).

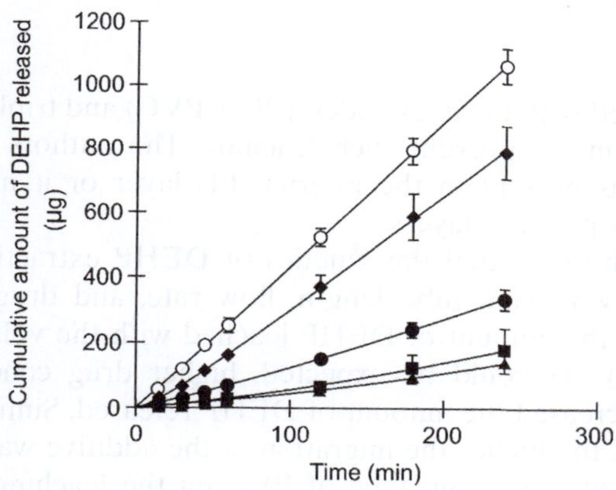

FIGURE 8.6 Effect of glucose concentration on the release of DEHP from PVC tubing into HCO60 solution: (○) water for injection (WFI), (◆) 10% glucose, (●) 30% glucose, (■) 50% glucose, (▲) 70% glucose. The concentration of HCO60 is 10 mg/mL. Data are expressed as mean ± SD ($n = 3$) [6].

polyethylene (PE). In spite of this arrangement, it has been shown that, depending on the preparation, DEHP from the PVC outer layer may be released into the infusion solution.

Boithias et al. [7] showed that DEHP is rapidly leached from PVC, coextruded, and also triple-layered IV tubing by action of an etoposide infusion solution. Table 8.3 presents the concentration of DEHP leached from tubing into etoposide infusion solutions as a function of time of exposure for both tubing lengths. The results showed that fast and considerable leaching of DEHP occurred even when the PVC component had no direct contact with the solution. DEHP leaching was greatest for the tubing made only of PVC,

TABLE 8.3 Concentration of DEHP Leached from Tubing into Etoposide Infusion Solutions as a Function of the Time of Exposure

	DEHP Concentration (μg/mL)			
Tubing Length and Type	0 h	2 h	4 h	6 h
25 cm				
Triple layered (PVC + EVA + PE)	<0.5	31.49 ± 3.78	54.83 ± 4.28	61.99 ± 1.25
PVC	<0.5	33.01 ± 1.87	55.46 ± 3.15	66.10 ± 1.23
Coextruded (PVC + PE)	<0.5	28.77 ± 2.33	48.13 ± 2.21	54.60 ± 2.53
50 cm				
Triple layered (PVC + EVA + PE)	<0.5	59.73 ± 3.04	85.01 ± 2.17	98.72 ± 2.33
PVC	<0.5	73.27 ± 0.96	117.53 ± 3.43	143.41 ± 11.39
Coextruded (PVC + PE)	<0.5	59.61 ± 1.49	82.47 ± 3.56	107.83 ± 9.68

Adapted from Reference 7.
EVA, ethylene vinyl acetate.

but it also occurred with the coextruded (PE + PVC) and triple-layered tubes, despite their claims to prevent such leaching. The authors concluded that either DEHP was present in the internal PE layer or it migrated rapidly through the other polymer layers.

The same authors studied the kinetics of DEHP extraction from triple-layered tubing by varying tube length, flow rate, and drug concentration. Figure 8.7 shows the amount of DEHP leached with the volume of solution infused (250 mL). As could be expected, higher drug concentrations and tubing lengths increased the amount of DEHP released. Similarly, the slower the flow rate was, the higher the migration of the additive was.

The influence of brand and type of PVC on the leaching of DEHP was evaluated using TPN solution and lipid emulsion in perfusion lines of different lengths [8, 9]. For lipid emulsion, not only pure PVC lines were tested but also PVC coextruded with PE and polyurethane (PU), and also a DEHP-free PVC line. TPN circulated through the line at a rate of 42 mL/h and the lipid emulsion at 1 mL/h, both for 24 hours. The temperature was kept between 22 and 28°C to reflect the environmental conditions of use. Despite being obtained from different manufacturers and having different lengths, the amount of DEHP leached from the PVC lines into the TPN solution was about 1.5 μg/mL DEHP (Table 8.4). By contrast, lipid emulsion leached between 13 and 107 μg/mL, despite the lines being shorter. Although TPN contains lipids (10 mL in 1000 mL), the pure lipid emulsion is much more effective in leaching DEHP than when mixed with the other constituents of TPN. Undoubtedly, the mixture lowers the formulation's lipophilicity and therefore the ability of extracting DEHP. Compared to pure PVC, PVC/PU and PVC/PE coextruded lines leached less DEHP. Surprisingly, PVC DEHP-free and PE lines also released DEHP, albeit in minute amounts.

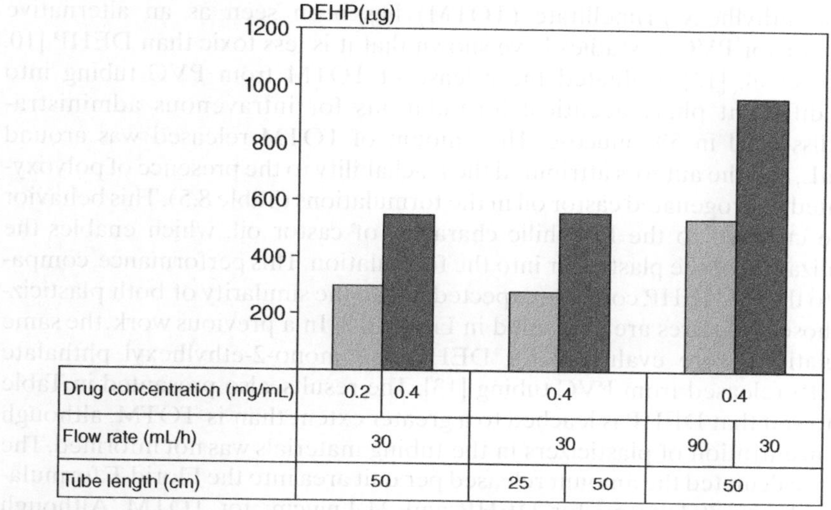

FIGURE 8.7 Cumulative amount of di-2-ethylhexylphthalate (DEHP) leached from polyvinyl chloride tubing after infusion of different drug concentrations, 0.2 and 0.4 mg/mL, different flow rates, 30 and 90 mL/h, and amount of DEHP leached from 25- and 50-cm tubing (adapted from Reference 7).

TABLE 8.4 Leaching of DEHP from Perfusion Lines from Different Materials and Lengths into TPN Solutions (1) and Lipid Emulsions (2) [8, 9]

Sample Code	Composition	Length (cm)	*n*	DEHP (μg/mL)[a]
Air VS (1)	PVC with DEHP	270	5	2.02 ± 0.57
IVAC (1)	PVC with DEHP	250	5	1.45 ± 0.10
Codan (1)	PVC with DEHP	240	5	1.51 ± 0.15
Dosi-flow (1)	PVC with DEHP	230	5	1.36 ± 0.20
Codan V8 (1)	PVC with DEHP	240	5	1.69 ± 0.66
OPVC10 (2)	PVC with DEHP	10	8	13.40 ± 1.18
CPV75 (2)	PVC with DEHP	75	8	39.48 ± 3.12
OPVC (2)	PVC with DEHP	150	8	73.88 ± 3.64
BPVC (2)	PVC with DEHP	150	10	106.70 ± 7.06
OPVCPU (2)	PVC with DEHP/PU	150	12	72.83 ± 2.86
NPVCPE (2)	PVC with DEHP/PE	150	14	41.64 ± 6.42
NDPVC (2)	PVC DEHP-free	150	12	0.23 ± 0.20
NPE (2)	PE	150	5	0.11 ± 0.05

[a] After discounting the residual DEHP contamination: emulsions 0.82 ± 0.42 μg/mL; TPN 0.05–0.69 μg/mL.

Tris(2-ethylhexyl)trimellitate (TOTM) has been seen as an alternative plasticizer for PVC as studies have shown that it is less toxic than DEHP [10, 11]. Ito et al. [12] evaluated the release of TOTM from PVC tubing into three different pharmaceutical formulations for intravenous administration, dissolved in 5% glucose. The amount of TOTM released was around 5 μg/mL, and the authors attributed the leachability to the presence of polyoxyethylated hydrogenated castor oil in the formulations (Table 8.5). This behavior can be credited to the lipophilic character of castor oil, which enables the solubilization of the plasticizer into the formulation. This performance, comparable to that of DEHP, could be expected due to the similarity of both plasticizers, whose structures are presented in Figure 8.8. In a previous work, the same formulations were evaluated for DEHP and mono-2-ethylhexyl phthalate (MEHP) released from PVC tubing [13]. The results, also presented in Table 8.5, showed that DEHP is leached to a greater extent than is TOTM, although the concentration of plasticizers in the tubing materials was not informed. The authors calculated the amount released per unit area into the Florid-F formulation, as being 307 ng/cm^2 for DEHP and 24.4 ng/cm^2 for TOTM. Although MEHP is a hydrolysis product of DEHP, the authors believed that it was already present in the PVC tubing since no plausible event (heat or the presence of enzymes) occurred to justify the hydrolysis reaction.

Wang and Storm [14] showed that DHEP and its derivatives are not the only plasticizers leached from PVC tubing. From three different PVC devices, one feeding tubing, and two suction catheters, 10 different compounds were extracted (see Table A.3, Appendix). Among them, butylhydroxyltoluene (BHT), which is an antioxidant, and bis (2-ethylhexyl) adipate (DOA) and di(isooctyl) phthalate (DOP) were identified in the extracts of all samples, but only DOA and DOP were quantified. Table 8.6 presents the amount of both plasticizers.

Another class of plasticizers, which is also used in medical devices made of PVC, is the class of adipates. Wang and Storm [15] investigated the stability and migration of polyadipates from PVC tubing (Unomedical phthalate-free feeding tubing, A/S Denmark) by aging the tubing in water and phosphate buffer at pH 1.7 at different temperatures. After 70 days of storage at 37°C, the weight loss was less than 0.5%, and only a small amount of adipic acid migrated into both the water and phosphate buffer. The aging in water was also carried out at 70 and 110°C. In this case, significant degradation was observed when the tubing was aged at 110°C. Besides adipic acid, 1,4-butanediol and monomers and oligomers of polyadipate were identified as migration products (see Table A.3, Appendix).

The leaching of polyadipate was evaluated in PVC nasogastric feeding tubes and was compared to the leaching of DEHP itself [16]. Although incubated for 28 days at 37°C, the feeding solutions (Fresenius Energy Fibre) were changed every day and analyzed. The highest leakage occurred in the second week for both plasticizer groups, but even after 4 weeks, there was still some leaching (Table 8.7). The load of polyadipate was at least 10 times lower than

TABLE 8.5 Release of DEHP, MEHP, and TOTM from PVC Tubing into Different Pharmaceutical Formulations for Intravenous Administration, Dissolved in 5% Glucose [12, 13]

Sample	Active Ingredient	Additives	Additive Conc. (mg/mL)	DHEP (μg/mL)	MEHP (μg/mL)	TOTM (μg/mL)
5% Glucose	Glucose	None		0.12	0.56	0.2
Prograf	Tacrolimus hydrate	Polyoxyethylated hydrogenated caster oil	200	4.60	0.39	5.8
		Ethanol	80% (v/v)			
Florid-F	Miconazole	Polyoxyethylated hydrogenated castor oil	1000	53.99	n.d.	4.3
Sandimmun	Cyclosporine	Polyoxyethylated hydrogenated castor oil	650	–	–	4.4
		Ethanol	32.9% (v/v)			
Lastet inj.		Polyethylene glycol 400 (PEG-400)	240	27.04	n.d.	–
		Polysorbate 80	32			
		Ethanol	n.i.			
		Citric acid	n.i.			

n.d., not detected; –, not evaluated; n.i., not informed.

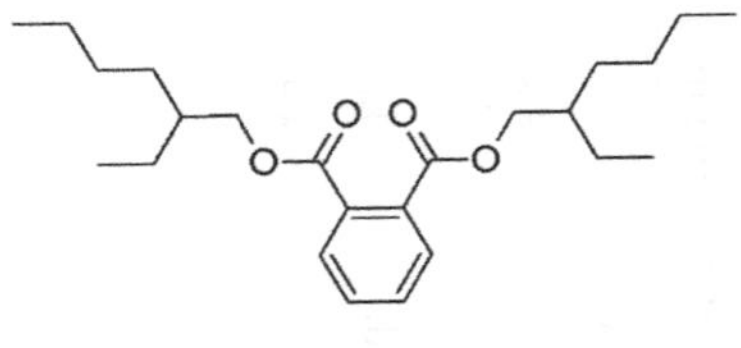

DEHP

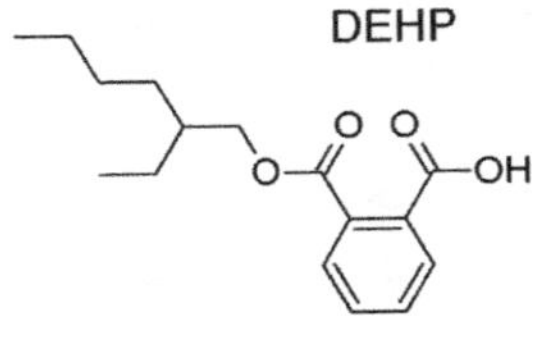

MEHP

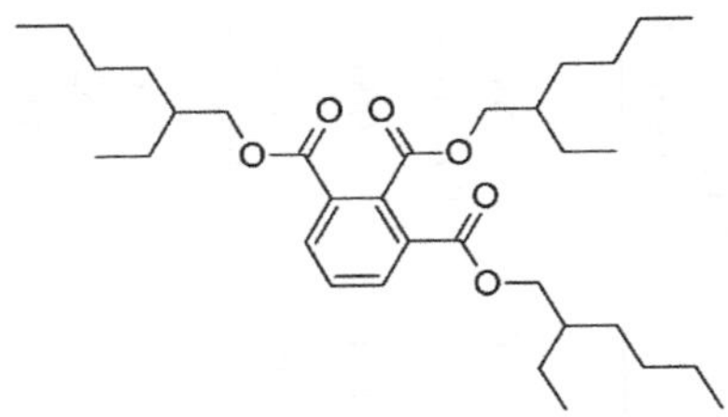

TOTM

FIGURE 8.8 Chemical structures of DEHP, MEHP, and TOTM.

TABLE 8.6 Bis (2-Ethylhexyl) Adipate (DOA) and Di(Isooctyl) Phthalate (DOP) Extracted from PVC Feeding Tubing and Suction Catheters

Device	DOA (mg/g)	DOP (mg/g)
Feeding tubing	3.9	0.30
Suction catheter 1	n.d.	287.3
Suction catheter 2	4.9	0.6

Adapted from Reference 14.
n.d., not detected.

TABLE 8.7 Cumulative Amount of Extracted Plasticizers from PVC Nasogastric Feeding Tubes into a Feeding Solution Kept at 37°C (n = 20)

Day	DEHP (μg)	Polyadipate (μg)
1–7	345.7 ± 111.7	25.0 ± 8.7
8–14	323.6 ± 63.3	30.3 ± 12.0
15–21	227.1 ± 161.9	20.1 ± 9.7
22–28	158.7 ± 118.0	5.1 ± 5.2
1–28 (total amount)	1063.4 ± 308.1	80.5 ± 30.4

Adapted from Reference 16.

that of DEHP, which suggests that it may be an alternative to reduce the DEHP intake from PVC medical devices.

Trimellitates and adipates are not the only alternative plasticizers for phthalates: citrates, dibenzoates, di-octyl terephthalate, and di-isononyl-cyclohexane-1,2-dicarboxylate may also be used for this purpose. The difference is that adipates, citrates, and dibenzoates are more water extractable than DEHP and are around 10–35% more expensive. Moreover, since all of them have had short use histories, human toxicology data are limited or lacking [17].

Although the liquid fluid in contact with a plastic material plays a key role in the leaching of polymer additives, surprisingly, migration may not require a liquid extracting medium to occur. Hill et al. [18] found several volatile compounds in the air passing through respiratory therapy tubing. In two different tubing materials, both made of PCV, more than 30 compounds were leached into the tubing's airspace. Table 8.8 shows the compounds found in two

TABLE 8.8 Volatile Compounds Collected by Direct Thermal Desorption of the PVC Tubing Material at 160°C, and Volatile Compounds Collected from the Air Flowing through the Hospital Tubing Manifold at Room Temperature [18]

	Compound Detected					
Compound	Direct Thermal Desorption	Air through Tubing 1	Control 1[a]	Direct Thermal Desorption	Air Through Tubing 2	Control 2[a]
Butylhydroxytoluene	x	x	–	x	x	–
Diethyl phthalate	x	x	–	x	x	–
Di-(2-ethylhexyl) phthalate	x	x	x	x	x	x
2-Ethylhexanol	x	x	x	x	x	x
p-Nonyl phenol	x	x	–	x	x	–
Benzoic acid	x	–	–	–	–	–
Ionol	x	–	–	x	–	–
Cumyl alcohol	x	–	–	x	–	–
7-Methyl tridecane	x	–	–	–	–	–
2,5-Dimethyl-3-hexanone	x	–	–	–	–	–
1,1-Dimethylethyl-4-methoxy phenol	x	x	–	x	x	–
Ethyl tetramethyl heptane	x	–	–	–	–	–
2-Pentyl furan	x	x	–	–	–	–
Tetradecane	x	–	–	–	–	–
3-Methyl-5-propyl nonane	x	–	–	–	–	–
3-Ethyl-5-methyl hexanone	x	–	–	–	–	–
2,6-Dimethyl octane	x	–	–	–	–	–
Acetophenone	x	x	–	x	x	x

(*Continued*)

TABLE 8.8 (*Continued*)

	Compound Detected					
Compound	Direct Thermal Desorption	Air through Tubing 1	Control 1[a]	Direct Thermal Desorption	Air Through Tubing 2	Control 2[a]
2,2,5-Trimethyl hexane	x	–	–	–	–	–
Pentadecane	x	x	–	–	x	–
Hexadecane	x	x	–	–	x	–
1,2,3-Propanetriol	x	x	–	–	x	–
Heptadecane	x	x	–	–	–	–
3-Methyl-3-hexanol	x	–	–	–	–	–
Hexanal	x	–	–	–	–	–
Decanal	x	x	–	x	x	–
2,2,9-Trimethyl decane	x	–	–	–	–	–
2-Ethyl hexanal	x	–	–	–	–	–
Dimethyl nonane	x	–	–	–	–	–
α-Methyl styrene	x	x	–	–	x	–
Styrene	x	x	–	–	–	–
3-Methyl nonane	x	–	–	x	–	–
Propanal	x	–	–	–	–	–
Cyclohexanone	x	–	–	–	–	–
2,2,6-Trimethyl octane	x	–	–	–	–	–
5-Methyl undecane	x	–	–	x	–	–
Acetone	x	x	x	x	x	x
Acetaldehyde	x	–	–	–	–	–

[a] Control: gas line using the same type of gas valve at the manifold but with no plastic tubing.

different sets of experiments: volatile compounds collected by direct thermal desorption of the tubing material at 160°C and volatile compounds collected from the air flowing through the hospital tubing manifold at room temperature. The results showed that DEHP was one among several other volatilized compounds. As the five first compounds listed in Table 8.8 were released from both tubing materials in higher concentrations (they are listed according to the amount released in the thermal desorption experiment), they were quantified in the air that flowed through the tubing lines (Table 8.9). Although DEHP is the most important additive in PVC, the compound volatilized to the greatest extent was BHT. Since no solvent was involved, the leaching may be related to the vapor pressure of the compounds, which is much higher for BHT (9.8×10^{-3} mmHg at 20°C) than for DEHP (6.2×10^{-8} mmHg at 25°C). Although not listed among the additives allowed by the British Pharmacopoeia (BP), BHT has been found in PVC for medical purposes and behaved as one of the potential leachables (see Table A.3, Appendix).

TABLE 8.9 Quantification of Volatile Compounds Released from PVC Tubing Used in Respiratory Therapy [18]

Tubing Lot No.	Compound	Tubing Section 1 (ng/L Air)	Tubing Section 2 (ng/L Air)	Control 1 (ng/L Air)[a]	Control 2 (ng/L Air)[a]
Y6H0127	Butylhydroxytoluene	104	92.7	n.d.	n.d.
	Di-(2-ethylhexyl) phthalate	<16	<16	<16	n.d.
Y6K0025	Butylhydroxyltoluene	87.1	60.7	n.d.	n.d.
	Di-(2-ethylhexyl) phthalate	<16	<16	<16	<16
	Diethyl phthalate	<4.5	<4.5	n.d.	n.d.
	2-Ethylhexanol	152	363	<6.9	<6.9
	p-Nonyl phenol	<7.3	n.d.	n.d.	n.d.

[a] Control: gas line using the same type of gas valve at the manifold but with no plastic tubing.

8.2 DELIVERY SYSTEMS MADE OF OTHER PLASTIC MATERIALS

Jenke et al. [19] studied the composition of extractables from plastic tubing used in pharmaceutical production facilities. Eight tubing materials made of silicone and rubber (Santoprene) were characterized by their extractable substances. The authors investigated not only organic but also inorganic extractables. Extracts were obtained by static experiments using both water and ethanol. The tubing was cut and autoclaved in water at 121°C for 1 hour, or lengths of tubing were filled with 100% ethanol and kept at 55°C for 24 hours. Metal determination was carried out by inductively coupled plasma atomic emission spectroscopy (ICP-AES), which included 29 elements. Besides the elements listed in Table 8.10, Be, Co, Cr, Cd, Se, V, Ge, Pb, and Bi were also measured but were not detected since they either presented a concentration not different from the blank sample or below the lowest quantity determinable (LQD). It is important to mention that the ICP-AES is not a very sensitive technique. LQD values for elements ranging between 0.01 and 0.1 mg/L are relatively high levels for contaminants such as Cd or Be. This means that these elements could possibly be present in the extracts but not detected using this technique. In general, all tested tubing materials presented metals that were extracted either by water or by ethanol, although water, due to its greater capability to interact with metallic ions, promoted higher extraction yields. All tubing materials contained extractable Ca, Mg, Zn, and B. Elevated concentrations of other metals were also extracted from samples 3 and 5, probably due to the tubing's reinforced embedded metallic wire.

TABLE 8.10 Levels of Total Carbon (Organic and Inorganic), and Silicone and Metals Extracted from the Tubing Materials with Water or Ethanol by Heating at 121°C for 1 Hour

Tubing Material	Total Inorganic Carbon (μg/g)	Total Organic Carbon (μg/g)	Si Ethanolic Extract (μg/g)	Si Aqueous Extract (μg/g)	Metal Concentration in Aqueous Extract		
					≤ 0.5 μg/mL	0.5–1.0 μg/mL	>1 μg/mL
Silicone 1	0	14.0	765	101	B, Mg, Zn		
Silicone 2	0.9	38.1	1360	<0.2	Ca	B, Mg, Mn Mo, Ti, Zr. Sn, Zn, Sb, Li, Ag, Ni	
Silicone 3 (embedded wire)	2.3	250	1860	66.0	Ca, Ba, Mn, Mg, Al, Cu		B, Fe, Zn,
Silicone 4	0.2	34.0	1120	130	Ca, B, Fe, Mg, Al, Zn		
Silicone 5 (embedded wire)	0	49.9	1300	87.3	Ca, B, Mn, Fe, Mg, Zn, Sb		
Silicone 6	0	46.9	739	<0.2	Mn, Fe, Mg, Zn, Cu, Sb, Ni	B	Ca
Santoprene 1	10.1	16.6	–[a]	<0.2	Ca, Ba, B, Mg, Zn,		
Santoprene 2	4.6	13.5	Not tested	0	Ca, Mo, Ti, Zr, Mn, Zn, Li, Ag, Ni		Mg

[a] Tube disintegration.

Adapted from Reference 19.

The authors also measured silicon and both organic and inorganic carbon. While inorganic carbon is indicative of the presence of carbonates (earth carbonates as additives), organic carbon is related to organic extractables. As could be expected, Santoprene (rubber) extracts contained higher concentrations of inorganic carbon and little, if any, extractable silicon. On the other hand, the amount of silicone in extracts of silicone polymers was very high. From Table 8.10, it is possible to see that these materials, though all made of silicone, are very different from each other. While about 100 μg/g Si was found in the aqueous extract from sample 1, no measurable Si was detected in the same extracts from samples 2 and 6. As could be expected, silicone extractables were much higher in ethanolic extracts, where 739 μg/g Si from sample 6 was the lowest amount leached.

The analysis of aqueous and ethanolic extracts by techniques such as liquid chromatography–mass spectrometry (LC-MS) and gas chromatography–mass spectrometry (GC-MS) showed that all silicon materials contained essentially the same peaks; that is, the same compounds were extracted. The distribution, however, varied from material to material. The primary organic extractables from silicone tubing were homologous series of silicone oligomers with the structural formula $[(CH_3)_2SiO]_n$. Direct matches with a mass spectrometry (MS) spectral library allowed the identification of oligomers with $n = 5$ to $n = 25$. However, while in sample 3 the peaks were at the lower range of n, the oligomer distribution for sample 2 was shifted toward higher molecular weights, which are less soluble and therefore less extractable. Figure 8.9 illustrates the differences in the ethanolic extracts of materials 2 and 3. Results of Santoprene extracts were quite different from those of silicone. GC-MS chromatograms for water extracts of Santoprene tubing showed peaks associated with C8 and C9 acids, phthalates, and other organic substances. Figure 8.10 shows the CG-MS chromatograms of static ethanol extracts of Santoprene tubing materials and Table 8.11 lists the compounds identified in Figure 8.9, based on matches of MS spectrum peaks.

Cyclodextrins (CDs) are potentially useful formulation components for poorly aqueous-soluble drugs. The hydrophobic interior of CD molecules provides an appealing environment for lipophilic drugs, while the exterior hydrophilic surface enables the aqueous solubility of the molecule. As CDs can solubilize liphophilic drugs, they can also potentially extract small molecules from production materials and administration devices.

The contact of a CD, sulfobutylether cyclodextrin (SBECD) with silicone tubing, promoted the migration of *o*-phenylphenol (*o*-PP) into the solution [20]. Extraction experiments conducted with silicone tubing from different suppliers and of different lots yielded *o*-PP levels between 0.007 and 3.89 μg/cm^2 (0.03–14.4 μg/g). The experimental conditions included rinsing the tubing with water for injection at 80°C and steam sterilizing at 121°C for 30 minutes prior to filling with the SBECD solution, which was then kept at room temperature for 24 hours. Nevertheless, rinsing and sterilizing the material prior to contact with the CD vehicle, even when repeated three times, reduced the

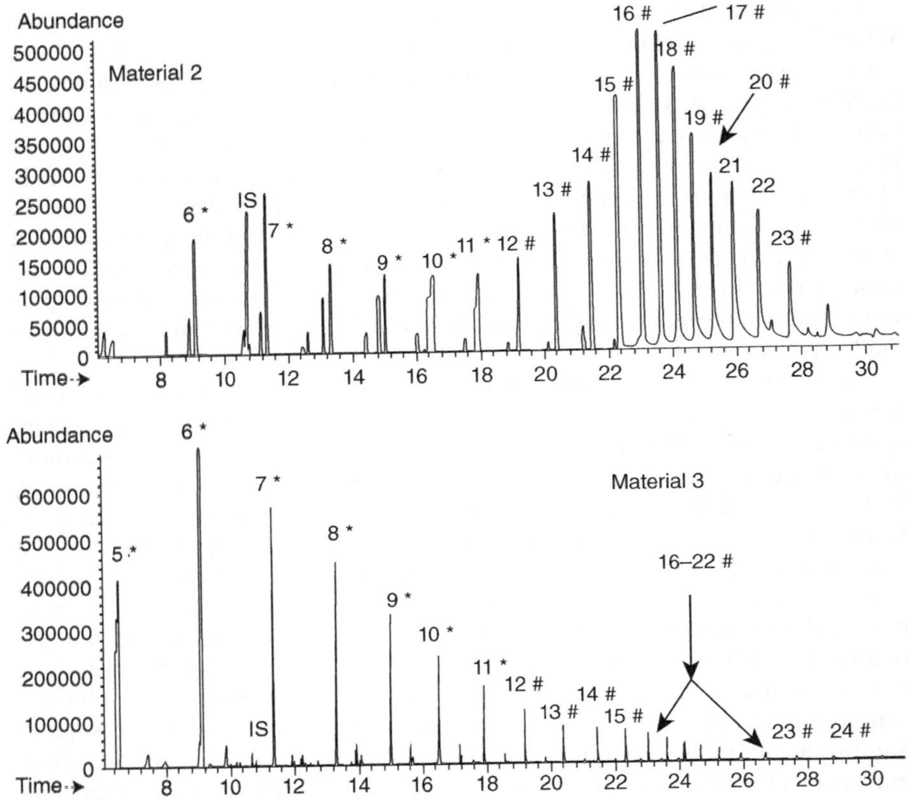

FIGURE 8.9 GC–MS chromatograms of the static ethanol extracts of silicone tubing materials (underivatized). The chromatograms from all the silicone materials were similar (same peaks but different relative sizes), and thus chromatograms from two different silicones are shown. A majority of the peaks are attributable to silicone oligomers. Peaks associated with cyclic oligomers are identified by the number of repeating units, *n*; for example, $[(CH_3)_2SiO]_n$. Peaks denoted with * produced exact compound matches versus a library of mass spectra, while peaks denoted by # produced a library match to the right compound class (cyclic oligomer) but wrong specific oligomer. Small peaks at 7.95, 10.32, and 11.93 minutes were linked to five to seven member linear silicone oligomers. IS, internal standard (dimethyl phthalate) [19].

available amount of *o*-PP by merely 21%. Moreover, in the experiment conducted with pure water, the maximum amount released, after triple extraction, was only 0.001 μg/cm². These data are shown in Table 8.12. The authors mentioned that, although *o*-PP is used in the rubber industry as an antimicrobial component, its presence in the silicone tubing has no obvious source.

Another CD, 2-hydroxypropyl-beta-cyclodextrin (HPβCD), is also able to extract *o*-PP from silicone tubing [21]. The experiments were conducted with

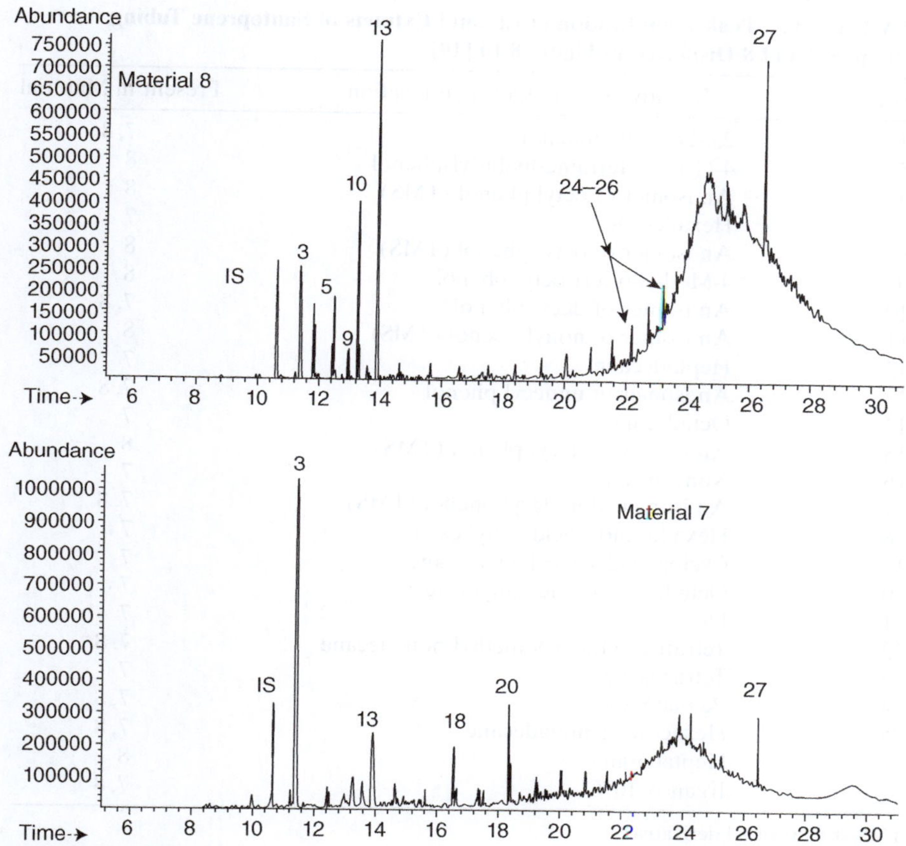

FIGURE 8.10 GC–MS chromatograms of static ethanol extracts of Santoprene tubing materials (underivatized). The chromatograms from these two Santoprene materials were quite different from those of the silicone materials (Figure 8.9). See Table 8–11 for the tentative peak identifications. Only those peaks with recorded spectral library matches are noted for each sample, although retention times and patterns may suggest some additional peak identifications [19]. IS, internal standard (dimethyl phthalate).

different concentrations of HPβCD and contact times, and the results showed that the extraction process is relatively slow, continuously increasing for 24 hours, but without an apparent relationship with the CD concentration. The large standard deviation observed among the replicates (Fig. 8.11) is attributed to an inhomogeneous distribution of *o*-PP in the tubing. Another experiment showed that the presence of a drug competing with *o*-PP for the place in the CD reduces the amount of the extractable compound. The same solution, containing 20% HPβCD and different concentrations of an active pharmaceutical ingredient (API), showed that the higher the API concentration, the

TABLE 8.11 Peak Identification in Ethanol Extracts of Santoprene Tubing Samples 7 and 8 Displayed in Figure 8.10 [19]

Peak #	Tentative Compound Identification	Present in Material
3	2,4-Di-t-butylphenol	7, 8
5	4-(1,1,3,3-Tetramethylbutyl)phenol	8
6	An isomer of octyl phenol (TMS)	8
7	Hexadecane	7
8	An isomer of octyl phenol (TMS)	8
9	4-Methyl-6-tert-octyl phenol	8
10	An isomer of decyl phenol	7, 8
11	An isomer of nonyl phenol (TMS)	8
12	Heptadecane	7
13	An isomer of undecyl phenol	7, 8
14	Octadecane	7
15	An isomer of decyl phenol (TMS)	8
16	Nonadecane	7
17	An isomer of undecyl phenol (TMS)	7, 8
18	Hexadecanoic acid, ethyl ester	7
19	Cyclohexadecane, heneicosane	7
20	Octadecanoic acid, ethyl ester	7
21	Docosane	7
22	Tetratriacontane, 9-methyl-nonadecane	7, 8
23	Tetracosane	7
24	Pentacosane	7, 8
25	Hexacosane, nonadecane	7, 8
26	Heptacosane	8
27	Irganox 1076	7, 8

TMS, trimethylsilyl derivative.

TABLE 8.12 Extraction of *o*-Phenylphenol from Silicone Tubing into a Cyclodextrin Vehicle Formulation after Sterilization and Different Periods of Contact at Room Temperature [20]

Conditions	μg *o*-PP/cm^2 tubing
Sterilization cycles before filling	
None	1.54
1	1.33
2	1.29
3	1.22
Extraction interval (h)	
0–24	0.0496
24–48	0.0103
48–72	0.0033
72–96	0.0010
0–24 (only water)	0.0010

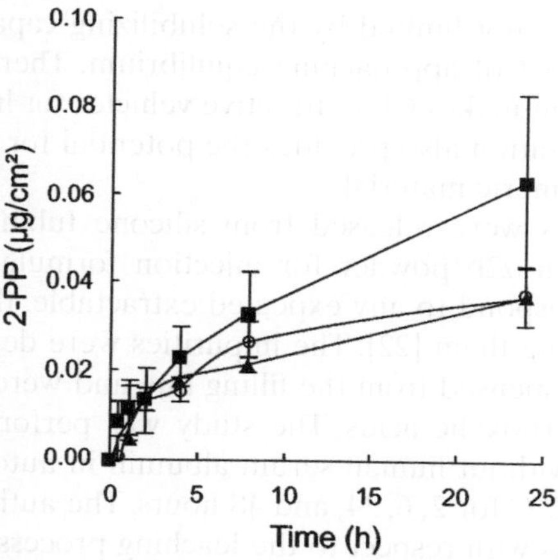

FIGURE 8.11 Extraction of 2-PP from silicone tubing as a function of contact time and HPβCD concentration, with 10% (w/v) HPβCD (■), 20% (w/v) HPβCD (▲), and 40% (w/v) HPβCD (○) [21].

TABLE 8.13 Extraction of *o*-PP from Silicone Tubing after 8 Hours' Incubation in Contact with Pure Water, 20% HPβCD Solution, and Ethanol

Silicone Tubing	Solution	API Concentration (mM)	*o*-PP Extracted (μg/cm² ± SD)
Lot 14936984	20% HPβCD	–	0.009 ± 0.001
Lot 16549274	Pure water	–	n.d.
	20% HPβCD	–	0.022 ± 0.010
	20% HPβCD	0.3	0.021
	20% HPβCD	3.1	0.007
	20% HPβCD	31	0.006
Autoclaved	20% HPβCD	–	0.0017 ± 0.002
	Ethanol	–	0.37 ± 0.14

n.d., not detected.

lower the amount of *o*-PP extracted. The results are presented in Table 8.13. The results in Table 8.13 also show that while in pure water no *o*-PP was extracted, in ethanol, the amount was approximately 15-fold higher than that in the CD solution.

Extraction rates found in this study are similar to those mentioned above for SBECD; comparable amounts of *o*-PP were extracted despite the different CD molecules, tubing brands, and lot-to-lot tubing variability. Autoclave sterilization did not reduce the *o*-PP released into the CD solutions. The maximum amount of *o*-PP obtained from these tubing samples was much lower than the solubility of *o*-PP in the CD solutions, which allows one to conclude that the amount extracted was not limited by the solubilizing capacity of the vehicle but rather by an effect of approaching equilibrium. Therefore, the physicochemical property that makes CDs attractive vehicles for lipophilic drug solubilization and stabilization also provides the potential for extraction of small molecules from polymeric materials.

Other extractables were released from silicone tubing into Intron-A, a lyophilized interferon-α2b powder for injection formulations. As the compounds did not correspond to any expected extractable, the authors set up a strategy for identifying them [22]. The impurities were detected in the prelyophilized solution dispensed from the filling line and were identified as halogenated biphenyl carboxylic acids. The study was performed by incubating Intron-A with and without human serum albumin in autoclaved and nonautoclaved tubing at 22°C for 2, 6, 24, and 48 hours. The authors established the following conclusions with respect to the leaching process:

- Higher levels of extractables were observed in untreated tubing versus treated tubing (five autoclaving cycles).
- The same level of extractables was observed in both medical- and technical-grade silicone tubing (RAU-SIK from Rehau Ltd.).
- The extraction process attained the highest rates over the first 2 hours.
- Previously used tubing leached lower or nondetectable levels of extractables.
- The presence of protein (albumin) facilitated the extraction of leachables.

Since the tubing is used in a dynamic process, the authors estimated that around 0.01–0.55 μg of the amount of leached impurities went into the first vials filled.

It is interesting to note that in tests carried out with a platinum-cured silicone tubing, these impurities were not leached. However, Pt-cured silicone tubing was the type of tubing that leached *o*-PP when in contact with cyclodextrins.

8.3 FILTERS

Filters are widely used in the pharmaceutical industry for the clarification of products, reduction of bioburden, and sterilization of thermosensitive drugs. Nevertheless, during filtration, formulation constituents can extract materials from the filter device. The extractable candidates depend on the combination of filter material, pharmaceutical product, and also processing parameters such as temperature and filtration rate. The materials used in these filters include surface-modified polyvinylidene fluoride (PVDF), polypropylene (PP), nylon,

TABLE 8.14 Extractables from Different Filter Brands

	Membrane					
Brand	Polyether-sulfone	Nylon	Cellulose Acetate	Cellulose Nitrate	Polyvinylidene Fluoride	Teflon
Nalgene	0.6%	0.8%	2.0%	2.3%	–	–
Pall	–	–	–	–	<5 mg/ cartridge (10 in.)	–
Millipore	–	–	–	–	<10 mg/ cartridge (0.18 m^2)	<15 mg/ cartridge (0.18 m^2)

Values furnished by the companies.

polysulfone, and regenerated cellulose (these polymers are fully described in the Appendix).

Because concerns on extractables and leachables have been increasing, some filter manufacturers inform the amount of extractables from their filtration devices. Table 8.14 presents the membranes of sterile filtration devices supplied by different companies and, when available, the amount of extractables.

Despite information given by the manufacturers, studies on filter extractables were conducted to investigate the action of different media on the filter material. Potential extractables from sterilizing grade filters are characterized using model solvents somewhat different from those used for plastics in general, probably due to the difficulties associated with the measurement of small amounts of extractables in the presence of high protein content or other complex matrices. Extraction in buffer solutions and techniques such as proton nuclear magnetic resonance spectroscopy (^{1}H-NMR) and Fourier transform infrared spectroscopy (FTIR), and total organic carbon (TOC) are employed for the purpose. In fact, extractables are not fully characterized, but, instead, the amount of nonvolatile residues (NVRs) or one specific component is determined.

In a study conducted with Durapore® cartridges, three different filters were analyzed using water, 0.01 M NaOH (pH 12), 0.01 M HCl (pH 2), and ethanol as extractor solutions [23]. Filter components were separately autoclaved and tested without previous flushing by soaking in solvent solutions for 20 hours at room temperature. The parameters evaluated were TOC and NVR, followed by the recording of FTIR and NMR spectra of the extractables. Although high-performance liquid chromatography ultraviolet detection (HPLC-UV) and liquid chromatography-nuclear magnetic resonance detection (LC-NMR) techniques have also been used, they did not add any information to the results obtained with FTIR and NMR. HPLC-UV presented lack of sensitivity, and

TABLE 8.15 Amounts of TOC and Nonvolatile Residues (NVRs) Recovered in Different Solvents from Different Filter Materials and the Identification of Extractables

Extraction Solvent	Filter Component: PVDF Hydrophilic	PVDF Hydrophobic	Cellulose	PP Substrate	PP Housing
Water					
NVR (mg)	21.6	8.5	12.3	8.5	0
TOC (mg)	7.2	0.5	7.3	0.3	0.1
0.01 M HCl					
NVR (mg)	38.2	5.9	17.3	13.1	2.6
TOC (mg)	14.8	0.3	11.8	0.4	0.3
NaOH 0.01 M					
NVR (mg)	n.a.	n.a.	n.a.	n.a.	n.a.
TOC (mg)	79.5	0	24.9	0	0
Ethanol					
NVR (mg)	38.5	65.1	16.3	40.5	0.5
TOC (mg)	n.a.	n.a.	n.a.	n.a.	n.a.
Compound identified	Oligomers Polyhydroxypropyl acrylate Propylene glycol	Oligomers	Glycerol Cellulose acetate Cellulose nitrate	Phenolic antioxidant	Oligomers

Adapted from Reference 23.
n.a., not analyzed.

LC-NMR only confirmed some of the NMR findings. Table 8.15 presents the results of TOC and NVR analysis as well as the extractables whose presence was confirmed in the different solvents and for the different filter materials. As could be expected, the hydrophobic PVDF filter leached much less TOC and NVR in aqueous extracts than did the hydrophilic PVDF filter. On the other hand, its ethanolic extract contained twice the amount of NVR of the hydrophilic PVDF filter extract. The amounts of NVR and TOC from the cellulose filter were lower than the amount leached by the hydrophilic PVDF filter in all extracts.

The same filter membrane, hydrophilic PVDF, was subjected to another evaluation of extractables using the intact cartridge (with PP housing) and several other buffer solutions [24]. The only extractables from PVDF filters were hydroxypropyl acrylate and propylene glycol. PVDF is made hydrophilic by cross-linking with hydroxypropyl acrylate, which forms the hydrophilic layer. In alkaline milieu, the polyhydroxypropyl acrylate suffers hydrolysis producing hydroxypropyl acrylate and propylene glycol. The results, summarized in Table 8.16, include those obtained using protein formulations,

TABLE 8.16 Amount of Propylene Glycol and TOC Extracted from a Durapore® Filter[a] by the Action of Different Solutions and Buffers, after 24 Hours' Exposure at Room Temperature

Extraction Media	pH	Propylene Glycol (mg/mL)	TOC (μg/mL)
Water	–	<0.1	1.18
Acetic acid	3	1.2	–
H_3PO_4	2	0.2	2.65
Na Acetate, 1% benzyl alcohol	5	<0.1	–
MOPSO + TMAC + 0.2% thiodiglycol	7.0	5.3	–
Tris + NaCl + EDTA + TMAC	7.0	<0.1	–
Tris + guanidine HCl	7.5	<0.1	–
Guanidine HCl + TMAC + HEPES	7.8	<0.1	–
Formulated proteins	5.5–6.5	<0.1	–
MOPSO	10.6	14	–
NH_4OH	11	11	17.8
Tris	11.1	8	–
NaOH	12.3	128	89.2

[a] Hydrophilic PVDF filter.
Adapted from Reference 24.
–, not evaluated; MOPSO, 3-(N-morpholinyl)-2-hydroxypropanesulfonic acid; TMAC, tetramethylammonium chloride; HEPES, N-(2-hydroxyethyl)piperazine-N′-(2-ethanesulfonic acid).

generated by recirculating the solutions through a Millidisk 40 filter, containing a hydrophilic PVDF membrane as well. Considerable amounts of extractables were only obtained in alkaline solutions, allowing one to conclude that pH is the only factor that governs the level of extractables from the PVDF filters.

Two representative components of single-use filtration systems were evaluated not only with respect to TOC and NVR but also organic compounds and metallic contaminants. Employing several analytical techniques, Ding and Nash [25] characterized extractables of a sterile connector and a filter capsule of an integrated single-use device for manufacturing biopharmaceuticals (Pall Kleenpak). Both parts were gamma-irradiated at 50 kGy prior to extraction, which was carried out in water and ethanol at room temperature. Extraction time was 24 hours for water and 4 hours for ethanol. The connector had a polycarbonate housing and an ethylene propylene O-ring; the filter incorporated a PVDF membrane, a PP membrane support, and capsule hardware. Very few compounds were extracted from the connector (Table 8.17). On the other hand, several organic and metallic impurities were found in both extracts of the filter. Among organic compounds, acids and alcohols were identified. The antioxidant, Irgafos 168, was present in the filter materials because its oxidation product was identified among the extractables. These compounds probably originated from the PP support since similar group of substances

TABLE 8.17 Extractables from an Integrated Single-Use System, Connector and Filter Used for Biopharmaceutical Manufacturing

	Connector				Filter Capsule			
TOC	0.09 ppm				8.92 ppm			
Extracts	Water		Ethanol		Water		Ethanol	
	Compound	Conc. (µg/L)	Compound	Conc. (µg/L)	Compound	Conc. (µg/L)	Compound	Conc. (µg/L)
NVR (FTIR analysis)	n.d.		n.d.		Primarily oligomers of acrylate	5.5 mg	Primarily oligomers of acrylate	56 mg
							Oligomers of polypropylene	
Organic acids	Benzoic acid	39	Benzoic acid	47	Benzoic acid	16	Oxalic acid	161
					Oxalic acid	461	Succinic acid	166
					Succinic acid	79	Stearic acid	1230
					Stearic acid	13	Malonic acid	103
							Lauric acid	234
							Myristic acid	85
							Palmitic acid	612
							1,4-Benzendicarboxylic acid	2240
							2-Ethylhexanoic acid	<1600
Other organics	n.d.		n.d.		n.d.		Acetal	850
							1,3-Di-tert-butylbenzene	<1600
							2,4-Di-tert-butylphenol	<1600
							Lauryl acetate	<1600
							Lauril acrylate	<1600
							1-Dodecanol	<1600
							1-Pentadecanol	<1600
							1-Heptadecanol	<1600
							Oxidized Irgafos	<200
Anions	n.d.		n.a.		Acetate	185	n.a.	
					Formate	8		
Metals	B, Na, Zn	<0.4	Ca	1.0	Na,	283	Na	157
					Ca	22.2	Ca	34.1
					Mg, Al, K, Cu	<7	Zn	2.3

Adapted from Reference 25.

n.a., not analyzed; n.d., not detected.

were also extracted from other PP made devices (see Table A.3, Appendix). Although titanium oxide is a component of both devices, connector and filter, titanium was not found among the metallic extractables.

CONCLUDING REMARKS

Devices responsible for the delivery of drugs and infusion solutions are made of practically the same polymers as plastic containers. As flexibility is a required feature, PVC and silicone are the chief materials. By contrast, filters, widely used in the pharmaceutical industry for clarification and sterilization, are made of rigid materials such as PP and PVDF. Nevertheless, the leachability of ingredients in both types of devices depends not only on the nature of the plastic material but also on the flow rate, temperature, presence of solvents, and time of contact.

REFERENCES

1. Loff S, Kabs F, Witt K, Sartoris J, Mandl B, Niessen KH, Waag KL. Polyvinylchloride infusion lines expose infants to large amounts of toxic plasticizers. J Pediatr Surg 35, 1775–1781, 2000.
2. Haishima Y, Seshimo F, Higuchi T, Yamazaki H, Hasegawa C, Izumi S, Makino T, Nakahashi K, Ito R, Inoue K, Yoshimura Y, Saito K, Yagami T, Tsuchiya T, Nakasawa H. Development of a simple method for predicting the levels of DEHP migrated from PVC medical devices into pharmaceutical solutions. Int J Pharm 298, 126–142, 2005.
3. Kambia K, Dine T, Gressier B, Bah S, Germe AF, Luyckx M, Brunet C, Michaud L, Gottrand F. Evaluation of childhood exposure to di(2-ethylhexyl) phthalate from perfusion kits during long-term parenteral nutrition. Int J Pharm 262(1-2), 83–91, 2003.
4. Loff S, Kabs F, Subotic U, Schaible T, Reinecke F, Langbein M. Kinetics of diethylhexyl-phthalate extraction from polyvinylchloride-infusion lines. JPEN J Parenter Enteral Nutr 26(5), 305–309, 2002.
5. Takehisa H, Naoko E, Masahiko S, Katsuhide T, Moriyuki O, Keizoh S, Mutsuko T, Kenji K, Shin'ichiro N, Toshio O. Release behavior of diethylhexyl phthalate from the polyvinyl-chloride tubing used for intravenous administration and the plasticized PVC membrane. Int J Pharm 297(1-2), 30–37, 2005.
6. Hanawa T, Endoh N, Kazuno F, Suzuki M, Kobayashi D, Tanaka M, Kawano K, Morimoto Y, Nakajima S, Oguchi T. Investigation of the release behavior of diethylhexyl phthalate from polyvinyl chloride tubing for intravenous administration based on HCO 60. Int J Pharm 267(1-2), 141–149, 2003.
7. Boithias S-B, Sautou-Miranda V, Bourdeaux D, Tramier V, Boyer A, Chopineau J. Leaching of DEHP from multilayer tubing into etoposide infusion solutions. Am J Health Syst Pharm 62, 182–188, 2005.

8. Loff S, Subotic U, Reinicke F, Wischmann H, Brade J. Extraction of Di-ethylhexylphthalate from perfusion lines of various material, length and brand by lipid emulsions. J Pediatr Gastroenterol Nutr 39(4), 341–345, 2004.
9. Loff S, Hannmann T, Subotic U, Reinecke FM, Wischmann H, Brade J. Extraction of diethylhexylphthalate by home total parenteral nutrition from polyvinyl chloride infusion lines commonly used in the home. J Pediatr Gastroenterol Nutr 47(1), 81–86, 2008.
10. Rathinam K, Srivastava SP, Seth PK. Hepatic studies of intraperitoneally administered TOTM and DEHP in rats. J Appl Toxicol 10, 39–41, 1990.
11. Christensson A, Ljunggren L, Nilsson-Thorell C, Arge B, Diehl U, Hagstam KE, Lundberg M. In vivo comparative evaluation of hemodialysis tubing plasticized with DEHP and TEHTM. Int J Artif Organs 14, 407–410, 1991.
12. Ito R, Miura N, Iguchi H, Nakamura H, Ushiro M, Wakui N, Nakahashi K, Iwasaki Y, Saito K, Suzuki T, Nakazawa H. Determination of tris(2-ethylhexyl) trimellitate released from PVC tube by LC-MS/MS. Int J Pharm 360(1-2), 91–95, 2008.
13. Ito R, Seshimo F, Miura N, Kawaguchi M, Saito K, Nakazawa H. High-throughput determination of mono- and di(2-ethylhexyl)phthalate migration from PVC tubing to drugs using liquid chromatography–tandem mass spectrometry. J Pharm Biomed Anal 39(5), 1036–1041, 2005.
14. Wang Q, Storm KB. Separation and analysis of low molecular weight plasticizers in poly(vinyl chloride) tubes. Polym Test 24(3), 290–300, 2005.
15. Wang Q, Storm BK. Migration of additives from poly (vinyl chloride) (PVC) tubes into aqueous media. Macromol Symp 225, 191–203, 2005.
16. Subotic U, Hannmann T, Kiss M, Brade J, Breitkopf K, Loff S. Extraction of the plasticizers diethylhexylphthalate and polyadipate from polyvinylchloride nasogastric tubes through gastric juice and feeding solution. J Pediatr Gastroenterol Nutr 44(1), 71–76, 2007.
17. Teknor Apex Company. PVC compounder introduces dedicated production lines. Plast Addit Comp 11(1), 9, 2009.
18. Hill SS, Shaw BR, Wu AHB. Plasticizers, antioxidants, and other contaminants found in air delivered by PVC tubing used in respiratory therapy. Biomed Chromatogr 17(4), 250–262, 2003.
19. Jenke DR, Story J, Lalani R. Extractables/leachables from plastic tubing used in product manufacturing. Int J Pharm 315, 75–92, 2006.
20. Zimmerman JA, Ballard JM, Wang H, Wu A, Gallagher KA. Extraction of o-phenylphenol from silicone tubing by a sulfobutylether cyclodextrin formulation. Int J Pharm 267(1-2), 113–120, 2003.
21. den Brok MW, van der Schoot SC, Nuijen B, Hillebrand MJ, Beijnen JH. 2-Hydroxypropyl-beta-cyclodextrin extracts 2-phenylphenol from silicone tubing. Int J Pharm 278(2), 303–309, 2004.
22. Chan TM, Pramanik B, Aslanian R, Gullo V, Patel M, Cronin B, Boyce C, McCormick K, Berlin M, Zhu X, Buevich A, Heimark L, Bartner P, Chen GD, Pu HY, Hegde V. Isolation, structural determination, synthesis and quantitative determination of impurities in Intron-A, leached from a silicone tubing. J Pharm Biomed Anal 49(2), 327–332, 2009.

23. Jiang TB, Phelan M. Extractables from filter components in model solvent streams for pharmaceutical filtration processes. PDA J Pharm Sci Technol 57(4), 287–299, 2003.
24. Kao YH, Bender J, Hagewiesche A, Wong P, Huang YF, Vanderlaan M. Characterization of filter extractables by proton NMR spectroscopy: studies on intact filters with process buffers. PDA J Pharm Sci Technol 55(5), 268–277, 2001.
25. Ding W, Nash R. Extractables from integrated single-use systems in biopharmaceutical manufacturing. Part I. Study on components (Pall Kleenpak™ connector and Kleenpak filter capsule). PDA J Pharm Sci Technol 63, 322–338, 2009.

9

MEDICAL DEVICES

According to regulatory agencies [1], medical devices are any instrument, apparatus, appliance, material, or other article, whether used alone or in combination, intended to be used for human beings for the purpose of diagnosis, prevention, monitoring, treatment or alleviation of diseases, modification of the anatomy, replacement of a physiological process, and control of conception.

Although not considered a medicinal product, and therefore subject to different legislations, medical devices may come into very close contact with the human body. In some cases, such as implants, the contact is perpetual, and while the required inertness may be satisfied by the implant material itself, this may not be the case for potential impurities contained within it.

In this chapter, the following medical devices will be considered:

- devices for medication and patient management
- dialyzers and other extracorporeal circuits (ECs)
- devices for administration of aerosolized drugs
- tissue substitutes.

A compilation on materials used to produce medical devices could not possibly cover all materials involved. In this chapter, the most commonly used

Sources of Contamination in Medicinal Products and Medical Devices, First Edition. Denise Bohrer.
© 2013 by John Wiley & Sons, Inc. Published 2013 by John Wiley & Sons, Inc.

materials are considered, and in so far as possible, the major constituents required for production were considered.

Most of the products are made of either natural or synthetic polymeric materials. Sources of contaminants are the processing of raw materials and additives in the manufacture of synthetic polymers and the isolation and purification of natural polymers. For synthetic polymers, a description of raw materials and other components is presented separately in the Appendix. Because many of them are used to make containers, all information related to synthetic polymers is grouped in this appendix. In this chapter, contamination arising from interactions between devices and body fluids or specific drugs is discussed.

Pharmacopeial compendia rule in different ways about polymeric material to be used for pharmaceutical purposes. While British Pharmacopoeia (BP) and Ph. Eur. limit the number of additives and their amount in different polymeric materials specifying only that "they be for parenteral and ophthalmic preparations" (see Table 6.10, Chapter 6), the United States Pharmacopeia (USP) does not specify polymer compositions but describes tests to be carried out with devices (made of polymers) according to the way they enter into contact with the human body. The USP classifies medical devices into three categories (*<1031> The Biocompatibility of Materials Used in Drug Containers*) [2]. These categories are *surface devices*, *external communicating devices*, and *implant devices*, and the appropriate procedure to investigate the biocompatibility of the material is defined by the duration of the contact between the device and the body. The contact is considered *limited* for less than 24 hours, *prolonged* for 24 hours to 30 days, or *permanent* for more than 30 days (Table 9.1). In Table 9.1, the plastic class designation for the three device categories

TABLE 9.1 USP Classification (⟨1031⟩) of Medical Devices According to the Nature and Extent of Contact and the Plastic Class Designation as a Function of the Use

		Plastic Class Designation according to the Contact time		
Device Category	Nature of Contact	Limited	Prolonged	Permanent
Surface devices	Skin	I	I	I
	Mucosal membrane	I	II	V
	Breached or compromised surface	III	V	VI
External communicating devices	Blood path indirect	IV	V	VI
	Tissues	IV	VI	VI
	Circulating blood	IV	VI	VI
Implant devices	Tissue or bone	–	–	VI
	Blood	–	–	VI

TABLE 9.2 Tests Required for the USP Classification of Plastics (⟨88⟩)

Extract Medium[a]	Animal	Administration Mode	Class I	II	III	IV	V	VI
0.9% NaCl	Mouse	Intravenous	x	x	x	x	x	x
	Rabbit	Intracutaneous	x	x	x	x	x	x
1 part 0.9% NaCl + 20 parts ethanol	Mouse	Intravenous		x	x	x	x	x
	Rabbit	Intracutaneous		x	x	x	x	x
Polyethylene glycol 400	Mouse	Intraperitoneal			x		x	x
	Rabbit	Intracutaneous					x	x
Vegetable oil (sesame or cottonseed)	Mouse	Intraperitoneal			x	x	x	x
	Rabbit	Intracutaneous				x	x	x
Implant strips of the material	Rabbit	Intramuscular				x		x

[a] Extraction: 25–120 cm^2 of the device (sheet, tubing) cut into small pieces (5 × 0.3 cm) or 0.1 g elastomer or 0.2 g other polymers/mL extraction fluid. Heating at 121°C for 60 minutes, or at 70°C for 24 hours, or at 50°C for 72 hours.

is also displayed. Although covering multiple purposes, most of these devices are made basically of either synthetic or natural polymeric materials, so the USP provides a plastic class designation, under the monograph *<88> Biological Reactivity Tests, In Vivo*. Plastics are numerically designed in categories related to the duration the device will be in contact with the body. The class category stipulates the tests that should be done. The procedures are based on the use of extracts obtained by contact of the plastic material with an extracting medium at one of three standard temperatures (50, 70, and 121°C). The extract is injected in mice or rabbits depending on the nature of the contact with the body and the reaction should be evaluated (Table 9.2). The extracting media are 0.9% NaCl, 0.9% NaCl/ethanol 1:20, polyethylene glycol 400, or vegetable oil.

Although these extraction media cover different solvent natures, considering chemical interactions (polar, apolar, hydrophilic, lipophilic), they actually do not include polymer specifications and limits for additives or impurities. Only under "<661> Containers-Plastics" are there limits for impurities such as heavy metals and nonvolatile residues for different and specific polymers, polyethylene, polypropylene, and polyethylene terephthalate (this topic is discussed in Section 6.2, Plastic Containers).

Impurities and contaminants may be introduced into medical devices not only from raw materials but also from substances added during processing of the product itself. A list of the main components and ancillary substances involved in the production of the materials used in medical devices considered in this chapter is displayed in Table 9.3.

TABLE 9.3 Main Constituents and Substances Involved in the Production of Common Medical Devices

Device	Characteristics	Main Components	Other Substances in the Product or Involved in the Production
Syringe	Drug administration	Polypropylene Glass	Silicon oil Elastomers
Access System	Blood circulation	PVC	
Infusion delivery system	Infusion solutions Parenteral nutrition Drug administration	PVC (bag and lines) Polycarbonate (burette)	
Catheter	Blood circulation	Polyurethane (PU) PVC	
Stent	Implantable mesh-like tube of thin wire	Nitinol (54.5% Ni, 45.5% Ti) Co/Cr alloy (40% Co, 20% Cr, 16% Ni, 7% Mo) Stainless steel 316L (62–72% Fe, 16–18% Cr, 10–14% Ni, 2–3% Mo, 2% Mn)	
Suture[a]	Absorbable	PLA and PLGA Catgut (collagen)	Dyes Chromium trioxide Magnesium stearate (cover) Silicone (cover)
	Nonabsorbable	Silk PET Polyamide 6 (nylon) Polyamide 6/6 PP Stainless steel 316L	 Silicone PTFE (cover)

[a] *Source*: http://www.covidien.com.

PLA, polylactic acid; PLGA, poly(lactide-co-glycolide acid); PET, polyethylene terephthalate; PP, polypropylene; PTFE, polytetrafluoroethylene.

9.1 GENERAL USE DEVICES

Syringes, catheters, sutures, gloves, and intravenous access systems are medical devices that come into very close contact with the human body. They are made of synthetic polymers (sutures and implants may also be made of natural and metallic components) and may have parts made of other materials, such as the syringes that may have a rubber sealing ring. Syringes may also be internally lubricated with silicone oil.

9.1.1 Medical Gloves

Because gloves have a general use in medical procedures and also in handling and compounding pharmaceutical products, and because they are made of polymeric materials similar to those used in other medical devices, this chapter records the main constituents of medical gloves and also potential contaminants.

Gloves can be made of different kinds of polymeric materials, which are considered in this book under rubber components (latex, nitrile, and neoprene), and under plastics (polyvinyl chloride [PVC]). Depending on the type, different additives must be present to form the material. The general constituents of these polymeric materials (monomers and usual additives) are given in the Appendix.

A comprehensive survey on specific components of rubber and plastic gloves was complied by Rose et al. [3] using information given by suppliers. The aim of the study was to review the presence of allergens in gloves since glove allergy plays a significant role in occupational medicine. Based on this resource, Table 9.4 presents these chemical components for different kinds of medical gloves. Among the chemical compounds used in the manufacture of gloves are several identified as extractables and leachables from other forms of the same polymer (Table A.3 in the Appendix). Bisphenol A (BPA), for example, is an antioxidant found in PVC gloves as well as in hemodialyzers (HDs).

The behavior of these additives is the same either when present in gloves or in other plastic materials. Table 9.5 presents additives released from gloves and the conditions used for their extraction. When comparing the compounds listed in Table 9.5 with those listed in Table A.3 in the Appendix, it becomes clear that the process is the same; migration occurs to a greater or lesser extent, depending on the extraction medium.

9.1.2 Syringes

The presence of silicone oil in products delivered through disposable syringes might have adverse consequences and has been seen as an undesirable contaminant. Silicone oil, used as a lubricant, is a syringe component permitted by pharmacopeial compendia, although no specific amounts or limits are set for its use.

Silicone oil release occurred in disposable insulin syringes causing the product to cloud and to deactivate [10–12]. Besides insulin, silicone oil was also released into intravitreal injection drugs leading to the formation of tiny silicone oil drops in the eyes of the patients. Bakri and Ekdawi [13] found silicone oil droplets in the vitreous cavity of all 15 patients examined after

TABLE 9.4 Components Present in the Different Types of Gloves [3]

Chemical Component	Glove Type			
	Latex	Nitrile	Neoprene	PVC
Zinc oxide	Yes	Yes	No	No
Titanium dioxide	Yes	Yes	No	No
Calcium nitrate	Yes	Yes	No	No
Fillers				
Calcium carbonate	Yes	No	No	No
Zinc sulfide	Yes	No	No	No
Barium sulfate	Yes	No	No	No
Plasticizers				
Naphthenic oil	Yes	No	No	No
Paraffinic process oil	Yes	No	No	No
Epoxy resins	No	No	No	Yes
Adipic polyester	No	No	No	Yes
Poly(adipic acid-co-1,2-propylene glycol)	No	No	No	Yes
Stabilizers				
Potassium oleate	Yes	No	No	No
Triethanolamine oleate	Yes	No	No	No
Casein	Yes	Yes	No	No
Di(2-ethylhexyl) phthalate	No	Yes	No	Yes
Dibutyl phthalate	No	No	No	Yes
Di-(n-octyl) tin-bis(2-ethylhexylmaleate)	No	No	No	Yes
Antioxidants				
Bisphenol A	No	No	No	Yes
Methylcyclohexyldimethylphenol	Unknown	Unknown	No	No
4,4′-Thiobis(6-tert-butyl-meta-cresol) (Lowinox® 44S36)	Yes	No	No	No
Butylated hydroxyanisole (BHA)	Yes	No	No	No
2-Tertiary-butyl-4-methylphenol	No	Yes	No	No
4,4′-Dihydroxydiphenyl ether	No	No	Yes	No
N-Isopropyl-N′-phenyl-paraphenylenediamine	Yes	Yes	No	No
N-Cyclohexyl-N′-phenyl-paraphenylenediamine	Yes	Yes	No	No
Accelerators				
2-Mercaptobenzothiazole	Yes	Yes	Yes	No
N-Cyclohexyl–2-benzothiazylsulfenamide	Yes	Yes	Yes	No
Morpholinylmercaptobenzothiazole	Yes	Yes	Yes	No
Dibenzothiazyldisulfide	Yes	Yes	Yes	No
Zinc diethyldithiocarbamate	Yes	Yes	Yes	No
Zinc dibutyldithiocarbamate	Yes	Yes	Yes	No
Zinc dimethyldithiocarbamate	Yes	Yes	Yes	No
1,3-Diphenylguanidine	Yes	Yes	Yes	No

(*Continued*)

TABLE 9.4 (*Continued*)

Chemical Component	Glove Type			
	Latex	Nitrile	Neoprene	PVC
Tetraethylthiuram disulfide	Yes	Yes	Yes	No
Tetramethylthiuram monosulfide	Yes	Yes	Yes	No
Tetramethythiuram disulfide	Yes	Yes	Yes	No
Dipentamethylenethiuram disulfide	Yes	Yes	Yes	No
Dibutyl thiourea	Yes	Yes	Yes	No
Diphenyl thiourea	Yes	Yes	Yes	No
Diethyl thiourea	Yes	Yes	Yes	No
Pigment				
Sulfur black	Yes	Yes	No	No
Pigment Yellow (CI 21290)	Yes	No	No	No
Pigment Blue 15:1 (CI 74160:3)	No	Yes	No	No
Pigment Green 7 (CI 74260)	No	Yes	No	No
Carbon black	Yes	No	No	No
Pigment Red 122 (CI 73915)	Yes	No	No	No
Disperse Yellow 134 (CI 47023)	No	No	No	Yes
Surfactant				
Fatty alcohol polyglycol ethers	Yes	Yes	No	No
Dowfax™ (mono- and di-alkyl diphenyl oxide disulfonate disodium salt)	No	Yes	No	No
Thickener				
Polyamide	Yes	Yes	No	No
Sodium carboxymethyl cellulose	Yes	Yes	No	No
Ammonia solution	Yes	Yes	No	No
Defoamers				
Blend metallic soaps/mild nonionic surfactants	Yes	No	No	No
Mineral oil	No	Yes	No	No
Tackifiers				
Terpene phenolic resin	No	Yes	No	No
Cross-linking agents				
Methylated melamine resin	No	Yes	No	No
Anticaking agents				
Silica with medium oil absorption	No	Yes	No	No
Emulsifiers				
Polyvinyl alcohol (PVA)	No	Yes	No	No
Promoters				
Methanol	Yes	Yes	No	No
Dispersers				
Xylene	Yes	No	No	No
Antimicrobials				
Formaldehyde	No	No	No	Yes
1,2 Benzisothiazolinone	No	No	No	Yes
Cetyl pyridinium chloride	Yes	No	No	No
Corn starch powder	Yes	Yes	No	No

TABLE 9.5 Additives Released from Different Kinds of Medical Gloves under Specific Conditions of Extraction

Kind of Glove	No. of Samples (Brands)	Extraction Conditions	Additive		Amount	Reference
PVC	16	Acetone	Bisphenol A	BPA	<0.001% in 15 brands 0.001% in 1 brand	[4]
PVC	1	Accelerated solvent extraction	Bisphenol A	BPA	Positive	[5]
Latex powder-free	19	Acetonitrile for 17 h at room temperature	2-Mercaptobenzothiazole	MBT	n.d. in 14 brands 0.01–0.17 mg/g glove in 5 brands	[6]
			Zinc dibutyldithiocarbamate	ZDBC	n.d. in 6 brands 0.36–2.63 mg/g glove in 13 brands	
			Zinc diethyldithiocarbamate	ZDEC	n.d. in 5 brands 0.38–4.99 mg/g glove in 14 brands	
Latex powdered	5	Acetonitrile for 17 h at room temperature	2-Mercaptobenzothiazole	MBT	n.d. in all brands	
			Zinc dibutyldithiocarbamate	ZDBC	0.81–4.59 mg/g glove in 5 brands	
			Zinc diethyldithiocarbamate	ZDEC	n.d. in 1 brand 1.62–5.17 mg/g glove in 4 brands	
Nitrile powder-free	14	Acetonitrile for 17 h at room temperature	2-Mercaptobenzothiazole	MBT	n.d. in 8 brands 0.01–1.31 mg/g glove in 6 brands	
			Zinc dibutyldithiocarbamate	ZDBC	n.d. in 11 brands 1.35–3.09 mg/g glove in 3 brands	

(Continued)

TABLE 9.5 (*Continued*)

Kind of Glove	No. of Samples (Brands)	Extraction Conditions	Additive		Amount	Reference
			Zinc diethyldithiocarbamate	ZDEC	n.d. in 7 brands 1.05–4.36 mg/g glove in 7 brands	
PVC	31	Water/methanol 3:1 overnight at room temperature	1,2-Benzisothiazolin-3-one	BIT	<2 ppm in 22 samples 3–9 ppm in 4 brands 16–26 ppm in 5 brands	[7]
PVC	4	Water	Formaldehyde		Positive for 4 brands	[8]
Latex	2	Water			Positive for 1 brand Negative for 1 brand	
Nitrile	3	Water			Positive for 1 brand Negative for 2 brand	
Latex	18	Acetone	Tetramethylthiuram-disulfide	TMTD	0.3 μmol/glove in 1 brand	[9]
			Tetrabutylthiuram-disulfide	TBTD	0.1–1.2 μmol/glove in 2 brands	
			Zinc-diethyldithiocarbamate	ZDEC	0.6–15.6 μmol/glove in 11 brands	
			Zinc-dibutyldithiocarbamate	ZDBC	0.4–8.9 μmol/glove in 9 brands	
			Zinc-dipentamethylen-dithiocarbamate	ZPC	0.5–6.7 μmol/glove in 7 brands	
			Zinc-diisobutyl-dithiocarbamate	ZDiBC	0.5–0.8 μmol/glove in 3 brands	
			Zinc-mercaptothiazole	ZMBT	<0.15–1.5 μmol/glove in 9 brands	
Synthetic	1	Acetone	Zinc-dibutyldithiocarbamate	ZBDC	3.5 μmol/glove	

n.d., not detected.

intravitreal drug injection, but no adverse effects were associated with the presence of these droplets.

In spite of these findings, contamination arising from the use of syringes seems to be much more related to its rubber parts, stoppers and plungers, than that to the syringe body itself. The polysorbate 80 present in one erythropoietin formulation stored in prefilled syringes (EPREX®) was responsible for the leaching of several compounds from the rubber stopper [14]. The leaching was verified only because of the incidence of antibody-mediated pure red cell aplasia in patients with chronic kidney disease who were treated with this formulation. The analysis and comparison of samples stored in prefilled syringes with uncoated and Teflon-coated rubber stoppers revealed that leachates were found only in formulations containing polysorbate 80 and stored in uncoated syringe stoppers. The leachates and their concentrations are presented in Table 9.6.

Table 9.6 also presents compounds leached into the drug stored in plastic syringes. Rubber stoppers of disposable syringes were responsible for the leaching of several compounds in contrast agent injections (Reference 15 in

TABLE 9.6 Concentration of Compounds Found in Different Injectable Formulations Stored in Disposable Plastic Syringe

Compound	Concentration	Reference
EPREX	(μg/mL)	[14]
4-tert-Amylphenol	0.046	
2-Chloro-4-tert-amylphenol	0.037	
2,2′-Methylene-bis-4-tert-amylphenol	0.243	
Vultac® 2 disulfide	0.778	
Vultac 2 trisulfide	0.235	
Vultac 2 tetrasulfide	0.142	
Vultac 2 pentasulfide	0.063	
Vultac 2 hexasulfide	0.024	
Contrast agent (diatrizoate meglumine)	(μg/g)[a]	[15]
Nonylphenol	1.2	
Unidentified phenol with 8-carbon side chain	1.2	
2-Mercaptobenzothiazole (MBT)	3.0	
Hydroxyethyl mercaptobenzothiazole	1.2	
Ethylene glycol	1.0	
Butylhydroxytoluene	1.2	
Xylene	0.9	
Tertiary butyl-2-benzothiazole sulfenylamide	0.7	
Digoxin (in propylene glycol and ethanol)	(μg/mL)	[16]
Hydroxyethyl mercaptobenzothiazole (HMBT)	2.83	
Water	(μg/mL)	[16]
Hydroxyethyl mercaptobenzothiazole (HMBT)	1.95	

[a] Calculated in relation to the actual surface area of the stopper in contact with the solution.

Table 9.6). The products found in a contrast agent (diatrizoate meglumine) and their concentrations after an overnight storage were listed. In a second investigation (4 years later), the contrast agents was analyzed, in three independent governmental laboratories, and 2-mercaptobenzothiazole (MBT) was found again. The amount of MBT detected by all laboratories was around 3.3 μg/mL. The author calculated that, at this concentration, the usual 100 mL dose of the contrast agent would deliver 0.33 mg of the toxic and allergenic substance into the patient's blood stream [17].

2-(2-Hydroxyethylmercapto) benzothiazole (HMBT) was identified in water and digoxin stored in disposable plastic syringes. After 5 minutes of storage, the amount of HMBT reached a mean of 1.95 μg in 1 mL water and 2.83 μg in the digoxin samples (three syringes were assayed for each sample) (Reference 16 in Table 9.6). The authors supposed that HMBT was formed during manufacture of the syringes as a result of a reaction between 2-mercaptobenzothiazole and ethylene oxide, used for sterilization.

An injection of dexamethasone sodium phosphate, after 18 months' storage in plastic Monoject syringes, presented contamination by rubber vulcanization accelerators MBT, tetramethyl thiuram disulfide (TMTD), and tetraethyl thiuram disulfide (TETD). These species were not quantified, but probably they were released from the syringe stopper since the analysis of the drug just after storage revealed no contamination by these substances [18].

9.2 EXTRACORPOREAL CIRCUITS

Extracorporeal circulation provides critical live support in the face of cardiopulmonary and renal failure. For patients with end-stage renal disease, renal replacement therapy is achieved by dialysis (hemodialysis or peritoneal dialysis) or kidney transplantation. The apparatus used to conduct hemodialysis consists of the dialyzer and the tubing for the transport of blood and dialysis solution.

Dialyzers are a capsule or a shell within which hollow fibers or parallel membrane plates are suspended in dialysate. The fibers or plates function as a semipermeable membrane along which blood and dialysate flow. By crossing this membrane, solutes and water move between a patient's intravascular compartment and the dialysate contained within the dialyzer.

While the dialyzer shell is made of polyurethane, polycarbonate (PC) or polystyrene, there are several types of fibers or membranes currently in use [19], which can be classified as cellulose-based and non-cellulose-based membranes. Besides the dialyzer, tubing lines for conducting both dialysate and blood complete the apparatus. In Table 9.7, the materials used to make the system are described.

BPA, a component of PC and polysulfone (PSf), is a leachable candidate of HDs since PC is the material employed to make the casing and PSf the

TABLE 9.7 Main Constituents and Substances Involved in the Production of Dialyzers

Dialyzer	Main Components	Abbreviation
Fiber or membrane	Cellulose	
	Cellulose acetate	CA
	Polyacrylonitrile	
	Cellulose triacetate	
	Cellulosynthetic (Hemophan)	
	Polysulfone	PSf
	Polyethercarbonate	PEC
	Polyamide	PA
	Polymethylmetacrylate	PMMA
Capsule	Polycarbonate	PC
	Polymethylmetacrylate	PMMA
	Polystyrene	PS
Tubing	Polyvinyl chloride	PVC

Source: Reference 19.

hollow fibers. Haishima et al. [20] showed that both parts release BPA when in contact with water, methanol, and a bovine serum solution used to simulate patients' blood. Water and methanol were used to provide the minimum and maximum release capability. For PC casings (without hollow fibers), 100 mL of methanol and water were used as solvents, and for the PSf hollow fibers, water and bovine serum solution (250 mL) circulated in the HD at 10 mL/min. Both experiments were carried out at room temperature for a period of 16 hours. Table 9.8 displays the HD features as well as the amount of BPA extracted in each experiment. The lower amount of BPA extracted by bovine serum from HD-I and HD-IV may be related to the materials other than PC and PSf used in these HDs. The hollow fiber in HD-I was made of CA and the casing in HD-IV was made of PS, and both are not supposed to have BPA in their constitution.

Besides the dialyzer itself, the tubing for the transport of blood and dialysis solution is also a source of contaminants for the circulating fluids. As it is usually made of PVC, the release of di-2-ethylhexylphthalate (DEHP) is the main concern.

Mettang et al. [21] evaluated the release not only of DEHP but also of mono-2-ethylhexyl phthalate (MEHP), 2-ethylhexanol (EH), and phthalic acid (PA) from *continuous ambulatory peritoneal dialysis* (CAPD) devices into the dialysis fluid. They observed that, besides DEHP, all of its derivatives were present in the dialysis fluid, and while the concentration of DEHP remained constant, the concentration of MEHP and EH decreased and the concentration of PA increased during the CAPD procedure.

TABLE 9.8 Amount of Bisphenol A Extracted from Different Parts of Hemodialyzers Made of Polycarbonate (PC), Polystyrene (PS), Polysulfone (PSf) and Cellulose Acetate (CA) [20]

	Material			BPA Released ng/Casing or ng/Module		
Hemodialyzer	Casing	Hollow Fiber	Part Tested	Water	Methanol	Bovine Serum
HD-I	PC		Casing	11.7	296.0	
HD-II	PC		Casing	13.7	345.0	
HD-I		CA	Fiber	34.1		196.1
HD-II		PSf	Fiber	141.8		2090
HD-III	PC	PSf	Entire module	31.0		1010
HD-IV	PS	PSf	Entire module	3.78		140.7

MEHP, EH, and PA are hydrolysis products of DEHP. The process may entail the following reactions:

$$\underset{\text{(DEHP)}}{C_6H_4(CO_2\ C_8H_{17})_2} + H_2O \rightarrow \underset{\text{(phthalic anhydride)}}{C_6H_4(CO)_2O} + 2\ \underset{\text{(EH)}}{C_8H_{17}OH}, \qquad (9.1)$$

$$\underset{\text{(phthalic anhydride)}}{C_6H_4(CO)_2O} + H_2O \rightarrow \underset{\text{(PA)}}{C_6H_4(COOH)_2}, \qquad (9.2)$$

and

$$\underset{\text{(DEHP)}}{C_6H_4(CO_2\ C_8H_{17})_2} + H_2O \rightarrow \underset{\text{(MEHP)}}{C_6H_4(CO_2\ C_8H_{17})COOH} + \underset{\text{(EH)}}{C_8H_{17}OH}. \qquad (9.3)$$

These DEHP derivatives could already be present as contaminants in the PVC bag or be formed during the autoclaving of the bags for sterilization. The higher water solubility of MEHP compared with DEHP (37 mg/L against 0.041 mg/L of DEHP) could also contribute to its increase in the CAPD fluid.

Kuhn-Velten et al. [22] detected only the release of DEHP from dialysis tubing, but the study was conceived to show the role of blood constituents in the migration. Since DEHP presents lipophilic properties, they assayed serum with different degrees of triglycerides and albumin. The results, presented in Table 9.9, allowed the authors to conclude that there is a relationship between the DEHP released and the content of triglycerides and albumin in the blood. Moreover, under physiological conditions, albumin extracts almost 18-fold more DEHP than do triglycerides (regarding their concentrations [TG and albumin] in human serum).

TABLE 9.9 DEHP Leached from PVC Hemodialysis Tubing into Serum and Samples with Different Concentrations of Triglycerides (TG) and Albumin

		Serum + TG TG Concentration (g/L)			Albumin in Phosphate Buffer Albumin Concentration (g/L)				
Sample	Serum	4.70	9.77	14.5	10	20	30	40	50
DEHP (μg/g)	24	25	39	27	9	18	19	21	35

Adapted from Reference 22.

TABLE 9.10 Amount of DEHP and TOTM Extracted from Different PVC Tubings during a Single Dialysis Session (4 Hours)

	DEHP Tubing	TOTM/DEHP Tubing	
	DEHP (mg)	DEHP (mg)	TOTM (mg)
Mean	122.9 ± 33.9	41.8 ± 4.5	75.1 ± 25.7
Range	55.0–166.2	37.5–49.2	47.6–125.7

The values were calculated from the measurement of the concentration of both plasticizers in the blood of 10 patients in each group.
Adapted from Reference 23.

Tris(2-ethylhexyl)trimellitate (TOTM) has been seen as an alternative plasticizer for PVC due to the toxicological concerns with DEHP (see Chapter 8, "Delivery Systems and Filters"). Kambia et al. [23] compared the leachability of DEHP and TOTM from PVC hemodialysis tubing into the blood of patients under hemodialysis treatment. For one group of patients (group 1), the PVC dialyzer tubing contained only DEHP, and for the other group (group 2), the PVC tubing contained both plasticizers TOTM and DEHP. The results (Table 9.10) showed that, although the amount of DEHP was higher in group 1 than in group 2, when both plasticizers were considered, the amount leached was practically the same. The sum was 116.9 mg TOTM + DEHP against 122.9 mg of DEHP only. As the tubing materials were not analyzed, it is not possible to affirm that TOTM was extracted to a lesser extent than DEHP because it is less extractable or because it was present in a lower concentration in the polymer. However, if TOTM is really less toxic than DEHP, the substitution may be advisable.

As shown under plastic containers, cyclohexanone may be an impurity in items made of PVC because of its use as a solvent sealer. Due to its high boiling point and relatively low vapor pressure, cyclohexanone may not evaporate, remaining in the polymeric material. Snell [24] evaluated the release of cyclohexanone from PVC dialysis tubing. Twenty sets of tubing from four different manufacturers were filled with water and allowed to stand for 24 hours. Twenty additional sets of hemodialysis tubing were rinsed with 1 L of 0.9%

TABLE 9.11 Release of Cyclohexanone from PVC Dialysis Tubing into Water after 24 Hours in Contact. Tubing of Four Different Manufacturers

		Cyclohexanone Concentration				
		First Set[a]		Additional Set[a]		
Manufacturer	Tubing Volume (mL)	Leach (mg/L)	Amount Leached (mg)	Rinse	Leach	Amount Leached (mg)
A	131	34.1 ± 4.4	4.47	0.099 ± 0.012	32.3 ± 4.4	4.24
B	134	43.7 ± 9.2	5.86	0.182 ± 0.028	41.5 ± 6.6	5.58
C	160	1.02 ± 0.23	0.16	~0.004	1.01 ± 0.21	0.16
D	102	1.58 ± 0.51	0.16	~0.009	2.68 ± 1.60	0.27

[a] Mean of 20 hemodialysis tubing sets.
Adapted from Reference 24.

TABLE 9.12 Cyclohexanone Leached into the Priming Fluid of Extracorporeal Circuits

Device	Cyclohexanone (μg/L)
Extracorporeal membrane oxygenation	2016
Cardiopulmonary bypass	452
Hemodialysis circuit	2100

Adapted from Reference 25.

NaCl solution before filling the tubes with water and allowing them to stand for 24 hours. The water volume ranged from 102 to 160 mL and permitted a calculation of the amount of cyclohexanone leached. Table 9.11 shows the mean concentration of cyclohexanone for each group. All sets released cyclohexanone into water, and rinsing the tubing with saline solution did not remove significant amounts of leachable cyclohexanone. There were great differences among the manufacturers. The cyclohexanone leached from tubing from manufacturers C and D was significantly less than from manufacturers A and B.

Extracorporeal membrane oxygenation (ECMO), as well as other ECs, may also leach cyclohexanone. The analysis of an ECMO, cardiopulmonary bypasses, and hemodialysis circuits priming fluids revealed that all of these devices released cyclohexanone [25]. As shown in Table 9.12, the amounts are quite high, mainly from the ECMO and hemodialysis circuits. The authors of this study hypothesized that cyclohexanone, delivered from these circuits, would induce cardiac insufficiency. They evaluated the cardiovascular function of laboratory animals after the administration of IV saline spiked with 210 μg/L cyclohexanone (the lowest concentration detected in EC fluids) and concluded that it induced significant adverse changes in all cardiovascular variables.

Karle et al. [26] evaluated the extraction rate of DEHP from PVC extracorporeal membrane oxygenation tubing and its relationship to time and

TABLE 9.13 Influence of Circuit Design on the Extraction Rate of DEHP from ECMO Tubing

	Circuit Characteristics							
					DEHP (μg/L)			
Circuit[a]	Surface Area (cm^2)	Diameter (cm)	Length (cm)	Circulating Volume (mL)	Saline	Albumin	Blood	Extraction Rates (μg/mL/h)
A	932	0.6	87.5	800	n.d.	n.d.	18.3	0.32 ± 0.12
B	1563	1.0	270	1000	n.d.	n.d.	21.8	0.57 ± 0.14
C (heparin bonded)	932	0.6	87.5	800	n.d.	n.d.	19.3	−0.2 ± 0.04

Each circuit was primed with a normal saline solution, 25% albumin solution, and then two units of human packed red blood cells, collected in citrate phosphate dextrose adenine-1 solution.
[a] Flow rate: 400 mL/min; each circuit contained a silicone membrane oxygenator (0.8 m^2), a heat exchanger that maintained the blood temperature at 36.5–37.0°C, and a 50-mL silicone reservoir (bladder).
Adapted from Reference 26.

circuit design. Three clinical circuit designs were studied (Table 9.13). Each circuit was primed with normal saline solution, 25% albumin, and then red blood cells in citrate solution. The DEHP concentrations during the normal saline and albumin prime were nondetectable, but at the end of the blood prime, the mean concentrations increased, as can be seen in Table 9.13. These values were considered zero point for calculations of the extraction rates for the following 48-hour circulations. While the extraction rates increased in circuits A and B, the concentration of DEHP in circuit C decreased at a rate of 0.2 μg/mL/h. The study demonstrated that DEHP leaches from standard ECMO circuits at a linear rate, which is dependent on the surface area of each circuit and that DEHP does not leach from heparin-coated tubing. This decrease was attributed to the heparin bonding in circuit C, which prevented DEHP accumulation.

Contact between blood and artificial surfaces in extracorporeal circulation produces adverse effects, such as red blood cell adhesion and destruction, platelet adhesion, and activation of the coagulation cascade. To improve hemocompatibility, heparin-coated PVC is commonly utilized in cardiopulmonary bypass procedures (as showed above). Although not designed to prevent the leaching of PVC additives, this coating alters the interaction blood-tubing material and therefore may change the release of DEHP.

Haishima et al. [27] analyzed the performance of heparin-coated tubing for ECMO not only with respect to the release of DEHP but also of MEHP. Two different heparin bindings, one covalent and one ionic, were evaluated and compared to non-heparin-coated tubing. The tubes, provided by two companies, were set at 37°C and 500 mL of bovine blood circulated into each circuit at a flow rate of 1.5 L/min for 6 hours. The results in Table 9.14 show that only the covalent bond type of coating may help in reducing the release. The total amount of DEHP plus MEHP leached from noncoated tubing reached 7.8–12.6 mg, and that from the heparin ionically bound tubing reached 10.6 mg,

TABLE 9.14 Amount of DEHP and MEHP Leached from Extracorporeal Membrane Oxygenation PVC Tubing, Heparin-Coated and Noncoated, into Bovine Blood after Circulating at 37°C [27]

	Concentration (μg/L)							
	DEHP				MEHP			
	Company I		Company II		Company I		Company II	
Circulation time (h)	Noncoated	Coated (Covalent Bound)	Noncoated	Coated (Ionic Bound)	Noncoated	Coated (Covalent Bound)	Noncoated	Coated (Ionic Bound)
1	4,528.4 ± 194.0	3,140.4 ± 429.7	5,626.2 ± 266.3	5,216.0 ± 185.3	221.3 ± 31.4	142.8 ± 19.0	303.4 ± 27.9	479.6 ± 5.2
3	8,240.1 ± 193.0	4,734.0 ± 365.1	13,062.6 ± 335.2	10,676.1 ± 687.0	302.0 ± 22.6	199.6 ± 13.1	497.7 ± 37.6	626.1 ± 10.7
6	15,503.0 ± 88.5	7,480.3 ± 376.2	24,792.9 ± 530.7	20,683.2 ± 1,212.2	416.1 ± 26.3	258.5 ± 24.3	713.6 ± 52.3	696.5 ± 21.7
Total amount (mg)[a]	7.8	3.7	12.6	10.6	–	–	–	–

[a] The amount of MEHP was computed together.

while only 3.7 mg migrated into the blood sample from the covalently bound heparin tubing.

9.3 DEVICES FOR ADMINISTRATION OF AEROSOLIZED DRUGS

The devices responsible for delivering a drug product to the lungs through the mouth or the nose are roughly classified into three categories: nebulizers, metered dose inhalers (MDIs), and dry powder inhalers (DPIs). Nebulizers produce a mist of drug-containing water droplets for inhalation. MDIs and DPIs are more complex devices and have generally three parts: a container, a valve, and an actuator, or mouthpiece.

An MDI delivers a specific amount of medication to the lungs and is the most commonly used type of inhaler. In MDIs, medication is stored in a pressurized canister that contains a propellant. The canister is attached to a plastic, hand-operated actuator. On activation, the MDI releases a fixed dose of medication in aerosol form. The aerosol is created when a valve is opened by pressing down on the propellant canister, allowing liquid propellant to spray out of the canister. The drug is usually contained in small particles suspended in the liquid propellant or the drug is dissolved in the propellant. In either case, the propellant evaporates rapidly as the aerosol leaves the device, resulting in small drug particles that are inhaled.

A DPI delivers medication to the lungs in the form of a dry powder and it does not contain propellants. In DPIs, the aerosol is a powder, contained within the device until it is inhaled. The therapeutic drug is manufactured in powder form as small powder particles. In many DPIs, the drug is mixed with lactose as a bulking agent to aid in powder uptake from the device during inhalation. The medication is commonly held in a capsule for manual loading.

The most crucial component of an MDI is its propellant. The propellant provides the force to generate the aerosol cloud and is also the medium in which the active component must be suspended or dissolved. Prior to the mid-1990s, MDIs used chlorofluorocarbons (CFCs) as propellants, but with the elimination of CFCs due to ozone depletion concerns, the propellants in new MDIs are hydrofluoroalkanes (HFAs), which do not result in ozone depletion. By 2010, CFCs used in medical aerosols are to be replaced by HFAs [28].

Because the propellant makes up at least 99% of most MDI formulations, they are the most important source of impurities. The Food and Drug Administration (FDA) provided guidance for the industry on the chemistry and manufacturing of MDIs and DPIs, including acceptance criteria for propellants. Specifications for pharmaceutical-grade CFCs and HFAs are presented in Table 9.15.

Another crucial impurity in MDIs and DPIs is particulate matter because of the impact of the deposition of foreign particles in the human respiratory tract. These particles may be derived from the active ingredient, excipients, container/closure components, and from the manufacturing process.

TABLE 9.15 Specifications for Pharmaceutical-Grade Chlorofluorocarbons (CFCs) and Hydrofluoroalkanes (HFAs)

Compound	CFC-11	CFC-12	CFC-114	HFA 227	HFA 134a
Formula	CCl_3F	CCl_2F_2	$CClF_2CClF_2$	C_3HF_7	$C_2H_2F_4$
Purity	99.8%	99.8%	99.8%	≥99.9%	≥99.99%
CFC-11		2000 ppm	500 ppm	–	5 ppm
CFC-12	2000 ppm		1000 ppm	–	100 ppm
CFC-114	40 ppm	150 ppm		–	5 ppm
HCFC (sum)	105 ppm	320 ppm	370 ppm	–	1590 ppm
Other CFC (sum)	150 ppm	485 ppm	620 ppm	–	60 ppm
Methyl chloride	10 ppm	40 ppm	–	–	–
Dichloromethane	50 ppm	10 ppm	–	–	–
Chloroform	20 ppm	10 ppm		–	–
Carbon tetrachloride	20 ppm	10 ppm	–	–	–
Total chloromethanes	50 ppm	50 ppm	–	–	–
Water	10 ppm	–	–	≤10 μg/g	≤10 μg/g
Nonvolatile matter	–	–	–	≤20 ppm	≤50 ppm
Acidity	2 ppm	–	–	≤0.1 μg/g	≤0.1 μg/g

Source: http://www.solvaychemicals.us: HFA propellants for medical use.

Additionally, the International Pharmaceutical Aerosol Consortium on Regulation and Science (IPAC-RS) [29] considers other sources of contaminants, such as heavy oils in the inner surface of the canister and extractables from elastomeric gaskets and seals and from plastic mouthpieces used in MDIs and DPIs. Labels, inks, and glues applied to containers fabricated from semipermeable plastic materials may also be sources of organic extractables and leachables. In MDIs, the leaching from elastomers might be even higher than in other container/closure systems because the propellant is under pressure and therefore has its solvent capability enhanced.

Nevertheless, contaminants in MDIs and DPIs have not been effectively evaluated. The presence of polynuclear aromatic hydrocarbons (PAHs) and N-nitrosamines in MDIs was investigated by Norwood et al. [30, 31]. While PAHs were detected at levels of 4–7 μg/canister, no N-nitrosamines at the trace level of 1 ng/canister were found among the analyzed samples.[1]

9.4 REPROCESSED MEDICAL DEVICES

The practice of reprocessing medical devices approved for sale as single-use devices has been adopted in many hospitals. By in-house recycling, hospitals

[1] The origin and presence of PAHs are discussed in Chapter 7, "Closures."

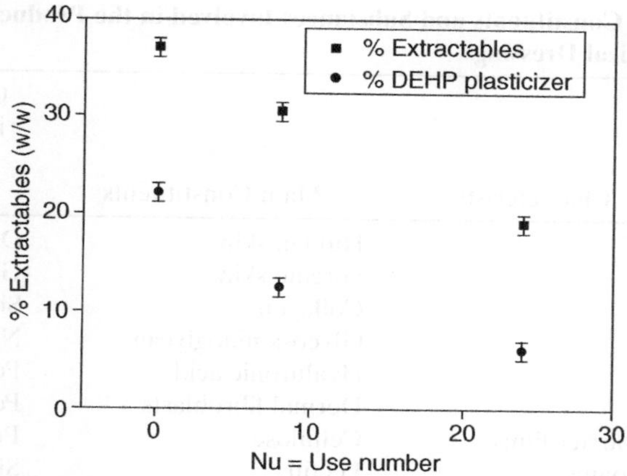

FIGURE 9.1 Squares: percentage of extractable material (w/w) obtained by supercritical fluid extraction, CO_2, 96°C, 48 MPa. Circles: percentage of DEHP plasticizer (w/w), separation, and quantification in the sample extracts performed off-line by gas chromatography [32].

can achieve substantial cost savings. Although the main concern about their reuse is the risk of infection and device malfunction, since they should be cleaned and sterilized upon every recycling, the potential amount of extractables remaining in the material may also change. From the point of view of extractables and leachables, recycling may help to reduce the release. A study with reused PVC catheters [32] showed that the total amount of extractables, and DEHP specifically, decreased with the number of reuses. Figure 9.1 shows that the amount of DEHP and the total extractables released after 10 reuses decreased by approximately 10 percentage points (pp).

9.5 TISSUE SUBSTITUTES

Probably, the most important issue dealing with tissue substitutes is sterility. The risk inherent to allografts (skin, bones, collagens) is disease transmission. Usually, products with proteic constitution are sensitive to heat and are, therefore, sterilized by techniques other than steam or dry heat. On the other hand, for devices made of synthetic materials, which come into intimate contact with the human body, sterilization is also an important issue since many of these cannot bear steam or dry heat sterilization without suffering permanent deformity. Options for sterilization methods include gas and ionizing radiation. Contamination arising from the effect of gas and radiation on natural and synthetic polymers is discussed in Chapter 12.

TABLE 9.16 Main Constituents and Substances Involved in the Production of Skin Substitutes and Surgical Dressings

Skin Substitutes and Surgical Dressings	Characteristics	Main Constituents	Other Substances in the Product or Involved in the Production
Grafts		Human skin Porcine skin	Dimethylsulfoxide Glycerol
Biological skin substitutes		Collagen Glycosaminoglycan Hyaluronic acid Dermal fibroblasts	Silicone Nylon Polyglycolic acid Polyglactin
Synthetic and semisynthetic dressings	Barrier films Foams Tissue adhesives	Cellulose Pectin Gelatin Alginates	Petroleum jelly Silicone sheet or gel Polyurethane film or foam
Synthetic dressings containing antimicrobial agent	Silver dressing	Rayon/polyester laminated between two silver-coated high-density polyethylene (HDPE) layers	Activated charcoal Silver
	Iodine dressing		Iodine
	Gentamicin dressing		Gentamicin

9.5.1 Skin Substitutes and Surgical Dressings

Wound coverings can be divided into two categories: permanent and temporary. Permanent coverings are called autografts, which are epidermis from healthy skin areas obtained from the patient himself. Temporary coverings include allografts (a skin not from the patient himself), biological skin substitutes, and synthetic or semisynthetic dressings [33]. Synthetic or semisynthetic dressings contain synthetic materials or natural nonbiologically active substances. Table 9.16 presents types of surgical dressings and skin substitutes and the materials involved in their production. Many dressings include substances with bioactive properties to promote healing and to control infection. These dressings may contain sustained-release antimicrobial agents such as iodine or silver compounds and gentamicin.

9.5.2 Hard Tissue Substitutes

There are several categories of bone graft substitutes, encompassing a variety of materials, both natural and synthetic. Many are formed from composites of one or more types of material, but they are usually built on a base material [34]. A classification of bone graft substitutes can be based on the nature of the materials employed. This classification and the materials most usually employed as hard tissue substitutes are presented in Table 9.17.

TABLE 9.17 Main Constituents and Substances Involved in the Production of Hard Tissue Substitutes

Hard Tissue Substitutes	Characteristics	Main Constituents
Allograft based		Demineralized bone matrix
Ceramic based		Calcium sulfate
	Phosphates	Tricalcium phosphate
		Synthetic hydroxyapatite
		Coralline
		Sea coral
	Bioactive glass	SiO_2, CaO, Na_2O, P_2O_5
Polymer based	Degradable polymers	Polylactic acid
		Poly(lactic-co-glycolic acid)
Metal-based	Alloys	Stainless steel 316L: (62–72% Fe, 16–18% Cr, 10–14% Ni, 2–3% Mo, 2% Mn)
		Cobalt-based Co/Cr alloy: (40% Co, 20% Cr, 16% Ni, 7% Mo)
		Titanium-based:
		Nitinol (54.5% Ni, 45.5% Ti)
		Ti6Al4V (90% Ti, 6% Al, 4% V)

9.5.3 Soft Tissue Substitutes

Together, dermal fillers and breast implants can be considered soft tissue substitutes, which may have reconstructive or cosmetic purposes. By definition, a dermal filler is a product that is injected or placed into the dermis [35, 36]. In current practice, several dermal fillers are available for use, in addition to subdermal fillers, which are those placed underneath the dermis in the subcutis. The origin of dermal fillers can be nonanimal, animal, and synthetic. Within these categories, collagen and silicone are the most popular materials (Table 9.18). Silicone breast implants have been available since the 1960s, and although they have evolved and improved, the silicone rubber shell covered with polyurethane foam continues to be basic to all implants. Although the prosthesis is mainly made of silicone, it is the thin polyurethane foam (1–2 mm) that stays in contact with the body and may cause the leaching of undesirable species.

This type of foam is manufactured from toluene diisocyanate (TDI), 2,4- and 2,6-TDI in a ratio of 4:1, a polyol, and water (see Table A.1, Appendix). In the production, water reacts with TDI to form carbon dioxide, which enhances the foaming process. However, the reaction of TDI with water also produces 2,4- and 2,6-toluene diamines (TDAs), which have been reported as contaminants in TDI-based polyurethane foams.

Benoit [37] conducted a careful study on breast implants in order to determine how much TDI and TDA contamination was already present in the PU foam and how much was formed by degradation under physiological

TABLE 9.18 Main Constituents and Substances Involved in the Production of Soft Tissue Substitutes

Soft Tissue Substitutes	Design	Main Constituents	Other Substances in the Product or Involved in the Production
Breast implant	Implantable	Silicone	Silica filler
		Polyurethane	
Dermal filler	Injectable	Bovine collagen	Glutaraldehyde
		Human-derived collagen	Lidocaine
		Porcine collagen	Phosphate
		Botolinum toxin	Sodium chloride
		Hyaluronic acid	D-ribose
		Polymethylmethacrylate	Glycerin
		Polylatic acid	Carboxymethylcellulose
		Calcium hydroxyapatite	Albumin

TABLE 9.19 Amount of 2,4-TDA, 2,6-TDA, TDI, and Toluene Isocyanate Amine (TIA) Found in the Extracts of the Polyurethane Foam Removed from New Breast Implants [37]

Sample Lot	2,6-TDA (μg/g)	2,4-TDA (μg/g)	TDI (μg/g)	TIA (μg/g)
A	2	18	270	64
A	31	288	477	453
A	5	31	367	113
B	13	75	128	124
B	18	137	1036	388
B	71	692	1013	806
B	6	49	433	160
C	6	40	119	219
D	29	231	373	399
D	21	194	868	419
D	36	360	920	780
E	21	181	347	328
E	34	310	556	537
E	10	117	621	291
E	22	199	475	395
Average	22 ± 17	195 ± 168	534 ± 290	354 ± 226

conditions. Before incubation at physiological conditions, the samples (PU foam removed from new breast implants) were cleaned by Soxhlet acetone extraction for a period of 148 hours. The result of the solvent analysis showed that the samples contained not only TDA but also TDI, polydiethyleneglycol adipate, a variety of polydimethylsiloxanes, and 2,3-dimethyl-6-t-butylphenol. Table 9.19 shows the amount of 2,4-TDA, 2,6-TDA, TDI, and toluene isocyanate amine (TIA) found in the extracts. The levels of contaminants varied widely from foam to foam, even within the same lot. The amount of 2,4-TDA, the toxic isomer, ranged from 18 to 692 μg/g, but the simultaneous presence of TDI and TIA could lead to the formation of more 2,4-TDA since both can react with water to form 2,4-TDA.

Although collagen is the most important dermal filler because it is the major structural component of the skin, other substances such as hyaluronic acid and polylatic acid have also been used as dermal fillers. Collagen may come from different origins. Bovine collagen was the first FDA-approved dermal filler, and since 2003, a newer bioengineered human collagen has also been approved for injection.

CONCLUDING REMARKS

Medical devices comprise several items used in the administration of drugs and in patients' handling. In this chapter, gloves, syringes, ECs, inhalers, and implants were considered. Most devices are made of the same polymers as plastic containers and closures, and therefore the same contaminants can be found among the leachables and extractables. Implants, on the other hand, besides synthetic polymers, encompass inorganic and natural materials. These substances carry their own impurities, which, depending on the environmental conditions, may migrate from the article into the surrounding tissue.

REFERENCES

1. MHRA. Medical Devices and Medicinal Products, United King Government, Bulletin No. 17, September 2008.
2. United States Pharmacopeia, USP 32 Chapter <1031> The biocompatibility of materials used in drug containers, medical devices and implants, 2009.
3. Rose RF, Lyons P, Horne H, Mark Wilkinson S. A review of the materials and allergens in protective gloves. Contact Dermatitis 61(3), 129–137, 2009.
4. Aalto-Korte K, Alanko K, Henriks-Eckerman M, Estlander T, Jolanki R. Allergic contact dermatitis from bisphenol A in PVC gloves. Contact Dermatitis 49(4), 202–205, 2003.
5. Matthieu L, Godoi AFL, Lambert J, van Grieken R. Occupational allergic contact dermatitis from bisphenol A in vinyl gloves. Contact Dermatitis 49(6), 281–283, 2003.
6. Depree GJ, Bledsoe TA, Siegel PD. Survey of sulfur-containing rubber accelerator levels in latex and nitrile exam gloves. Contact Dermatitis 53(2), 107–113, 2005.
7. Aalto-Korte K, Ackermann L, Henriks-Eckerman M, Välimaa J, Reinikka-Railo H, Leppänen E, Jolanki R. 1,2-Benzisothiazolin-3-one in disposable polyvinyl chloride gloves for medical use. Contact Dermatitis 57(6), 365–370, 2007.

8. Pontén A. Formaldehyde in reusable protective gloves. Contact Dermatitis 54(5), 268–271, 2006.
9. Knudsen BB, Hametner C, Seycek O, Heese A, Koch H-U, Peters K-P. Allergologically relevant rubber accelerators in single-use medical gloves. Contact Dermatitis 43(1), 9–15, 2000.
10. Bernstein RK. Clouding and deactivation of clear (regular) human insulin: association with silicone oil from disposable syringes? Diabetes Care 10(6), 786–787, 1987.
11. Chantelau E, Berger M, Böhlken B. Silicone oil released from disposable insulin syringes. Diabetes Care 9(6), 672–673, 1986.
12. Baldwin RN. Contamination of insulin by silicone oil: a potential hazard of plastic insulin syringes. Diabetic Medicine 5(8), 789–790, 1988.
13. Bakri SJ, Ekdawi NS. Intravitreal silicone oil droplets after intravitreal drug injections. Retina (Philadelphia, Pa.) 28(7), 996–1001, 2008.
14. Sharma B, Bader F, Templeman T, List P, Ryan M, Heavner GA. Technical investigation into the cause of the increased incidence of anti-body-mediated pure red cell aplasia associated with EPREX®. European Journal of Hospital Pharmacy 5, 86–91, 2004.
15. Hamilton G. Contamination of contrast agent by rubber components of 50-mL disposable syringe. Radiology 152, 539–540, 1984.
16. Petersen MC, Vine J, Ashley JJ, Nation RL. Leaching of 2-(2-hydroxyethylmercapto) benzothiazole into contents of disposable syringes. Journal of Pharmaceutical Sciences 70(10), 1139–1143, 1981.
17. Hamilton G. Contamination of contrast agent by MBT in rubber seals. Canadian Medical Association Journal 136(10), 1020–1021, 1987.
18. Lau DW, Law S, Walker SE, Iazzetta J. Dexamethasone phosphate stability and contamination of solutions stored in syringes. PDA Journal of Pharmaceutical Science and Technology 50(4), 261–267, 1996.
19. Daugirdas JT, Ing TS. *Handbook of Dialysis*, 2nd ed. Little, Brown, Boston, 1994.
20. Haishima Y, Hayashi Y, Yagami T, Nakamura A. Elution of bisphenol-A from hemodialyzers consisting of polycarbonate and polysulfone resins. Journal of Biomedical Materials Research 58(2), 209–215, 2001.
21. Mettang T, Thomas S, Kiefer T, Fischer FP, Kuhlmann U, Wodarz R, Rettenmeier AW. The fate of leached di(2-ethylhexyl)phthalate in patients undergoing CAPD treatment. Peritoneal Dialysis International 16(1), 58–62, 1996.
22. Kuhn-Velten WN, Kramer U, Susanto F, Schram J, Reinauer H. Differential contribution of albumin and lipids to the hydrophobic extraction of bis-(2-ethylhexyl) phthalate from hemodialysis tubing by human serum: analysis by gas chromatography—stable isotope dilution mass spectrometry. International Journal of Environmental Analytical Chemistry 80(4), 245–256, 2001.
23. Kambia K, Dine T, Azar R, Gressier B, Luyckx M, Brunet C. Comparative study of the leachability of di(2-ethylhexyl) phthalate and tri(2-ethylhexyl) trimellitate from haemodialysis tubing. International Journal of Pharmaceutics 229(1-2), 139–146, 2001.
24. Snell RP. Gas chromatographic determination of cyclohexanone leached from hemodialysis tubing. Journal of AOAC International 76(5), 1127–1132, 1993.

25. Thompson-Torgerson CS, Champion HC, Santhanam L, Harris ZL, Shoukas AA. Cyclohexanone contamination from extracorporeal circuits impairs cardiovascular function. American Journal of Physiology. Heart and Circulatory Physiology 296(6), H1926–H1932, 2009.
26. Karle VA, Short BL, Martin GR, Bulas DI, Getson PR, Luban NL, O'Brien AM, Rubin RJ. Extracorporeal membrane oxygenation exposes infants to the plasticizer, di(2-ethylhexyl)phthalate. Critical Care Medicine 25(4), 696–703, 1997.
27. Haishima Y, Matsuda R, Hayashi Y, Hasegawa C, Yagami T, Tsuchiy T. Risk assessment of di(2-ethylhexyl)phthalate released from PVC blood circuits during hemodialysis and pump–oxygenation therapy. International Journal of Pharmaceutics 274(1-2), 119–129, 2004.
28. Noakes T. Medical aerosol propellants. Journal of Fluorine Chemistry 118, 35–45, 2002.
29. Leachables and Extractables Technical Team. Leachables and Extractables Testing: Points to Consider, 2001. Available at http://www.ipacrs.com/PDFs/Points_to_Consider_FINAL.PDF (accessed November 2011).
30. Norwood DL, Paskiet D, Ruberto M, Feinberg T, Schroeder A, Poochikian G, Wang Q, Deng TJ, DeGrazio F, Munos MK, Nagao LM. Best practices for extractables and leachables in orally inhaled and nasal drug products: an overview of the PQRI recommendations. Pharmaceutical Research 25(4), 727–739, 2008.
31. Norwood DL, Mullis JO, Feinberg TN, Davis LK. N-nitrosamines as "special case" leachables in a metered dose inhaler drug product. PDA Journal of Pharmaceutical Science and Technology 63(4), 307–321, 2009.
32. Granados DL, Jimenez A, Cuadrado TR. Assessment of parameters associated to the risk of PVC catheter reuse. Journal of Biomedical Materials Research 58(5), 505–510, 2001.
33. Dieckmann C, Renner R, Milkova L, Simon JC. Regenerative medicine in dermatology: biomaterials, tissue engineering, stem cells, gene transfer and beyond. Experimental Dermatology 19(8), 697–706, 2010.
34. Laurencin C, Khan Y, El-Amin SF. Bone graft substitutes. Expert Review of Medical Devices 3(1), 49–57, 2006.
35. Bray D, Hopkins C, Roberts DN. A review of dermal fillers in facial plastic surgery. Current Opinion in Otolaryngology & Head and Neck Surgery 18(4), 295–302, 2010.
36. Klein AW, Elson ML. The history of substances for soft tissue augmentation. Dermatologic Surgery 26(12), 1096–1105, 2000.
37. Benoit FM. Degradation of polyurethane foams used in the Même breast implant. Journal of Biomedical Materials Research 27(10), 1341–1348, 1993.

PART II

PHYSICAL CONTAMINATION

10

PARTICULATE MATTER

Particulate contamination can be classified as intrinsic or extrinsic, depending on its origin. Intrinsic contaminants arise from the manufacturing, packaging, transport, and storage of solutions; extrinsic contaminants are introduced or generated during drug reconstitution and administration to the patient.

The presence of particulate matter, while undesirable in any pharmaceutical product, is truly a problem in intravenous and ophthalmic preparations. The most common particulates in intravenous preparations are glass fragments, from the opening of glass ampoules, particles from rubber stoppers and intravenous equipment, and particles from plastic syringes.

Pharmacopeial compendia prescribe the examination of particulate contamination for injections and infusion solutions and consider particulate contamination as the presence of extraneous mobile undissolved particles, other than gas bubbles, unintentionally present in the solution. They limit the number of particles according to their size and the volume of the preparation. Pharmacopeial count limits for particulates in parenteral solutions are given in Table 10.1. The limit depends on the method used for determination and also on the volume of the sample. Two different procedures for determination are generally proposed, a light obscuration particle count test (LO) and a microscopic particle count test (M), since neither is applicable to all kinds of samples.

Generally, particulate matter is eliminated by filtration, but the word "elimination" or even the concept of "being substantially free of…," as appears in

Sources of Contamination in Medicinal Products and Medical Devices, First Edition. Denise Bohrer.
© 2013 by John Wiley & Sons, Inc. Published 2013 by John Wiley & Sons, Inc.

TABLE 10.1 Limits for Particulate Contamination in Infusion Solutions Established by Pharmacopeial Compendia

Compendia	Volume	Particle Size (μm)	LO Limit	M Limit
BP	SVP	≥10	6000/container	3000/container
		≥25	600/container	300/container
	LVP	≥10	25/mL	12/mL
		≥25	3/mL	2/mL
USP	SVP	≥10	6000/container	–
		≥25	600/container	–
	LVP	≥10	25/mL	12/mL
		≥25	3/mL	2/mL
JP	SVP	≥10	6000/container	3000/container
		≥25	600/container	300/container
	LVP	≥10	25/mL	12/mL
		≥25	3/mL	3/mL
Ph. Int.	SVP, LVP	–	One or more particles in more than one container	–

The limits are related to the method used for the determination.
LO, light obscuration particle count test; M, microscopic particle count test; SVP, small-volume parenteral (≤100 mL); LVP, large-volume parenteral (>100 mL); JP, Japanese Pharmacopeia; Ph. Int., International Pharmacopoeia.

some compendia, are levels practically impossible to reach. In fact, compliance with pharmacopeial prescriptions is limited to the removal of particles above a certain size, since the elimination of particles becomes increasingly more difficult the smaller the particle is. The difficulty to eliminate particles seems to increase in an exponential fashion as the size decreases [1].

It is possible to display the relationship between size and number of particles graphically. Bikhazi et al. [2] reported a logarithmic relationship between the number of particles and their diameter in a solution. Thus, using the limits prescribed by the British Pharmacopoeia (BP), the graphical representation would be as shown in Figure 10.1.

The author, however, considers that if the number/size distribution of particles follows a random pattern, graphically represented in Figure 10.1, it cannot be considered contamination. On the other hand, if the solution no longer contains random identities but rather is dominated by particles such as starch grains from a stopper composition, glass shards, from a delaminating container, skin fragments, or clothing fibers, a nonrandom behavior would occur and be defined as "contamination." In other words, only positive deviation from the log–log particle-size distribution curve, constructed using pharmacopeial limits, for example, would be considered contamination. Figure 10.2 graphically displays the particulate matter found in 5% dextrose solutions. The

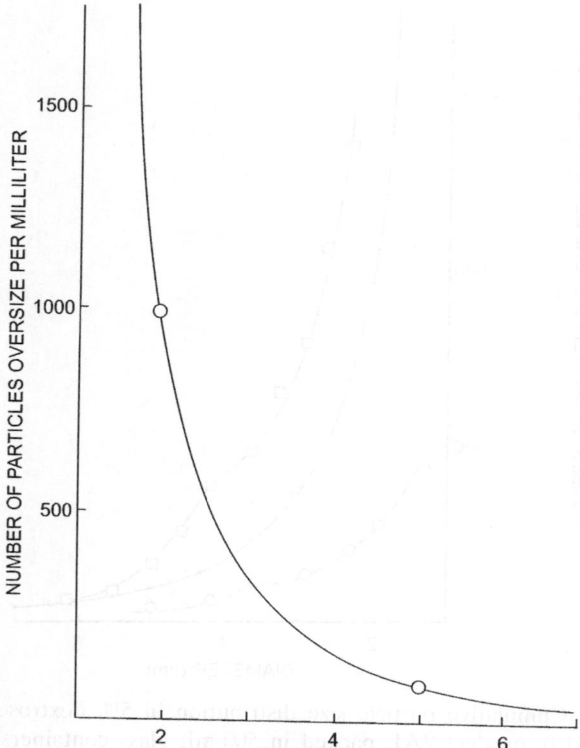

FIGURE 10.1 Coordinate plot of log of number of particles oversize per milliliter versus log particle diameter for BP specifications [2].

smooth curve represents BP specifications and the other two represent the samples. One product is within the specifications (2G) and the other is not (2AL). According to the authors, while only carbon black particles with sizes between 1 and 3 μm were identified in the 2G sample, carbon black, starch grains, lacquer flakes, and rubber with sizes up to 100 μm were found in the 2AL sample [2].

Techniques for counting particulate matter allow a discrimination of the particles by sizes even lower than the pharmacopeial limits and, though not prescribed, it is also possible to characterize particulate matter by its composition. Techniques such as optical microscopy and scanning electron microscopy (SEM) give information on particle shape and texture, while the elemental composition can be determined through the use of energy-dispersive X-ray analysis (EDXRA). The elemental composition allows a determination of the nature, and therefore the origin, of the particulate matter. Figure 10.3 displays the EDXRA spectra of particles of different origins [3]. Although glass and

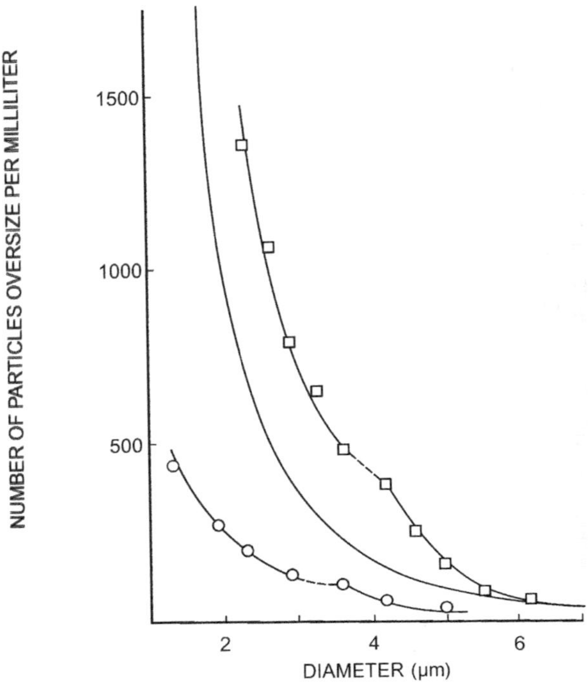

FIGURE 10.2 Cumulative particle size distribution in 5% dextrose solutions for injection. Key: (□), product 2AL packed in 500-mL glass containers with lacquer-coated rubber stopper; (○), product 2G packed in 500-mL plastic bags; and smooth curve, BP specification [2].

rubber particles present similar spectra (Fig. 10.3a,b), displaying mainly silicon and aluminum, the presence of zinc and sulfur distinguishes the spectrum obtained from a rubber particle. Zinc oxide and sulfur are the main constituents of rubber and are not ordinary glass ingredients (see Chapter 7, "Closures"). Figure 10.3c displays the spectrum of a stainless steel particle exhibiting mainly iron, and Fig. 10.3d shows the spectrum of a barium sulfate particle. Type I glass often contains a small percentage of barium oxide, which has been shown to leach into the product as barium ions. If the formulation contains a sulfate salt, or is stabilized by bisulfite, barium sulfate is able to form a fine precipitate.

Despite the existence of multiple sources of particulate matter, and the presence of materials such as cotton and rayon fibers [4], globules of silicon grease, and even hair [5], glass fragments, and rubber cores are the most frequently found materials.

The opening of glass ampoules is an important source of particulate matter in products for infusion. Since the 1970s, there have been reports of

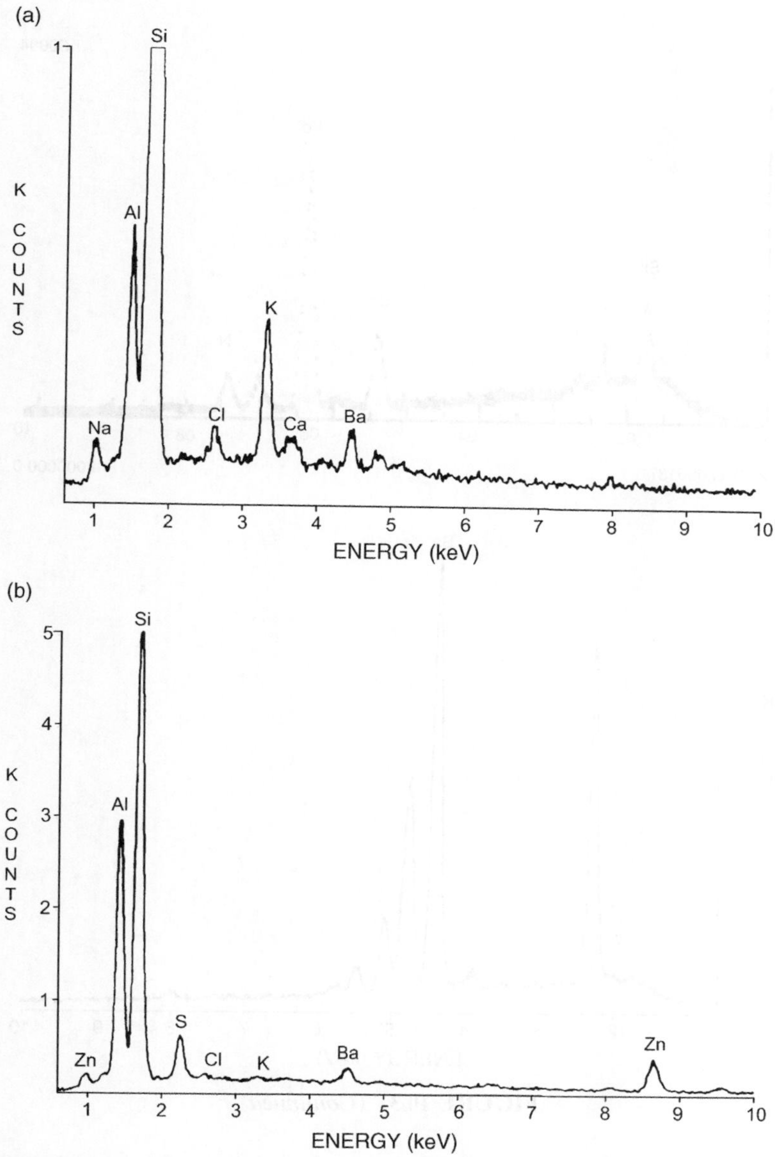

FIGURE 10.3 Energy-dispersive X-ray analysis (EDXRA) spectra of particles of (a) glass, (b) rubber, (c) stainless steel, and (d) barium sulfate [3].

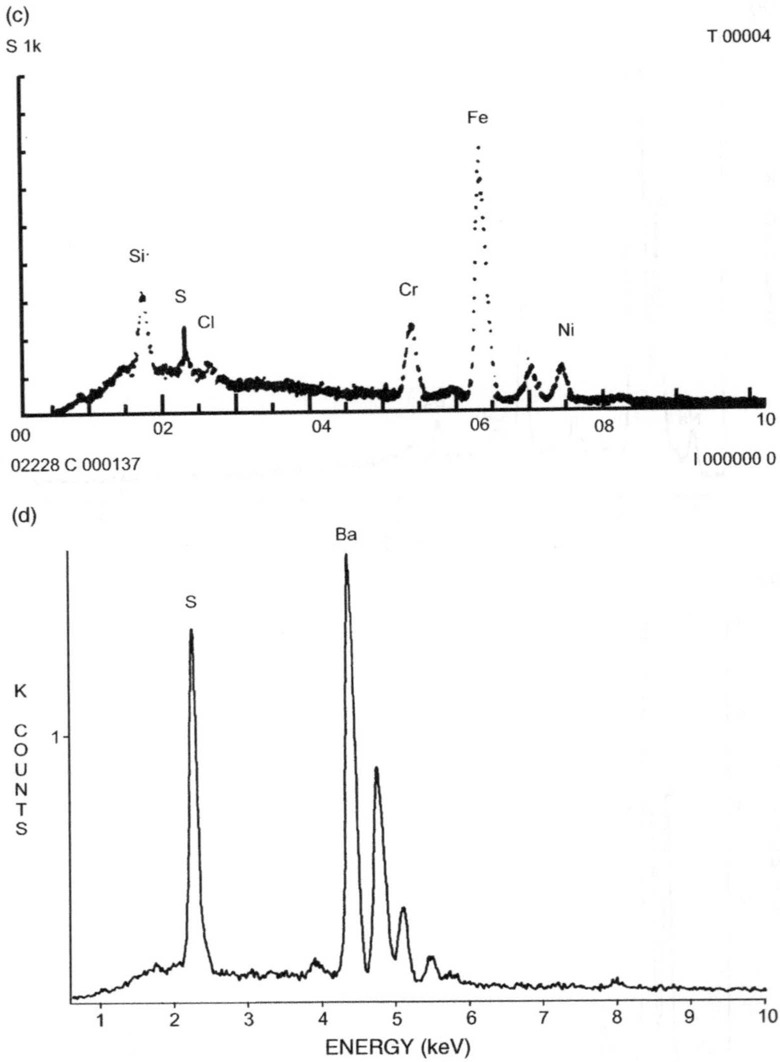

FIGURE 10.3 (*Continued*)

contamination of glass particles in ampoules generated by fragmentation of glass during ampoule opening. Table 10.2 presents several cases and studies on the presence of fragments in glass ampoules.

From the study conducted by Carbone-Traber and Shanks (Reference [11] in Table 10.2), it is possible to see that 1-mL ampoules had significantly fewer particles than either 5- or 10-mL ampoules. The authors supposed that the reason is the smaller diameter of the opening, which is 5 mm in the 1-mL

TABLE 10.2 Particles Found in Glass Ampoules

Product or Ampoule Features	Volume	No. of Samples	Finding	Year	Reference
Furosemide		–	Fragments >5 μm	1972	[6]
Furosemide	2 mL	–	Several hundred particles after opening	1972	[7]
Anesthetic	2 mL	A number of samples	Fragments 3–4 mm	1973	[8]
Local anesthetic	–	1	Minute particles after opening	1974	[9]
Lidocaine HCl	–	1	5–10 particles 20–100 μm long	1974	
Spinal-22		1	Numerous smaller particles		
Water for injection	10 mL	5 (single lot)	All ampoules contained visible glass particles	1985	[10]
Not identified	1 mL	10 (single lot)	14 ± 4 particles (average)	1986	[11]
Not identified	5 mL	10 (single lot)	28 ± 6 particles (average)	1986	
Not identified	10 mL	10 (single lot)	32 ± 6 particles (average)	1986	
Transparent ampoules metal etched	10 mL	20 (single lot)	45.9 ± 15.4 particles (average)	1989	[12]
Transparent ampoules chemically scored	10 mL	20 (single lot)	3.2 ± 0.9 particles (average)	1986	
Amber ampoules metal etched	10 mL	20 (single lot)	6.0 ± 1.7 particles (average)	1986	
Amber ampoules chemically scored	10 mL	20 (single lot)	3.1 ± 0.6 particles (average)	1986	
Propofol	20 mL	1	4 particles 2.5–17.0 mm long present before opening	1990	[13]
Etomidate emulsion	–	9	Total of 9 particles computing all samples	2003	[14]
Different medications from operating theaters	varied	510	Total of 113 particles computing all samples	2003	
Different medications from hospital pharmacy	1 mL	18	4 glass particles (10–50 μm)	2004	[15]
Different medications from hospital pharmacy	2 mL	18	71 glass particles (5–60 μm)	2004	
Atropine, neostigmine, ephedrine, adrenaline	1 mL	30	1 glass particle	2008	[16]
Fentanyl, suxamethonium, ranitidine, metoclorpramid, frusemide	2 mL	30	2 particles	2008	
Atracurium, dopamine, dobutamine	5 mL	30	3 glass particles	2008	
Calcium chloride, glyceryl trinitrate, dextrose 50%	10 mL	30	5 glass particles	2008	

TABLE 10.3 Glass Particles in Ampoules Larger and Smaller Than 2 mL Metal Etched and Color Scored with Different Wall Thickness

	Wall Thickness			
	0.70–0.80 mm		0.80–0.90 mm	
Ampoule Type	No. Ampoules	No. Particles (Average)	No. Ampoules	No. Particles (Average)
Metal etched				
≤2 mL	162	0.14	20	0.35
>2 mL	13	0.46	55	0.47
Color scored				
>2 mL	–	–	93	0.24

Adapted from Reference 14.

ampoule in contrast to 7 and 9 mm in the 5- and 10-mL ampoules, respectively. In another investigation [14], ampoules were pooled into sizes larger or smaller than 2 mL, and the results, presented in Table 10.3, confirmed the finding that smaller ampoules produce fewer particles. The study also measured the wall thickness of the ampoules, and those with thicker walls produced a greater number of particles, although for ampoules larger than 2 mL, no significant difference was found. Among the 510 ampoules collected from operating theaters and analyzed in this study, 92 were color scored and 417 were metal etched. The comparison between color-scored and metal-etched ampoules of the same size showed the number of particles to be greater in metal-etched ampoules. A similar result was found by Sabon et al. [12], who evaluated transparent and amber ampoules, as well as metal etched and color scored ampoules. The results, also presented in Table 10.2, showed that the number of particles is higher in metal-etched ampoules when compared with either transparent or amber chemically scored ampoules.

Roseman et al. [17] observed flakes in solutions of unopened glass ampoules (Type I glass ampoules). They observed that all flakes had similar characteristics (colorless, platelike, <1 μm in thickness), but their sizes ranged from a few micrometers to a hundred micrometers in length. The elemental analysis of four flakes gave the results presented in Table 10.4. Because the flakes presented the same elements of glass, the authors broke a glass ampoule and analyzed their fragments (results also presented in Table 10.4). Since the compositions of the flakes and glass fragments were similar to each other, the authors believed that the flakes originated in the ampoule due to an attack on the glass surface by some chemical substance. It has been shown that, besides strong alkali or hydrofluoric acid, other substances such as EDTA and other organic acids such as citric acid, as well as phosphate salts, are able to attack glass. In this case, the existence of such particulates in glass ampoules was not generated by the breaking of the ampoule but was attributed to a delamination process that could occur just above the bottom of the ampoule, a region of

TABLE 10.4 Elemental Analysis of Flakes Found in a Solution Stored Glass Ampoule and of Glass Fragments Resulting from a Broken Glass Ampoule

	Composition (%)				
Element	Flake 1	Flake 2	Flake 3	Flake 4	Broken Glass[a]
Silicon	25–30	35–45	30–35	25–30	30–40
Aluminum	10–12	3–5	8–10	10–11	2–5
Potassium	4–6	n.d.	3–5	3–4	n.d.
Calcium	1–2	n.d.	n.d.	1–2	n.d.
Boron	3–8	2–6	3–8	Present	4–10
Sodium	1–3	1–2	1–3	1–2	4–6

All ampoules were of Type I glass.
[a] Mean of four samples; n.d., not detected.
Adapted from Reference 17.

thermal stress because of the intense heat used to form the bottom of the container.

If glass particles are originated by fragmentation during the breaking of glass ampoules, rubber particles or "coring," as they are called in the medical jargon, are generated during the insertion of a needle through the stopper of a medication vial. The rubber is shaved when the needle is introduced through the stopper. Several reports on the occurrence of coring have been published in recent years. Table 10.5 lists some of these reports and their findings. From the reports by Seto et al. [20] and Riess and Strong [25], pictures show the presence of particles clearly visible to the naked eye (Figs. 10.4 and 10.5).

Asakura et al. [27] investigated the occurrence of coring in insulin cartridge formulations for pen-type injectors. The pen-type insulin injector has been developed for patients to mount and self-inject insulin from a cartridge several times. For this investigation, three samples were collected from 30 pen-type injectors: the first two insulin units delivered (sample A), the second two insulin units delivered (sample B), and the insulin solution remaining in the cartridge (sample C). The result of the analysis is presented in Table 10.6. It was found that coring occurred at the very high ratio of 97% of the cartridges. Based on the size and shape of the particles, the conclusion was that they originated from the needle shaving off rubber particles from the stopper when it is mounted on the injector. As the 31-gauge needle of the injector has an inner diameter of approximately 0.15 mm, it is wide enough to let the particles found in samples A and B to pass through.

The number of particles generated by glass and plastic ampoules after ampoule opening was counted and compared [28]. The graphics in Figure 10.6 show that plastic ampoules produced a significantly lower number of particles (of all sizes) than did glass ampoules. This result is expected since plastic is not a rigid material, and the possibility of fragmentation by opening is smaller than it is for glass.

TABLE 10.5 Particles Originated from Rubber Parts of Vials and Syringes

Product Features	Volume	No. Samples	Method of Investigation	Finding	Year	Reference
Diprivan vial	50 mL	1	–	Foreign body due to coring from the rubber cap	1996	[18]
Spinal anesthesia glass syringe	–	1	Spinal tray was opened and bupivacaine solution was drawn with the disposable glass syringe	Numerous foreign bodies floating	1994	[19]
IV solutions in glass bottles with rubber bungs	500 mL	50	Seals were pierced once with a IV infusion set	1–5 small fragments visible to the naked eye in all bottles (see Fig. 10.4)	1998	[20]
Diprivan 1% vial	50 mL	1	–	Two fragments due to coring from the rubber cap	2001	[21]
Propofol vial	50 mL	100	Drawn propofol into a syringe from the vial using a 18G injection needle and observation of the syringes	18 syringes presented rubber fragments	2001	[22]
Propofol syringe	–	1	Aspiration of lidocaine into propofol from an Abboject® syringe	Foreign body sheared from the syringe plunger	2006	[23]
Botulinum toxin A injections	2 mL	1	2 mL saline introduced into the Botox bottle	Rubber core from the stopper	2007	[24]
Propofol vial	20 mL	1	Drawn propofol through a blunt plastic cannula from the sealed vial	Gray rubber core ca. 0.4 × 1.5 mm (see Fig. 10.5)	2008	[25]
Prefilled glass syringe containing a biological drug product	–	–	The particle originated from a polymeric pin used during syringe manufacture	Black particle ~300 microns visually observed containing glass fibers, Nylon-6, and Nylon-MXD6	2010	[26]

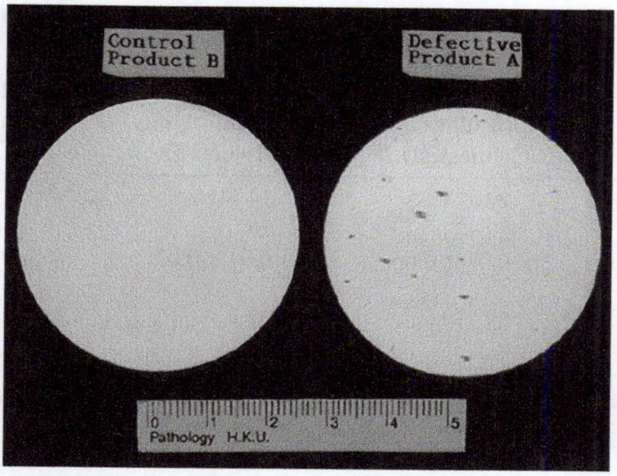

FIGURE 10.4 Membrane filters of two samples of IV solutions from different manufacturers stored in glass bottles (500 mL) with rubber bungs [20].

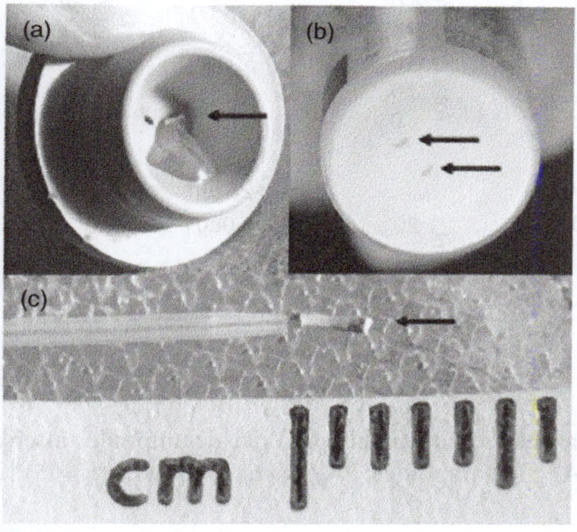

FIGURE 10.5 (a) A blunt plastic cannula pushed through the rubber stopper of a propofol vial cut out a single rubber core (arrow). (b) Two rubber cores (arrows) on the bottom of the vial only visible when inspected from below. (c) Rubber core (arrow) identical to the one in our patient in relation to the used 24-gauge angiocath [25].

The container/closure system of iodine-based contrast media was evaluated and compared as a source of particulate matter [29]. The products were packed in glass ampoules and in rubber-capped vials. Ampoules were opened in the conventional manner; however, the vials were opened so as not to pierce the rubber membrane by removing the cap at the same time as the metal ring. In both cases, the total volume, 20 mL, was emptied into the counting cuvette.

TABLE 10.6 Coring Occurrence in Insulin Pen-Type Injector after the First and Second Injection [27]

	Sample A[a] (First 2 Insulin Units Injected)	Sample B[a] (Second 2 Insulin Units Injected)	Sample C[a] (Solution Remaining in the Cartridge)
Coring occurrence	22 (73%)	14 (47%	29 (97%)
Rubber pieces	29	26	58
Particle size (mm)	0.07 ± 0.06	0.04 ± 0.02	0.18 ± 0.15
Particle shape			
Aggregate	22	22	25
Needle-like	7	4	32
Spiral	0	0	1

[a] $n = 30$.

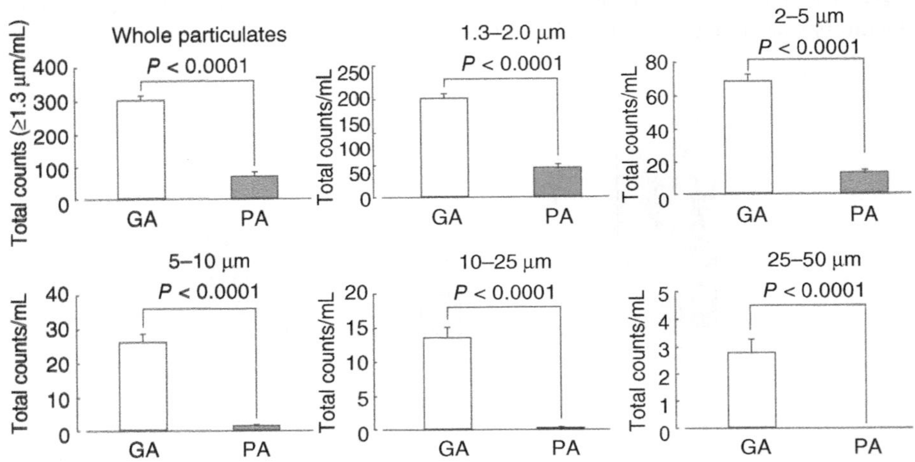

FIGURE 10.6 Particle counts in glass and plastic ampoules after opening according to particle size. GA, glass ampoules; PA, plastic ampoules ($n = 15$) [28].

The results showed that, for most products, the higher number of particles was found in the glass ampoules (Table 10.7), which is in agreement with other studies that demonstrated glass to be more prone than other materials to generate particles due to fragmentation. However, one must keep in mind that the vials were not emptied by conventionally penetrating the membrane but rather by removing the entire cap intact. Therefore, the results do not represent an authentic scenario.

Foroni et al. [30] measured and characterized inert particles in individual solutions and in the final ternary mixture prepared by the sterile transfer technique. The results show that the concentration of particles in the components packaged in glass ampoules, KCl and disodium phosphate, was higher

TABLE 10.7 Number of Particles in Contrast Media Packaged in 20-mL Ampoules and in Rubber-Capped Vials

Sample	Type of Package (*n* = 10)	Particle Count/mL[a]					
		5–10 μm		10–30 μm		>30 μm	
A	Ampoule	258	320	79	120	1	13
	Vial	90	61	11	45	0	5
B	Ampoule	136	310	73	125	1	8
	Vial	107	323	25	46	1	5
C	Ampoule	301	302	171	131	3	8
	Vial	119	369	29	183	1	3
D	Ampoule	337	370	88	322	1	13
	Vial	36	138	6	47	1	4
E[b]	Ampoule	394	450	124	320	1	9

[a] Results: first column measured with a Coulter counter, second column measured with scanning electron microscope. [b]Vial not available.
Adapted from Reference 29.

than in the other components packaged in plastic containers. However, when the components were transferred into the bag and their total volume was computed, lipid emulsions, amino acids, 10% KCl, and 30% dextrose contributed most to increase the number of particles. In addition, particles composed of silica, aluminum, and sodium were identified as contaminants in the KCl solution, and particles containing sulfur and silicate were identified as contaminants in all samples after the transfer to bags. The authors consider the presence of sulfur to be related to rubber particles and attribute the presence of silicate to contact with talcum from gloves used by manipulators during the manufacturing of the bags.

Oie and Kamiya [31] evaluated particulate contamination in samples of admixed and unadmixed parenteral nutrition (PN) solution bags from 10 hospitals in Japan. Seven samples were used as controls since they had not been mixed with ampoules or vials (unadmixed samples). The size and number of particles were measured using a particle counter, and the identification of elements was carried out by SEM coupled to energy dispersion spectroscopy. The authors collected the residual volume of the samples (10–60 mL) after their usage at patients' beds. The results are presented in Table 10.8. Admixed solutions presented a higher number of particles than unadmixed solutions, and solutions prepared with four or more glass ampoules presented a higher number of particles than those prepared with three or fewer ampoules. The scanning electron micrograph of two types of particles found on the filter after filtration (0.22-μm membrane filter) of 50 mL of a standard admixture containing glucose, electrolytes, amino acids, and vitamins are presented in Figure 10.7. Their compositions suggest that they are particles of glass (Fig. 10.7a) and of rubber (Fig. 10.7b).

TABLE 10.8 Mean and Range of Particles Found in Admixed Parenteral Nutrition Solutions and Antibiotics Classified According to Their Size

			Particle Size										
			>1.3 μm		>5 μm		>10 μm		>25 μm		>50 μm		
Sample	Volume (mL)	n	Mean	Range	Mean	Range	Mean	Range	Mean	Range	Mean	Range	Reference
Unadmixed samples[a]	10–60	7	62.7	8–146	1.7	0–4	0.4	0–2	–	–	0	–	[31]
Admixed samples[a]	10–60	192	960.9	30–9,539	42.8	0–587	6.4	1–146	–	–	0.09	0–1	[31]
Solutions mixed using 1–3 glass ampoules[a]	10–60	129	862.1	30–5,707	31.3	0–176	4.4	0–24	–	–	0.1	0–1	
Solutions mixed using 4–13 glass ampoules[a]	10–60	63	1,163.4	142–9,539	66.2	4–587	10.6	1–146	–	–	0.08	0–1	
Electrolytes + KCl glass ampoule	22	60	758.4	388–1,560	20.9	5–44	9.9	2–43	–	–	2.9	0–9	[32]
Electrolytes + KCl prefilled syringes	22	63	158.6	98–299	6.5	1–22	1.8	0–11	–	–	0		
Electrolytes + NaCl glass ampoule	22	66	736.1	379–1,009	23.9	9–65	10.6	2–28	–	–	2.5	0–6	

Electrolytes + NaCl prefilled syringes	22	15	179.2	128–245	7.6	3–16	3.2	0–7	–	–	0		
Saline + dobutamine glass ampoules	10	30	1,884.5	1,001–3,890	29.4	15–56	11.1	2–30	–	–	3.2	1–8	
Dobutamine prefilled syringes	10	10	178.9	160–199	2.6	0–6	0	–	–	–	0		
Amino acid/ dextrose plus vitamins (pediatric)	10	6	–	–	151	±24	44	±13	13	±5	4	±3	[33]
Lipid emulsion plus trace elements, vitamins (pediatric)	35	8	–	–	5,881	±1,538	766	±317	729	±406	418	±193	
Claforan (antibiotic)	1-g vial	3	539		–		12		3		1		[34]
Cefantral (antibiotic)	1-g vial	3	16,728		–		87		32		9		
Taxim (antibiotic)	1-g vial	3	17,362		–		152		80		16		

[a] Residual volume collected.

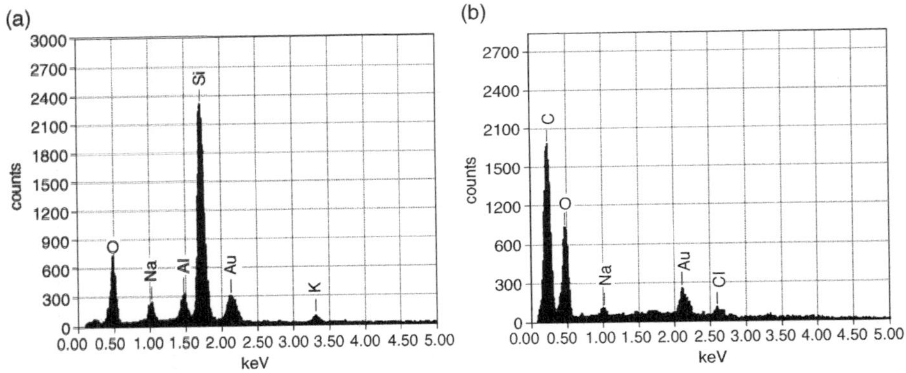

FIGURE 10.7 Identification of two types of particles by scanning electron microscopy coupled to energy dispersion spectroscopy. (a) suggests glass particles and (b) suggests rubber particles [31].

Also presented in Table 10.8 are the results of the study conducted by Yorioka [32] comparing the contribution of glass ampoules and prefilled syringes for the number of particles in final admixed infusions. The number of particles was higher in fluids admixed with KCl, NaCl, and dobutamine in glass ampoules than those admixed with these products stored in syringes. Moreover, glass fragments were only identified in the samples prepared with glass ampoules, supporting the assumption that glass particulate matter is largely introduced by ampoule breaking.

Table 10.8 also presents the results of two other studies on particle counting, one dealing with admixtures for pediatric administration [33] and the other with antibiotics [34]. The elevated number of particles in pediatric PN may be implicated, according to the authors, in the pathogenesis of pulmonary hypertension and granulomatous pulmonary arteritis. They calculated, based on the results presented in Table 10.8, that a 3000-g infant fed parenterally according to standard nutritional regimens would receive 37,000 particles between 2 and 100 μm in 1 day's feed, of which 80% were derived from the fat emulsion. In the study with antibiotics, a significant difference in the number of particles was found among the samples. Since these samples had the same active ingredient, the authors hypothesized that the difference may be related to the fact that the same overall quality and purity might not be maintained during drug production once the patent has expired.

Ball et al. [35] evaluated the influence of administration sets on the number of particles delivered to patients by analyzing samples collected from the lines immediately before the connection to the patient. All-in-one PN solutions for adult patients and two-in-one mixtures with a separate lipid emulsion in 50-mL syringes for pediatric patients were sampled. From the adult solutions, the first 30 mL and the subsequent 100 mL were collected, and from the pediatric

TABLE 10.9 Particle Counts per Milliliter in Two Successive Fractions Collected from Administration Sets of Adult and Pediatric Parenteral Nutrition Solutions

	Particle Counts/mL				
Sample[a]	>40 μm	>25 μm	>10 μm	>5 μm	3–5 μm
All-in-one adult					
First 30 mL	0.06 ± 0.45	1.06 ± 0.39	0.96 ± 0.45	1.40 ± 0.73	3.47 ± 1.24
Second 100 mL	0.15 ± 0.09	0.24 ± 0.14	0.29 ± 0.07	0.42 ± 0.12	0.90 ± 0.41
Two-in-one pediatric					
First 50 mL	0.16 ± 0.23	0.38 ± 0.26	1.43 ± 1.09	11.77 ± 7.38	7.59 ± 2.56
Second 50 mL	0.16 ± 0.17	0.34 ± 0.21	0.80 ± 0.40	9.32 ± 3.74	5.72 ± 2.67
Lipid emulsion syringe					
First 5 mL	0.44 ± 0.37	0.76 ± 0.34	1.27 ± 1.23	8.98 ± 4.63	16.72 ± 10.85
Second 5 mL	0.42 ± 0.67	0.69 ± 0.42	1.35 ± 1.08	9.83 ± 6.73	15.55 ± 12.68

[a] $n = 20$.

Adapted from Reference 35.

TABLE 10.10 Particulate Matter Present in Saline Solutions Stored in Different Container Materials and Particulate Matter Aggregated to These Solutions by Handling with Syringes and Tubing Connections

Sample	Identification	Material	Particulate (≥5 μm)
1	100-mL saline bag	Polypropylene	12 ± 3.3
2	500-mL saline bag	EVA	1 ± 0.5
3	10-mL disposable syringe filled with sample 1	Polypropylene	164 ± 1.4
4	500-mL saline bag with tubing connection	EVA	22 ± 5.9
5	500-mL saline bag with tubing connection previously washed with 20 mL water	EVA	No particles observed

Volume analyzed: 5 mL.

Adapted from Reference 36.

solutions, first and second aliquots of 50 mL were collected. Two samples of 5 mL each were taken from the lipid emulsion syringes. Table 10.9 lists the particle counts and their size range in the first and second samples drawn from the infusion sets. It is possible to identify a "washout" effect, with the first samples more contaminated than the second samples.

This washout effect caused by sample transferring was also demonstrated in the study conducted by Kuramoto et al. [36]. In Table 10.10, it is possible to see that particulate matter is aggregated to saline solutions merely by transferring it with a syringe or a connecting tube, and that previously washing the administration set with water removed practically all particulate matter ≥5 μm.

An example of the introduction of particulate matter due to sample handling is the contamination of radiograph contrast media caused by transferring the contrast from the vial to a disposal syringe used in automatic injectors. Radiograph contrast media should be injected at a controlled high speed to generate a constant image. Sendo et al. [37] demonstrated that iodine-based contrast media interact with the silicone oil, which coats the inner surface of syringes, generating immiscible oil droplets dispersed in the formulation. Nonionic contrast media, iohexol, ioversol, iopamidol, and the ionic medium, ioxaglate, were analyzed by particle counting before and after transfer into a disposable syringe. The results, a mean of three determinations with vials of the same lot, showed that all contrast media presented considerably higher counts in the syringe than in the vial (Fig. 10.8). The particle counts increased in the order ioxaglate > ioversol > iohexol > iopamidol and exceeded the limit prescribed by pharmacopeias for particles ≥10 μm in two samples, ioxaglate and ioversol. Particles obtained by drawing contrast medium and water into disposable syringes were analyzed by X-ray emission spectrometry. While both samples presented a peak corresponding to silicone, particles from the contrast medium also contained iodine and sulfur (Fig. 10.9). The authors suggested that contrast media also interact with the rubber plunger of the syringe and leach the sulfur present in the particle.

In a subsequent study [38], the authors compared the number of particles in the same type of contrast medium, but packaged in different containers: large-volume prefilled syringes and vials. The results showed that vial preparations of iohexol and ioversol had lower levels of particulate contamination, whereas for ioxaglate, the syringe presented lower contamination (Table 10.11). In this case, the chemical analysis of the particulate revealed that the elements present in the ioxaglate vial (aluminum, calcium, sulfur, titanium, and silicon) matched those present in the rubber stopper made of butyl rubber. On the other hand, the elements in the particulate from the ioversol syringe were released from the rubber piston and silicone oil coating on the inner surface of the syringe.

Silicone oil was also released in insulin packaged in disposable syringes [39, 40] and intravitreal injection drugs [41]. These studies, however, did not consider silicone droplets as particulate matter. Even so, their presence caused insulin inactivation and the formation of tiny silicone oil drops in the eyes of the patients, although no adverse reaction was observed.

The presence of particulate matter in contrast media used for angiography has been a concern since the 1970s because of the risk of tissue damage, coronary vasoconstriction, and embolism. This may be the reason that iodinated contrast media are one of the top 10 drugs responsible for anaphylaxis or anaphylactoid reactions [42, 43].

In 1983, Hamilton [44, 45] raised the possibility that the use of plastic syringes for injecting contrast media intravenously could be the cause of an

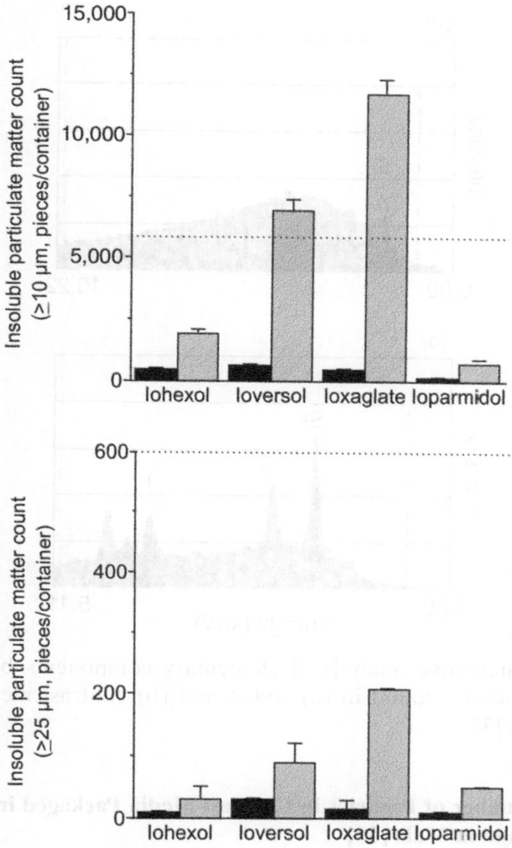

FIGURE 10.8 Particulate matter counts in contrast media before and after transferring to a disposable syringe. Vial preparation (black), after transfer (gray), United States Pharmacopeia (USP) acceptable limit (dashed line) [37].

increased incidence of adverse reactions. The analysis of the material revealed the presence of several contaminants originated by the contact of the contrast medium with the rubber stopper of a 50-mL syringe (these substances are listed in Table 9.6 in Chapter 9, Section 9.1.2, Syringes). The investigation did not include particle counting or silicone oil, although syringe lubricant was mentioned as a possible course of action. This led to a proposal for a general recommendation for procedures requiring injection of an angiograph contrast agent [45], which included "Inspect plastic syringes before use for visible evidence of crystal or globules on or around the plunger." Based on the results mentioned above, there is a chance that silicone oil droplets were among the contaminants and contributed to the adverse reactions observed in the patients.

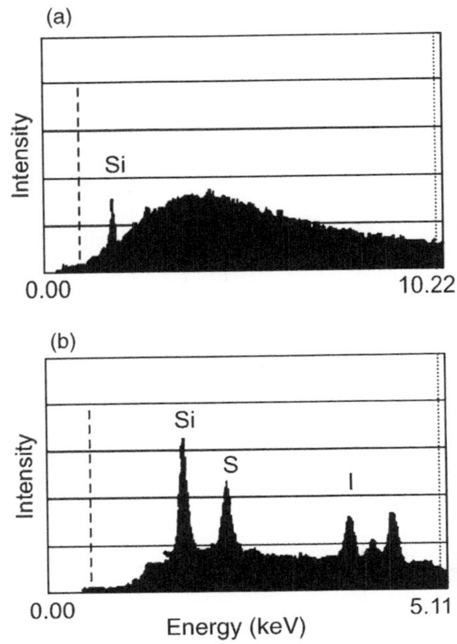

FIGURE 10.9 Qualitative analysis of elementary components by X-ray emission spectrometry of particles found in (a) water, and (b) contrast medium, drawn into disposable syringes [37].

TABLE 10.11 Number of Particles in Contrast Media Packaged in Large-Volume Syringes (20 mL) and in Vials [38]

Contrast Medium	Type of Package	Particle Count/mL			
		>2 μm	>5 μm	>10 μm	>25 μm
Iohexol	Syringe	11 ± 2	4 ± 1	1 ± 1	0
	Vial	6 ± 1	4 ± 1	3 ± 0	2 ± 0
Ioversol	Syringe	123 ± 20	41 ± 11	16 ± 9	0
	Syringe (angiography)	116 ± 42	43 ± 19	17 ± 9	0
	Vial	7 ± 1	5 ± 1	5 ± 1	2 ± 1
Ioxaglate	Syringe	9 ± 5	3 ± 2	1 ± 0	0
	Vial	78 ± 11	15 ± 1	3 ± 1	0

Another source of particulate that also originates from the packaging material but does not correspond to pieces of the container itself, is the leaching of additives from plastic containers with subsequent precipitation due to reaction with a formulation constituent or due to the pH of the solution.

Solutions with relatively elevated pH levels that come into contact with polymers containing stearate salts of calcium and zinc, as is the case of polyvinyl

TABLE 10.12 Particle Counts Before and After In-Line Filtration of Three Lots of ViaSpan Solutions (Filter Pore Size: 40 μm)

	Particle Counts per Milliliter					
	5–10 μm		>10–25 μm		>25 μm	
Sample Lot[a]	Before	After	Before	After	Before	After
1	210 ± 56	48 ± 15	151 ± 56	23 ± 9	68 ± 31	<1 ± 0
2	86 ± 32	16 ± 5	32 ± 9	5 ± 2	5 ± 3	<1 ± 0
3	131 ± 18	20 ± 4	57 ± 9	4 ± 1	21 ± 4	<1 ± 0

[a] $n = 8$.

Adapted from Reference 46.

chloride (PVC) and elastomers, promote the migration of these salts into the solution. The extent of the subsequent precipitate formation is dependent on the solution pH. Precipitation of small quantities of calcium, magnesium, and zinc fatty acid salts led to the formation of visible and subvisible particulates in a solution for storage and flushing of organs for transplantation (ViaSpan, DuPont Merck Pharmaceutical) [46]. These particles were originated from the PVC bags where the formulation was stored. Probably, the release of these cations and fatty acids was stimulated by the pH and composition of the solution. The solution, buffered at pH 7.4, contained several organic ingredients. ViaSpan manufacturers evaluated in-line filtration (40 μm) in order to reduce particulate contamination, which proved to be efficient in reducing particle counts to less than one-fourth for all particle sizes (Table 10.12).

A similar effect was observed by Jenke [47], who evaluated the influence of pH on the extraction of zinc, calcium, stearic acid, and palmitic acid from a synthetic polyisoprene rubber. It was observed that, at a high pH, stearic acid and zinc were extracted in great enough amounts to precipitate zinc stearate (these results are presented in Chapter 7, "Closures").

Equally related to precipitation, but not to packaging materials, is the physical incompatibility associated with nutrient admixtures for total parenteral nutrition (TPN). The most known case is the precipitation of calcium phosphate. The issue of incompatibility is more likely in solutions for neonates than in solutions for adults because of the higher calcium concentration and the elevated room temperature at administration (37°C). Parikh et al. [48] demonstrated that the final amino acid concentration is also a factor for admixture stability. The compatibility of high concentrations of organic calcium and inorganic phosphates in neonatal TPN solutions containing amino acids and dextrose, but not lipids, was tested. Nine bags with the same calcium and phosphate concentrations but with different concentrations of amino acids, dextrose, and cysteine were aseptically prepared under laminar airflow with the aid of a compounding device and were stored at 23–27°C. Particle counts were assessed

TABLE 10.13 Particle Counts in TPN Admixtures Containing Different Concentrations of Amino Acids, Dextrose, and Cysteine, after 24 Compounding at Room Temperature

	Bag Composition[a]			Particle Counts/mL (Mean ± SD)		
Bag	Amino Acids (g/dL)	Dextrose (g/dL)	Cysteine HCl (mg/dL)	≥5 μm	≥10 μm	≥25 μm
1	0.5	5.0	20	350 ± 306	81 ± 74	7 ± 6
2	0.5	5.0	20	182 ± 52	41 ± 9	1 ± 2
3	1.0	5.0	40	184 ± 86	30 ± 15	3 ± 3
4	1.5	5.0	60	126 ± 56	19 ± 11	1 ± 1
5	1.5	10.0	60	126 ± 49	22 ± 18	1 ± 2
6	3.0	5.0	120	178 ± 92	39 ± 30	4 ± 2
7	3.0	10.0	120	106 ± 43	24 ± 9	2 ± 2
8	4.0	5.0	160	205 ± 94	41 ± 17	4 ± 2
9	4.0	10.0	160	214 ± 137	53 ± 53	3 ± 3

[a] Other components: calcium gluconate (60 mg/dL), disodium phosphate (46.5 mg/dL), potassium acetate (39 mg/dL), magnesium sulfate (4.8 mg/dL), trace elements (2.2 mL/dL), zinc sulfate (0.2 mg/dL), selenious acid (2.2 μg/dL), multivitamins (2.2 mL/dL), heparin (50 units/dL).
Adapted from Reference 48.

within 1, 6, 24, and 30 hours after compounding. Table 10.13 shows the results of particle counting 24 hours after compounding. In bags 3–9, no significant changes in measured particle counts occurred. On the other hand, in bags 1 and 2, with the lowest concentration of amino acids, a significantly higher number of particles was observed. Moreover, when the particle numbers were considered over time (from 0 to 30 hours after compounding), there was a significant increase in particles in these samples, different from the others, where the number of particles remained in the same range. Figure 10.10 displays the increased particle counts after 30 hours for admixtures 1 and 2. This effect is due to the association of ionized amino acids with calcium and phosphate ions to form soluble complexes, thus making them unavailable to interact with one another to produce insoluble particles.

A singular case of contamination by particulate matter is the formation of particulate matter as a consequence of metallic tungsten contamination. Tungsten microparticles may be introduced into prefilled syringes during the creation of the needle hole. In their turn, these microcontaminants may interact with therapeutic proteins to produce visible particles. During syringe manufacture, a hot tungsten pin is used to create the needle exit hole, and, depending on the process, the temperature can reach 1000–1200°C. In temperatures above 500°C, metallic tungsten oxidizes and forms tungsten oxide, which sublimes above 800°C. Therefore, sublimated oxide may be deposited in the cooler parts of the syringe. The product is then potentially exposed to tungsten contamination upon the filling of the syringe. When the product contains

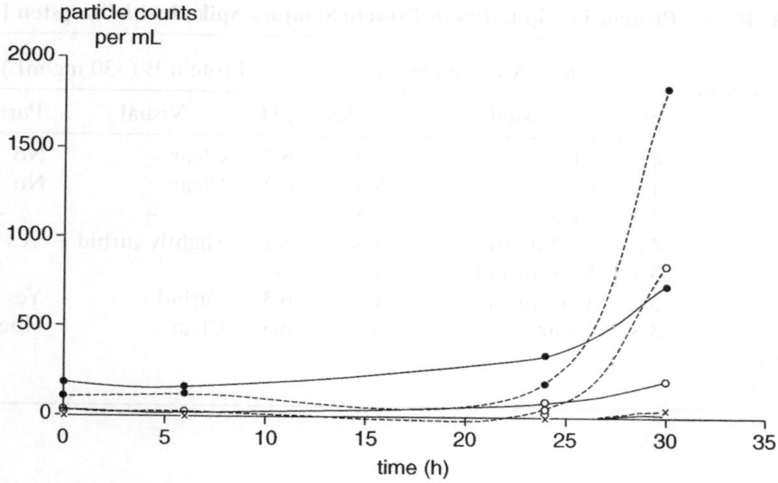

FIGURE 10.10 Particle numbers counted over time (from 0 to 30 hours after compounding) for admixtures 1 and 2 in Table 10.13. Particles ≥5 μm (•), particles ≥10 μm (○), particles ≥25 μm (x). Solid line: sample 1; dashed line: sample 2 (adapted from Reference 48).

protein as an ingredient, the problem can be twofold: Besides the contamination by tungsten, there is the risk of protein precipitation. Sodium tungstate is a usual protein precipitant in clinical chemical analysis, and, therefore, it is highly probable that precipitation will occur upon filling the syringe with a protein-containing formulation. Bee et al. [49] evaluated this possibility and concluded that pH is an important factor in this process since below pH 6, tungsten oxide is soluble and, therefore, available to react with proteins. The precipitation of a monoclonal antibody (mAb) (formulated at 0.02 mg/mL) in a pH 5 buffered solution occurred in the presence of only 3 ppm of sodium tungstate.

In another study [50], tungsten pins used in syringe manufacturing were placed in glass vials with buffers used in protein formulations (buffer A, pH 4.0, and buffer B, pH 6.3). The vials, kept at 45°C, were incubated for 12 hours and were submitted to sonication for 1.5 hours before and after incubation. The extracts were centrifuged and the supernatant analyzed for tungsten concentration determination. The results showed concentrations as high as 4971 mg/L in buffer A and 3801 mg/L in buffer B. These extracts were used to spike protein formulations at tungsten levels displayed in Table 10.14. The spiked protein solutions were filled into Luer-type syringes to provide a typical exposure to silicone oil. The results, presented in Table 10.14, revealed that the tungsten released by the pin at concentrations over 100 mg/L promoted protein aggregation and consequently particulate matter in both buffed formulations.

TABLE 10.14 Protein Precipitation in Protein Samples Spiked with Tungsten [50]

Tungsten Concentration (mg/L)	Protein A (<1 mg/mL)			Protein B (>30 mg/mL)		
	pH	Visual	Particles	pH	Visual	Particles
0	4.0	Clear	No	6.2	Clear	No
1	4.0	Clear	No	6.2	Clear	No
10	4.0	Clear	No	–	–	–
100	4.0	Slightly turbid	Yes	6.2	Slightly turbid	Yes
500	3.9	Very turbid	Yes	–	–	–
1000	3.8	Very turbid	Yes	6.3	Turbid	Yes
1000 (only buffer, no protein)	3.8	Clear	No	6.3	Clear	Fiberlike

The evidence that protein aggregation is caused by tungsten is that turbidity did not occur in samples containing only the buffer.

The benefits of using IV in-line filters have been investigated by several authors. The Center for Disease Control and Prevention recommends the filtration of infusates during manufacture instead of by delivering (in-line filters) as the most cost-effective and practical way to remove particulates. In terms of removing particulate matter, in-line filters are very effective, as can be seen in Table 10.12 for ViaSpan samples. On the other hand, filtration of infusates during manufacture instead of during administration may not be as effective because it does not encompass particulates introduced by handling and administration sets.

Hellinger et al. [51] examined the particle content of a solution used for organ preservation and in heart surgery before and after handling for clinical application (infusion line kit) and also after a terminal in-line filtration (0.2 μm). The particles in the original commercial fluid, a histidine-tryptophane-α-ketoglutarate (HTK) solution, and those retained in the filter were analyzed to determine their origin. Figure 10.11 shows the number of particles in the fluid after each step of the procedure. Permissible levels of particles, with a distribution of diameters between 0.2 and 20 μm, were found in the fluid. A significant additional increase in the particle count was observed when the fluid was administered through the infusion kit, despite including a 270-μm pore size in-line filter. Finally, the filtration of the HTK solution by a terminal 0.2-μm in-line filter significantly reduced the number of particles. The particle analysis revealed aluminum and silicon in the original HTK solution, from storage in glass bottles, and chloride in the fluid after infusion through the kit, probably from the PVC lines. The release of particles from the infusion kit was presumably a consequence of mechanical alterations of the tubes, such as abrasion or clamping. Nevertheless, an effective reduction in the particle counts was achieved through terminal 0.2-μm filtration.

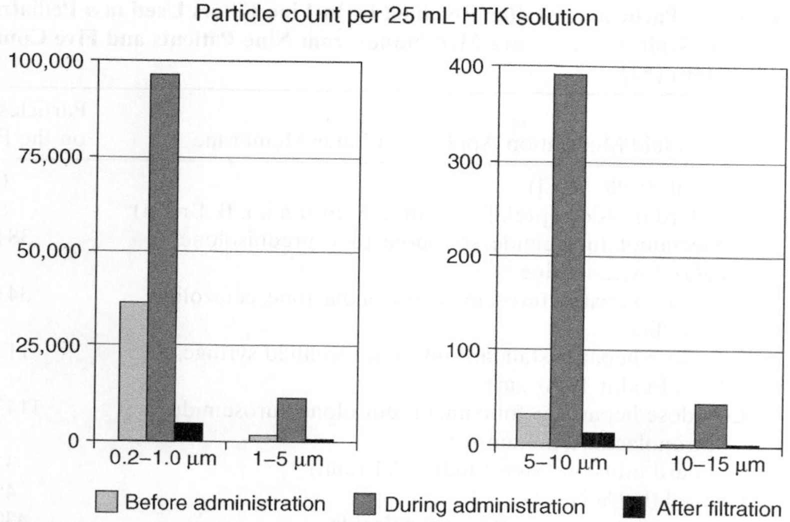

FIGURE 10.11 Particle counts in the original HTK cardioplegic solution, after administration using an infusion kit and after terminal in-line filtration (adapted from Reference 51).

A study conducted by Jack et al. [52] analyzed the particulate matter retained by in-line filters used in a pediatric intensive care unit. Twenty filter membranes from nine patients and five controls (saline solution) were examined for particle counting and particle constitution. The results of particle counting, presented in Table 10.15, gave an average number of 550 particles/cm^2 for the fluids and of 45.6 particles/cm^2 for the controls. Particle size ranged from 5 to >100 µm with the majority between 5 and 50 µm. Energy dispersion microscopy analysis detected silicon, carbon, oxygen, sodium, chlorine, potassium, calcium, magnesium, iron, and aluminum in the filter material, where carbon and oxygen probably originated from the membrane itself. Although in this case it was not possible to affirm that particles were introduced during administration, the number of particles retained by the filters demonstrated that there was contamination despite stringent infusion regiments. Moreover, the number of particles originated in admixtures and the number in the controls and in standard infusions suggest a correlation between the complexity of infusion regiment and the particle load.

The major problem with particulate contamination in infusion fluids is not related to the composition itself (since they are mostly pieces of the container and therefore innocuous elements) but to the potential of each particle to cause adverse reactions. The probability of an adverse reaction occurring is proportional to the number (and size) of particles introduced in the circulatory system. Turco and Davis [53] suggested that responses resulting from the infusion of particles included physical occlusion, inflammatory responses,

TABLE 10.15 Particulate Matter Retained by In-Line Filters Used in a Pediatric Intensive Care Unit. Twenty Filter Membranes from Nine Patients and Five Controls (Saline Solution) [52]

Filter	Fluid/Medication Applied via Filter Membrane	Particles/cm^2 on the Filter
1	Control (0.9% NaCl)	6
2	Standard infusion (prefilled syringe, Sterofundin, B. Braun)	8
3	Paracetamol, furosemide, spironolactone, prednisolone, cefazolin, ranitidine	384
4	Low dose heparin, furosemide, spironolactone, cefazolin, ranitidine	343
5	Low dose heparin, standard infusion (prefilled syringe, Sterofundin, B. Braun)	113
6	Low dose heparin, cefotaxim, prednisolone, furosemide, spironolactone, metamizole	1143
7	Standard infusion (Sterofundin, B. Braun)	37
8	Control (0.9% NaCl)	45
9	Parenteral nutrition, potassium chloride	430
10	Low-dose heparin, potassium chloride, insulin	238
12	Parenteral nutrition, potassium chloride, high-dose heparin	1424
13	Glucose 40%, potassium chloride	97
14	Standard infusion (Sterofundin, B. Braun)	124
15	Standard infusion (Sterofundin, B. Braun), magnesium aspartate, piritramide	392
16	Standard infusion (Sterofundin, B. Braun), colloidal volume therapy, magnesium aspartate, potassium chloride	521
17	Paracetamol, cefotaxim, prednisolone, piritramide	407
18	Low dose heparin, cefotaxim, prednisolone	279
19	Cefotaxim, prednisolone, pantoprazol	277
20	Parenteral nutrition, potassium chloride, insulin	1021
21	4-Hydroxy butyric acid (GABA)	1993
22	Low-dose heparin, potassium chloride, ampicillin, meropenem, acyclovir, phenobarbital	1344
23	Clonidine, fentanyl, midazolam	424
24	Control (0.9% NaCl)	93
25	Control (0.9% NaCl)	46

neoplastic responses, and antigenic responses. Animal studies have shown that the distribution of particles in tissues is related to their diameter. Particles larger than 8 μm are trapped in the lung capillaries; 3- to 6-μm particles are lodged in the spleen and hepatic lymph nodes; and 1-μm particles remain in the liver [54].

Particle composition can also be a concern when the patient has latex allergy. At least two reports have confirmed allergic reactions caused by the administration of latex originated in the syringes where the medication was prepared. In both cases, the rubber stopper of plastic syringes leached latex particles into the preparate [55, 56].

CONCLUDING REMARKS

The major problem with particulate contamination in infusion fluids is related to the potential of each particle to cause adverse reactions. Although not fully demonstrated, particulate matter might be associated with tissue damage, coronary vasoconstriction, and embolism.

Despite materials such as cotton, nylon fibers, and even hair having been found in infusion fluids, containers and delivery systems are the main sources of particulate matter. Glass ampoules and rubber closures pose major hazards for intravenous infusions. In glass ampoules, due to its nature, glass may fragment in microscopic and macroscopic particles, which are aspirated through the needle. In rubber closures, during the insertion of a needle, the rubber is shaved, thus introducing rubber particles into the vial.

The admixing of PN solutions leads to an increase in particulate matter, not only from ampoule opening or stoppers perforation but also from administration sets and sample handling. Even if filters are used during the manufacture of infusion solutions, bags and administration sets contribute significantly to the particle load administered, and, therefore, in-line filters used during administration are more effective in reducing particulate matter contamination.

REFERENCES

1. Groves MJ. Particulate contamination in parenterals: current issues. Boll Chim Farm 130, 347–354, 1991.
2. Bikhazi AB, Shiatis JA, Haddad AF. Quantitative estimation of particulate matter in pharmaceutical preparations intended for intravenous administration. J Pharm Sci 66, 181–186, 1977.
3. DeLuca PP, Knapp JZ. Particulate matter. In: *Pharmaceutical Dosage Forms—Parenteral Medications*, Vol. 3, Avis KE, Lieberman HA, Lachman L (eds.). Marcel Dekker, New York, 1993, pp. 117–230.
4. Garvan JM, Gunner BW. Particulate contamination of intravenous fluids. Br J Clin Pract 25(3), 119–121, 1971.
5. Wood RW. Foreign bodies in spinal syringes. Anesth Analg 81, 211, 1995.
6. Turco S, Davis NM. Glass particles in intravenous injections. N Engl J Med 287(23), 1204–1205, 1972.
7. Turco S, Davis NM. Preventing the injection of glass particles with furosemide solution. Hosp Pharm 7, 423–424, 1972.
8. Katz H, Borden H, Hirscher D. Glass-particle contamination of color-break ampules. Anesthesiology 39(3), 354, 1973.
9. Furgang FA. Glass particles in ampules. Anesthesiology 41(5), 525, 1974.

10. Shaw NJ, Lyall EG. Hazards of glass ampoules. Br Med J (Clin Res Ed) 291(6506), 1390, 1985.
11. Carbone-Traber KB, Shanks CA. Glass particle contamination in single-dose ampules. Anesth Analg 65(12), 1361–1363, 1986.
12. Sabon RL, Cheng EY, Stommel KA, Hennen CR. Glass particle contamination: influence of aspiration methods and ampule types. Anesthesiology 70(5), 859–862, 1989.
13. Finkelstein A, Lokhandwala BS, Pandey NS. Particulate contamination of an intact glass ampule. Anesthesiology 73(2), 362–363, 1990.
14. Lye ST, Hwang NC. Glass particle contamination: is it here to stay? Anaesthesia 58(1), 93–94, 2003.
15. Preston ST, Hegadoren K. Glass contamination in parenterally administered medication. J Adv Nurs 48(3), 266–270, 2004.
16. Zabir AF, Choy CY, Rushdan R. Glass particle contamination of parenteral preparations of intravenous drugs in anaesthetic practice. South Afr J Anaesth Analg 14(3), 17–19, 2008.
17. Roseman TJ, Brown JA, Scothorn WW. Glass for parenteral products: a surface view using the scanning electron microscope. J Pharm Sci 65, 22–29, 1976.
18. Sakai T, Koh H, Matsuki A. A case of the foreign body due to "coring." Masui 45(12), 1533–1535, 1996.
19. Cohen S. A foreign body in a spinal syringe. Anesth Analg 79, 1214–1215, 1994.
20. Seto WH, Ching PT, Kumana CR. Rubber emboli. Eur J Clin Pharmacol 53(5), 313–315, 1998.
21. Adachi Y, Takigami J, Watanabe K, Satoh T. A case of coring on using a 1% Diprivan vial. Masui 50(6), 635–636, 2001.
22. Shiroyama K. The incidence of "coring" during aspiration of propofol from a 50-mL vial. J Anesth 15(2), 120, 2001.
23. Cohen S, Prieto N, Hunter CH. A foreign body in propofol syringe. Anesth Analg 103(3), 784, 2006.
24. Beer K. Potential foreign body emboli associated with botulinum toxin A injections. J Drugs Dermatol 6(2), 220–221, 2007.
25. Riess ML, Strong T. Near-embolization of a rubber core from a propofol vial. Anesth Analg 106(3), 1020–1021, 2008.
26. Nashed-Samuel Y, Torraca G, Liu D, Fujimori K, Zhang Z, Wen ZQ, Lee H. Identification of an extraneous black particle in a glass syringe: extractables/leachables case study. PDA J Pharm Sci Technol 64(3), 242–248, 2010.
27. Asakura T, Seino H, Nozaki S, Abe R. Occurrence of coring in insulin vials and possibility of rubber piece contamination by self-injection. Yakugaku Zasshi 121(6), 459–463, 2001.
28. Kawasaki Y. Study on insoluble microparticulate contamination at ampoule opening. Yakugaku Zasshi 129(9), 1041–1047, 2009.
29. Winding O. Intrinsic particles in angiographic contrast media. Radiology 134, 317–320, 1980.
30. Foroni LA, Rochat MH, Trouiller P, Calop JY. Particle contamination in a ternary nutritional admixture. J Parenter Sci Technol 47, 311–314, 1993.

31. Oie S, Kamiya A. Particulate and microbial contamination in in-use admixed parenteral nutrition solutions. Biol Pharm Bull 28, 2268–2270, 2005.
32. Yorioka K, Oie S, Oomaki M, Imamura A, Kamiya A. Particulate and microbial contamination in In-use admixed intravenous infusions. Biol Pharm Bull 9(11), 2321–2323, 2006.
33. Puntis JW, Wilkins KM, Ball PA, Rushton DI, Booth IW. Hazards of parenteral treatment: do particles count? Arch Dis Child 67(12), 1475–1477, 1992.
34. Lehr HA, Brunner J, Rangoonwala R, Kirkpatrick CJ. Particulate matter contamination of intravenous antibiotics aggravates loss of functional capillary density in postischemic striated muscle. Am J Respir Crit Care Med 165(4), 514–520, 2002.
35. Ball PA, Bethune K, Fox J, Ledger R, Barnett M. Particulate contamination in parenteral nutrition solutions: still a cause for concern? Nutrition 17, 926–929, 2001.
36. Kuramoto K, Shoji T, Nakagawac Y. Usefulness of the final filter of the IV infusion set in intravenous administration of drugs—contamination of injection preparations by insoluble microparticles and its causes. Yakugaku Zasshi 126(4), 289–295, 2006.
37. Sendo T, Otsubo K, Hisazumi A, Aoyama T, Oishi R. Particle contamination in contrast media induced by disposable syringes. J Pharm Sci 84(12), 1490–1491, 1995.
38. Sendo T, Hirakawa M, Yaginuma M, Aoyama T, Oishi R. Quality evaluation of radiographic contrast media in large-volume prefilled syringes and vials. Acad Radiol 5(6), 444–447, 1998.
39. Chantelau E, Berger M, Böhlken B. Silicone oil released from disposable insulin syringes. Diabetes Care 9(6), 672–673, 1986.
40. Baldwin RN. Contamination of insulin by silicone oil: a potential hazard of plastic insulin syringes. Diabet Med 5(8), 789–790, 1988.
41. Bakri SJ, Ekdawi NS. Intravitreal silicone oil droplets after intravitreal drug injections. Retina 28(7), 996–1001, 2008.
42. Morcos SK. Acute serious and fatal reactions to contrast media: our current understanding. Br J Radiol 78(932), 686–693, 2005.
43. Wang DY, Forslund C, Persson U, Wiholm BE. Drug-attributed anaphylaxis. Pharmacoepidemiol Drug Saf 7(4), 269–274, 1998.
44. Hamilton G. Adverse reactions to intravenous pyelography contrast agents. Can Med Assoc J 129, 405–406, 1983.
45. Hamilton G. Contamination of contrast agent by rubber components of 50-mL disposable syringe. Radiology 152, 509–510, 1984.
46. Mueller BJ, Guessford SA, Chen TT, Karczmarcyk RM, Schreiber MA, Ycas JW, Gorko MA, Repta AJ. Effect of inline filtration on ViaSpan cold-storage solution. Am J Health Syst Pharm 55(3), 266–271, 1998.
47. Jenke D. Extraction of stearate salts from plastic materials used in pharmaceutical applications. PDA J Pharm Sci Technol 64, 200–210, 2010.
48. Parikh MJ, Dumas G, Silvestri A, Bistrian BR, Driscoll DF. Physical compatibility of neonatal total parenteral nutrient admixtures containing organic calcium and inorganic phosphate salts. Am J Health Syst Pharm 62(11), 1177–1183, 2005.
49. Bee JS, Nelson SA, Freund E, Carpenter JF, Randolph TW. Precipitation of a monoclonal antibody by soluble tungsten. J Pharm Sci 98(9), 3290–3301, 2009.

50. Jiang Y, Nashed-Samuel Y, Li C, Liu W, Pollastrini J, Mallard D, Wen ZQ, Fujimori K, Pallitto M, Donahue L, Chu G, Torraca G, Vance A, Mire-Sluis T, Freund E, Davis J, Narhi L. Tungsten-induced protein aggregation: solution behavior. J Pharm Sci 98(12), 4695–4710, 2009.
51. Hellinger A, Piotrowski J, Konerding MA, Burchard WG, Doetsch N, Peitgen K, Erhard J, Reidemeister JC. Impact of particulate contamination in crystalloid cardioplegic solutions: studies by scanning and transmission electron microscopy. Thorac Cardiovasc Surg 45(1), 20–26, 1997.
52. Jack T, Brent BE, Boehne M, Müller M, Sewald K, Braun A, Wessel A, Sasse M. Analysis of particulate contaminations of infusion solutions in a pediatric intensive care unit. Intensive Care Med 36(4), 707–711, 2010.
53. Turco SJ, Davis NM. Detrimental effects of particulate matter on the pulmonary circulation. J Am Med Assoc 217, 81–82, 1971.
54. Hearse DJ, Sonmez B, Saldanha C, Braimbridge MV, Maxwell MP, Erol C. Particle-induced coronary vasoconstriction in the rat heart: pharmacological investigation of underlying mechanisms. Thorac Cardiovasc Surg 34, 316–325, 1986.
55. Vassallo SA, Thurston TA, Kim SH, Todres ID. Allergic reaction to latex from stopper of a medication vial. Anesth Analg 80, 1057–1058, 1995.
56. Kearns CF, Norris A. Latex allergy and plastic syringes. Anesth Analg 82, 429, 1996.

PART III

MICROBIOLOGICAL CONTAMINATION

11

MICROBIOLOGICAL AND ENDOTOXIN CONTAMINATION

To determine sources of the microbial contamination of pharmaceutical products, the following two scenarios were considered: the contamination of the product itself and the contamination acquired through the administration or use of the product.[1] Intrinsic or extrinsic contamination results from microorganism access while manufacturing or handling products.

The relevance of microbial contamination is related to the sterility level of the manufactured product. The restrictions for the preparation of nonsterile products are not as strongly enforced as for sterile products, where viable microorganisms must be absent; therefore, compendia have been generated to identify nonsterile preparations and to set the limits of microorganisms according to the origin of the raw materials and the methods used for preparation and administration. Table 11.1A–C display the limits set by three pharmacopeias: United States Pharmacopeia (USP), European Pharmacopeia (Ph. Eur.), and Japanese Pharmacopeia. Despite differences in limits, all compendia establish that pathogens, such as *Escherichia coli* and *Pseudomonas aeruginosa*, should be absent from all manufactured products.

Jimenez [1] used an interesting strategy to evaluate microorganism prevalence and origins in pharmaceutical products: identifying microorganism

[1] Although endotoxin is not a microorganism, because of its origin, contamination by endotoxins is discussed in this chapter. The topic of endotoxins is also addressed in Chapter 13.

Sources of Contamination in Medicinal Products and Medical Devices, First Edition. Denise Bohrer.
© 2013 by John Wiley & Sons, Inc. Published 2013 by John Wiley & Sons, Inc.

TABLE 11.1A Microbiological Quality of Pharmaceutical Preparations According to the British Pharmacopoeia (BP)

Category	Preparation Requirements	General Limits Total Viable Aerobic Count per Gram or per Milliliter	Specific Limits Per gram or per Milliliter
1	Sterile	Absent	–
2	Topical use Transdermal patches	$<10^2$ $<10^2$ per patch	Absence of *Pseudomonas aeruginosa* Absence of *Staphylococcus aureus*
3a	Oral and rectal administration	$<10^3$ aerobic bacteria $<10^2$ fungi	Absence of *Escherichia coli*
3b	Oral administration Containing raw materials of natural origin	$<10^4$ aerobic bacteria $<10^2$ fungi	$<10^2$ enterobacteria Absence of *E. coli* Absence of *S. aureus* Absence of *Salmonella* (10 g or 10 mL)
4a	Herbal drugs Boiling water added	$<10^7$ aerobic bacteria $<10^5$ fungi	$<10^2$ *E. coli*
4b	Herbal drugs Boiling water not added	$<10^5$ aerobic bacteria $<10^4$ fungi	$<10^3$ enterobacteria Absence of *E. coli* Absence of *Salmonella* (10 g or 10 mL)

British Pharmacopoeia—Appendix XVI-D.

contamination in recalled product. All products recalled by the Food and Drug Administration (FDA) from 1998 to 2006 because of microbial contamination were assessed, which allowed for an evaluation of the raw materials and manufacturing processes as sources of contamination. The research showed that *Burkholderia cepacia* was the most prevalent microorganism in both sterile and nonsterile products, followed by the *Pseudomonas* spp. in nonsterile products; notably, microorganisms were identified in only 23% of the recalled sterile products (Table 11.2). Both bacterial species are gram-negative human pathogens associated with water contamination. As these types of bacteria were uncovered in the bioburden of ultrapure water (UPW) systems (see Table 11.9, Section 11.1), they are capable of growing even under minimal nutrient supply. Therefore, slight deficiencies in water sanitization could support their development. However, the presence of mold and yeast, which

TABLE 11.1B Acceptance Criteria for Microbiological Quality of Nonsterile Dosage Forms According to the USP

Preparation Requirements	Total Aerobic Microbial Count (cfu/g or cfu/mL)	Total Combined Yeasts/Molds Counts (cfu/g or cfu/mL)	Microorganisms Absent
Nonaqueous preparations for oral use	10^3	10^2	*Escherichia coli*
Aqueous preparations for oral use	10^2	10^1	*E. coli* *Pseudomonas aeruginosa* *Staphylococcus aureus*
Rectal use	10^3	10^2	–
Oromucosal use	10^2	10^1	*S. aureus* *P. aeruginosa*
Gingival use	10^2	10^1	*S. aureus* *P. aeruginosa*
Cutaneous use	10^2	10^1	*S. aureus* *P. aeruginosa*
Nasal use	10^2	10^1	*S. aureus* *P. aeruginosa*
Auricular use	10^2	10^1	*S. aureus* *P. aeruginosa*
Vaginal use	10^2	10^1	*S. aureus* *P. aeruginosa* *Candida albicans*
Transdermal patches	10^2	10^1	*S. aureus* *P. aeruginosa*
Inhalation use	10^2	10^1	*S. aureus* *P. aeruginosa* Absence of bile-tolerant gram-negative bacteria

United States Pharmacopeia—⟨1111⟩.

were uncovered in 19% and 5% of the recalled nonsterile and sterile products, respectively, indicates deficient environmental control.

The data in Table 11.2 show that the microorganism contamination is different in sterile and nonsterile products, which is probably due to their distinct origin. Nonsterile products include mouthwash, creams, and supplements, while sterile products are infusion solutions or drugs, such as antibiotics and anesthetics. The microorganisms uncovered from recalled sterile products

TABLE 11.1C Acceptance Criteria for Microbiological Quality of Nonsterile Dosage Forms According to the JP

Preparation Requirements	Total Aerobic Microbial Count (cfu/g or cfu/mL)	Total Combined Yeast/Mold Counts (cfu/g or cfu/mL)	Microorganisms Absent
Raw materials	≤1000	≤100	Not specified
Inhalation (liquid)	≤20	≤20	*P. aeruginosa* *Staphylococcus aureus*
Inhalation (powder)	≤100	≤50	*P. aeruginosa* *S. aureus*
Nasal	≤100	≤50	*S. aureus* *P. aeruginosa*
Vaginal	≤100	≤50	*S. aureus* *E. coli* *Candida albicans*
Optic or topical (including transdermal patches)	≤100[a]	≤50[a]	*S. aureus* *P. aeruginosa*
Rectal	≤1000	≤100	Not specified
Oral (solid)	≤1000	≤100	*E. coli*
Oral (liquid)	≤100	≤50	*E. coli*
Herbal drugs			Microorganisms with specific limits
Boiling water added (category 1)	10^7	10^4	
	10^2		*E. coli*
	Not detected		*Salmonella*
Other preparations (category 2)	10^5	10^3	
	10^3		Enterobacteria and other gram-negative bacteria
	Not detected		*Salmonella*
	Not detected		*E. coli*

Japanese Pharmacopeia.
[a] Per patch.

did not correspond to those typically associated with the parenteral formulation contaminations or nosocomial infections. This chapter presents a discussion of the evidence that the main sources of microbial contamination are extrinsic and associated with practices that allow the microbes to enter the product.

TABLE 11.2 FDA Recalls from 1998 to 2006 Because of the Microbial Contamination of Sterile and Nonsterile Products

Recall Cause	Number of Occurrences: Nonsterile Formulations	Sterile Formulations	Sterile Medical Devices
Lack of sterility assurance	–	67	86
Microbial contamination	26	9	8
Mold and yeast	25	6	4
Burkholderia cepacia	29	4	1
Pseudomonas spp.	31	1	0
Aspergillus spp., *Penicillium* spp.	1	2	1
Mycobacterium chelonae	0	3	0
Methylobacterium spp.	0	3	0
Enterobacter spp.	3	0	0
Bacillus spp.	3	1	0
Staphylococcus spp.	3	0	0
Ralstonia pickettii	2	1	0
Salmonella spp.	2	0	0
Klebsiella pneumoniae	2	0	0
Candida spp.	2	0	0
Stenotrophomonas maltophilia	0	1	0
Fusarium spp.	0	1	0
Acinetobacter baumanii	1	0	0
Aeromonas sobria	1	0	0
Citrobacter freundii	1	0	0
Flavimonas oryzihabitans	1	0	0

Adapted from Reference 1.

11.1 WATER

By far, the most relevant "raw material" for the pharmaceutical industry is water. Although the compendia describe several types of water, the minimum purity level in this context is that of "purified water." Table 11.3 lists the biological parameters of the different types water described in pharmacopeial compendia.

To become a raw material for pharmaceutical purposes, the water must be processed. A usual source is the drinking water from municipal supplies, which should comply with governmental regulations. Table 11.4 shows the microbiological quality of drinking water according to regulatory agencies. The purification processes that are suitable for producing water for pharmaceutical purposes are listed in Table 11.5. Theoretically, with the exception of

TABLE 11.3 Biological Parameters for Different Types of Water Established in the Pharmacopeial Compendia

	USP		BP		JP	
Type of Water	Microorganisms	Bacterial Endotoxins	Microorganisms	Bacterial Endotoxins	Microorganisms	Bacterial Endotoxins
Water	–	–	–	–	<100 general bacteria/mL none coliform bacilli	–
Purified water	$<10^2$ cfu/mL	<0.25 EU/mL	$<10^2$ cfu/mL	<0.25 IU/mL	No limits	No limits
Highly purified water	–	–	<10 cfu/mL	<0.25 IU/mL	–	–
Sterile purified water	Absent	–	–	–	–	
Sterile water for injection	Absent	<0.25 EU/mL	Absent	<0.25 IU/mL	Absent	<0.25 EU/mL
Bacteriostatic water for injection	Absent	<0.5 EU/mL	–	–	–	–
Sterile water for irrigation	Absent	<0.25 EU/mL	–	–	–	–
Sterile water for inhalation	Absent	<0.5 EU/mL	–	–	–	–
Water for hemodialysis	$<10^2$ cfu/mL	2 USP EU/mL	$<10^2$ cfu/mL	<0.25 IU/mL	–	–

TABLE 11.4 Microbiological Quality of Drinking Water According to Regulatory Agencies

Parameter	U.S. EPA[a]	EU	Japan
Colony count at 22°C	500/mL	100/mL	100/mL
Colony count at 37°C	–	20/mL	–
Escherichia coli (total coliforms)	5%[b]	0/250 mL	Absent
Enterococci	–	0/250 mL	–
Pseudomonas aeruginosa	–	0/250 mL	–
Clostridium perfringens	–	0/100 mL	–
Cryptosporidium	>99% removal	–	–
Giardia lamblia	>99.9% removal	–	–
Legionella	zero	–	–
Viruses	>99.99% removal	–	–

[a] Standards that should be achieved.
[b] *No more than 5.0% samples total coliform-positive in a month.*
–, no limits established.
U.S. EPA: http://water.epa.gov/drink/contaminants/basicinformation/index.cfm.
EU: http://ec.europa.eu/environment/water/water-drink/index_en.html.
Japan: http://www.nilim.go.jp/lab/bcg/siryou/tnn/tnn0264pdf/ks0264011.pdf.

TABLE 11.5 Water Purification Systems and the Microbial Limits

Purification System	Microorganisms Level (cfu/mL) or Baseline Removal	Endotoxins (EU/mL) or Baseline Removal
Filtration	50% bacteria	Zero removal
Distillation	<10	<10
Reverse osmosis	<10	<10
Deionization	Zero removal	Zero removal
Membrane filtration	<10	<10

Millipore Catalog for water purification.
Guidelines for Drinking-Water Quality, 4th ed. World Health Organization, 2011.

deionization, the processes are able to reduce the water bioburden to less than 10 cfu/mL, fulfilling the specifications for purified water but not for sterile water or water for injection (WFI). In fact, a combination of processes is performed to produce sterile water.

Among all types of water, water for hemodialysis is the most vulnerable to microbiological contamination, exceeding the prescribed limits because of the large volume required for each patient and the production at local facilities, hospitals, and clinics. Water purification for hemodialysis includes filtration and either ion exchange (IE) or reverse osmosis (RO). Some facilities perform IE followed by RO to achieve a better water quality. After the introduction of RO in the purification of water for hemodialysis, purification improvements would seem to solve contamination problems. Studies have shown, however,

TABLE 11.6 Microbiological Quality of Water in Hemodialysis Centers

No. of Centers	No. of Samples	Water Treatment	Treated Water (cfu/mL)	Country	Reference
9	18	RO	64.2 ± 173.6	Japan	[2]
85	85	RO (90.5%) IE (9.5%)	63.07 ± 162.85	Greece	[3]
3	82	RO	58 (2–341)	United States	[4]

TABLE 11.7 Microbiological and Endotoxin Parameters of Water for Hemodialysis Collected in Different Dialysis Centers after Purification. Samples Incubated at 22°C

	Microbiological Parameter							
	Raw Water		Ion Exchanger		Reverse Osmosis		UV Disinfection	
Center	cfu/mL	EU/mL	cfu/mL	EU/mL	cfu/mL	EU/mL	cfu/mL	EU/mL
1	250	2.2	250	2.8	50	0	–	–
2	<10	9.2	–	–	20	0	–	–
3	<10	9.2	–	–	130	0	–	–
4	<10	5.7	155	>9.49	0	0	0	0
5	<10	5.8	150	>9.49	120	0	0	0
6	<10	0.6	<10	0.2	<10	0	–	–

Adapted from Reference 5.

that inadequate maintenance and the occurrence of biofilms[2] on the membranes of RO systems could be causes for the failure to control microbial bioburden and the presence of endotoxins.

The American Association for the Advancement of Medical Instrumentation (AAMI) sets the standard for the water quality for hemodialysis, and the limits are the same as those described in the USP monograph (Table 11.3). Table 11.6 displays the bioburden of water for hemodialysis produced in different centers that used the RO purification system. The range shows that the bioburden can exceed the limit of 100 cfu/mL, although according to RO system manufacturers, the maximum bioburden would be 10 cfu/mL.

Table 11.7 shows the results of a study comparing different treatment combinations adopted by various hemodialysis centers. All of the centers received their water from public suppliers: centers 1, 4, 5, and 6 purified their water using IE followed by RO, and in centers 4 and 5, the water was additionally disinfected by UV irradiation. Two samples were collected in center 2, where single RO systems were installed for each dialysis machine. IE, a system

[2] Biofilm formation is addressed in Section 11.6.

that is able to remove ionic species, was not satisfactory for the removal of microbial and endotoxin contamination. A reliable water treatment involved the combination of more than one purification system (centers 4 and 5). RO alone, or in combination with IE, was able to eliminate endotoxins but not microbial contamination (centers 1, 2, and 3). Nevertheless, when combined with UV irradiation, IE followed by RO completely eliminated the water bioburden.

The analysis of industrial pure water produced in plants using either IE or RO revealed that a complex microbial community is able to grow in both systems. Bereschenko et al. [6] monitored the *in situ* initiation and development of microbial biofilm layers on the surfaces of fresh and chemically treated RO membranes. Bohus et al. [7] characterized the bacterial community structure of the UPW and the biofilms formed in pipelines and on the resins of an IE water plant purification. The phylogenetic groups isolated from the water and sites of collection in both plants (IE and RO) were practically the same (Table 11.8).

In the IE plant [7], the number of viable aerobic heterotrophic bacteria in the exhausted bed samples was two times higher than in the regenerated bed samples, indicating that the microorganisms were enriched on the resin surface during the 2 weeks of continuous use. The β-proteobacteria was the most abundant group present followed by the α-proteobacteria. Moreover, because the IE resins were cross-linked with divinylbenzene and the carbon steel surfaces were covered with a butyl-type rubber, the authors supposed that these

TABLE 11.8 Phylogenetic Groups Isolated from Water and Sites of Collection in Two Plants for the Production of Pure Water

Ion Exchange Water Plant		Reverse Osmosis Water Plant	
Collection Site	Microorganism	Collection Site	Microorganism
Water from exhausted IE bed	α-Proteobacteria β-Proteobacteria γ-Proteobacteria	Feed water	β-Proteobacteria γ-Proteobacteria Acinetobacteria
Water from regenerated IE bed	β-Proteobacteria γ-Proteobacteria Low G + C Gram+ *Chromobacterium* sp.	Noncleaned membrane after 6 months	α-Proteobacteria β-Proteobacteria γ-Proteobacteria
Inner surface of feed pipelines	*P. necessarius* *Sphingomonas* sp. *Blastochloris* sp.	Membrane cleaned weekly fter 6 months	α-Proteobacteria Acinetobacteria
Inner surface of product pipelines	*Blastochloris* sp.		
Resin	*Methylosinus* sp.		

Purification systems consisted of ion exchange (IE) and reverse osmosis (RO).
Adapted from References 6 and 7.

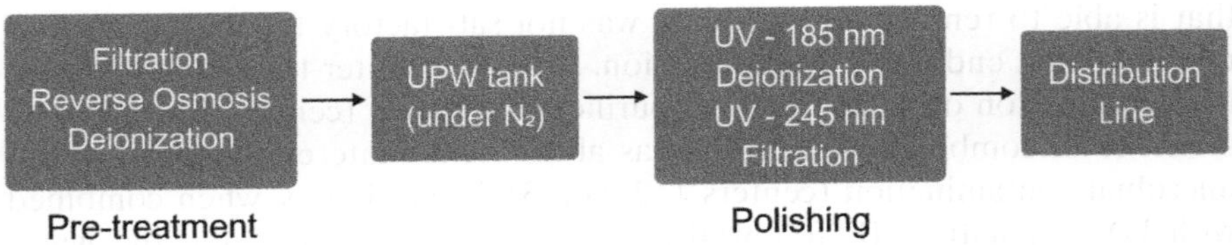

FIGURE 11.1 Schematic presentation of the typical UPW production showing the main components common to all six UPW systems. Incoming water samples were used as controls (adapted from Reference 8).

polymers could provide a reasonable range of nutrients to support an evolving complex microbial community. The authors concluded that the detected microorganisms could enter the purification system from the crude water or from the chemicals used for resin regeneration. Although the cells or their DNA could survive the drastic physicochemical effects of water treatment, the larger populations seemed to arise from true planktonic species adapted to the low nutrient environment.

The microbial ecology of the RO water plant [6] demonstrated that conventional cleaning with toxic chemicals (sodium bisulfite and acid detergent descaler) affected the occurrence of biofouling but was not effective in actually cleaning the system because the microbial colonization on collapsed biofilms occurred directly after chemical cleaning. The colonization process involves a similar set of events as for clean surfaces: the initiation of early biofilm structures and a spatiotemporal development into a multispecies slime layer with a complex three-dimensional structure. Therefore, the failure to remove established biofilms from RO membranes is the primary reason for the limited effect of conventional chemical treatment on the prevention or elimination of biofouling in RO water purification plants.

Kulakov et al. [8] demonstrated that even UPW may present microbial contamination. Six industrial systems for UPW production, which included filtration, RO, IE, and UV irradiation at 245 and 185 nm, were investigated (Fig. 11.1). The isolated bacteria were mostly gram-negative, and some groups, including *Ralstonia pickettii*, *Pseudomonas saccharophilia*, *Bradyrhizobium* sp., and *Stenotrophomonas* sp., seemed to be indigenous for UPW production systems (Table 11.9). The authors concluded that these bacteria represent the most important communities in polishing and distributing parts of UPW systems. Because of the presence of some nitrogen-fixing strains (e.g., *Bradyrhizobium* and *P. saccharophilia*), the authors speculated that the widespread use of nitrogen gas in UPW production systems (to reduce O_2 and CO_2, usual sources of oxygen and carbon) may contribute to bacterial contamination. Other studies have also found representatives of the genus *Pseudomonas*, *R. pickettii*, and *Burkholderia* in distilled water [9] and in UPW [10] production systems.

TABLE 11.9 Bacterial Strain Identification on the Basis of 16S rRNA Gene Analysis in Tested Ultrapure Water Systems[a]

UPW System	Location of Sampling Ports	Nearest Neighbor(s) of Main Strain in BLAST Search of GenBank
1	Before UV 245 nm	*Ralstonia pickettii* *Bradyrhizobium* sp. *Flavobacterium* sp. *Burkholderia* sp. *Stenotrophomonas* sp. *Mycobacterium* sp. *Bacillus* sp. Other
1	Distribution line	*R. pickettii* *Bradyrhizobium* sp. *Pseudomonas saccharophilia* Other
2	Distribution line	*Bradyrhizobium* sp. Other
3	Distribution line	*R. pickettii* *Bradyrhizobium* sp. Other
4	Distribution line	*P. saccharophilia* *Sphingomonas* sp. Other
5	Distribution line	*R. pickettii*
6	Distribution line, storage tank, and before UV 245 nm	*Pseudomonas fluorescens* *R. pickettii* *P. saccharophilia* *Bradyrhizobium* sp. *Sphingomonas* sp. Other

[a] Bacterial strains isolated by growth on R2A media were identified on the basis of 16S rRNA gene sequence.
Adapted from Reference 8.

11.2 RAW MATERIALS

Raw materials afford a wide spectrum of inorganic and organic compounds in dry, aqueous, and nonaqueous systems. The contamination of raw materials is more challenging in nonsterile products because the presence of the bioburden in the raw material has a reasonable probability of remaining in the final product. Depending on their nature, raw materials are more or less prone to present microbial contamination. Untreated products of natural origin may present a high level of contamination, whereas products of synthetic origin are usually free of contamination.

TABLE 11.10 General Classification of Raw Material Used in Pharmaceuticals and Their Ability to Support Microbial Growth

Raw Material	Performance
Water	Highly favorable to microbial growth
Acids, alkalis, salts	Extreme pH prevent growth and survival Dry salts may contain low levels
Oils, waxes, paraffin	Anhydrous forms do not support growth
Fatty acids, alcohols, esters	Unless water is present, do not support growth
Surfactants, emulsifiers	Support microbial growth when supplied as aqueous solutions
Talc, clays	Subject to contamination
Proteins, starches, botanicals, gums, resins	Subject to contamination and microbial growth
Humectants	Usually neat forms do not support growth
Colors, pigments	Subject to contamination
Preservatives, antioxidants, chelating agents	Most do not support growth
Fragrances, essential oils	Usually neat forms do not support growth

Adapted from Reference 11.

The bioburden may either increase or decrease during the various stages of manufacturing. Microorganisms (spores) that remain dormant in dry forms of raw materials may be allowed to grow during manufacture when they contact water. However, the microorganisms present in raw materials may be killed, as the steps of processing may include, for example, the addition of acids or decoction.

Orth [11] categorized raw materials and their ability to support microbial growth. Table 11.10 lists the various categories and the general uses of the raw materials. Dry products do not support microbial growth, and substances such as alcohols, surfactants, and essential oils may present antimicrobial properties.

Raw materials of a natural origin, such as plants and animals, are subject to contamination during growth, harvest or slaughter, and processing as a result of the exposition to soil, water, and air. The contamination of botanical materials was evaluated by Razem and Katusin-Razem [12] to define dose requirements for microbial decontamination by irradiation. They concluded that the degree of contamination depends on the available surface; thus, flowers and leaves are approximately 100 times more contaminated than fruits and seeds. Moreover, the degree of processing and the amount of water also influence microbial bioburden; therefore, crude herbs become more contaminated than extracts. Table 11.11 presents the upper and lower limits of contamination for botanical materials.

TABLE 11.11 Typical Microbial Contamination Levels in Botanical Materials [12]

Material	Contamination Level (cfu/g)
Flowers and leaves	10^4–10^8
Fruits and seeds	10^2–10^6
Liquid extracts	10^4–10^6
Dry extracts	10^2–10^5
Herbal ointments	10^4–10^7

De la Rosa et al. [13] analyzed 115 samples of excipients from different origins, such as animal, mineral, and vegetable. The results in Table 11.12 show that 70% of the analyzed samples presented less than 100 cfu/g of viable aerobic bacteria and that 95% of the samples presented less than 100 cfu/g of molds and yeasts. Arabic and tragacanth gums, both of vegetable origin, were identified as the most contaminated products, and pregelatinized starch and microcrystalline cellulose were the least contaminated excipients. The results indicated that the microbiological quality of these raw materials, with the exception of five samples (two samples of corn starch and three samples of lactose), was suitable and within the limits prescribed by the pharmacopeial compendia.

The microbial contamination of some raw materials used for preparing large-volume parenterals is presented in Table 11.13. All raw materials were relatively free of microorganisms with the exception of mannitol, which presented endotoxins. The large number of samples, from different suppliers and lots, facilitated the confirmation that inorganic species do not favor microbial growth. This finding is also consistent with the data shown in Tables 11.12 and 11.13, where the nonmineral samples presented higher contamination levels than the mineral samples.

Although sodium bicarbonate did not present elevated microbial contamination in the solid form, the bicarbonate solution used to prepare dialysate supported a considerable amount of bacterial and endotoxin growth. An investigation of three hemodialysis centers during a period of 1 year showed that the mean bacterial and endotoxin concentrations in bicarbonate solutions used to prepare dialysates were substantially high at 130,000 cfu/mL and 6.300 pg/mL, respectively. The most prevalent bacteria were *Pseudomonas* spp., *Xanthomonas maltophilia*, and *Alcaligenes dentrificans*. Figure 11.2 shows the microbial and endotoxin levels not only in bicarbonate solutions but also in water and dialysates during the period of the study. Although the amount of bioburden present in the water samples was usually below the AAMI recommendation level (<100 cfu/mL), most of the dialysates (prepared with concentrated sodium bicarbonate) presented contamination above the 2.000-cfu/mL limit.

TABLE 11.12 Microbial Contamination of Raw Material Excipients

Raw Material	Lactose	Talc	Pregelatinized Starch	Tragacanth Gum	Gelatin	Arabic gum	Corn Starch	Microcrystalline Cellulose
No. of samples	36	27	3	1	8	18	19	3
No. of lots	20	19	3	1	8	18	10	3
				Number of Samples Contaminated				
Bacteria (cfu/g)								
<100	33	18	3	0	6	3	14	3
100–1000	3	9	0	0	2	7	4	0
>1000	0	0	0	1	0	8	1	0
Yeast (cfu/g)								
<100	35	27	3	0	7	17	18	3
>100	1	0	0	1	1	1	1	0

Adapted from Reference 13.

TABLE 11.13 Microbial and Endotoxin Contamination of Raw Materials Used for Preparing Large-Volume Parenterals

Raw Material	NaCl	KCl	$CaCl_2.2H_2O$	$NaHCO_3$	Glucose	Fructose	Sorbitol	Mannitol	Glycine
No. of samples[a]	43	30	15	23	39	15	22	22	14
	Number of Samples Contaminated								
Bacteria (cfu/10 g)									
<1	12	11	6	7	3	6	6	0	4
1–10	26	17	8	14	16	9	12	13	7
11–100	5	2	1	2	12	0	4	9	2
101–1000	0	0	0	0	8	0	0	0	1
Yeast (cfu/10 g)									
<1	21	16	12	16	5	7	12	2	10
1–10	19	13	3	7	23	7	9	13	3
11–100	0	1	0	0	9	1	0	7	1
101–1000	0	0	0	0	2	0	0	0	0
Endotoxin (ng/g)									
0.05	0	0	0	0	17	9	10	10	0
0.05–0.5	43	30	15	23	22	5	11	3	14
0.5–5	0	0	0	0	0	1	0	6	0
5–50	0	0	0	0	0	0	0	3	0

[a] From different lots and suppliers.
Adapted from Reference 14.

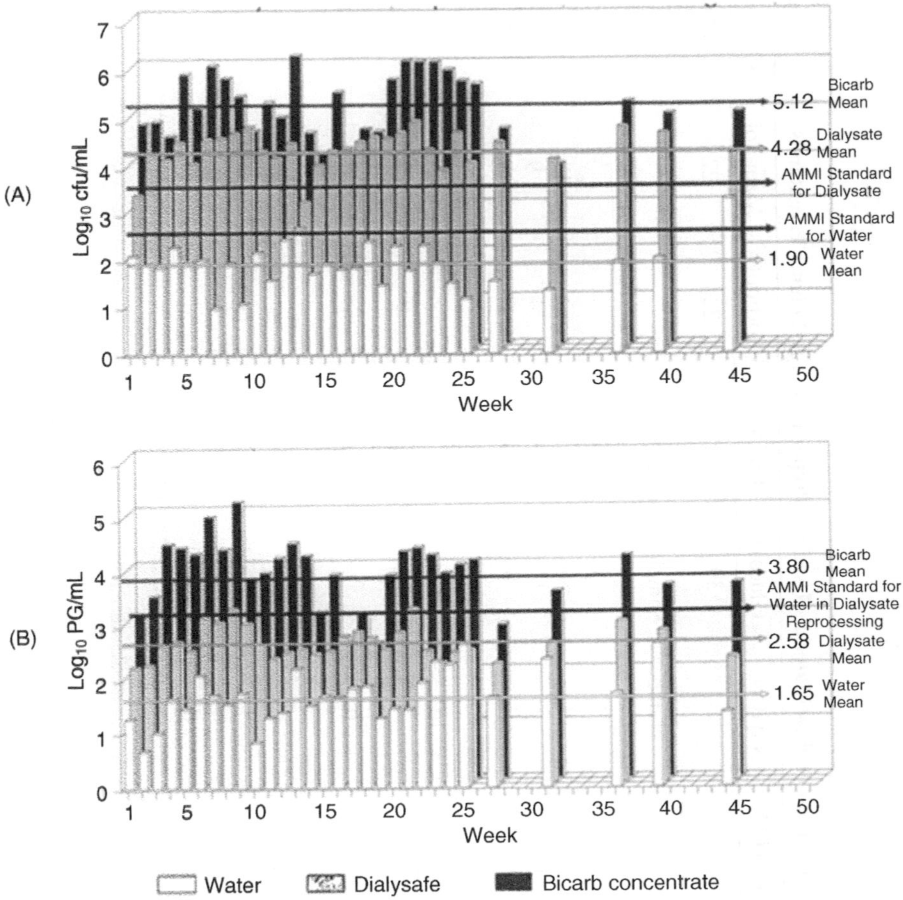

FIGURE 11.2 Mean microbial concentrations (A) and endotoxin concentrations (B) in dialysis fluids by week at three hemodialysis centers evaluated for a period of 1 year [15].

Gelatin has many pharmaceutical applications but presents high microbial bioburden because of its animal origin. Table 11.14 shows the results of a broad survey on gelatin contamination. Samples of 75 gelatin batches from six production plants in different countries were examined. The most frequent microorganisms were *Bacillus licheniformis*, which was found in all production plants, followed by *Bacillus fumarioli* and *Bacillus cereus*, present in three plants, *Bacillus coagulans*, present in two plants, and *Bacillus gelatini*, present in one plant.

TABLE 11.14 Origin and Cultures Isolates Found in Gelatin Samples

Production Plant	Location	Raw Material	No. of Batches
A	Belgium	Porcine skin	23
B	France	Bovine bone	20
C	France	Porcine bone	7
D	France	Bovine skin	2
		Porcine skin	2
E	Argentina	Bovine skin	16
F	United States	Porcine skin	5

Microorganisms Found	Frequency (Batches)	Microorganisms Found	Frequency (Batches)
Bacillus cereus	11	*Anoxybacillus contaminans*	1
Bacillus coagulans	15	*Anoxybacillus flavithermus*	2
Bacillus fumarioli	24	*Brevibacillus agri*	5
Bacillus amyloliquefaciens	5	*Brevibacillus borstelensis*	1
Bacillus licheniformis	44	*Geobacillus stearothermophilus*	5
Bacillus pumilus	3	*Enterobacter sakazakii*	2
Bacillus sonorensis	1	*Staphylococcus epidermidis*	1
Bacillus subtilis	3	*Staphylococcus homints*	1
Bacillus gelatini	9	*Staphylococcus pasteuri*	1
Bacillus thermoamylovorans	1	*Staphylococcus* sp.	4

Adapted from Reference 16.

11.3 STERILE PRODUCTS

Sterility is defined as the total absence of all viable life forms. Parenteral and ophthalmic products are expected to be sterile. Any microorganisms may be opportunistically pathogenic if administered parenterally to debilitated or immunocompromised patients or if applied to susceptible tissues such as the eye.

Sterile pharmaceutical products manufactured under aseptic conditions from presterilized ingredients require the addition of an antimicrobial preservative. Terminally sterilized products require the addition of preservatives only when formulated for multiple uses, which refer to any kind of vial that was used more than once among different patients. The preservative is added to prevent the proliferation of microorganisms or to protect against microbial entry during usage.

Assuming that industrially manufactured sterile products are delivered under strict control standards and regulations, the contamination of intravenous preparations is the result of lapses in aseptic techniques (compounding pharmacies or healthcare settings) associated with favorable conditions for the growth of microorganisms. In contrast to pharmaceutical industries, compounding pharmacies and healthcare settings adhere to different quality control standards, which may not be routinely followed.

11.3.1 Single- and Multiple-Dose Products

One of the main problems regarding the inability to adhere to standards is the repeated use of single-dose (without preservatives) parenteral medications as multiple doses or for multiple patients because these aliquots appear to contain quantities in excess of those required to treat a single patient. Even multidose medications are equally subject to mishandling. Table 11.15 displays the contamination levels of different products in single- and multiple-dose vials collected in hospital wards. The methods for handling these samples for administration and the viable microorganisms presented in these samples are also presented in Table 11.15. The findings reveal different rates of contamination. These differences may be attributed, at least in part, to the difficulties involved in the evaluation of the methodical details, such as the procedure for sample collecting, types of medication, time and temperature of storage, and environmental conditions.

Motamedifar and Askarian ([17]; see Table 11.15) reported that the mean contamination rates among all wards involved in the study was 5.6%, and because approximately 44.5% of the samples were contaminated with *Staphylococcus epidermidis*, the authors attributed the bacterial contamination to poor aseptic techniques associated with the multiple use of preservative-free solutions. Nogler-Semenitz et al. [18] and Mattner and Gastmeier [19] reported a similar conclusion, although the contamination rates were relatively lower in both studies.

The results uncovered by Driver et al. ([20]; see Table 11.15) showed that drugs prepared in syringes and stored in an operating room did not support microbial growth, even after 8 days at room temperature. Among the researched drugs, three contained antimicrobial agents and the others were known to possess bactericidal or fungicidal properties. However, although contained in multidose vials, all of the syringes were prepared at once, which reinforces that the contamination source is not the multidose vial itself but the way it is handled. Microorganisms could enter the vials as a consequence of inserting subsequent needles from syringes that have already been used on patients. Figure 11.3 demonstrates how a microorganism (hepatitis C virus [HCV] in the picture) may enter a multidose vial.

Another approach to evaluate the microbial growth in multidose vials is to inoculate the drugs with pathogens to simulate clinical conditions and to determine the microbial viability. Table 11.16 presents the results of four studies on microbial viability in drugs, with or without preservatives, used in intensive care treatments. For testing, the drugs were used in concentrated or diluted (with saline or dextrose) forms, inoculated with common nosocomial isolates, and incubated at room temperature (21 or 25°C) for 48 hours. The

TABLE 11.15 Contamination Rates and Viable Microorganisms Found in Multiple-Dose Vials Collected in Hospital Wards

Product Type	Sample Collection or Handling	Samples	Preservative	No. of Samples Analyzed	Percentage Contaminated	Microorganism Found (Samples not Specified)	Reference
Multidose	Open vials collected in 37 different wards, stored at room temperature, and used for more than one patient	Saline Potassium chloride Sodium bicarbonate Calcium carbonate Insulin WFI	No No No No Yes No	Total 637	0% ICU to 7.6% in pediatric unit	*Staphylococcus epidermidis* *Bacillus coagulans* *Pseudomonas maltophilia* α-Hemolytic streptococci *Actinomyces viscosus* *Acinetobacter* sp. *Streptococcus viridans* *Micrococcus* sp. *Corynebacterium* sp. *Serratia marcescens* β-Hemolytic streptococci	[17]
Multidose	Open vials collected in 18 different wards	Saline Heparin	No Yes	Total 68	4.4%	Spore-forming bacteria Coagulase-negative *Streptococcus*	[18]
Single dose		Saline WFI	No No	Total 17	5.9%	Coagulase-negative *Streptococcus*	[18]

(*Continued*)

TABLE 11.15 (*Continued*)

Product Type	Sample Collection or Handling	Samples	Preservative	No. of Samples Analyzed	Percentage Contaminated	Microorganism Found (Samples not Specified)	Reference
Multidose		Heparin + saline	Yes	11	0%	–	[18]
Multidose	Open vials collected in 47 different wards	Saline	No	total 227	0.9%	*S. epidermidis*	[19]
		Heparin	Yes				
		Glucose	No				
		WFI	No				
		Insulin	Yes				
		Others[a]	Yes				
Multidose	Drugs stored in syringes after dawning into sterile syringes and diluting in operating room under normal clinical practices Kept at room temperature. Sterility tested at days 4 and 8	Atropine	Yes	126	0%	–	[20]
		Ephedrine	None	126	0%		
		Lidocaine	None	126	0%		
		Oxytocin	Yes	126	0%		
		Succinylcholine	Yes	126	0%		
		Thiopental	None	126	0%		
		Sterile water	None	126	0%		

[a] Cathecolamines, atropine, antibiotics, midazolam, furosemide, iopromide, and others.

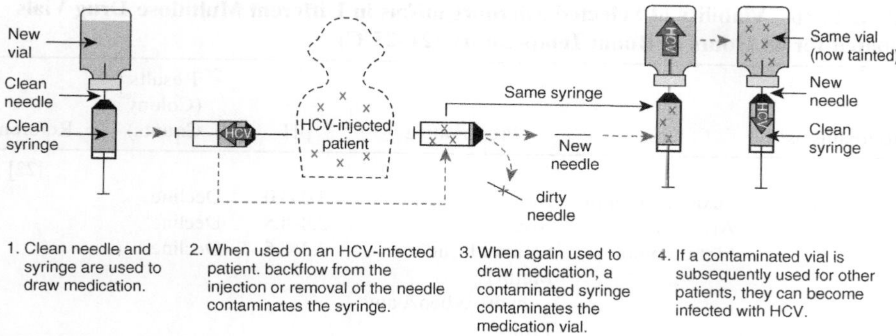

FIGURE 11.3 Unsafe injection practices and circumstances that likely resulted in transmission of hepatitis C virus (HCV) [21].

results showed that, when present, preservatives may be responsible for reducing colony counts.

Another factor that affects microbial proliferation is the environmental pH. Most bacteria grow best within a pH range of 6.5–7.5. A low pH could also function as a sufficient preservative. Formulations with a low pH tend to show a more rapid decline in colony counts, as occurred with the use of suxamethonium, atropine, and metaraminol in study I displayed in Table 11.16. The decreased colony counts in nonpreserved methohexital sodium and sodium thiopental preparations used in study III were likely due to the high pH because both presented a pH > 10. Only three medications in Table 11.16 supported the proliferation of microorganisms: propofol, lidocaine, and insulin. However, lidocaine and insulin contained preservatives. In addition to the absence of preservatives, propofol presented a pH in the 7.0–8.5 range. Figure 11.4 shows the results from experiments evaluating microorganism growth in propofol samples. Propofol proved to be an excellent culture medium for most inoculated microorganisms (see Section 11.3.3).

The action of preservatives is restricted by sample dilution. Dilution could reduce the amount of preservative to a less than optimal concentration and the pH to a range that permits microbial growth. In study I in Table 11.16, the bactericidal drug effect was diminished by dilution, but no significant differences were observed when dextrose or saline were used as diluents.

11.3.2 Parenteral Nutrition (PN)

PN formulations are susceptible to microbial growth because their components, amino acids, glucose, and lipids, support fungal and bacterial growth. Nevertheless, the intrinsic contamination of PN solutions is rare. However, extrinsic contamination, arising from preparation and handling and through the use of catheters for central venous access of PN infusions, is a common occurrence. Thomas et al. [26] surveyed the literature concerning the

TABLE 11.16 Viability of Selected Microorganisms in Different Multidose Drug Vials Incubated for 48 Hours at Room Temperature (21–25°C)

Microorganisms[a]	Drug	Preservative	pH	Results (Colony Counts)	Reference
Study I					[22]
S. epidermidis	Suxamethonium	None	3.0–5.0	Decline	
	Atropine	None	2.8–4.5	Decline	
	Metaraminol	Sodium bisulfite Methyl hydroxybenzoate Propyl hydroxybenzoate	3.2–3.5	Decline	
	Ephedrine	None	4.5–7.0	Decline	
Study II					[23]
S. aureus	Midazolam	EDTA, benzyl alcohol	3	Decline	
E. coli	Propofol	None	7.0–8.5	Proliferation	
E. faecalis	Morphine	–	2.5–6	Decline	
P. aeruginosa	Fentanyl citrate	None	4–7.5	Decline	
C. albicans	Pethidine	–	–	Decline	
	Bupivacaine	Methylparaben	4–6.5	Decline	
	Atracurium besylate	Benzyl alcohol	3.25–3.65	Decline	
	Vecuronium bromide	Mannitol	4	Decline	
	Adrenaline	–	–	Decline	
	Dopamine	Sodium metabisulfite	3.3	Decline	
	Dobutamine	Sodium bisulfite	2.5–5.5	Decline	
	Noradrenaline	Sodium metabisulfite	3–4.5	Decline	
	Isoprenaline	Sodium metabisulfite	2.5–4.5	Decline	
	Glyceryl trinitrate	–	–	Decline	
	Sodium nitroprusside	None	3.5–6	Decline	
Study III					[24]
S. aureus *S. epidermidis*	Lidocaine	Methylparaben Sodium metabisulfite	6.52	Proliferation	
P. aeruginosa *C. albicans*	Succinylcholine	Methylparaben Propylylparaben	3.8	Decline	
E. coli	Insulin	Phenol	6.3	Proliferation	
A. calcoaceticus Enterococcus	Procainamide	Sodium bisulfite Benzyl alcohol	4.3	Decline	
E. agglomerans	Potassium chloride	None	6.6	Survive	
E. cloacae	Sodium heparine	Benzyl alcohol	6.64	Survive	

TABLE 11.16 (*Continued*)

Microorganisms[a]	Drug	Preservative	pH	Results (Colony Counts)	Reference
S. marcescens	Methohexital sodium	None	10.69	Decline	
K. pneumoniae	Sodium thiopental	None	10.55	Decline[b]	
P. mirabilis					
P. cepacia					
Study IV					[25]
S. aureus	Iopromide	None	6.5–7.7	Decline	
S. epidermidis	Sodium meglumine	None	6.5–7.7	Decline	
S. cohnii	Diatrizoate				
P. aeruginosa					
C. albicans					

[a] All microorganisms were incubated in all drugs.
[b] Survival of *C. albicans* for 48 hours.
–, not informed.

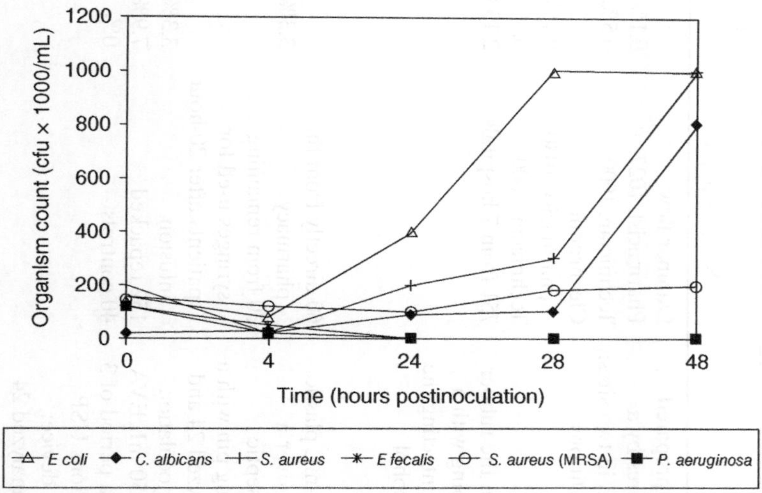

FIGURE 11.4 Growth of different microorganisms in propofol samples after inoculation at 22°C [23]. MRSA, methicillin-resistant *Staphylococcus aureus*.

contamination rates of IV admixing infusions according to the environment where the samples were prepared. The rate of contamination was higher in admixtures prepared in wards or nursing units than in those prepared in a laminar flow workbench (7.87% and 2.77%, respectively). However, the same authors compared the laminar flow workbench and a class 1000 cleanroom for the preparation of admixtures. The results, which are presented in Table 11.17

TABLE 11.17 Microorganism Growth in Parenteral Infusion Solutions Prepared under Different Conditions and in Different Hospitals

Procedure	No. of Samples	Contaminated (%)	Microorganisms Found	Reference
Samples(simulated products) prepared following identical procedures by a pharmacist and a technician in two sites: Laminar flow standard conditions Class 1000 cleanroom	Laminar flow Pharmacist: 1027 Technician: 1000 Cleanroom pharmacist: 1030 Technician: 1000	 0.1% 0.5% 0.1% 0.6%	Coagulase-negative *Staphylococcus* *Bacillus* sp. *Corynebacterium* sp.	[26]
0.5–1 mL infusate drawn from the rubber IV injection port of the tubing with a disposable syringe after disinfecting the injection port with 70% alcohol	751 from 7 hospitals	2.13%	*Klebsiella pneumoniae* *Acinetobacter baumannii* *Enterobacter cloacae* *Serratia marcescens* Coagulase-negative *Staphylococcus* *Bacillus* sp.	[27]
Lipid emulsion repacked in 5-mL plastic syringes by six pharmacists over a period of 1 month under aseptic conditions using a dispensing pin with a Safsite valve. Samples analyzed 24 and 48 hours and 7 days after inoculation	60 directly from the pharmacy 30 from remaining syringes used for patients after 20-hour infusion	3.3% at day 7 3.3% at day 7	Gram-positive rods	[28]
Lipid emulsion repacked in 250-mL EVA bags by a pharmacist over a period of 3 weeks under aseptic conditions USP ⟨797⟩ using a compounding device; samples kept at 22°C and analyzed 24 and 48 hours after inoculation	152 repacked 40 controls	7.9% 0%	Gram-positive cocci Gram-positive rods Yeast	[29]

Lipid emulsion for neonatal patients collected after 24 infusions directly from original container and after 12 hours' infusion from repacked in syringes; samples handled under class 5 hood	60 in the original container 90 repacked into syringes 60 drawdown of the original container	0% 3.3% 0%	Coagulase-negative *Staphylococcus* *Klebsiella oxytoca* *Citrobacter freundii*	[30]
All-in-one formulations (glucose, amino acids, lipids, vitamins, electrolytes, and trace elements) compounded under aseptic conditions, incubated at 30–35°C for 48–72 hours	1294	4.6%	*Staphylococcus epidermidis* *Enterobacter cloacae*	[31]
Admixed and nonadmixed i.v. solutions for pediatric patients	750 from 8 wards	5.1% ad-mixed 1.3% nonadmixed	*Enterobacter* sp. *Klebsiella* spp.	[32]
Standard crystalloid solutions: saline, glucose, Ringer's lactate	264 in use from 24 to 185 hours 261 controls (unopened flasks)	2.7% 3.0%	Coagulase-negative *Staphylococcus*	[33]
1 mL infusate collected at the time of recruitment (1) and after 72 hours from the administration set (2), with a disposable syringe	621 samples 1 472 samples 2	0.6% 1.3%	Coagulase-negative *Staphylococcus* *Enterobacter agglomerans* *Achromobacter piechaudii* *Acinetobacter haemolyticus* *Acinetobacter lwoffi* *Pseudomonoas aeruginosa* *Chrysobacterium meningosepticum* Nitrite-*Flavobacterium* Group CDCII	[34]
1.5 mL infusate collected with a disposable syringe from the infusion sets of patients with reported bacteremia	384	2%	Coagulase-negative *Staphylococcus* *E. cloacae* *Acinetobacter lwoffi* *Pseudomonoas aeruginosa* *Enterobacter aerogenes* *Pseudomonoas stutzeri* *Pseudomonoas fluorescens*	[35]

suggested that when proper aseptic techniques are adopted, there is no significant difference between the environments.

Table 11.17 also displays the results of the contamination of parenteral infusion solutions in different hospitals. The results substantiate the idea that intrinsic contamination is rare, but extrinsic contamination may be a serious problem, as contamination rates of up to 8% were reached. Among the components of PN, lipid emulsions supported the most bacterial growth.

Because lipid emulsion is an important part of neonatal PN therapy, and daily doses of lipid emulsion can be as low as 2 mL infused over 20 hours, Reiter ([28]; see Table 11.17) investigated the sterility of intravenous fat emulsions in plastic syringes used in neonatal intensive care units (NICUs). A total of 90 samples were cultured: 60 samples were obtained from syringes provided by the pharmacy that were prepared under aseptic conditions (normal), and 30 samples were obtained from syringes randomly collected among those used by patients after the 20-hour infusion (the surplus emulsion remaining in the syringes was analyzed). All samples were negative for bacteria growth at 24 and 48 hours, but in three samples (one from pharmacy and two from patients), gram-positive rods grew at day 7.

A similar study was conducted to evaluate the sterility of pediatric lipid emulsions repacked by an automated compounding device ([29]; see Table 11.17). In this case, higher contamination rates were observed. A total of 152 bags were prepared using the compounding device, and bacterial growth occurred in 7.9% (12 bags). The authors concluded that the use of the compounding device should not be recommended because contamination was observed even though all activities had been performed in compliance with the USP <797>chapter. No positive bacterial counts were observed in the control samples.

Because lipid emulsions are favorable for microbial growth, Didier et al. [36] inoculated total PN containing a lipid emulsion with 12 bacterial species that were most likely to cause nosocomial infections. The samples were incubated for 5 days at 4, 25, and 35°C. None of the species grew at 4°C, but at 25 and 35°C, the growth of *Candida albicans* and *Staphylococcus saprophyticus* was observed. The authors attribute this lack of growth to the slightly acidic pH (between 5.2 and 6.6), which is a major inhibitory influence for microorganism growth. However, when pure lipid emulsions were used, increased microorganism growth was observed. Figure 11.5 shows the bacterial growth, endotoxin production, and pH change in pure lipid emulsions inoculated with *Enterobacter cloacae* and incubated at 5 and 25°C for 24 hours. These results show that while the bacteria survived at 5°C, no proliferation or endotoxin production was detected, and the pH remained unchanged. In contrast, at 25°C, the organism rapidly proliferated, reaching concentrations of $>10^7$ cfu/mL in 24 hours. Endotoxin was detected after proliferation reached 10^2 cfu/mL, while the pH decreased when proliferation reached 10^7 cfu/mL.

A better evaluation of the role of temperature in microbial growth can be obtained by comparing the data in Table 11.18, where samples of glucose and

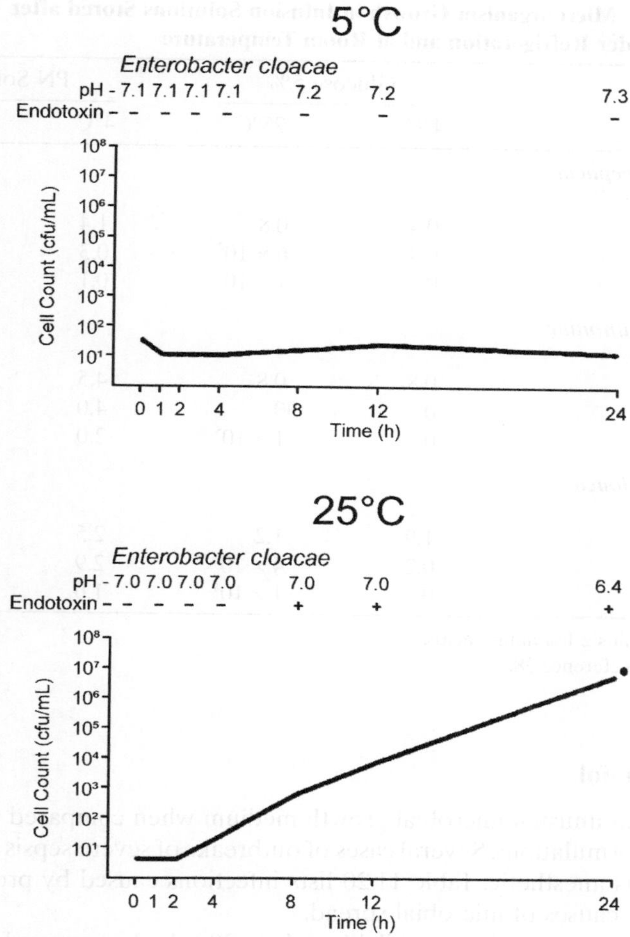

FIGURE 11.5 Bacterial growth, endotoxin production, and pH change in lipid emulsion at 5 and 25°C. Adapted from Reference 37.

standard PN solutions were inoculated with three different bacteria. While most of the microorganisms did not proliferate in both solutions at 4°C, their growth reached significant densities at room temperature.

Pure lipid emulsions support microbial growth due to the favorable pH (approximately 7.5), osmolarity (250–350 mOsm/L), and glycerol content (2.25–2.50%), irrespective of the final lipid concentration. However, when used in total nutrient admixtures, the lipid emulsion becomes a poor growth medium because these features are lost [39]. Table 11.19 presents the growth of *C. albicans* in PN solutions with different compositions. The data suggest that the presence of lipids, electrolytes, and vitamins did not increase microbial growth.

TABLE 11.18 Microorganism Growth in Infusion Solutions Stored after Inoculation under Refrigeration and at Room Temperature

	Glucose 5%		PN Solution[a]	
Product	4°C	25°C	4°C	25°C
Pseudomonas cepacia				
cfu/mL on day:				
0	0.4	0.8	1.4	1.8
2	0.4	6×10^4	0.3	50
7	0	1×10^6	0.1	5×10^7
Klebsiella pneumoniae				
cfu/mL on day:				
0	0.8	0.8	4.5	7.0
2	0	30	4.0	5×10^4
7	0	1×10^5	2.0	2×10^7
Enterobacter cloacae				
cfu/mL on day:				
0	1.9	1.2	2.5	2.6
2	0.2	4×10^2	2.9	8×10^3
7	0	1×10^4	1.0	3×10^7

[a] 10% glucose plus 2.1% amino acids.
Adapted from Reference 38.

11.3.3 Propofol

Propofol is an unusual microbial growth medium when compared with other lipid-based formulations. Several cases of outbreaks of severe sepsis are associated with this anesthetic. Table 11.20 lists infections caused by propofol and the probable causes of microbial spread.

Before 1991, propofol was available only in 20-mL glass ampoules. In 1991, propofol became available in 50- and 100-mL rubber-topped vials. Despite recommendations that the doses be prepared just prior to administration and the infusion be completed within 6 hours after container opening, the use of larger vials increased the risk of extrinsic contamination because the larger volumes could be used for an extended period of time and for more than one patient. Table 11.21 presents the results of microbial growth in syringes used in different scenarios where anesthetic procedures using propofol were conducted. The rates of contamination reached 8.8%, and *S. epidermidis* was the most prevalent pathogen.

These investigations showed that even under standard clinical conditions, contamination can occur at any stage of sample handling and that even short periods of time between drawing up and anesthesia induction were not able to avoid bacterial growth. Figure 11.6 shows the microbial growth in propofol samples during a period of 48 hours. The graph shows that bacterial growth

TABLE 11.19 Microbial Growth in PN Solutions with Different Compositions after 48 Hours of Inoculation with *Candida albicans* at 25°C

	Components (%)								
Sample	Glucose	Aminoacids	Lipids	Electrolytes	Trace Elements	Vitamins	pH	cfu	Reference
1	20	4.25	–	–	–	–	–	6×10^2	[38]
2	12.5	1.5	1.5	–	–	–	6.2	1×10^3	[40]
3	25	1.5	1.5	–	–	–	6.1	1×10^3	[40]
4[a]	12.5	4.3	5	Yes	–	–	5.7	1×10^3	[41]
5	5	3	5	Yes	Yes	Yes	5.75	$>1 \times 10^2$	[42]
6	10	3	5	Yes	Yes	Yes	5.74	$>1 \times 10^2$	[42]
7	25	3	5	Yes	Yes	Yes	5.65	$>1 \times 10^2$	[42]
8	18.3	2.5	3.3	Yes	Yes	Yes	6.1	4×10^2	[43]
9	8.3	2.9	3.3	Yes	Yes	Yes	5.6	1.6×10^4	[43]
10	17.6	5	4	Yes	Yes	Yes	5.6	1.6×10^5	[36]
11	3.3	2.8	4.3	Yes	Ycs	Yes	6.5	1.3×10^2	[44]
12	16.7	2.8	3.3	Yes	Yes	Yes	6.4	8×10^1	[44]

[a] 37°C.

TABLE 11.20 Outbreaks Associated with the Anesthetic Agent Propofol and Possible Sources of Contamination

Microorganism Outbreak	No. of Patients	Positive Cases (%)	Period	Reference	Sources of Contamination
Hepatitis C virus	6	100%	7–10/2007	[48]	Inappropriate reuse of syringes Use of medication vials intended for single use on multiple patients
Klebsiella pneumoniae and *Serratia marcescens*	35	20%	09/2008	[49]	Positive cultures in Vial of propofol in use Syringe with propofol
Staphylococcus aureus	668	2.4%	4/1990	[45]	
Candida albicans	364	1.1%	4/1990	[45]	Presumable sources of contamination for these outbreaks:
S. aureus	56	13%	5/1990	[45]	-Not disinfecting the rubber stopper of vials before use
Enterobacter agglomerans	18	22.2%	8/1990	[45]	-Prepare multiple syringes at on time for use throughout the day
S. marcescens	360	1.7%	9-10/1992	[45]	-Reuse of syringes for different patients
None identified	463 50	0.6% 2.0%	12/1992	[45]	-Transfer prepared syringes between rooms -Fail to wear gloves during preparing or administering propofol
S. marcescens	36	19.4%	10/1999	[46]	
S. aureus	9	56%	6/1996	[47]	
Hepatitis C virus	16	100%	3/1998	[21]	

TABLE 11.21 Results of Prospective Studies on the Contamination of Propofol Infusions in ICUs and Operating Theaters

Samples	Study Design	Number of Samples	Culture Positive	Microorganism Identified	Reference
Propofol administered to patients in an ICU setting	Samples prepared in 20-mL syringes by bedside nursing and administered using a syringe driver	302	18 (5.9%)	*Staphylococcus epidermidis*	[50]
Propofol administered to patients in an hospital	1.0 mL propofol collected from remnants syringes and ampoules immediately after administration	100	3 (3%)	*S. epidermidis* *Pseudomonas aeruginosa*	[51]
Propofol administered to patients in operating theaters	Elapsed time between drawing up and induction:			*Bacillus* sp., Diptheroid, *Micrococcus* sp., coagulase-negative *Staphylococcus*	[52]
	<10 min	80	5 (6.3%)		
	11–69 min	140	10 (7.1%)		
	≥60 min	34	5 (6.3%)		
Propofol administered to patients in operating theaters	Drawing up propofol following:			*S. epidermidis*	[53]
	Manufacturer's handling protocol (group I): Disinfecting vials and ampoules before filling syringes; Replacing empty syringes; Discarding all materials at the end of surgery	160	14 (8.8%)		
	Modified handling protocol (group II): Refilling empty syringes; Renewing only the infusion line to the patient	160	13 (8.1%)		

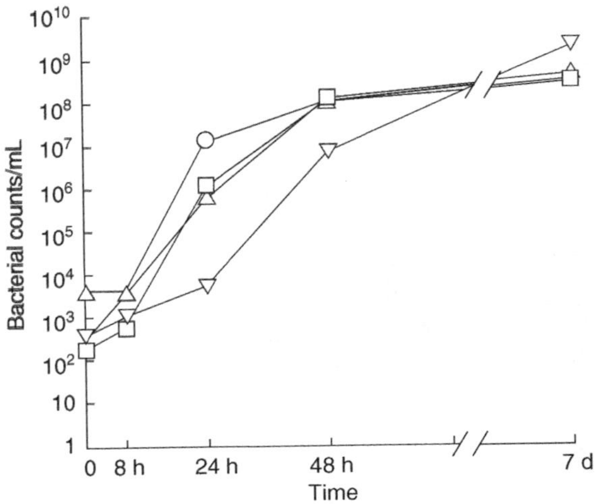

FIGURE 11.6 Viability of *Serratia marcescens* at 30°C in soybean oil (○) and propofol (△), and of *Burkholderia cepacia* at 30°C in soybean oil (□) and propofol (▽) [54].

occurred 4 hours after inoculation and subsequently increased rapidly, justifying the recommendation on the propofol label that administration should commence promptly and be completed within 6 hours after the vials have been opened.

Tessler et al. [55] obtained similar results by comparing propofol with Intralipid 10% incubated with *Staphylococcus aureus*, *Moraxella osloensis*, and *C. albicans*. The growth rates were similar; a 10% growth rate was observed after 8 hours, which is an expected result because both media contain 10% soybean oil. The authors also showed that the lipid vehicle is responsible for microbial growth because pure 2,6-diisopropylphenol (propofol) was bactericidal. No microorganism growth was observed in this medium.

A comparison of the results in Tables 11.15, 11.17, and 11.21 reveals that the contamination rates in the propofol samples are higher than in the other formulations. This is likely because the propofol samples are usually prepared in intensive care unit (ICU) settings for patient sedation or in operating theaters for patient anesthesia, under conditions that are not as controlled as in a compounding pharmacy, where PN and other kinds of samples are usually prepared.

To reduce the risk of bacterial contamination, some manufacturers incorporate antimicrobial additives into propofol formulations. The compositions of some commercially available propofol formulations are displayed in Table 11.22. Two formulations, Diprivan and Propofol, present antimicrobial agents. The effects of EDTA on microbial growth are shown in Figure 11.7, where two propofol formulations, with and without 0.005% EDTA, were cultured

TABLE 11.22 Composition of Commercial Propofol Formulations

Company	Commercial Name	Ingredients	Concentration (mg/mL)	Antimicrobial
AstraZeneca Pharmaceuticals	Diprivan	Soybean oil	100.0	EDTA 0.005%
		Glycerol	22.5	
		Egg lecithin	12.0	
Teva Parenteral Medicines	Propofol	Soybean oil	100.0	Sodium metabisulfite 0.25 mg/mL
		Glycerol	22.5	
		Egg yolk phospholipid	12.0	
Fresenius Kabi	Propoven	Soybean oil	100.0	–
		Glycerol	22.5	
		Egg yolk phospholipid	12.0	
B. Braun	Lipuro	Soybean oil	50.0	–
		Glycerol	–	
		Medium-chain triglycerides	–	
		Egg lecithin	–	
		Sodium oleate	–	

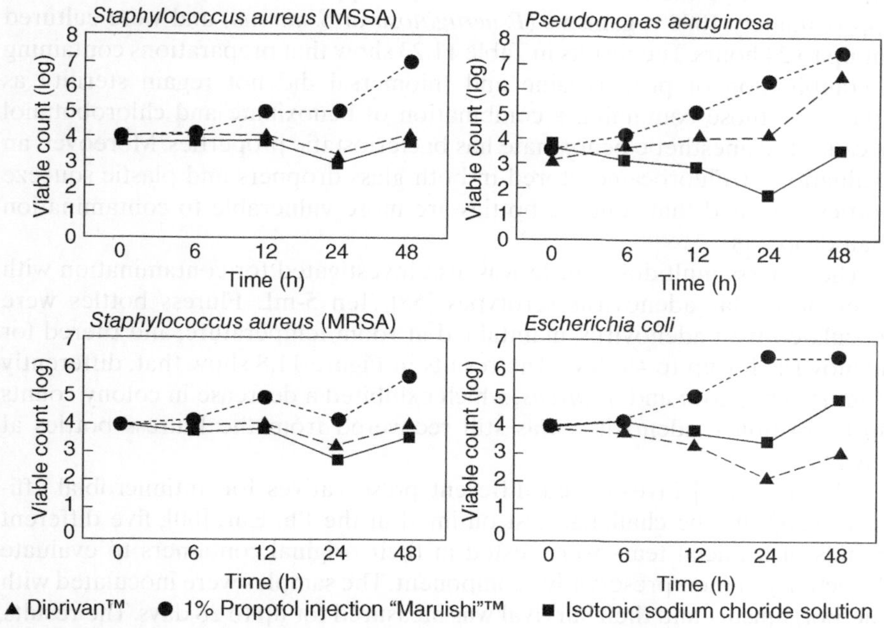

FIGURE 11.7 Microorganisms growth curve in Diprivan™, 1% propofol injection (Maruishi™), and in isotonic sodium chloride solution after inoculation with *Staphylococcus aureus*, *Escherichia coli*, and *Pseudomonas aeruginosa*, at 22.5°C [56].

with *S. aureus*, *E. coli*, and *P. aeruginosa* at 22.5°C. The addition of EDTA to propofol retarded the growth of microorganisms commonly associated with nosocomial infections.

11.3.4 Ophthalmic Products

Eyedrops are at risk of contamination with pathogenic microorganisms if they are preservative-free because most products are supplied in multiple application containers. These containers are usually droppers (with a pipette) or squeeze bottles. During administration, the pipette attached to the cap of the bottle comes completely out of the container, exposing the open contents of the bottle directly. Spillage can also increase the incidence of contamination. Poor technique in administering the drops is another risk factor for contamination, especially because most patients self-administer the drops. Even preserved multiple-use containers may accidentally become contaminated through applicators or caps, which may not recover sterility after an application because they do not contact the preserved solution. For these reasons, the occurrence of bacterial ocular infections transmitted by contaminated eyedroppers has been documented.

Duffner et al. [57] evaluated the ability of four commercially available ophthalmic preparations containing preservatives to regain sterility. Fluorescein–anesthetic solutions and their dropper tips were inoculated with approximately 1000 organisms (*P. aeruginosa* or *S. aureus*) and were cultured for up to 24 hours. The results in Table 11.23 show that preparations containing a combination of proparacaine and thiomersal did not regain sterility as quickly as those containing a combination of benoxinate and chlorobutanol because the anesthetic benoxinate has bacteriostatic properties. Moreover, an evaluation of Fluorocaine, stored in both glass droppers and plastic squeeze bottles, revealed that squeeze bottles are more vulnerable to contamination than droppers.

The Fluress multidose bottle was also investigated for contamination with common ocular adenovirus serotypes [58]. Ten 5-mL Fluress bottles were inoculated with adenovirus, maintained at room temperature, and titered for adenovirus for up to 49 days. The results in Figure 11.8 show that, differently from *P. aeruginosa* and *S. aureus*, which exhibited a decrease in colony counts up to 24 hours, adenovirus was still recovered from the Fluress bottles at day 28.

Charnock [59] investigated different preservatives for antimicrobial efficacy. Adopting the challenge test outlined in the Ph. Eur. [60], five different brands of artificial tears were tested in their original containers to evaluate the activity of their preservative component. The samples were inoculated with microorganisms, and their survival was measured for up to 28 days. The results, presented in Table 11.24, demonstrated that only samples containing benzalkonium chloride/EDTA satisfied the major criteria for antimicrobial preservation for all microorganisms tested.

TABLE 11.23 Mean Colony Counts after the Inoculation of Solutions and the Contamination of the Bottle or Pipette Tips of Ophthalmic Fluorescein Formulations with *P. aeruginosa* and *S. aureus*

Product	Fluress	Fluorpro	Fluorocaine	Fluorocaine
Dispenser	Dropper	Squeeze Bottle	Squeeze Bottle	Dropper
Constituents	Benoxinate hydrochloride 0.4% Chlorobutanol 1% Povidone Boric acid	Proparacaine hydrochloride 0.5% Thimerosal 0.01% Povidone Glycerin	Proparacaine hydrochloride 0.5% Thimerosal 0.01% Povidone Glycerin	Proparacaine hydrochloride 0.5% Thimerosal 0.01% Povidone Glycerin
		Solution		
P. aeruginosa				
1 min	0	TNTC	TNTC	TNTC
15 min	0	6.7	5.7	3.3
2 h	0	0	0	0.5
24 h	0	0	0	0
S. aureus				
1 min	0	17.3	26.8	2.0
15 min	0	11.0	13.3	2.5
2 h	0	10.8	6.3	3.0
24 h	0	0	0	0
		Bottle Tip or Pipette Tip		
P. aeruginosa				
1 min	0	105.0	80.6	23.0
15 min	0	105.0	94.0	11.6
2 h	0	8.8	42.6	0
24 h	0	0	0	0
S. aureus				
1 min	0.8	TNTC	TNTC	TNTC
15 min	0	4.8	TNTC	11.3
2 h	0	7.5	40.0	0.8
24 h	0	0	0	0

Counts were measured after 1 and 5 minutes, and 2 and 24 hours after inoculation and contamination.
TNTC, too numerous to count.
Adapted from Reference 57.

Oldham and Andrews [61] used the same Ph. Eur. test for the efficacy of antimicrobial preservation to establish the inherent antimicrobial activity of unpreserved eyedrops. Preservative-free multidose eyedrops are available for patients susceptible to adverse reactions from antimicrobial agents. Twenty-one commonly used eyedrops were inoculated with selected microorganisms.

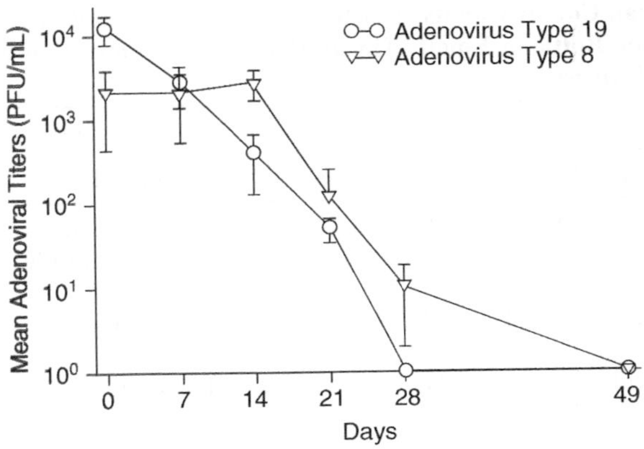

FIGURE 11.8 The decrease of adenovirus types 8 (▽) and 19 (○) titers over time in multidose bottles of Fluress [58].

Two, atropine and chloramphenicol, were incubated at 20–25°C, as stipulated, or under refrigeration (2–8°C). All others were maintained only under refrigeration for 28 days. Table 11.25 presents the results for atropine and chloramphenicol at both temperatures, where lower rates of replication of microorganisms can be observed in the refrigerated samples. The results for the other samples, displayed in Table 11.26, revealed that most samples have their own inherent antimicrobial activity. Most of the eyedrops tested had less inherent antimicrobial action against *S. aureus*, which is a problem that is also encountered with preserved products (see Tables 11.23 and 11.24), suggesting that *S. aureus* could be a recurrent microorganism in eyedrops.

Table 11.27 shows the spontaneous microbial growth in different ophthalmic products, including preserved and nonpreserved eyedrops. Cultures were harvested after different periods of use from different parts of containers collected either from patients or in ophthalmologic clinics.

Rahman et al. ([62]; see Table 11.27) investigated the sterility of 10 different preservative-free eyedrops on either day 3 or 7 of use. None of the antibiotic eyedrop formulations showed signs of contamination, but the overall incidence of contamination in the nonantibiotic formulations was 19%. Hypromellose was the most contaminated, followed by prednisolone and acetylcysteine. These results are similar to those obtained by Oldham and Andrews [61] (Table 11.25), demonstrating the vulnerability of these formulations to microbial growth when unpreserved.

Kim et al. ([63]; see Table 11.27) compared preserved and preservative-free multidose artificial tears packed in reclosable containers after three or more uses in an interval of 10 hours. While 2.4% of the preservative-free bottles were contaminated, the preserved solutions presented no contamination. The

TABLE 11.24 Mean Colony Counts after the Inoculation of Artificial Tears Containing Different Preservatives with *P. aeruginosa*, *S. aureus*, and *C. albicans* ($T = 22 \pm 2°C$)

Product	1	2	3	4	5[b]
Preservative system	Benzalkonium chloride 0.01% EDTA 0.05%	Chlorobutanol 0.5%	Purite, stabilized oxychloro complex[a] 50 ppm	Sodium silver chloride complex 0.001%	Methyl-ethyl-, propyl-paraben
		Pseudomonas aeruginosa			
Time after inoculation					
0.25 h	n.d.	12 ± 9.9	100 ± 28	52 ± 47	47 ± 22
1 h	n.d.	0.62 ± 0.20	93 ± 25	36 ± 27	33 ± 31
6 h	n.d.	n.d.	54 ± 32	4.0 ± 1.4	2 ± 1.6
24 h	n.d.	n.d.	n.d.	n.d.	n.d.
7 days	n.d.	n.d.	n.d.	n.d.	n.d.
28 days	n.d.	n.d.	n.d.	n.d.	n.d.
		Staphylococcus aureus			
Time after inoculation					
0.25 h	n.d.	98 ± 13	94 ± 3.5	84 ± 12	102 ± 13
1 h	n.d.	86 ± 2.8	89 ± 3.5	97 ± 20	86 ± 4.6
6 h	n.d.	88 ± 12	58 ± 29	94 ± 25	74 ± 18
24 h	n.d.	3.0 ± 2.6	n.d.	20 ± 15	6.5 ± 0.71
7 days	n.d.	n.d.	n.d.	0.005 ± 0.007	n.d.
28 days	n.d.	n.d.	n.d.	n.d.	n.d.
		Candida albicans			
Time after inoculation					
0.25 h	1.7 ± 0.14	111 ± 0.71	93 ± 8.0	110 ± 6.5	109 ± 2.1
1 h	0.11 ± 0.13	104 ± 6.4	109 ± 6.4	94 ± 12	94 ± 23
6 h	n.d.	118 ± 19	121 ± 16	95 ± 11	93 ± 0.71
24 h	n.d.	86 ± 12	88 ± 7.8	65 ± 25	83 ± 2.1
7 days	n.d.	0.25 ± 0.35	n.d.	2.5 ± 2.1	18 ± 31
28 days	n.d.	n.d.	n.d.	n.d.	2.3 ± 4.0

[a] Consists of an equilibrium mixture of oxychloro species: 99.5% chlorite, 0.5% chlorate, and trace amounts of chlorine dioxide.
[b] Undeclared concentration.
Adapted from Reference 59.

authors concluded that the advanced age of the users and fingertip contact were significant risk factors for microbial contamination.

Nentwich et al. ([64]; see Table 11.27) analyzed eyedrops collected in the clinical settings of a hospital. Among 101 samples, only 10 (belonging to mydriatics) were not preserved. The overall contamination rate was 6%, but while

TABLE 11.25 Microorganism Growth in Unpreserved Atropine and Chloramphenicol Eyedrops Stored under Refrigeration and at Room Temperature after Inoculation in Accordance with the European Pharmacopeia Test for Efficacy of Antimicrobial Preservation [61]

	Atropine 1%		Chloramphenicol 0.5%	
Product	2–8°C	25°C	2–8°C	25°C
	Pseudomonas aeruginosa (Log Reduction)			
Time after inoculation				
6 h	NR	2.54	1.00	0.57
24 h	NR	4.51	1.43	0.88
7 days	NR	NR	NR	NR
28 days	NR	NR	NR	NR
	Staphylococcus aureus (Log Reduction)			
Time after inoculation				
6 h	0.63	0.44	0.83	0.49
24 h	0.88	0.75	0.95	0.69
7 days	NR	NR	3.05	NR
28 days	NR	NR	NR	NR
	Candida albicans (Log Reduction)			
Time after inoculation				
6 h	0.44	0.53	0.51	0.25
24 h	1.00	0.78	0.82	0.64
7 days	1.81	1.08	2.73	1.20
28 days	NR	4.96	NR	NR
	Aspergillus niger (Log Reduction)			
Time after inoculation				
6 h	0.69	0.66	0.81	1.12
24 h	1.05	0.76	1.93	1.34
7 days	2.51	3.52	2.17	3.49
28 days	NR	3.29	3.22	3.31

NR, not recovered.

1% of the solutions presented microbial growth, 5% of the tips were contaminated.

Medication used to treat glaucoma ([68]; see Table 11.27), even containing antimicrobial agents, presented elevated microbial contamination rates as compared to the other products listed in Table 11.27. In this study, eyedrops were administered by the patients themselves, and the authors presumed that they may inadvertently touch their eyelids because 20% of the bottle tips were contaminated as compared to 8% of the bottle contents. Moreover, the most frequent organisms were *S. epidermidis*, *S. aureus*, and *Corynebacterium* spp.,

TABLE 11.26 Results of Unpreserved Eyedrops Stored at 2–8°C for up to 28 Days According to the Efficacy of Antimicrobial Preservation Test of the European Pharmacopeia [61]

Product	Conc. (%)	Log Reduction			
		C. albicans	*A. niger*	*P. aeruginosa*	*S. aureus*
Homatropine	1.0	Pass A	Pass A	Pass A	Pass B
Phenylephrine	10.0	Pass B	Pass A	Pass A	Pass B
Chloraphenicol	0.5	Pass A	Pass A	Pass B	NR
Gentamicin	1.5	<1	Pass A	Pass A	Pass A
Gentamicin	0.3	<1	Pass A	Pass A	Pass A
Cyclopentolate	1.0	Pass B	Pass B	Pass A	NR
Atropine	1.0	Pass B	Pass A	Pass A	NR
Pilocarpine	0.5	Pass B	Pass B	Pass A	NR
Pilocarpine	1.0–4.0	<1	<1	Pass A	NR
Cefuroxime	5.0	Pass B	<1	Pass B	>1
Acetylcysteine	5.0	Pass A	Pass B	Pass A	>2
Citrate	6.5	Pass B	Pass B	Pass A	>1
Hypromellose	0.3	Pass A	Pass A	Pass A	<1
Potassium ascorbate	–	Pass B	Pass B	Pass A	NR
Sodium chloride	0.9	Pass B	Pass B	>3	>3
Sodium chloride	5.0	Pass A	Pass B	Pass A	NR
Dexamethasone	0.1	<1	Pass B	Pass B	NR
Povidone	0.5	Pass B	Pass B	>4	>1
Prednisolone	0.01	Pass B	Pass B	<2	>1
Sodium bicarbonate	2.0	Pass A	>1	>3	<1
Sodium cromoglycate	2.0	Pass B	Pass B	<1	<1
Penicillin	0.3	<1	Pass B	>3	Pass A

The data show the results for day 28 for bacteria and fungi
Criteria: bacteria: criterion A log reduction 3 (24 hours), NR (28 days), criterion B log reduction 1 (24 hours), NI (28 days); fungi: criterion A log reduction 2 (7 days), NI (28 days), criterion B log reduction 1 (14 days), NI (28 days).
NR, no recovery; NI, no increase.

all of which are part of the resident skin and normal eye flora. Høvding and Sjursen ([70]; see Table 11.27) obtained similar results. The microorganisms detected in this study were also similar to the normal conjunctival and skin flora, suggesting that they originated from contaminated dropper tips.

Botulinum toxin is used in ophthalmology for a variety of ocular problems. Because usual doses are less than the contents in the bottle, they are shared among patients and remain in use for more than the 4-hour period recommended by the manufacturers. The study conducted by Menon and Murray ([66]; see Table 11.27) showed that, although botulinum toxins (specifically, Botox and Dysport) are preservative-free preparations, they remained sterile for 4 hours at room temperature and with further storage at 3–5°C

TABLE 11.27 Contamination of Preserved and Nonpreserved Eyedrops

Samples	Preservative	Sample	Part Analyzed	*n*	Contamination Incidence	Microorganisms Isolated[a]	Reference
Multidose eyedrops	No	Gentamicin	Solution	20	0%	1, 2, 3, 4, 5, 6, 7	[62]
	No	Cefuroxime	Solution	18	0%		
	No	Acetylcysteine	Solution	18	11.1%		
	No	Prednisolone	Solution	14	21.4%		
	No	Dexamethasone	Solution	7	0%		
	No	Vancomycin	Solution	6	0%		
	No	Hypromellose	Solution	5	60%		
	No	Amphotercin	Solution	3	0%		
	No	Cyclosporin	Solution	2	0%		
	No	Fluconazole	Solution	2	0%		
Multidose	No	Artificial tears	Solution	207	2.4%	2, 8	[63]
	Yes	Artificial tears	Solution	35	0%		
Multidose eyedrops	Yes	Mydriatics	Solution and tip	56	2% solution, 5% tip	3, 9, 19, 20, 21	[64]
	Yes	Tetracaine	Solution and tip	10	0% solution, 10% tip		
	Yes	Antibiotics	Solution and tip	18	0% solution, 0% tip		
	Yes	Steroids	Solution and tip	14	0% solution, 7% tip		
	Yes	others	Solution and tip	3	0% solution, 0% tip		
Single-dose vial	No	Bevacizumab	Solution	12	0%		[65]
Single-dose bottle	No	Botulinum toxin	Solution	11	0%		[66]

Multidose solution	No	Moxifloxacin	Solution	61	0%	2	[67]
			Cap	61	0%		
			Tip	61	0%		
			Thread	61	1.6%		
Medication to treat glaucoma	Yes	Miotics	Solution	67	15%	1, 4, 10, 11, 12, 13, 14, 15, 16, 17	[68]
	Yes	β blockers	Solution	75	27%		
	Yes	Sympathomimetics	Solution	41	17%		
	Yes	Steroids and antibiotics	Solution	11	18%		
Bottles of solution	No	Boric acid	Solution	42	2.4%	1	[69]
			Inner cap	42	40.5%		
			Bottle edge	42	14.3%		
Multidose eyedrops	Yes	Not specified	Bottle	638	1.9%	2, 10	[70]
			Tip	180	21.1%		
			Pipette	196	0%		

[a] Microorganisms identified: *S. aureus* (1), coagulase-negative *Staphylococcus* (2), *Bacillus* spp. (3), *Serattia* spp. (4), *Klebsiella oxytoca* (5), *Enterobacter cloacae* (6), alpha *Streptococcus* (7), *Acinetobacter* sp. (8), *Staphylococcus epidermidis* (9), *Moraxella* sp. (10), *Enterobacter aerogenes* (11), *Flavobacterium* spp. (12), *Pseudomonas* spp. (13), *Proteus mirabilis* (14), *Corynebacterium* spp. (15), *Propionibacterium acnes* (16), *Streptococcus viridans* (17), micrococci (18), *Micrococcus* spp. (19), *Haemophilus* spp. (20), gram-negative rod (21), gram-positive bacillus (22).

for at least 5 days. The botulinum toxin itself likely presents antimicrobial properties.

Mason et al. ([67]; see Table 11.27) conducted a similar investigation using Vigamox (moxifloxacin), a preservative-free antibiotic eyedrop solution. The contamination rates were evaluated after clinical use for preoperative and postoperative prophylaxis for cataract surgery. The samples were collected from three different scenarios: bottles used by patients at home, bottles used pre- and postoperative for 3 and 7 days, respectively, and bottles used for 1 day in a surgical holding on multiple patients, being the contamination of the solution itself and the contamination of caps, tips, and threads evaluated. From a total of 63 bottles, only one (1.6%) presented contamination on the bottle thread. This low rate of organism recovery is likely due to the ability of the antibiotic to preserve itself.

An analogous situation was observed with the use of bevacizumab. The smallest container of bevacizumab is a 4-mL single-dose vial containing 100 mg. Because the intravitreal administration of this vascular endothelial growth factor for the treatment of neovascular age-related macular degeneration consumes only 0.05–0.10 mL, it is a common practice to fractionate the vial into smaller amounts. Chen et al. ([65]; see Table 11.27) investigated the sterility of bevacizumab by aseptically piercing the vials (12 vials) approximately 10 times in 6 months and storing them at 4°C during the entire period of study. No growth was detected in any sample, showing that, even without preservatives, bevacizumab could remain sterile, providing that aseptic withdrawal of the solution is observed.

Boric acid solution is used as an ophthalmologic antiseptic in different circumstances, mainly at home and usually by a number of patients. Although the content of boric acid bottles used by patients at home presented low contamination rates, the frequency of contaminated caps and bottle edges was high. After interviewing the users, Jose et al. ([69]; see Table 11.27) concluded that inappropriate bottle handling was the cause of the contamination: contact of the bottle cap with hands or with a contaminated surface where the cap was deposited during instillation. The authors supposed that the low contamination in the solution itself is possibly associated with the antiseptic characteristics of boric acid.

Figure 11.9 displays flasks used for ophthalmic preparations including those designed to reduce microbial proliferation as a result of cap contamination: cap attached to the bottle and flask inverted. By inverting the flask, the tip is permanently in contact with the solution and, consequently, with the preservative.

Contact lenses (CLs), lens care solutions, and lens storage cases present the highest contamination rates among ophthalmic products. Szczotka-Flynn et al. [71] reviewed the incidence of microbial contamination of these products and concluded that more than half of the lenses routinely harbored microorganisms, including potentially pathogenic species. In storage cases, the

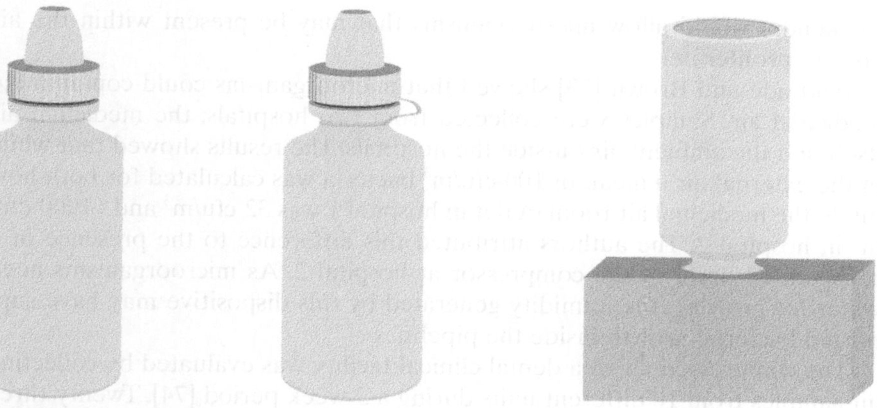

FIGURE 11.9 Flasks for ophthalmic preparations: conventional and options to reduce cap contamination.

incidence of positive microbial bioburden is also greater than 50%. All types of care solutions can become contaminated, including up to 30% of preserved products. The frequency of storage case contamination is higher than the frequency of lens contamination, and unlike CL contamination, which is almost exclusively bacterial, the bioburden in cases or in internal wells usually involves a mixture of bacteria, fungi, and protozoa. The incidence of positive microbial bioburden within storage cases ranges from 24% to 81%. Biofilms are the major factor responsible for transferring resistant organisms from the lens case to the lens surface. It is well known that biofilms are much more resistant to antimicrobial agents than the free-swimming organisms (see Section 11.6, Biofilms), and therefore, once formed, the removal is challenging.

The widespread contamination of CL and storage cases establishes a basis for microbial keratitis and ocular inflammation. Insufficient hand washing and the use of unpreserved lens care solutions are sources of contamination and the production of ocular infectious diseases.

11.4 MEDICINAL GASES

Although medicinal gases are classified as pharmaceutical products and are included in pharmacopeia compendia, medicinal gas monographs do not include microbial specifications. The most vulnerable medicinal gas to present microbiological contamination is compressed air. Contamination can arise from the atmosphere itself, into the compressor, and from the distribution system [72]. Moreover, the accumulation of condensed water in a compressed air tank and the absence of activities for long periods (overnight or over weekends) would allow microorganisms that may be present within the air lines to proliferate.

Andrade and Brown [73] showed that microorganisms could contaminate medicinal air. Samples were collected from two hospitals: the medicinal air itself and the ambient air outside the hospitals. The results showed that while in the external air a mean of 100-cfu/m^3 bacteria was calculated for both hospitals, the medicinal air room outlet in hospital 1 was 32 cfu/m^3 and >1000 cfu/m^3 in hospital 2. The authors attributed this difference to the presence of a refrigerated dryer in the compressor at hospital 2. As microorganisms need water for growing, the humidity generated by this dispositive may have supported bacterial growth inside the pipelines.

The compressed air of a dental clinical facility was evaluated by collecting air samples from 10 different units during a 3-week period [74]. Twenty-three of the 99 samples (23%) were positive for bacterial contamination. The number of colonies varied among the dental units from 1 to 150, and only one unit was free of contamination. The authors demonstrated that periods of inactivity may be responsible for higher microbial growth. Among the air samples collected from the same lines early in the morning and late in the afternoon, 45% of the morning samples were positive for bacterial isolates, while all of the afternoon samples were negative.

Conte et al. [75] investigated microbial growth in different locations of a compressed air supply at a dental clinic. The results in Table 11.28 show that microbial growth occurred in all the parts of the system. Although the contamination of compressed air was not high (~22 cfu/m^3), the colony counts were noticeably high in vacuum hoses and in the water condensed inside the tank. The warm moist environment in these parts was responsible for supporting the growth of microorganisms and their probable spread into the delivered air.

Although the data on the contamination of medicinal air did not reveal noteworthy levels, compressors and pipelines may develop and transport microorganisms and, therefore, may become a potential source of pathogens.

11.5 MEDICAL DEVICES

The unsafe reuse of syringes and failure to adhere to endoscope reprocessing guidelines are the leading causes of microbial transmission and have been associated with numerous outbreaks of bacterial infections [76]. Catheters have also been associated with infections; although the catheter itself does not cause these infections, the migration of skin organisms at the insertion site or direct contamination of the catheter or catheter hub by contact with the hands, contaminated fluids, or other contaminated devices are major causative factors. Gloves, nebulizers, anesthesia circuits, and even inanimate surfaces, such as keyboards, telephones, and monitors, are also sources of microbial transmission.

TABLE 11.28 Microbial Growth in Different Parts of a Compressed Air System

	Location				
Microorganism	Outside Air (cfu/m^3)	Ambient Air (cfu/m^3)	Inner Surface Vacuum Hose (cfu/in^2)	Condensation in the Compressed air Storage Tank (cfu/mL)	Air Supply (cfu/m^3)
Bacteria					
Acinetobacter calcoaceticus	–	78			–
Bacillus	11	22			22
Bacillus cereus	33				–
Coagulase-negative *Staphylococcus*	–	11			22
Staphylococcus aureus	11	–			22
Moraxella osloensis	–	–			22
Gram-negative	–	33	>3,280,000[a]	32,472[a]	–
Fungi					
Aspergillus niger	11	11	–	–	–
Aspergillus fumigatus	307	–	–	–	–
Penicillium	–	11	–	–	–
Rhodotorula	–	–	20–19,680	–	–
Acremonium	–	–	120	–	–

[a] Gram-negative and other bacteria.
Adapted from Reference 75.

11.5.1 Syringes

Syringes are a vehicle for microbial transmission when the same syringe is used for more than one patient. Even changing the needle, the use of vials, bottles, or bags of intravenous solution as common sources of supply for more than one patient allow microbial access to the syringe, as demonstrated in Figure 11.3.

Table 11.29 shows some examples of how the use of syringes may or may not lead to microbial contamination. Studies were designed to evaluate microbial accessibility during syringe preparation under real or simulated situations. The results show that when no precautions are taken (syringes left without caps, coming in contact with fingers, or being reused, even after changing the needle), the contamination is high (from 1% to 67% of the samples). Seventy-five percent of the bulb-syringe irrigators flushed with sterile isotonic saline samples stored in a residential bathroom become contaminated in 4 weeks.

TABLE 11.29 Microbial Contamination during Syringe Preparation

Study Design	Conditions	Number of Samples Analyzed	Contamination Incidence	Reference
Syringes filled with tryptone soya broth from 100-mL vials	Prepared under class 5 laminar-airflow hood in a class 6 cleanroom	100 simple filling	0%	[77]
		100 syringe left without cap	0%	
		100 syringe tip in contact with fingers	0%	
		100 syringe tip in contact with object	0%	
	Prepared under class 7 drug preparation area of an operating room	100 simple filling	0%	
		100 syringe left without cap	1%	
		100 syringe tip in contact with fingers	24%	
		100 syringe tip in contact with object	3%	
	Uncontrolled decentralized pharmacy in a ward	100 simple filling	0%	
		100 syringe left without cap	1%	
		100 syringe tip in contact with fingers	10%	
		100 syringe tip in contact with object	67%	
1-mL syringes filled with tryptone soya broth under aseptic technique	Stored at room temperature (22°C) for 6 months	40	0%	[78]
	Stored in a refrigerator (5°C) for 6 months	40	0%	
	Stored in a freezer (–20°C) for 6 months	40	0%	
Syringes prepared with tryptone soya broth from:	Prepared by intensive care units nurses in usual circumstances	650 prepared from 10-mL ampoules	22%	[79]
10-mL ampoules	No special conditions or precautions	100 prepared from 50-mL vial	2%	
50-mL vials (rubber cap)	Prepared by pharmaceutical technicians using gloves, hair cap, surgical mask and under laminar-airflow	100 prepared from 10-mL ampoules	1%	
		100 prepared from 50-mL vial	0%	

Syringe reuse changing needle	5-mL syringes filled with saline. Three mL injected through the injection site of a tubing connected to a unit of blood running at maximum flow through a 16-gauge needle. Needle syringe was removed and replaced. The remaining fluid in the syringe was tested for blood.	100	34%	[80]
Pen-like injectors for insulin	Cartridge contamination due to blood reflux	146	4.1%	[81]
Syringe filling with anesthetics	Ten trainee anesthetics drew up syringes without taking any aseptic precaution			[82]
	Propofol	10	20%	
	Midazolam	10	0%	
	Thiopentone	10	20%	
	0.9% saline	10	20%	
Bulb-syringe irrigators were flushed twice a day with flushing solutions and stored for 4 weeks in a residential bathroom	Sterile isotonic saline	8	75%	[83]
	Hypertonic saline (table salt + sterile water)	8	0%	
	Baking soda-saline (table salt + baking soda + sterile water)	8	12.5%	

TABLE 11.30 Microbiological Contamination of Automatic Syringe Injectors after Successive Injections of Saline Solution

Injection Sequence	Contamination
1, 2, 3, 4	No
5	*Candida* spp.
6	Nonhemolytic *Streptococcus* spp.
7	Coagulase-negative *Staphylococcus*
8	Coagulase-negative *Staphylococcus*
Exchange of Syringes	
9, 10, 11, 12, 13	No
14	Coagulase-negative *Staphylococcus*
15	Coagulase-negative *Staphylococcus*
16	*Bacillus* spp.
17	No

Adapted from Reference 84.

However, syringes filled with soya broth under aseptic techniques remained uncontaminated even after storage at room temperature for 6 months.

Buerke et al. [84] showed that even the optimization of environmental conditions does not prevent contamination from multiple-use syringes. Automatic injectors with contrast agent and saline were evaluated after each injection under normal hygienic and intensive hygienic conditions. Each simulation was performed eight times. The results in Table 11.30 show that samples collected after the fourth probe of each trial under normal hygienic conditions were not sterile. In a second investigation [85], the authors evaluated microbial contamination following two protocols: In one, the syringes were used for one patient and then discharged, and in the other, the saline syringes were used for four injections (times of reuse with no contamination according to first trial; Table 11.30). Even when maintaining only the first four syringe reuses, contamination was not entirely prevented, and while no contamination was observed in any of the 60 injections in the first protocol, two syringes revealed contamination with coagulase-negative *Streptococcus* in the second protocol.

11.5.2 Endoscopes

Endoscopy-related infection occurs through the spread of microorganisms through contaminated equipment as a result of deficient disinfection/sterilization procedures [86].

Table 11.31 displays the microbial contamination levels of endoscopes randomly collected in hospitals or endoscopy centers and cleaned according to standard protocols. The devices were collected immediately after use but before cleaning and immediately after cleaning, but before disinfection or

TABLE 11.31 Microbial Contamination and Endotoxin Levels of Endoscopes Randomly Collected Immediately after Use but before Cleaning and Immediately after Cleaning but before High-Level Disinfection or Sterilization

		Before Cleaning		After Cleaning		
Instrument	*n*	Bacterial Contamination (cfu/device)	Endotoxins (EU/device)	Bacterial Contamination (cfu/device)	Endotoxins (EU/device)	Reference
Bronchoscopes	20	1.02×10^4–1.95×10^7	1.13×10^1–6.46×10^3	6.76×10^3–5.13×10^5	1.25×10^1–8.94×10^2	[87]
Duodenoscopes	20	0–2.82×10^7	6.28–2.71×10^3	4.68×10^3–2.19×10^5	0.10–9.61	
Colonoscopes	21	5.25×10^5–2.82×10^8	8.92×10^1–1.89×10^6	1.48×10^3–4.07×10^4	0.96–8.79×10^1	
Sinuscope	1	1.19×10^4	–	9.72×10^2	–	[88]
Irrigation forceps	1	8.88×10^3	–	6.19×10^2	–	
Cystoscope sheath	2	3.5×10^1	–	1.9×10^1–4.6×10^1	–	
Adaptador resectoscope	1	3.76×10^3	–	2.4×10^1	–	
Tissue extractor	10	1.2×10^1–3.88×10^3	–	3.3×10^1–1.62×10^4	–	
Hook blade w/irrigant	2	2.6×10^1–1.40×10^2	–	1.2×10^1–2.5×10^1	–	
Ureteroscope	1	1.55×10^2	–	7	–	
Colonoscope surface	10	1.2×10^4–1.5×10^6	–	8.2×10^2–9.5×10^4	–	[89]
Colonoscope channel	10	1.3×10^7–2.0×10^{10}	–	1.3×10^3–4.3×10^5	–	

–, not measured.

TABLE 11.32 Microbial Identification in Endoscopes after Different Stages of Cleanness

	Bioburden Identification			
Device	Before Cleaning	After Cleaning	After Disinfecting	Reference
Colonoscope surface	*Staphylococcus* sp. *Streptococcus* sp. *E. coli* Bacteroides *Flavomonas*	*Staphylococcus* sp. *Pseudomonas* sp. *Xanthomonas* sp. *Klebsiella* sp.	–	[89]
Colonoscope channel	*E. coli* *Bacteroides* *Staphylococcus* sp.	*Pseudomonas* sp. *Xanthomonas* sp. *Aeromonas* sp. *Enterococcus* sp. *Staphylococcus* sp.	–	
Other endoscopes	*Staphylococcus* sp. *Micrococcus* sp. Diphtheroids *Bacillus* sp. *Pseudomonas* sp.	*Staphylococcus* sp. *Stenotrophomonas* sp. *Bacillus* sp. *Micrococcus* sp. Diphtheroids *Pseudomonas* sp.	–	[88]
Bronchoscopes and gastroenterological endoscopes	–	–	*Staphylococcus aureus* Enterobacteria *Pseudomonas aeruginosa* *Pseudomonas* sp. *Stenotrophomonas maltrophilia.* *Acinetobacter* sp. *Candida* sp	[90]

–, not evaluated.

sterilization. The results show that although the bioburdens were reduced, manual cleaning is not capable of eliminating microorganisms. The same result occurred with endotoxins; although the levels were reduced through cleaning, the total elimination was not achieved.

Table 11.32 displays microorganisms observed before and after endoscope cleaning. The microorganisms present before cleaning were those associated with the intestinal tract, in the case of colonoscopes, and with the hospital environment, in the case of other endoscopes. By contrast, the microorganisms present on the devices after cleaning were predominately gram-positive cocci and bacilli, which are part of the normal microbiota of the skin, and

Pseudomonas, *Stenotrophomonas*, *Xanthomonas*, and *Aeromonas*, which are common waterborne microorganisms.

In another study [91], sampling was carried out in two stages: collecting the last rinse water from the endoscope inside the washer–disinfector (anterograde sampling) and flushing the endoscope channel with sterile water outside the washer–disinfector (retrograde sampling). The results showed that, while in anterograde sampling contamination was confirmed in 7.9% of the samples, for retrograde sampling, the level of contaminated samples reached 31%. In addition to gram-positive cocci, Enterobacteriaceae species and gram-negative nonfermenters were found in both samples, and 2.8% of anterograde-sample and 25% of retrograde-sample cultures gave positive results for fungi, specifically the *Candida* sp. Although during endoscope reprocessing healthcare workers wore gloves, which cultured negative for yeast, the authors considered the hands to be the source of the contamination because yeast is usually present on the hands of most people [92]. Scanning electron microscopy (SEM) of endoscopes after reprocessing revealed the incomplete removal of debris on the interior part of the endoscope, where fibrin and biofilms containing rods and cocci could be seen (Fig. 11.10).

11.5.3 Other Devices

Table 11.33 presents studies designed to show how contamination is spread through gloves and nebulizers. Gloves used by healthcare workers are important sources of microbial contamination. Gloves, collected after routine patient care in two hospitals, revealed that 18–70% of them presented contamination with pathogenic bacteria. Table 11.33 also shows that endotoxins may be a contaminant in new gloves made of latex because they were observed in more than 20% of the analyzed samples [96].

Nebulizers are frequently contaminated with microbes from the usual upper respiratory tract flora, and when not properly cleaned or disinfected, may be a vehicle for the passage of microorganisms between users. In three independent studies on the contamination of nebulizers ([97–99]; see Table 11.33), up to 58% of the analyzed nebulizers and inhalers presented microbial contamination after a single use.

Anesthesia circuits are also potential media for pathogenic organism transfer between patients. Carbon dioxide absorbers, namely, soda lime-based cartridges, could be an ideal culture medium (warm and humid) for microorganism growth in the respiratory circulation system of anesthetic devices. Although soda lime has a bactericidal effect, the survival of *P. aeruginosa*, *S. aureus*, and *Mycobacterium tuberculosis* in soda-lime solutions and the capability of microorganisms to cross the soda-lime cartridge [100] have been reported. Langevin et al. [100] showed that once nebulized into the expiratory limb of an anesthesia breathing circuit, nearly 100% of the bacteria traversed the soda-lime canister to reach the inspiratory limb, as long as the fresh gas flow was not

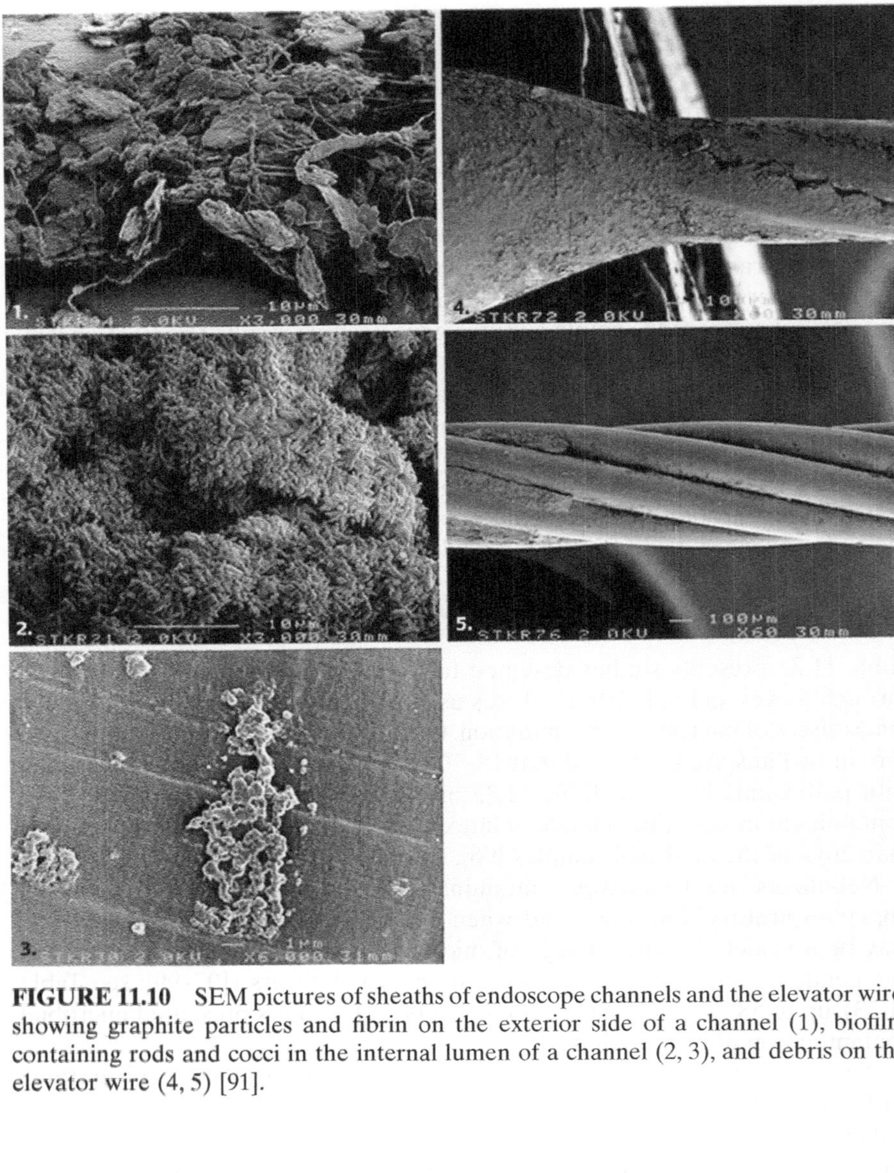

FIGURE 11.10 SEM pictures of sheaths of endoscope channels and the elevator wire, showing graphite particles and fibrin on the exterior side of a channel (1), biofilm containing rods and cocci in the internal lumen of a channel (2, 3), and debris on the elevator wire (4, 5) [91].

interrupted. Nebulized bacteria were eliminated only when the fresh gas flow was discontinued for more than 1 hour.

Arai and Azevedo [101] evaluated the contamination of reprocessed and non-reprocessed parts of anesthesia circuits before each of 56 anesthetic procedures. Reprocessed parts were washed with tap water, submersed in 1% hypochlorite or 2% glutaraldehyde solution for 30 minutes, washed again with water, and then dried with compressed air. The results, displayed in Table 11.34,

TABLE 11.33 Microorganism Contamination Spread through Gloves and Nebulizers

Device	Study Design	Conditions	No. of Samples	Microorganism	Contamination Incidence	Reference
Gloves	Healthcare workers (HCW) caring for patients colonized or infected	Gloves sampled after HCW completed patient care activities	137	Methicillin-resistant *Staphylococcus aureus* Vancomycin-resistant enterococci	17.5%	[93]
Gloves	HCW caring for patients colonized or infected	Culture samples from HCWs' hands and gloves after patient care	131	Vancomycin-resistant enterococci	52% After touching environment only 70% After touching environment and patients	[94]
Gloves	Prospective observational study of HCW caring for patients colonized or infected with pathogenic bacteria	Five wards: three intensive care units and two medical wards	120	Pathogenic bacteria	18.3% Because used gloves were not removed before performing care activities that necessitated strict aseptic precautions	[95]
Gloves	Latex gloves	1 g cut into 5-mm pieces + 10 mL sterile water for 24 h at 50°C	5	Endotoxins	20% Pyrogen test positive	[96]
Inhalers	Metered dose inhaler Easi-Breathe Turbohaler	Swabs taken from the inside of mouthpiece	31	Upper respiratory tract flora Coagulase-negative *Staphylococcus* *S. aureus*	58%	[97]

(*Continued*)

TABLE 11.33 (*Continued*)

Device	Study Design	Conditions	No. of Samples	Microorganism	Contamination Incidence	Reference
Inhalers	Face mask and spacer device	Sterile cotton-tipped swabs rotated around the inner surface	30	*Pseudomonas* spp. *Klebsiella* *Propeus* *Enterobacter* *Escherichia coli* *S. aureus* Gram-negative *staphylococcus* *Staphylococcus epidermidis* *Enterococcus* *Bacillus* spp. *Corynebacterium* *Sarcinia lutea* *Candida albicans*	40% spacer reservoirs 26.7% masks	[98]
Nebulizers	Single-patient nebulizer before cleaning		44	*Pseudomonas aeruginosa* *S. aureus* *Haemophilus* Yeast	17% 14% 1% 6%	[99]

TABLE 11.34 Number of Reprocessed Parts (Branches) and Collector Jars and Canisters Contaminated in Anesthesia Circuits before Anesthetic Procedures and Microorganisms Found

Microorganism	Inspiratory Branches ($n = 56$)	Expiratory Branches ($n = 56$)	Canisters ($n = 56$)	Collector Jars ($n = 11$)	Soda Lime ($n = 56$)
Candida sp.	4	7	6	1	0
Dermatophytus sp.	5	6	3	1	0
Penicillium sp.	5	1	2	0	0
Aspergilium sp.	0	1	1	0	0
Other fungi	2	2	1	0	0
S. aureus	1	2	0	0	0
S. epidermidis	0	1	0	1	0
S. saprophyticus	4	2	0	1	0
A. baumannii	0	0	0	1	0
No growth	35	34	43	6	56

Adapted from Reference 101.

TABLE 11.35 Contamination Rates of Anesthesia Circuits before and after Cleaning Procedures

Parts	Before		Adopted cleaning Procedure	After		Reference
	Total	Contaminated		Total	Contaminated	
Flow control knobs	78	14	Machine wiped with a detergent	77	5	[102]
Handset	56	25	Additional cleaning using 70%	50	3	[103]
Keypad	56	26	Isopropyl alcohol	50	2	

showed that the level of contaminated reprocessed parts (branches) reached 39.3%, and collector jars and canisters reached 45% and 23%, respectively. Although microorganisms did not grow in soda lime, it did not inhibit their survival inside the circuit.

Two studies evaluated microorganism growth in parts of reusable analgesia and anesthesia infusion pumps that did not directly contact the patient but were commonly touched by the anesthetist during induction of anesthesia.The contamination rates are presented in Table 11.35. In both studies, measures for reducing contamination, such as cleaning with detergent or alcohol, helped to reduce the bioburden.

Oliveira and Damasceno [104] surveyed the occurrence of contamination on other inanimate surfaces of common use in hospitals. Twenty-one locations were sampled and the presence of bacteria was detected on the monitors, bed grids, faucets, telephones, keyboards, and other objects. There was a prevalence of methicillin-resistant *S. aureus*, *Clostridium difficile*, *Acinetobacter baumannii*, and vancomycin-resistant *Enterococcus*. Using molecular typification, similarities were detected among the strains collected from these objects and those isolated from colonized or infected patients. Thus, these authors concluded that there is a need for the identification and control of sources of pathogens in the entire hospital environment.

11.6 BIOFILMS

A wide range of microorganisms persist as surface-attached communities surrounded by matrix materials and extracellular polymeric substances (EPSs), including polysaccharides, nucleic acids, and proteins. These colonies, which interact to form different microbial species, are called biofilms. Biofilm formation is an effective survival strategy, and conventional methods for bacterial cleaning are often ineffective for biofilm populations because of the existence of the matrix barrier. The degree of roughness of a surface plays an important role in biofilm adhesion. Rough surfaces are more favorable to biofilm formation than smooth surfaces. Once a biofilm has formed, the resultant structure is highly viscoelastic and behaves in a rubberlike manner that is difficult to penetrate. Biofilms are found in aquatic environments, tissues, and medical biomaterials and devices. It has been shown that biofilms transmit infectious diseases due to their formation in indwelling medical devices. The most susceptible devices to biofilm formation and the respective microorganisms that form the biofilms are listed in Table 11.36 [105].

Indwelling medical devices are often primarily colonized by a single bacterial species, but as the biofilm develops, multispecies consortia quickly develop. Machado et al. [106] evaluated the presence and adherence of microorganisms to central venous catheters used for PN using SEM. The formation of biofilms in the internal and external lumen occurred in almost all catheters used for more than 48 hours. These authors suggest that the biofilms were probably constituted of the colonizing microorganism and proteins of the host (fibronectin and fibrinogen). Figure 11.11 displays SEM micrographs of catheters where coccid structures are adhered to the lumen of a catheter and colonies of *C. albicans* and *S. epidermidis* were observed.

Biofilm-related diseases may be caused by the resistance of biofilms to antimicrobial agents and the detachment of cells or cell aggregates, which, because of their altered phenotype as compared to corresponding planktonic

TABLE 11.36 Devices Susceptible to Biofilm Formation and the Respective Microorganisms

Device	Common Biofilm Bacterial Species
Sutures	*Staphylococcus epidermidis* and *S. aureus*
Arteriovenous shunts	*S. epidermidis* and *S. aureus*
Schleral buckles	Gram-positive cocci
Contact lens	*P. aeruginosa* and gram-positive cocci
Urinary catheter	*E. coli* and other gram-negative rods
Endotracheal tubes	A variety of bacteria and fungi
Hickman catheters	*S. epidermidis* and *C. albicans*
Central venous catheters	*S. epidermidis* and others
Mechanical heart valves	*S. aureus* and *S. epidermidis*
Vascular grafts	Gram-positive cocci
Biliary stent blockage	A variety of enteric bacteria and fungi
Orthopedic devices	*S. aureus* and *S. epidermidis*
Penile prostheses	*S. aureus* and *S. epidermidis*

Adapted from Reference 105.

cells[3] [105], could overcome the host immune system. The structure of biofilms and the nature of the EPS confer an inherent resistance to antimicrobial agents, for example, antibiotics, disinfectants, or sterilant agents. The molecules of the antimicrobial agents must diffuse through the biofilm matrix to inactivate the microorganisms, and in some cases, the activity of the antimicrobial agents is greatly reduced. Table 11.37 presents the effectiveness of selected antibiotics in the presence of staphylococcal slime. While pefloxacin, teicoplanin, and vancomycin demonstrated the greatest decrease in efficacy, the other agents were only moderately affected.

Biofilms, because of their antimicrobial resistance and irreversible attachment to surfaces, represent a major challenge to clinicians and healthcare workers in dealing with nosocomial infections. Bacterial biofilms are a problem, as they have definite infection implications, particularly for immunocompromised patients.

11.7 DIALYSIS CIRCUITS

Bacterial contamination is a potential threat in dialysis circuits. Although the methods used to purify the water and dialysate must follow strict quality standards, intermittent hemodialysis and hemofiltration are vulnerable procedures because of their multiple-component circuitry. Kanagasundaram et al. [108] demonstrated that a significant bacterial contamination was present in the dialysate circuitry of bicarbonate-based continuous hemodialysis. Samples of

[3] Planktonic cells are the free-swimming cells.

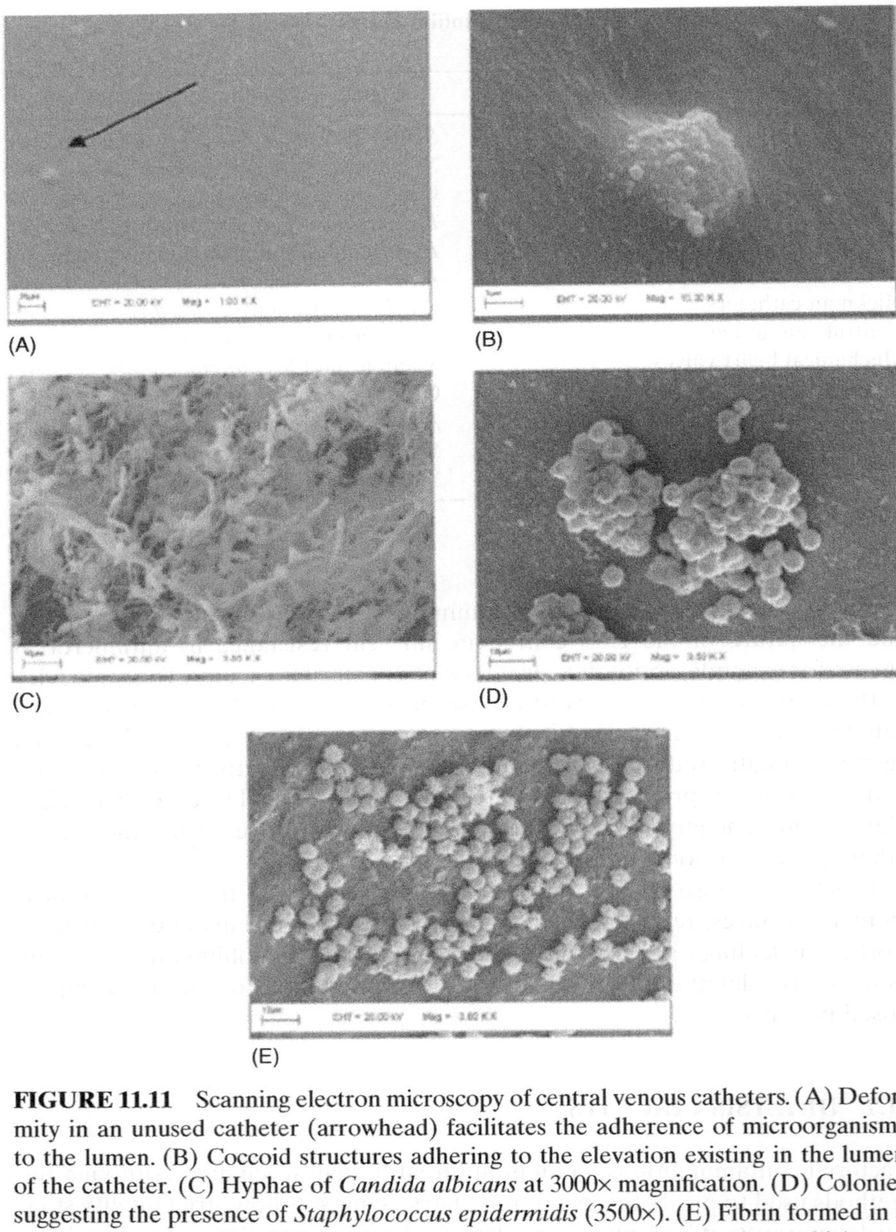

FIGURE 11.11 Scanning electron microscopy of central venous catheters. (A) Deformity in an unused catheter (arrowhead) facilitates the adherence of microorganisms to the lumen. (B) Coccoid structures adhering to the elevation existing in the lumen of the catheter. (C) Hyphae of *Candida albicans* at 3000× magnification. (D) Colonies suggesting the presence of *Staphylococcus epidermidis* (3500×). (E) Fibrin formed in a catheter after less than 48 hours of use and lymphocytes from a patient, which may suggest an acute inflammatory response (3000×). The dimensions of the structures analyzed, in micrometer, are indicated by the bar. EHT, extra-high tension; Mag, magnification [106].

TABLE 11.37 Effectiveness of Antimicrobial Activity of Selected Antibiotics for Staphylococcal Slime

Antimicrobial Agent	Reduction in Effectiveness[a] (%)	Antimicrobial Agent	Reduction in Effectiveness[a] (%)
Cloxacillin	3.7	Gentamicine	9.8
Amoxicillin/clavulanic acid	5.3	Tobramycin	16
Imipenem	7.0	Netilmicin	11
Cefpirome	11	Amikacin	12
Erythromycin	2.3	Isepamicin	10
Roxithromycin	0.94	Ofloxacin	13
Clindamycin	1.4	Ciprofloxacin	9.5
Rifampin	0.99	Pefloxacin	30
Fusidic acid	4.4	Daptomycin	9.1
Trimethoprim/sulfamethoxazole	4.8	Teicoplanin	52
Doxycyline	10	Vancomycin	63

[a] Calculated as the average of the percent difference in the zone diameter produced by each antimicrobial agent alone and the control.
Adapted from Reference 107.

the dialysate were drawn from four points along the dialysate circuit in 18 therapies (mean duration 119.5 ± 72.0 hours), and contamination developed in 6 therapies. The authors attributed this contamination to breaches in aseptic techniques during the connection of the dialysate bag with components within the circuit.

Contamination of the dialysate induces the development of biofilm within the dialysis machine fluid pathway. Because biofilms are ubiquitous in flowing aqueous environments and may develop on virtually all surfaces immersed in natural aqueous environments, they may also be found within dialysis machines. Additionally, biofilms grow particularly rapidly in flowing systems where a regular nutrient supply is provided, as in the case of the dialysis fluids.

Man et al. [109] evaluated biofilm formation on silicon tubing in different dialysis circuits during a period of 2–4 years. They observed that the biofilms detached from tubing samples drawn from the water, bicarbonate, and fresh dialysate paths within the dialysis machines contained approximately 1.10^3–1.10^6 bacteria/cm^2, yet only some living bacteria were found. Endotoxin levels ranged from 1 to 12 EU/cm^2. In contrast, in the dialysate fluid, no viable bacteria were observed, and the endotoxin content was below the detection level. The results are presented in Table 11.38.

Marion-Ferey et al. [110] evaluated biofilm formation in silicone tubing removed from dialysis circuits, which were submitted to one of the following protocols: no decontamination or decontamination with 3% citric acid at 37°C for 5 minutes and autoclaving for 30 minutes or a chemical product (Dialox 0.3%, Actril 0.3%, or bleach 0.25%). The results in Table 11.39 showed that

TABLE 11.38 Living Bacteria, Total Bacteria, and Endotoxins on the Biomass of Various Tubing Samples and in Dialysate Fluid [109]

Tubing Sample	System 1 (10,732-h Usage)			System 2 (18,194-h Usage)			System 3 (19,192-h Usage)		
	Living Bacteria (cfu/cm^2)	Total (Bacteria/ cm^2)	Endotoxin (EU/cm^2)	Living Bacteria (cfu/cm^2)	Total (Bacteria/ cm^2)	Endotoxin (EU/cm^2)	Living Bacteria (cfu/cm^2)	Total (Bacteria/ cm^2)	Endotoxin (EU/cm^2)
Water path	32	2×10^5	2.95	22	2×10^5	9.82	15	1×10^4	12.31
Bicarbonate path	21	6×10^4	1.73	18	4×10^5	1.95	11	1×10^3	0.61
Dialysat path	10	9×10^4	1.47	14	8×10^5	11.14	12	7×10^4	4.67
Dialysate fluid	0	–	<0.005	0		<0.005	0		<0.005

TABLE 11.39 Characteristics of the Tubing Samples Removed from Dialysis Systems and Qualitative and Quantitative Analysis of the 13 Clinical Samples [110]

Sample	Center	Removing Site	Total Running Time (Months)	Decontamination[a]	Microscopic Observation	Total Bacteria/ cm^2	Culturable Bacteria (cfu/cm^2)	Endotoxin (EU/cm^2)
1	B	Control (new, unused)	0	None	Clean	180	<1	<0.005
2	B	Control (new, unused)	0	None	Clean	115	<1	<0.005
3	C	Connection loop-dialysis machine	12	None	Mature biofilm	5.3×10^6	9.7×10^5	354
4	C	Connection loop-dialysis machine	12	None	Mature biofilm	1.2×10^3	1.6×10^3	115
5	C	Connection loop-dialysis machine	12	None	Mature biofilm	1.3×10^5	358	89
6	A	Inside dialysis machine	12	C	Microcolonies + crystals	4550	11	0.3
7	A	Inside dialysis machine	12	C	Mature biofilm	3.0×10^5	133	3.7
8	A	Inside dialysis machine	12	C	Microcolonies	2.0×10^4	81	1.9
9	A	Exit dialysis machine	12	A	Crystal + isolated bacteria	2130	4	0.05
10	A	Exit dialysis machine	12	A	Crystal + isolated bacteria	1335	8	0.01
11	A	Exit dialysis machine	12	C	Microcolonies + isolated bacteria	6400	38	0.2
12	A	Exit dialysis machine	12	A	Crystal + isolated bacteria	530	7	0.1
13	A	Exit dialysis machine	12	C	Microcolonies + crystals	4000	13	0.2

[a] A: citric acid 3%, 5 minutes + autoclaving 121°C, 30 minutes; C: chemical decontamination (Dialox 0.3%, Actril 0.3%, or bleach 0.25%).

only the two controls were completely clean. Crystals were consistently present on the inner surface of the tubing with isolated adherent bacteria or microcolonies. The results indicate that the method of decontamination influences the amount of biofilm formation. The tubing treated with citric acid had less contamination. Conversely, the tubing removed from the connection between the water distribution loop and the dialysis machine was contaminated with a thick, extensive, mature biofilm.

In another study [111], 24 continuous veno-venous hemofiltration replacement fluid circuits were evaluated at the completion of therapy at an ICU from 23 patients (two samples were collected from one patient). The dialysis tubing was also analyzed. The results, displayed in Table 11.40, revealed that only 7 of the 24 investigated circuits were free from microbial contamination. The

TABLE 11.40 Microbiological Results by Individual Hemofiltration Circuit [111]

Circuit Number	Circuit Life Span (h)	Infusate Culture Isolate	Colony Counts (cfu/mL)	Endotoxin Counts (EU/mL)	Biofilm (+/)
1	6	*Bacillus cepacia, Micrococcus*	1	<0.01	+
2	5	*Pantoea* sp.	1	<0.01	
3	36			0.012	+
4	43	Gram-positive rods, gram-negative rods	>100 >100	0.01	+
5	47	Coagulase-negative *Staphylococcus*	1	0.01	+
6	9	Mold species, Coagulase-negative *Staphylococcus*	1	<0.01	+
7	38	*Pseudomonas fluorescens*	>10	0.014	
8	21			<0.01	+
9	11			0.01	
10	60			<0.01	+
11	86			<0.01	+
12	4			<0.01	
13	79	Not identified	1	<0.01	+
14	39	*Klebsiella*	1	0.061	
15	10			<0.01	
16	18			<0.01	+
17	29	Gram-negative rods	3	<0.01	
18	52			0.011	
19	28			0.014	
20	26			0.012	+
21	56			<0.01	
22	64			<0.01	+
23	26			<0.01	
24	36			<0.01	+

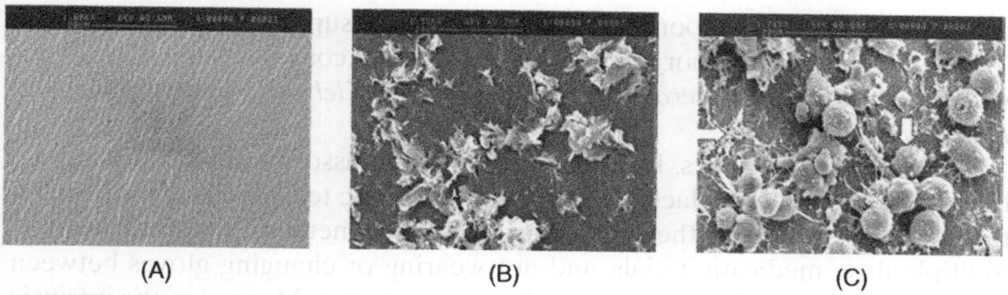

FIGURE 11.12 (A) Intermediate power electron microscopy image (circuit number 15) without evidence of biofilm. (B) Intermediate power electron microscopy image (circuit number 3) indicating the presence of biofilm fragments on the tubing surface. (C) Intermediate power electron microscopy image of tubing (circuit number 20) showing the presence of biofilm and bacteria. Circuits are identified in Table 11.40 [111].

microbiological limits for water for hemofiltration are <0.1 cfu/mL and <0.03 EU/mL endotoxin (Ph. Eur.). According to these standards, 37.5% of the circuits failed for microbial contamination, and 4% failed for the presence of endotoxin. Although the internal tubing cultures were negative, electron microscopy revealed that 13 of the 24 collected tubing samples presented biofilms. The mean circuit life span of those positive for biofilms was 40.9 hours but ranged from 6 to 86 hours, and four of the biofilm-positive samples were found in tubing harvested before 24 hours had elapsed. Figures 11.12A–C show electron microscopy images of three tubing samples, indicating no evidence of biofilm, biofilm fragments, and a well-established biofilm, respectively. The presence of biofilm is not necessarily associated with a positive fluid culture or tubing, as demonstrated in this study. The authors speculated that contamination may have occurred through nonsterile connections between replacement fluid bags and the replacement fluid circuit. The conclusion that the most vulnerable parts of dialysis circuits are the connections is consistent with other studies.

11.8 NOSOCOMIAL INFECTIONS

Nosocomial infections are infections acquired in healthcare settings. They may result from the medical care or contact with pathogens from another patient in a hospital or clinic. According to the National Nosocomial Infections Surveillance (NNIS) system (system established in 1970 to collect data on nosocomial infections in hospitals in the United States) among several device-associated infections, ventilator-associated pneumonia, central line-associated bloodstream infections (BSIs), and urinary catheter-associated infections are the most prevalent [112]. Catheter-related BSI occurs in 1.3–26.2% of patients with central venous catheters used to administer PN. Because of their nutritional components, PN solutions can support microbial growth. The most common microorganisms causing BSI are coagulase-negative *Staphylococcus*, *S. aureus*, *Enterococcus*, *Candida* spp., *Klebsiella pneumoniae*, and *P. aeruginosa*.

Nosocomial outbreaks, however, are not only associated with the device but are also related to a lack of adherence to aseptic techniques. Unsafe injection practice, primarily the reuse of syringes and needles, contamination of multiple-dose medication vials, and not wearing or changing gloves between patients are the usual causes of microbial transmission. Moreover, the intrinsic contamination of drugs, infusion solutions, and even disinfectant solutions, although to a lesser extent, may also count for triggering nosocomial infections.

Hepatitis B virus (HBV) and HCV are also considered nosocomial infections. The transmission of HBV and HCV may occur in long-term care facilities, mainly those for the regular monitoring of blood glucose, from percutaneous or mucosal exposures to blood or other body fluids from an infected patient or healthcare worker. Table 11.41 lists outbreaks of HBV and HCV and other pathogens originated by the most varied sources, from intrinsically contaminated soap to the reuse of the materials for blood sampling. The data were collected from the last 20 years, and the reports were selected to present a demonstrative panel of the sources of microbial contamination in healthcare settings.

Vonberg and Gastmeier [139] reviewed outbreaks related to contaminated drugs, which occurred from 1990 to 2006. The data are summarized in Table 11.42. Blood products responded for 26.6% of the outbreaks, followed by heparin/sodium chloride solutions with 23.4%. Seventy-four percent of the contaminated drugs had been administered intravenously, and of those, 49.2% corresponded to multidose vials. Excluding blood products, intrinsically contaminated products occurred in 21.1% of the cases, and the extrinsic contamination originated in hospital pharmacies and in ICUs/operating rooms corresponded to 65.3% and 12.6%, respectively.

Interestingly, the number of outbreaks caused by disinfectants and propofol in the period surveyed is practically the same, seven and six, respectively, but because of the widespread use of disinfectants and the strict use of propofol, the number of patients involved in the disinfectant-related outbreaks was 622 as compared to the 53 in propofol outbreaks. However, while the mortality rate was 4.3% in disinfectant-related outbreaks, this number reached 13.8% in propofol-related outbreaks (data not shown).

Because of their nature and use, blood products are rarely taken as multiuse drugs, and therefore, the label recommendation of "any unused portion must be discarded" is usually followed. Nevertheless, intrinsically contaminated blood products were responsible for 27.4% of the outbreaks listed in Table 11.42 attaining 16.5% of the patients.

Blood preparations, if contaminated, would allow microorganisms to proliferate rapidly. Obayashi et al. [54] evaluated the microbial viability in 12

TABLE 11.41 Outbreaks of HBV and HCV and Other Bloodstream Infections and the Sources of Transmission

Local or Procedure	Source	Microorganism	No. Patients Contaminated	Year	Ref.
NICU	Commercial mineral water used to prepare milk	*P. aeruginosa*	20	2010	[113]
NICU	Bottle soap dispenser	*Serratia marcescens*	5	2009	[114]
Endoscopy	Reuse of single-use vial of propofol for multiple patients	HBV HCV	6 HCV 5HBV 1 HBV and HCV	2007	[115]
Blood glucose monitoring at two assisted living facilities	Contamination of glucose meters, diabetic care supplies, or hands of nurses were implicated as sources	HBV	8 out of 109 (facility A) 7 out of 21 (facility B)	2007/2008	[116]
Injection	Prefilled heparin syringe from a single manufacturer Prefilled saline syringe from a single manufacture	*S. marcescens*	162	2007	[117]
Cataract surgery in two hospitals	Intrinsically contaminated ophthalmic solution	*Burkholderia cepacia* *P. aeruginosa*	4 hospital A 2 hospital B	2005	[118]
NICU	Non-antimicrobial soap	*S. marcescens*	9	2006	[119]
Infusion	Admixture prepared with a contaminated lot of magnesium sulfate from a compounding pharmacy	*S. marcescens*	10	2005	[120]
Pain management procedures	Reuse of single-dose medication vials Common needle and syringe to access multiple medications	*S. marcescens*	7	2005	[121]
NICU	In-use PN solutions	*S. marcescens*	7	2005	[122]

(*Continued*)

TABLE 11.41 (*Continued*)

Local or Procedure	Source	Microorganism	No. Patients Contaminated	Year	Ref.
Oxygen delivery reusable device	Contamination during calibration with unfiltered water Contamination of vapor transfer cartridge during patient use Failure to remove organisms during routine decontamination	*Ralstonia mannitolilytica*	38	2005	[123]
Myocardial perfusion	Syringes used to add saline to radipharmaceutical left in workstation for several preparations	HCV	16 out of 90	2004	[124]
Mechanical ventilation	Nebulizers used for multiple patients without disinfection	*S. marcescens*	71	1999/2000	[125]
General pediatric ward	Intrinsic contamination of 0.05% oxymetazoline hydrochloride nasal spray	*B. cepacia*	5	2003/2004	[126]
NICU	Prepared milk from multidose powder	*P. aeruginosa*	30	2004	[127]
Blood glucose monitoring	Staff not wearing gloves and hand hygiene not performed between patients	HBV	8 out of 22	2004	[128]
Blood glucose monitoring	Insulin and multidose medication not labeled with patient names Reuse of needle or lancet Fail to change gloves between patients Spring-loaded barrel and glucometer not routinely cleaned between patients	HBV	15 out of 160	2003	[128]
Blood glucose monitoring	Glucometer not routinely cleaned between patients	HBV	11 out of 192	2003	[128]
Injection	Injections of atropine, dexamethasone, and vitamin B_{12} drawn from multiple-dose vials into one syringe	HBV	38 out of 1042	2002	[129]
Chemoterapy	Use the same syringe to draw blood from patients' central venous catheters and to draw catheter-flushing solution from 500-cc saline bags that were used for multiple patients	HCV	99 out of 486	2002	[129]

Pain remediation	A single needle and syringe used to administer each of three sedation medications Versed® (midazolam HCl, fentanyl, and propofol) to up to 24 sequentially treated patients at each clinic session.	HCV and HBV	69 HCV out of 793 31 HBV out of 793	2002	[129]
General pediatric ward	Extrinsic contamination of ranitidine multidose from an automatic compounding machine used for up to 48 h at room temperature	*Enterobacter cloacae*	4	2001	[130]
Endoscopy	Extrinsic contamination of multiple-dose anesthesia vial	HCV	19 out of 1315	2001	[129]
Surgical intensive care unit	Tap water from the main faucet in the ICU used for taken oral medication	*S. marcescens*	19	2000/2001	[131]
NICU	Individual bottles of 1% chlorxylenol soap used by HCWs	*S. marcescens*	32	1994/1995	[132]
Blood glucose monitoring	Lanced device end cal not changed between patients	HBV	9 out of 67	1996	[133]
Blood glucose monitoring	Staff not wearing gloves and hand hygiene not performed between patients Used and unused lanced caps placed in the same box	HBV	3	1996	[133]
NICU	0.25% disinfectant solution (8.0 g/dL formaldehyde, 8.0 g/dL glyoxal, 4.5 g/dL glutaral) used for disinfection of surfaces, laminar air systems and infusion pumps	*Klebsiella oxytoca*	28	1996/1999	[134]
Pediatric intensive care unit	Breaks in aseptic techniques Reduced hand washing frequency among HCW due to hand dermatitis	*S. marcescens*	14 out of 25	1995	[135]
Hemodialysis	Staff handled blood contaminated items without changing gloves or washing hands	HBV	14 out of 20	1994	[136]
Hemodialysis	Extrinsic contamination of multiple-dose heparin vial	HBV	7 out of 131	1994	[136]
Anesthesia	Extrinsic contamination of fentanyl by a healthcare worker involved with illicit narcotic use	*S. marcescens*	26 out of 90	1998/1999	[137]
Hospitalized patients	4% Chlorhexidrine solution without alcohol diluted to 2% with nonsterile distilled water distributed in reusable, nonsterile plastic containers for several cleaning procedures	*S. marcescens*	151	1991/1992	[138]

TABLE 11.42 Distribution of Contaminated Substances Causing Hospital-Acquired Infections, Number of Patients Involved, and Most Frequent Contaminating Pathogens [139]

Group I			Group II		
Contaminated Product	No. of Outbreaks	No. of Patients	Contaminated Product	No. of Outbreaks	No. of Patients
Red blood cells	14	39	Heparin/sodium chloride	30	451
Clotting factor concentrates	10	173	Total parenteral nutrition	9	109
Other blood products	10	121	Disinfection substances	7	622
			Propofol	6	53
			Albuterol	4	143
			Ultrasound gel	4	36
			Fentanyl	3	15
			Ranitidine	2	50
			Others	25	413
Pathogens			Pathogens		
Yersinia enterocolitica			*Burkholderia cepacia*		
Serratia spp.			*Serratia* spp.		
Hepatitis A virus			*Enterobacter* spp.		
Hepatitis C virus			*Ralstonia pickettii*		
Parvovirus B19			*Staphylococcus aureus*		
			Klebsiella spp.		
			Pseudomonas spp.		
			Acinetobacter spp.		
			Candida spp.		
			Hepatitis B virus		
			Hepatitis C virus		

blood products. The samples were incubated with strains of *Serratia marcescens* and *B. cepacia* at 7 and 30°C for up to 7 days. The results showed that the bacteria did not proliferate at 7°C but proliferated rapidly just after inoculation at 30°C. The bacterial growth at 30°C can be seen in Figure 11.13. In four products, however, *B. cepacia* did not grow. Although the growth of *S. marcescens* was not inhibited in these samples, the authors attributed this occurrence to the antibacterial action of the added glycine because all four products contained glycine as a protein stabilizer (see Table 11.43). These results indicate that blood products, as lipid emulsions, are able to support microbial growth soon after the microorganisms enter the product package.

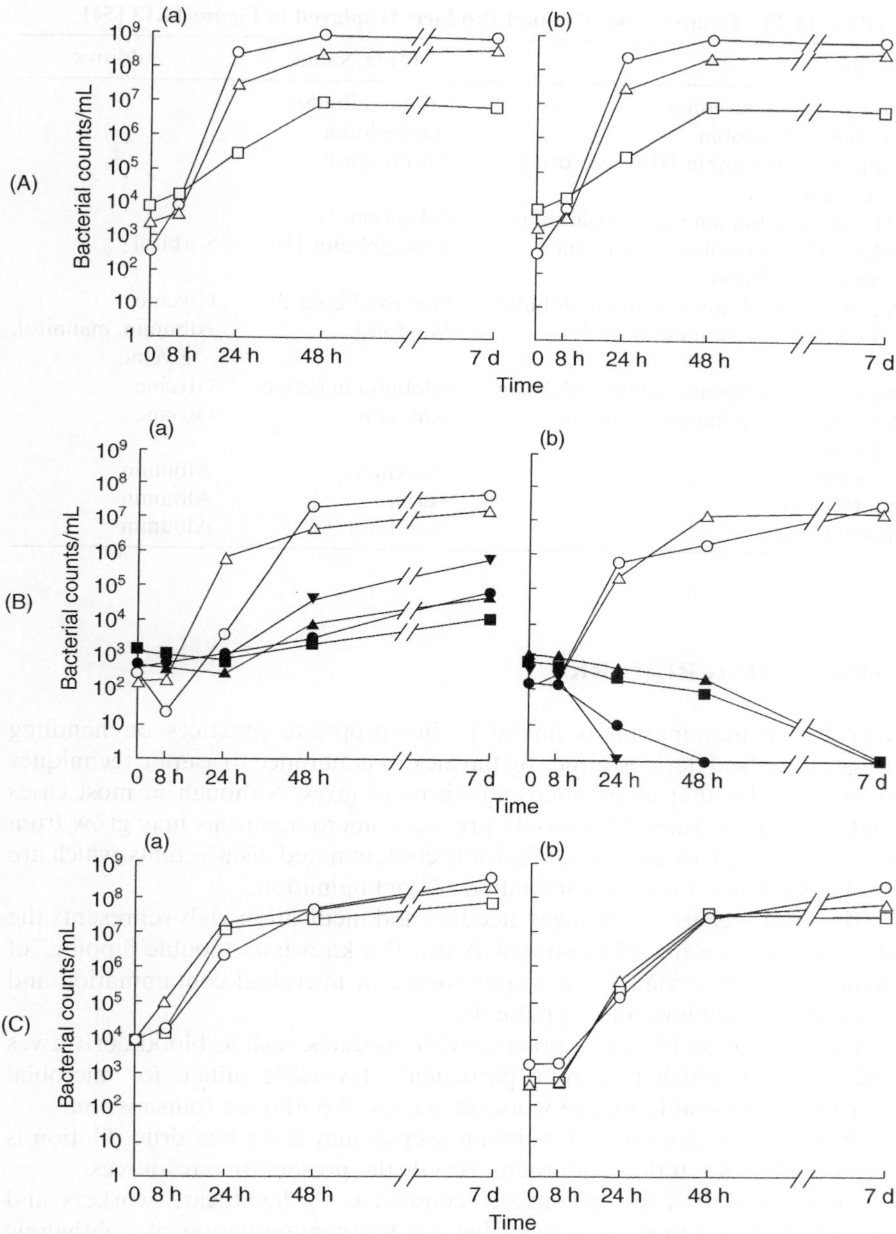

FIGURE 11.13 Viability of *Serratia marcescens* (a) and *Burkholderia cepacia* (b) at 30°C in (A) human serum albumin (○), human haptoglobin (△), and human antithrombin III freeze-dried concentrated (□); in (B) polyglobin-N (○), venoglobulin-IH (△), gamma-venin P (●),venilon-I (▲), globulin injection (■), and ahlbulin (▼); and in (C) urokinase (○), interferon-β (△), and interferon-α (□) [54].

TABLE 11.43 Composition of Blood Products Displayed in Figure 11.13 [54]

Product	Trade Name	Additive
Human serum albumin	Sekijuji albumin	–
Human haptoglobin	Haptoglobin	–
Human antithrombin III freeze-dried concentrated	Anthrobin P	–
pH 4-treated human immunoglobulin	Polyglobin-N	
Polyethylene glycol-treated human immunoglobulin	Venoglobulin-IH	Sorbitol
Pepsin-treated human immunoglobulin	Gamma-Venin P	Glycine
Sulfonated human immunoglobulin	Venilon-I	Albumin, mannitol, glycine
Pepsin-treated human immunoglobulin	Globulin injection	Glycine
Antilymphocyte immunoglobulin, equine	Ahlbulin	Glycine
Urokinase	Urokinase	Albumin
Interferon-β	Feron	Albumin
Interferon-α	Sumiferon	Albumin

CONCLUDING REMARKS

Microbial contamination is linked to inappropriate practices on handling drugs or medical devices, either by the lack of adherence to aseptic techniques or by attitudes that allow microorganisms to grow. Although in most cases contamination results from unsafe practices, microorganisms may grow from unexpected sources, such as intrinsically contaminated disinfectants, which are products intended for the elimination of contamination.

The improper use of syringes, needles, and medication vials represents the highest threat for microbial spread. A practice known as "double dipping" of single or multiuse vials is the major source of microbial contamination and consequent infections among patients.

The association of unsafe practice with products, such as blood derivatives and propofol, which present a particularly favorable milieu for microbial growth, is responsible for the worse scenarios of pathogen transmission.

Preservatives designed to maintain asepsis may fail when drug dilution is performed or when the bioburden exceeds the preservative resources.

Unsafe practices are potentially conducted by healthcare workers and patients. Patients may be responsible for the contamination of ophthalmic drops by touching the eyelids or bottle droppers. In this context, the design of a container may also influence contamination because bottles with tips attached would not allow their deposition on surfaces during medicine application.

Biofilms seem to be the utmost challenge for the prevention of microbial development. The biofilm structure allows for its attachment to surfaces and

TABLE 11.44 Intrinsic versus Extrinsic Contamination of Sterile Products

	Intrinsic[a]		Extrinsic[b]	
	Microorganism	Frequency (%)	Microorganism	Frequency (%)
Achromobacter piechaudii			x	1.1
Actinomyces viscosus			x	1.1
Acinetobacter spp.			x	6.5
Aspergillus spp. *Penicillium* sp.	x	8.7		
Bacillus spp.	x	4.3	x	6.5
Burkholderia cepacia	x	17.4		
Candida spp.			x	1.1
Chrysobacterium meningosepticum			x	1.1
Citrobacter freundii			x	1.1
Coagulase-negative *Staphylococcus*			x	21.5
Corynebacterium spp.			x	
Diphtheroid			x	1.1
Enterobacter spp.			x	9.7
Flavobacterium spp.			x	2.2
Fusarium spp.	x	4.3		
Haemophilus spp.			x	1.1
HCV			x	2.2
Klebsiella spp			x	5.4
Methylobacterium spp.	x	13.0		
Micrococcus spp.			x	3.2
Mold and yeast	x	26.1	x	1.1
Moraxella sp.			x	2.2
Mycobacterium chelonae	x	13.0		
Propionibacterium acnes			x	1.1
Proteus mirabilis			x	1.1
Pseudomonas aeruginosa	x	4.3	x	3.2
Pseudomonas spp.			x	4.3
Ralstonia pickettii	x	4.3		
Serratia marcescens			x	7.5
Stenotrophomonas maltophilia	x	4.3		
Staphylococcus aureus			x	6.5
Streptococcus spp.			x	5.4

Frequency: calculated from cumulative entries in the selected tables.

[a] Data from Table 11.2.

[b] Data from Tables 11.15, 11.17, 11.20, 11.21, 11.27.

hinders the penetration of antimicrobial agents; thus, once settled, biofilm removal is not attained through usual cleaning/disinfecting/sterilizing practices. Biofilms are ubiquitous, growing in RO membranes and catheters, endoscopes, hemodialysis circuits, and other reusable devices.

Table 11.44 displays the microorganisms discussed throughout this chapter and the frequency that they occurred. By comparing the microorganisms responsible for the recall of sterile products to the microorganisms spread through the administration of products in clinics and hospitals, it is clear that they have different origins. Intrinsic contamination is not the major source of pathogens but rather the environment of the medical facility and the negligent behavior of healthcare workers.

REFERENCES

1. Jimenez L. Microbial diversity in pharmaceutical product recalls and environments. J Pharm Sci Technol 61, 383–399, 2007.
2. Oie S, Kamiya A, Yoneda I, Uchiyama K, Tsuchida M, Takai K, Naito K. Microbial contamination of dialysate and its prevention in haemodialysis units. J Hosp Infect 54(2), 115–119, 2003.
3. Arvanitidou M, Spaia S, Katsinas C, Pangidis P, Constantinidis T, Katsouyannopoulos V, Vayonas G. Microbiological quality of water and dialysate in all haemodialysis centres of Greece. Nephrol Dial Transplant 13(4), 949–954, 1998.
4. Pegues DA, Oettinger CW, Bland LA, Oliver JC, Arduino MJ, Aguero SM, McAllister SK, Gordon SM, Favero MS, Jarvis WR. A prospective study of pyrogenic reactions in hemodialysis patients using bicarbonate dialysis fluids filtered to remove bacteria and endotoxin. J Am Soc Nephrol 3(4), 1002–1007, 1992.
5. Vorbeck-Meister I, Sommer R, Vorbeck F, Hörl WH. Quality of water used for haemodialysis: bacteriological and chemical parameters. Nephrol Dial Transplant 14(3), 666–675, 1999.
6. Bereschenko LA, Prummel H, Euverink GJ, Stams AJ, van Loosdrecht MC. Effect of conventional chemical treatment on the microbial population in a biofouling layer of reverse osmosis systems. Water Res 45(2), 405–416, 2011.
7. Bohus V, Tóth EM, Székely AJ, Makk J, Baranyi K, Patek G, Schunk J, Márialigeti K. Microbiological investigation of an industrial ultrapure supply water plant using cultivation-based and cultivation-independent methods. Water Res 44(20), 6124–6132, 2010.
8. Kulakov LA, McAlister MB, Ogden KL, Larkin MJ, O'Hanlon JF. Analysis of bacteria contaminating ultrapure water in industrial systems. Appl Environ Microbiol 68(4), 1548–1555, 2002.
9. Kayser WV, Hickman KC, Bond WW, Favero MS, Carson LA. Bacteriological evaluation of an ultra-pure water-distilling system. Appl Microbiol 30(4), 704–706, 1975.
10. Patterson MK, Husted GR, Rutkowski A, Mayette DC. Isolation, identification and microscopic properties of biofilms in high-purity water distribution systems. Ultrapure Water 8, 18–23, 1991.

11. Orth DS. *Handbook of Cosmetic Microbiology*. Marcel Dekker, New York, 1993, pp. 15–54.
12. Razem D, Katusin-Razem B. Dose requirements for microbial decontamination of botanical materials by irradiation. Radiat Phys Chem 63, 697–701, 2002.
13. de la Rosa MC, Medina MR, Vivar C. Microbiological quality of pharmaceutical raw materials. Pharm Acta Helv 70(3), 227–232, 1995.
14. Baggerman C, Kannegieter LM. Microbiological contamination of raw materials for large-volume parenterals. Appl Environ Microbiol 48(3), 662–664, 1984.
15. Gordon SM, Oettinger CW, Bland LA, Oliver JC, Arduino MJ, Aguero SM, McAllister SK, Favero MS, Jarvis WR. Pyrogenic reactions in patients receiving conventional, high-efficiency, or high-flux hemodialysis treatments with bicarbonate dialysate containing high concentrations of bacteria and endotoxin. J Am Soc Nephrol 2(9), 1436–1444, 1992.
16. De Clerck E, Vanhoutte T, Hebb T, Geerinck J, Devos J, De Vos P. Isolation, characterization, and identification of bacterial contaminants in semifinal gelatin extracts. Appl Environ Microbiol 70(6), 3664–3672, 2004.
17. Motamedifar M, Askarian M. The prevalence of multidose vial contamination by aerobic bacteria in a major teaching hospital, Shiraz, Iran 2006. Am J Infect Control 37(9), 773–777, 2009.
18. Nogler-Semenitz E, Lass-Flörl C, Nogler M, Speer G, Dierich MP. Bacterial contamination of solutions for parenteral administration for single- and multiple-dose vials after multiple use in the hospital. Wien Med Wochenschr 157(15–16), 398–401, 2007.
19. Mattner F, Gastmeier P. Bacterial contamination of multiple-dose vials: a prevalence study. Am J Infect Control 32(1), 12–16, 2004.
20. Driver RP, Jr., Snyder IS, North FP, Fife TJ. Sterility of anesthetic and resuscitative drug syringes used in the obstetric operating room. Anesth Analg 86(5), 994–997, 1998.
21. Centers for Disease Control and Prevention (CDC). Acute hepatitis C virus infections attributed to unsafe injection practices at an endoscopy clinic—Nevada, 2007. MMWR Morb Mortal Wkly Rep 57, 513–517, 2008.
22. Wu C, Engler C, Norton R. Growth of *Staphylococcus epidermidis* in anaesthetic resuscitative drugs: implications for potential contamination. Anaesth Intensive Care 33(1), 69–72, 2005.
23. Graystone S, Wells MF, Farrell DJ. Do intensive care drug infusions support microbial growth? Anaesth Intensive Care 25(6), 640–642, 1997.
24. Highsmith AK, Greenhood GP, Allen JR. Growth of nosocomial pathogens in multiple-dose parenteral medication vials. J Clin Microbiol 15, 1024–1028, 1982.
25. Dominik RH, Segebade IE, Taenzer V. Risk of microbial contamination of iodinated contrast media on multiple use of large-volume bottles. Eur J Radiol 19(3), 198–205, 1995.
26. Thomas M, Sanborn MD, Couldry RIV. admixture contamination rates: traditional practice site versus a class 1000 cleanroom. Am J Health Syst Pharm 62(22), 2386–2392, 2005.
27. Macías AE, Muñoz JM, Bruckner DA, Galván A, Rodríguez AB, Guerrero FJ, Medina H, Gallaga JC, Cortés G. Parenteral infusions bacterial contamination in

a multi-institutional survey in Mexico: considerations for nosocomial mortality. Am J Infect Control 27(3), 285–290, 1999.

28. Reiter PD. Sterility of intravenous fat emulsion in plastic syringes. Am J Health Syst Pharm 59(19), 1857–1859, 2002.
29. Ybarra JV, Rose WE, Curtis CS, Sacks GS. Sterility of pediatric lipid emulsions repackaged by an automated compounding device. JPEN J Parenter Enteral Nutr 35(3), 391–394, 2011.
30. Crill CM, Hak EB, Robinson LA, Helms RA. Evaluation of microbial contamination associated with different preparation methods for neonatal intravenous fat emulsion infusion. Am J Health Syst Pharm 67(11), 914–918, 2010.
31. Montejo O, Cardona D, Sánchez F, Rigueira AI, Coll P, Bonal J. Microbiological quality control study of "all-in-one" total parenteral nutrition admixtures. JPEN J Parenter Enteral Nutr 24(3), 183–186, 2000.
32. Muñoz JM, Zapién R, Ponce-De-León S, Alvarez JA, Mosqueda JL, Gallaga JC, Macías AE. Endemic intravenous fluid contamination in pediatric wards. Rev Invest Clin 61(5), 378–382, 2009.
33. Rickard CM, Vannapraseuth B, McGrail MR, Keene LJ, Rambaldo S, Smith CA, Ray-Barruel G. The relationship between intravenous infusate colonisation and fluid container hang time. J Clin Nurs 18(21), 3022–3028, 2009.
34. Macías AE, Ponce-De-León S, Huertas M, Maravilla E, Romero C, Montoya TG, Muñoz JM, Lopez-Vidal Y. Endemic infusate contamination and related bacteremia. Am J Infect Control 36(1), 48–53, 2008.
35. Macías AE, Huertas M, de Leon SP, Munoz JM, Chavez AR, Sifuentes-Osornio J, Romero C, Bobadilla M. Contamination of intravenous fluids: a continuing cause of hospital bacteremia. Am J Infect Control 38(3), 217–221, 2010.
36. Didier ME, Fischer S, Maki DG. Total nutrient admixtures appear safer than lipid emulsion alone as regards microbial contamination: growth properties of microbial pathogens at room temperature. JPEN J Parenter Enteral Nutr 22(5), 291–296, 1998.
37. Jarvis WR, Highsmith AK. Bacterial growth and endotoxin production in lipid emulsion. J Clin Microbiol 19(1), 17–20, 1984.
38. Weil DC, Arnow PM. Safety of refrigerated storage of admixed parenteral fluids. J Clin Microbiol 26(9), 1787–1790, 1988.
39. Driscoll DF. Sterility of fat emulsion in plastic syringes. Am J Health Syst Pharm 60(2), 197, 2003.
40. Mershon J, Nogami W, Williams JM, Yoder C, Eitzen HE, Lemons JA. Bacterial/fungal growth in a combined parenteral nutrition solution. JPEN J Parenter Enteral Nutr 10(5), 498–502, 1986.
41. Gilbert M, Gallagher SC, Eads M, Elmore MF. Microbial growth patterns in a total parenteral nutrition formulation containing lipid emulsion. JPEN J Parenter Enteral Nutr 10(5), 494–497, 1986.
42. Rowe CE, Fukuyama TT, Martinoff JT. Growth of microorganisms in total nutrient admixtures. Drug Intell Clin Pharm 21(7–8), 633–638, 1987.
43. Scott EM, Gorman SP, Wyatt TD, Magill EA. Growth of microorganisms in total parenteral nutrition mixtures and related clinical observations. J Clin Hosp Pharm 10(1), 79–88, 1985.

44. D'Angio R, Quercia RA, Treiber NK, McLaughlin JC, Klimek JJ. The growth of microorganisms in total parenteral nutrition admixtures. JPEN J Parenter Enteral Nutr 11(4), 394–397, 1987.

45. Bennett SN, McNeil MM, Bland LA, Arduino MJ, Villarino ME, Perrotta DM, Burwen DR, Welbel SF, Pegues DA, Stroud L, Zeitz PS, Jarvis WR. Postoperative infections traced to contamination of an intravenous anesthetic, propofol. N Engl J Med 333(3), 147–154, 1995.

46. Henry B, Plante-Jenkins C, Ostrowska K. An outbreak of *Serratia marcescens* associated with the anesthetic agent propofol. Am J Infect Control 29(5), 312–315, 2001.

47. Kuehnert MJ, Webb RM, Jochimsen EM, Hancock GA, Arduino MJ, Hand S, Currier M, Jarvis WR. *Staphylococcus aureus* bloodstream infections among patients undergoing electroconvulsive therapy traced to breaks in infection control and possible extrinsic contamination by propofol. Anesth Analg 85(2), 420–425, 1997.

48. Massari M, Petrosillo N, Ippolito G, Solforosi L, Bonazzi L, Clementi M, Manzin A. Transmission of hepatitis C virus in a gynecological surgery setting. J Clin Microbiol 39(8), 2860–2863, 2001.

49. Muller AE, Huisman I, Roos PJ, Rietveld AP, Klein J, Harbers JB, Dorresteijn JJ, van Steenbergen JE, Vos MC. Outbreak of severe sepsis due to contaminated propofol: lessons to learn. J Hosp Infect 76(3), 225–230, 2010.

50. Webb SA, Roberts B, Breheny FX, Golledge CL, Cameron PD, van Heerden PV. Contamination of propofol infusions in the intensive care unit: incidence and clinical significance. Anaesth Intensive Care 26(2), 162–164, 1998.

51. Soong WA. Bacterial contamination of propofol in the operating theatre. Anaesth Intensive Care 27(5), 493–496, 1999.

52. McHugh GJ, Roper GM. Propofol emulsion and bacterial contamination. Can J Anaesth 42(9), 801–804, 1995.

53. Lorenz IH, Kolbitsch C, Lass-Flörl C, Gritznig I, Vollert B, Lingnau W, Moser PL, Benzer A. Routine handling of propofol prevents contamination as effectively as does strict adherence to the manufacturer's recommendations. Can J Anaesth 49(4), 347–352, 2002.

54. Obayashi A, Oie S, Kamiya A. Microbial viability in preparations packaged for single use. Biol Pharm Bull 26(5), 667–670, 2003.

55. Tessler M, Dascal A, Gioseffini S, Miller M, Mendelson J. Growth curves of *Staphylococcus aureus*, *Candida albicans*, and *Moraxella osloensis* in propofol and other media. Can J Anaesth 39(5-1), 509–511, 1992.

56. Jansson JR, Fukada T, Ozaki M, Kimura S. Propofol EDTA and reduced incidence of infection. Anaesth Intensive Care 34(3), 362–368, 2006.

57. Duffner LR, Pflugfelder SC, Mandelbaum S, Childress LL. Potential bacterial contamination in fluorescein-anesthetic solutions. Am J Ophthalmol 110(2), 199–202, 1990.

58. Kowalski RP, Romanowski EG, Waikhom B, Gordon YJ. The survival of adenovirus in multidose bottles of topical fluorescein. Am J Ophthalmol 126(6), 835–836, 1998.

59. Charnock C. Are multidose over-the-counter artificial tears adequately preserved? Cornea 25(4), 432–437, 2006.
60. European Pharmacopoeia. Appendix XVI C. Efficacy of antimicrobial preservation, 7th ed. 2011.
61. Oldham GB, Andrews V. Control of microbial contamination in unpreserved eyedrops. Br J Ophthalmol 80(7), 588–591, 1996.
62. Rahman MQ, Tejwani D, Wilson JA, Butcher I, Ramaesh K. Microbial contamination of preservative free eyedrops in multiple application containers. Br J Ophthalmol 90(2), 139–141, 2006.
63. Kim MS, Choi CY, Kim JM, Chung HR, Woo HY. Microbial contamination of multiply used preservative-free artificial tears packed in reclosable containers. Br J Ophthalmol 92, 1518–1521, 2008.
64. Nentwich MM, Kollmann KH, Meshack J, Ilako DR, Schaller UC. Microbial contamination of multi-use ophthalmic solutions in Kenya. Br J Ophthalmol 91(10), 1265–1268, 2007.
65. Chen YH, Wu PC, Shiea J, Lo LH, Wu YC, Kuo HK. Evaluation of the sterility, stability, and efficacy of bevacizumab stored in multiple-dose vials for 6 months. J Ocul Pharmacol Ther 25(1), 65–69, 2009.
66. Menon J, Murray A. Microbial growth in vials of Botulinum toxin following use in clinic. Eye 21(7), 995–997, 2007.
67. Mason BL, Alfonso EC, Miller D. In-use study of potential bacterial contamination of ophthalmic moxifloxacin. J Cataract Refract Surg 31(9), 1773–1776, 2005.
68. Geyer O, Bottone EJ, Podos SM, Schumer RA, Asbell PA. Microbial contamination of medications used to treat glaucoma. Br J Ophthalmol 79, 376–379, 1995.
69. Jose AC, Castelo Branco B, Ohkawara LE, Yu MC, Lima AL. Use conditions of boric acid solution in the eye: handling and occurrence of contamination. Arq Bras Oftalmol 70(2), 201–207, 2007.
70. Høvding G, Sjursen H. Bacterial contamination of drops and dropper tips of in-use multidose eyedrop bottles. Acta Ophthalmol (Copenh) 60(2), 213–222, 1982.
71. Szczotka-Flynn LB, Pearlman E, Ghannoum M. Microbial contamination of contact lenses, lens care solutions, and their accessories: a literature review. Eye Contact Lens 36(2), 116–129, 2010.
72. Moss E. Medical gas contamination: an unrecognized patient danger. J Clin Monit 11(1), 73–76, 1995.
73. Andrade CM, Brown T. Microbial contamination of central supply systems for medical air. Braz J Microbiol 34(1), 29–32, 2003.
74. Hubar JS, Pelon W, Gardiner DM. Evaluation of compressed air used in the dental operatory. J Am Dent Assoc 133(7), 837–841, 2002.
75. Conte M, Lynch RM, Robson MG. Microbiological contamination of compressed air used in dentistry: an investigation. J Environ Health 64(4), 15–20, 2001.
76. Perz JF, Thompson ND, Schaefer MK, Patel PR. US outbreak investigations highlight the need for safe injection practices and basic infection control. Clin Liver Dis 14(1), 137–151, 2010.
77. Stucki C, Sautter AM, Favet J, Bonnabry P. Microbial contamination of syringes during preparation: the direct influence of environmental cleanliness and risk manipulations on end-product quality. Am J Health Syst Pharm 66(22), 2036, 2032.

78. Gonzalez S, Miller D, Murphy SP. Maintenance of sterility in 1 mL polypropylene syringes. Am J Health Syst Pharm 64(18), 1962–1964, 2007.

79. van Grafhorst JP, Foudraine NA, Nooteboom F, Crombach WHJ, Oldenhof NJJ, van Doorne H. Unexpected high risk of contamination with staphylococci species attributable to standard preparation of syringes for continuous intravenous drug administration in a simulation model in intensive care units. Crit Care Med 30(4), 833–836, 2002.

80. Trépanier CA, Lessard MR, Brochu JG, Denault PH. Risk of cross-infection related to the multiple use of disposable syringes. Can J Anaesth 37(2), 156–159, 1990.

81. Sonoki K, Yoshinari M, Iwase M, Tashiro K, Iino K, Wakisaka M, Fujishima M. Regurgitation of blood into insulin cartridges in the pen-like injectors. Diabetes Care 24(3), 603–604, 2001.

82. Magee L, Godsiff L, Matthews I, Farrington M, Park GR. Anaesthetic drugs and bacterial contamination. Eur J Anaesthesiol Suppl 12, 41–43, 1995.

83. Williams GB, Ross LL, Chandra RK. Are bulb syringe irrigators a potential source of bacterial contamination in chronic rhinosinusitis? Am J Rhinol 22(4), 399–401, 2008.

84. Buerke B, Mellmann A, Stehling C, Wessling J, Heindel W, Juergens KU. Microbiologic contamination of automatic injectors at MDCT: experimental and clinical investigations. AJR Am J Roentgenol 191(6), 283–287, 2008.

85. Buerke B, Puesken M, Mellmann A, Seifarth H, Heindel W, Wessling J. Microbiologic contamination and time efficiency of use of automatic MDCT injectors with prefilled syringes: results of a clinical investigations. AJR Am J Roentgenol 194(2), 299–303, 2010.

86. ASGE. Standards of practice committee Infection control during GI endoscopy. Gastrointest Endosc 67(6), 781–790, 2008.

87. Alfa MJ, Degagne P, Olson N. Worst-case soiling levels for patient-used flexible endoscopes before and after cleaning. Am J Infect Control 27(5), 392–401, 1999.

88. Chan-Myers H, McAlister D, Antonoplos P. Natural bioburden levels detected on rigid lumened medical devices before and after cleaning. Am J Infect Control 25(6), 471–476, 1997.

89. Chu NS, McAlister D, Antonoplos PA. Natural bioburden levels detected on flexible gastrointestinal endoscopes after clinical use and manual cleaning. Gastrointest Endosc 48(2), 137–142, 1998.

90. Saliou P, Garlantézec R, Baron R, Quiot JJ, Cholet F, Le Floch MF, Lejeune B. Microbiological investigation of endoscopes at Brest Hospital over a period from 2007 to 2009. Pathol Biol 59(2), 88–93, 2011.

91. Buss AJ, Been MH, Borgers RP, Stokroos I, Melchers WJ, Peters FT, Limburg AJ, Degener JE. Endoscope disinfection and its pitfalls-requirement for retrograde surveillance cultures. Endoscopy 40(4), 327–332, 2008.

92. Bonassoli LA, Bertoli M, Svidzinski TI. High frequency of *Candida parapsilosis* on the hands of healthy hosts. J Hosp Infect 59(2), 159–162, 2005.

93. Snyder GM, Thom KA, Furuno JP, Perencevich EN, Roghmann MC, Strauss SM, Netzer G, Harris AD. Detection of methicillin-resistant *Staphylococcus aureus* and

vancomycin-resistant *Enterococci* on the gowns and gloves of healthcare workers. Infect Control Hosp Epidemiol 29(7), 583–589, 2008.

94. Hayden MK, Blom DW, Lyle EA, Moore CG, Weinstein RA. Risk of hand or glove contamination after contact with patients colonized with vancomycin-resistant *Enterococcus* or the colonized patients' environment. Infect Control Hosp Epidemiol 29(2), 149–154, 2008.
95. Girou E, Chai SHT, Oppein F, Legrand P, Ducellier D, Cizeau F, Brun-Buisson C. Misuse of gloves: the foundation for poor compliance with hand hygiene and potential for microbial transmission? J Hosp Infect 57(2), 162–169, 2004.
96. Haishima Y, Murai T, Nakagawa Y, Hirata M, Yagami T, Nakamura A. Chemical and biological evaluation of endotoxin contamination on natural rubber latex products. J Biomed Mater Res 55(3), 424–432, 2001.
97. Harkins KJ. Bacterial contamination of inhalers. J Hosp Infect 43(3), 321, 1999.
98. Cohen HA, Cohen Z, Kahan E. Bacterial contamination of spacer devices used by children with asthma. J Am Med Assoc 290(2), 195–196, 2003.
99. Vassal S, Taamma R, Marty N, Sardet A, Athis P, Brémont F, Dalphin ML, Plésiat P, Rault G, Thubert J, Dominique S, Lemeland JF, Derelle J, Blech MF, Roussey M, Perrin M, Sautegeau A. Microbiologic contamination study of nebulizers after aerosol therapy in patients with cystic fibrosis. Am J Infect Control 28(5), 347–351, 2000.
100. Langevin PB, Rand KH, Layon AJ. The potential for dissemination of *Mycobacterium tuberculosis* through the anesthesia breathing circuit. Chest 115(4), 1107–1114, 1999.
101. Arai LA, Azevedo RB. Contamination of anesthesia circuits by pathogens. Rev Bras Anestesiol 61(1), 50–59, 2011.
102. Rothwell M, Pearson D, Wright K, Barlow D. Bacterial contamination of PCA and epidural infusion devices. Anaesthesia 64(7), 751–753, 2009.
103. Baillie JK, Sultan P, Graveling E, Forrest C, Lafong C. Contamination of anaesthetic machines with pathogenic organisms. Anaesthesia 62(12), 1257–1261, 2007.
104. Oliveira AC, Damasceno QS. Surfaces of the hospital environment as possible deposits of resistant bacteria: a review. Rev Esc Enferm USP 44(4), 1118–1123, 2010.
105. Costerton JW, Stewart PS, Greenberg EP. Bacterial biofilms: a common cause of persistent infections. Science 21, 1318–1322, 1999.
106. Machado JD, Suen VM, Figueiredo JF, Marchini JS. Biofilms, infection, and parenteral nutrition therapy. JPEN J Parenter Enteral Nutr 33(4), 397–403, 2009.
107. Souli M, Giamarellou H. Effects of slime produced by clinical isolates of coagulase-negative *Staphylococci* on activities of various antimicrobial agents. Antimicrob Agents Chemother 42, 939–941, 1998.
108. Kanagasundaram NS, Larive AB, Paganini EP. A preliminary survey of bacterial contamination of the dialysate circuit in continuous veno-venous hemodialysis. Clin Nephrol 59(1), 47–55, 2003.
109. Man N-K, Degremont A, Darbord J-C, Collet M, Vaillant P. Evidence of bacterial biofilm in tubing from hydraulic pathway of hemodialysis system. Artif Organs 22(7), 596–600, 1998.

110. Marion-Ferey K, Enkiri F, Pasmore M, Husson GP, Vilagines R. Methods for biofilm analysis on silicone tubing of dialysis machines. Artif Organs 27(7), 658–664, 2003.

111. Moore I, Bhat R, Hoenich NA, Kilner AJ, Prabhu M, Orr KE, Kanagasundaram NS. A microbiological survey of bicarbonate-based replacement circuits in continuous veno-venous hemofiltration. Crit Care Med 37(2), 496–500, 2009.

112. National Nosocomial Infections Surveillance (NNIS) report, data summary from October 1986-April 1997, issued May 1997. A report from the NNIS System. Am J Infect Control 25(6), 477–487, 1997.

113. Naze F, Jouen E, Randriamahazo RT, Simac C, Laurent P, Blériot A, Chiroleu F, Gagnevin L, Pruvost O, Michault A. *Pseudomonas aeruginosa* outbreak linked to mineral water bottles in a neonatal intensive care unit: fast typing by use of high-resolution melting analysis of a variable-number tandem-repeat locus. J Clin Microbiol 48(9), 3146–3152, 2010.

114. Buffet-Bataillon S, Rabier V, Bétrémieux P, Beuchée A, Bauer M, Pladys P, Le Gall E, Cormier M, Jolivet-Gougeon A. Outbreak of *Serratia marcescens* in a neonatal intensive care unit: contaminated unmedicated liquid soap and risk factors. J Hosp Infect 72(1), 17–22, 2009.

115. Gutelius B, Perz JF, Parker MM, Hallack R, Stricof R, Clement EJ, Lin Y, Xia GL, Punsalang A, Eramo A, Layton M, Balter S. Multiple clusters of hepatitis virus infections associated with anesthesia for outpatient endoscopy procedures. Gastroenterology 139(1), 163–170, 2010.

116. Counard CA, Perz JF, Linchangco PC, Christiansen D, Ganova-Raeva L, Xia G, Jones S, Vernon MO. Acute hepatitis B outbreaks related to fingerstick blood glucose monitoring in two assisted living facilities. J Am Geriatr Soc 58(2), 306–311, 2010.

117. Blossom D, Noble-Wang J, Su J, Pur S, Chemaly R, Shams A, Jensen B, Pascoe N, Gullion J, Casey E, Hayden M, Arduino M, Budnitz DS, Raad I, Trenholme G, Srinivasan A. Multistate outbreak of *Serratia marcescens* bloodstream infections caused by contamination of prefilled heparin and isotonic sodium chloride solution syringes. Arch Intern Med 169(18), 1705–1711, 2009.

118. Sunenshine R, Schultz M, Lawrence MG, Shin S, Jensen B, Zubairi S, Labriola AM, Shams A, Noble-Wang J, Arduino MJ, Gordin F, Srinivasan A. An outbreak of postoperative gram-negative bacterial endophthalmitis associated with contaminated trypan blue ophthalmic solution. Clin Infect Dis 48(11), 1580–1583, 2009.

119. Rabier V, Bataillon S, Jolivet-Gougeon A, Chapplain JM, Beuchée A, Bétrémieux P. Hand washing soap as a source of neonatal *Serratia marcescens* outbreak. Acta Paediatr 97(10), 1381–1385, 2008.

120. Sunenshine RH, Tan ET, Terashita DM, Jensen BJ, Kacica MA, Sickbert-Bennett EE, Noble-Wang JA, Palmieri MJ, Bopp DJ, Jernigan DB, Kazakova S, Bresnitz EA, Tan CG, McDonald LC. A multistate outbreak of *Serratia marcescens* bloodstream infection associated with contaminated intravenous magnesium sulfate from a compounding pharmacy. Clin Infect Dis 45(5), 527–533, 2007.

121. Cohen AL, Ridpath A, Noble-Wang J, Jensen B, Peterson AM, Arduino M, Jernigan D, Srinivasan A. Outbreak of *Serratia marcescens* bloodstream and central

nervous system infections after interventional pain management procedures. Clin J Pain 24(5), 374–380, 2008.

122. Arslan U, Erayman I, Kirdar S, Yuksekkaya S, Cimen O, Tuncer I, Bozdogan B. *Serratia marcescens* sepsis outbreak in a neonatal intensive care unit. Pediatr Int 52(2), 208–212, 2010.

123. Jhung MA, Sunenshine RH, Noble-Wang J, Coffin SE, St John K, Lewis FM, Jensen B, Peterson A, LiPuma J, Arduino MJ, Holzmann-Pazgal G, Atkins JT, Srinivasan A. A national outbreak of *Ralstonia mannitolilytica* associated with use of a contaminated oxygen-delivery device among pediatric patients. Pediatrics 119(6), 1061–1068, 2007.

124. Patel PR, Larson AK, Castel AD, Ganova-Raeva LM, Myers RA, Roup BJ, Farrell KP, Edwards L, Nainan O, Krick JP, Blythe D, Fiore AE, Roche JC. Hepatitis C virus infections from a contaminated radiopharmaceutical used in myocardial perfusion studies. J Am Med Assoc 296, 2005–2011, 2006.

125. Takahashi H, Kramer MH, Yasui Y, Fujii H, Nakase K, Ikeda K, Imai T, Okazawa A, Tanaka T, Ohyama T, Okabe N. Nosocomial *Serratia marcescens* outbreak in Osaka, Japan, from 1999 to 2000. Infect Control Hosp Epidemiol 25(2), 156–161, 2004.

126. Dolan SA, Dowell E, Lipuma JJ, Valdez S, Chan K, James JF. An outbreak of *Burkholderia cepacia* complex associated with intrinsically contaminated nasal spray. Infect Control Hosp Epidemiol 32(8), 804–810, 2011.

127. Sánchez-Carrillo C, Padilla B, Marín M, Rivera M, Cercenado E, Vigil D, Sánchez-Luna M, Bouza E. Contaminated feeding bottles: the source of an outbreak of *Pseudomonas aeruginosa* infections in a neonatal intensive care unit. Am J Infect Control 37(2), 150–154, 2009.

128. Center for Disease Control. Transmission of Hepatitis B Virus among persons undergoing blood glucose monitoring in long-term-care facilities. MMWR Morb Mortal Wkly Rep 54(09), 220–223, 2005.

129. Center for Disease Control. Transmission of hepatitis B and C viruses in outpatient settings—New York, Oklahoma, and Nebraska, 2000–2002. MMWR Morb Mortal Wkly Rep 52(38), 901–906, 2003.

130. Selenic D, Dodson DR, Jensen B, Arduino MJ, Panlilio A, Archibald LK. *Enterobacter cloacae* bloodstream infections in pediatric patients traced to a hospital pharmacy. Am J Health Syst Pharm 60(14), 1440–1446, 2003.

131. Horcajada JP, Martínez JA, Alcón A, Marco F, De Lazzari E, de Matos A, Zaragoza M, Sallés M, Zavala E, Mensa J. Acquisition of multidrug-resistant *Serratia marcescens* by critically ill patients who consumed tap water during receipt of oral medication. Infect Control Hosp Epidemiol 27(7), 774–777, 2006.

132. Archibald LK, Corl A, Shah B, Schulte M, Arduino MJ, Aguero S, Fisher DJ, Stechenberg BW, Banerjee SN, Jarvis WR. *Serratia marcescens* outbreak associated with extrinsic contamination of 1% chlorxylenol soap. Infect Control Hosp Epidemiol 18(10), 704–709, 1997.

133. Center for Disease Control. Nosocomial hepatitis B virus infection associated with reusable fingerstick blood sampling devices. MMWR Morb Mortal Wkly Rep 46(10), 217–221, 1997.

134. Reiss I, Borkhardt A, Füssle R, Sziegoleit A, Gortner L. Disinfectant contaminated with *Klebsiella oxytoca* as a source of sepsis in babies. Lancet 356(9226), 310, 2000.

135. Manning ML, Archibald LK, Bell LM, Banerjee SN, Jarvis WR. *Serratia marcescens* transmission in a pediatric intensive care unit: a multifactorial occurrence. Am J Infect Control 29(2), 115–119, 2001.

136. Center for Disease Control. Outbreaks of hepatitis B virus infection among hemodialysis patients. MMWR Morb Mortal Wkly Rep 45(14), 285–289, 1996.

137. Ostrowsky BE, Whitener C, Bredenberg HK, Carson LA, Holt S, Hutwagner L, Arduino MJ, Jarvis WR. *Serratia marcescens* bacteremia traced to an infused narcotic. N Engl J Med 346(20), 1529–1537, 2002.

138. Vigeant P, Loo VG, Bertrand C, Dixon C, Hollis R, Pfaller MA, McLean AP, Briedis DJ, Perl TM, Robson HG. An outbreak of *Serratia marcescens* infections related to contaminated chlorhexidine. Infect Control Hosp Epidemiol 19(10), 791–794, 1998.

139. Vonberg RP, Gastmeier P. Hospital-acquired infections related to contaminated substances. J Hosp Infect 65(1), 15–23, 2007.

PART IV

MISCELLANEOUS

12

CONTAMINATION FROM STERILIZATION PROCEDURES

Sterilization and disinfection are essential for ensuring that medicines and medical devices do not transmit infectious pathogens to patients. Sterilization should destroy or eliminate all forms of microbial life, including fungi, viruses, all forms of bacteria, and their spores. Disinfection, on the other hand, is performed on hard inanimate surfaces and objects to destroy or irreversibly inactivate infectious fungi and bacteria but not necessarily their spores.

Sterilization can be achieved by applying the proper combinations of heat, chemicals, irradiation, high pressure, and filtration. Methods for eliminating microorganisms from pharmaceutical products and medical devices are listed in Table 12.1. There are a few disinfectants (hydrogen peroxide, formaldehyde, and peracetic acid [PA]), depending on the application conditions, that kill spores. These disinfectants are called chemical sterilants. Because these agents are in contact with humans or animals, they are considered drugs and are approved and regulated by agencies, such as the Food and Drug Administration (FDA). Choosing the appropriate method of sterilization or disinfection depends on the necessary level of microorganism elimination and the features of the product.

Sterilization is not necessary for all items related to patient care. Thus, based on an item's intended use, healthcare policies establish when sterilization or disinfection is required, and the universe of items that need to be sterilized or

Sources of Contamination in Medicinal Products and Medical Devices, First Edition. Denise Bohrer.
© 2013 by John Wiley & Sons, Inc. Published 2013 by John Wiley & Sons, Inc.

TABLE 12.1 Sterilization and Disinfection Agents

Method	Agent	Action		Use	Materials' Compatibility	Residuals	
		Sterilant	Disinfectant			Sterilizing agent	Decomposition Products
Steam	Water + heat	x		Drugs	Poor		
Dry heat	Heat	x		Drugs	Poor		
Ionizing radiation	γ-Ray	x		Drugs	Poor		x
	Electron beam			Medical devices			
Gas	Ethylene oxide	x		Medical devices Reusable medical devices	Good	x	x
Filtration	Pore barrier	x		Drugs: liquids that would be damaged by heat, irradiation, or chemical sterilization	–		
Chemical	Hydrogen peroxide	x (gas)	x (liquid)	Medical devices Reusable medical devices	Fair	x	x
	Formaldehyde	x (gas)	x (liquid)	Hemodialyzers, hemodialysis machines Reusable medical devices	Good	x	
	Peracetic acid	x	x	Hemodialyzers Reusable medical devices	Fair	x	
	Glutaraldehyde or glutaraldehyde/phenol/phenate		x	Endoscopes, dialyzers, anesthesia, and respiratory equipment	Good	x	
	Ortho-phthalaldehyde		x	Reusable medical devices	Good	x	
	Isopropyl alcohol		x	Medication preparation areas	Good	x	
	Ethyl alcohol		x	Medication preparation areas	Good	x	
	Hypochlorite		x	Hemodialysis machines Surfaces		x	

Adapted from Reference 1.

disinfected is vast. Selection of a sterilizing or disinfecting procedure should be performed in accordance with the product to be processed. Drugs and single-use devices are generally industrially sterilized by dry or steam heat, ethylene oxide (EtO), or radiation. On the other hand, reusable devices are usually processed in local facilities. Because of the necessary structure, multiple-use devices are sterilized mostly using chemical agents. Therefore, residuals of chemical disinfectants or sterilants are more frequently found in reusable medical devices.

12.1 RESIDUALS FROM RADIATION STERILIZATION

Photons or subatomic particles in movement can transfer their energy to the molecules of a medium, inducing molecular excitation or ionization. When the particles or photons carry high energy, ionization of atoms or molecules takes place and the phenomenon is called *radiolysis*. Radiolysis is the result of exposition of the matter to *ionizing radiation*.

Ionizing radiation consists of subatomic particles or photons that are energetic enough to detach electrons from atoms or molecules, ionizing them. The occurrence of ionization depends on the energy of the impinging individual particles or waves and not on their number. An intense flood of particles or waves will not cause ionization if these particles or waves do not carry enough energy to be ionizing. Ionizing particles are *protons*, *electrons*, and *neutrons*. The ability of photons to ionize an atom or molecule depends on their wavelength. Radiation on the short wavelength end of the electromagnetic spectrum—*X-rays*, *gamma rays*, and *cosmic rays*—is ionizing. The different types of electromagnetic radiations and their correspondent energies and wavelengths are shown in Figure 12.1.

Gamma rays and electrons (beta particles) are used in the sterilization of devices and drugs. The absorbed radiation dose generally is expressed in radiation absorbed doses (*rads*). One rad is equivalent to an absorbed energy of 100 erg/g of material. The currently used International System of Units (SI) unit for radiation absorbed doses, however, is the *gray*, which is equivalent to an energy absorption of 1 J/kg. One gray is equivalent to 100 rad, and 25 kGy, a common radiation dose for sterilization, is equivalent to 2.5 Mrad.

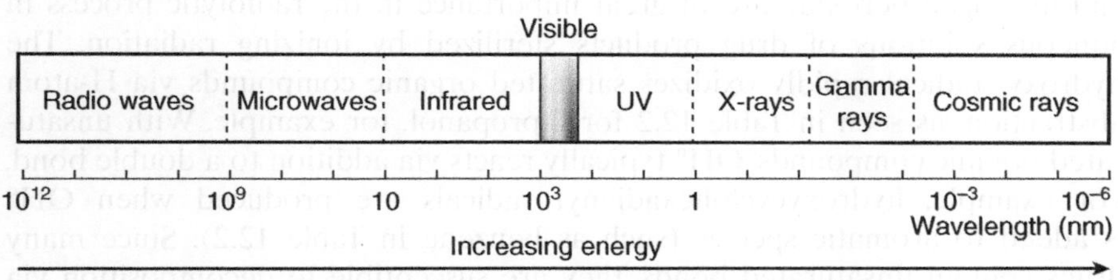

FIGURE 12.1 The electromagnetic spectrum.

In both gamma- and electron-beam (e-beam) radiation sterilization processes, available oxygen is quickly consumed, contributing to the formation of oxidation products. In the case of e-beam processing, the time of energy application is so short that before more oxygen can permeate into the material from its external surface, the application of radiant energy has been terminated. On the other hand, in gamma radiation, the application of ionizing energy continues over a much longer period of time, allowing reactants, such as oxygen, to permeate back into depleted areas of the material, resulting in a greater degree of damage.

12.1.1 Radiolysis of Water

Radiolysis of water is the main reaction that occurs by the action of ionizing radiation on pharmaceutical products since most liquid or even solid preparations contain water as a solvent. Bombarding liquid water with high-energy photons or subatomic particles causes excitation and ionization of the water molecules. In a few picoseconds, the excited molecules are transformed into radical and ionic products according to the following equations [2]:

$$H_2O + \text{ionizing radiation} \rightarrow H_2O^\circ \rightarrow H^\circ + OH^\circ \quad (12.1)$$

$$H_2O + \text{ionizing radiation} \rightarrow H_2O^+ + e^- \quad (12.2)$$

$$e^- \rightarrow e^-_{aq} \quad (12.3)$$

$$H_2O^+ \rightarrow H^+_{aq} + OH^\circ \quad (12.4)$$

As these reactions show, the primary species formed are hydroxyl radicals (OH°), hydrogen atoms (H°), and solvated electrons (e^-_{aq}). These species are able to react with each other or with any surrounding molecule. While the hydrogen atom and the hydrated electron are reducing species, the hydroxyl radical is a powerful oxidant. Table 12.2 presents reactions occurring between these primary species and between these species and other selected molecules, which may be present in pharmaceutical preparations.

Another oxidizing species produced by the hydroxyl radical is hydrogen peroxide. Because of their high yields and their reactivity, the hydroxyl radical and hydrogen peroxide are of great importance in the radiolytic process in aqueous solutions of drug products sterilized by ionizing radiation. The hydroxyl radical rapidly oxidizes saturated organic compounds via H-atom abstraction, as seen in Table 12.2 for 2-propanol, for example. With unsaturated organic compounds, OH° typically reacts via addition to a double bond. For example, hydroxycyclohexadienyl radicals are produced when OH° is added to aromatic species (such as benzene in Table 12.2). Since many drugs contain unsaturated bonds, they are susceptible to decomposition via

TABLE 12.2 Reactions between the Primary Species Generated by Radiolysis of Water and between these Species and Other Selected Molecules

Hydrogen Atom	Solvated Electron	Hydroxyl Radical
$H^\circ + H_2O \rightarrow H_2 + OH^\circ$	$e^-_{aq} + H_2O \rightarrow H^\circ + OH^-$	$OH^\circ + OH^\circ \rightarrow H_2O_2$
$H^\circ + H^\circ \rightarrow H_2$	$e^-_{aq} + HO^\circ \rightarrow OH^-$	$OH^\circ + OH^- \rightarrow H_2O + O^{\circ -}$
$H^\circ + OH^- \rightarrow H_2O + e^-_{aq}$	$e^-_{aq} + H^+ \rightarrow H^\circ$	
$H^\circ + O_2 \rightarrow HO_2^\circ$	$e^-_{aq} + O_2 \rightarrow O_2^-$	
$H^\circ + Cu^{2+} \rightarrow Cu^+ + H^+$	$e^-_{aq} + Cu^{2+} \rightarrow Cu^+$	$OH^\circ + Cu^{2+} \rightarrow [CuOH]^{2+}$
$H^\circ + (CH_3)_2CH(OH) \rightarrow H_2 + (CH_3)_2C^\circ HO$		$OH^\circ + (CH_3)_2CH(OH) \rightarrow H_2O + (CH_3)_2C^\circ HO(85.5\%) + (C^\circ H_2)(CH_3)CHOH(13.3\%)$
$H^\circ + C_6H_6 \rightarrow$ cyclohexadienyl radical		$OH^\circ + C_6H_6 \rightarrow$ hydroxycyclohexadienyl radical
$H^\circ + C_6H_6OH \rightarrow$ hydroxycyclohexadienyl radical		

Adapted from Reference 2.

radiolysis this way. Of the reactions listed in Table 12.2, the most likely to occur in aqueous preparations is reaction with oxygen (O_2). Oxygen solubility in water, at 25°C and under atmospheric pressure, is 8.9 mg/L, which is a considerable amount. Because of this, the hydroperoxy radical (HO_2°), formed by the reaction of O_2 with hydrogen, accounts for the increased oxidant power of irradiated water.

12.1.2 Effect of Ionizing Radiation on Drug Products

Sterilization using gamma rays is an alternative method for sterilizing thermosensitive drugs. Although radiosterilization does not modify drug activity, it can produce new radiolytic species.

Ionizing radiation for sterilization purposes is mainly used for parenterals, ophthalmic preparations, and topical ointments. Another specific application of radiation is the sterilization of complex drug delivery systems.

A drug may contain organic and inorganic constituents, which react differently to ionizing radiation. While ionization caused by radiation usually does not lead to permanent alterations in the structure of inorganic compounds, the effect can be much more destructive on organic compounds. Moreover, there is a great difference in irradiating drugs in solid state (dry state) or in solution, since solutions are usually aqueous and radiolysis of water increases the radiation effects. Among inorganic species, metals and crystalline solids are less vulnerable to the effects of radiation because of the fixed position of the atoms in the crystalline lattice. The electronic excitation affects the physical properties of the crystal rather than the chemical behavior.

The release of electrons causes disruption in covalent bonds, and consequent atomic abstraction, radical recombination, and addition to π-bonds. These reactions will lead to irreversible changes in the structure of organic molecules. Nevertheless, the irradiation of aromatic compounds results in considerably lower degradation than does irradiation of aliphatic compounds of similar function and molecular structure. This has been attributed to the stabilizing effect of the benzene ring. The delocalized π-orbitals are able to withstand the excitation energy without permitting the molecule to dissociate.

In general, degradation can be minimized by carrying out the irradiation of drugs in solid state and for liquids at low temperature. For radiosensitive drugs, the presence of radioprotective excipients may reduce the radiation effect. Recently, radiation has been used in the sterilization of encapsulated drugs, the so-called drug delivery systems. As encapsulation is usually carried out before sterilization, studies on these systems' stability must encompass both drug and polymer behavior against radiation. An appraisal of radiolysis products of some drugs and drug release systems are listed in Table 12.3.

12.1.3 Polymers in Drug Delivery

Controlled drug delivery occurs when a polymer, whether natural or synthetic, is carefully combined with a drug or other active agent in such a way that the active agent is released from the material in a predesigned manner. This can be in the form of a formulation or device that provides timely release of the therapeutic agent or delivers the therapeutic agent to a desired body location.

Besides direct delivery at the site where the drug action is needed, the main advantage of this system is the possibility of an effective stabilization of the drug since the polymer can protect the drug from the physiological environment. It is required, for example, in the delivery of labile drugs such as proteins as well as for prolonged drug availability or sustained delivery of drugs for chronic indications requiring frequent injections.

Drug delivery is employed in systems as simple as solid oral dosage forms (coating) as well as more complex systems such as the tissue engineering. Specific properties are needed for the polymeric matrix in order to either allow or prevent the release of the drug in the gastrointestinal tract, or to allow the release of a drug into a tissue at the correct time.

The release of drug molecules from a delivery device is controlled by various mechanisms, and the most widely used are diffusion, dissolution, osmosis, and ion exchange [26]. The polymeric matrix will be chosen depending on the desired mode of release. For diffusion-controlled delivery, drugs are loaded in a water-insoluble polymer matrix, through which the drug diffuses.

TABLE 12.3 Radiation Sterilization of Selected Drugs, Degradation Products, and Applicability of the Technique

Drug	Physical Form	Radiation (kGy)	Degradation Products	Observation	Reference
Metronidazole	Solid	10–25	2-Methyl-5-nitroimidazole 2-Methyl imidazole	Feasible	[3]
Trifluorothymidine	Solid	25–100	Trifluoromethyluracil 5-Carboxy 2-deoxyuridine 5-Carboxi uracil 5-Hidroxydifluoromethyl 5-Carboxydeoxyuridine	Reasonably stable, 0.5% degradation at 25 kGy	[4]
Isoproterenol	Solid	10–25	Stable free radicals	Drug purity decreased 0.19% at 25 kGy	[5]
Formoterol	Solid	5–30	Few number of free radicals	Minor differences in chromatograms of irradiated nonirradiated samples	[6]
Ritodrinehydrochloride	Solid	0–50	Radiolytic products	Degradation of 2.8% at 25 kGy	[7]
Nifedipine	Solid	10–100	Free radicals	Degradation of 1.3% at 25 kGy	[8]
Felodipine	Solid	10–100	–	No degradation	[8]
Nitrendipine	Solid	10–100	Free radicals	Degradation of 1.1% at 25 kGy	[8]
Nimodipine	Solid	10–100	–	No degradation	[8]
Sulfacetamide sodium	Solid	5–50	–	No degradation	[9]
Ketoprofen	Solid	25–200	Benzophenon and acetyl benzophenon at radiation doses >1000 kGy	Feasible at 25 kGy	[10]
Sulfatiazole	Solid	5–50	–	No detectable changes	[11]
Cyclophosphamide	Solid	5–50	Three decomposition products	Degradation >2% at 30 kGy	[12]
Doxorubicin	Solid	5–50	–	No detectable changes	[13]

(*Continued*)

TABLE 12.3 (*Continued*)

Drug	Physical Form	Radiation (kGy)	Degradation Products	Observation	Reference
Orciprenaline	Solid	25	Stable degradation products	Slight increase in impurities	[13]
Fenoterol	Solid	25	Stable degradation products	Slight increase in impurities	[13]
Bupivacaine	Solid	25	–	3% decrease of drug content	[14]
Salbutamol	Solid	25	Several degradation products	Degradation <2% at 25 kGy	[15]
Salbutamol	Aqueous solution	25	Several degradation products	Several degradation products	[15]
Dopamine	Solid	10–25	–	Degradation of 1.1% at 25 kGy	[15]
Norepinephrine	Solid	10–25	–	Degradation of 0.03% at 25 kGy	[16]
Paclitaxel		25 and 75	Benzaldehyde Benzoic acid	Not defined	[17]
Cortisone acetate	Aqueous solution	25	Not identified	High sensitivity to radiation even at low doses	[18]
Metoclopramide	Solid	15	Several degradation products	Not feasible	[19]
Cladribine/PLGA	Solid	10–200	3.5% 2-Chloroadenine irradiation at 100 kGy	Feasible at 10–25 kGy	[20]
Morphine	Solution	25 ($T = -80°C$)	Not identified	Freezing improves the stability not affected	[21]
Heparin	Solution	25	Not identified	Freezing improves the stability not affected	[21]
Gentamicin collagen/ PLGA	Solid	–	Free radicals detected	Disappeared after 4 weeks Feasible	[22]
Fluconazole /PLA		25	–	Chemical stability not affected	[23]
17β-estradiol	Solid	26.6	–	Degradation of 0.4%	[24]
17β-Estradiol/PLG		26.6	9,11-Dehydroestradiol PLG conjugates	Up to 5% drug degradation depending on drug loading	[24]
Captopril		6.9–34.8	–	No degradation	[25]
Captopril/PLGA		6.9–34.8	Oxidation to disulfide	–	[25]

Dissolution-controlled delivery systems are prepared using water-soluble or biodegradable polymers, and the release is controlled by the dissolution of the polymeric matrix. Environment-sensitive polymers possess properties that respond to environmental changes such as pH and temperature. In this case, the polymeric structure will be selected to support the desired polymer reaction. A polymeric structure sensitive to pH changes, for example, should be a polyelectrolyte such as poly(acrylic acid) (PAA) and poly(*N*,*N*′-diethylaminoethyl methacrylate) (PDEAEM), which dissociate or not according to pH changes.

Potential disadvantages of the use of polymeric materials are toxicity or nonbiocompatibility of the materials used and the formation of undesirable by-products of degradation. Unexpected degradation products may occur during sterilization procedures, with possibly deleterious effects on both drug and polymer. Polymeric materials used in drug delivery system are presented in Table 12.4.

12.1.3.1 Polylactide (PLA) and Poly(lactide-co-glycolide) (PLGA) The most commonly used polymers in drug delivery systems are the poly(lactic-co-glycolic acid) (PLGA) and the poly(lactic acid) (PLA). Important characteristics of these polymers are biodegradability, biocompatibility, and low immunogenicity and toxicity [27].

PLGA is synthesized by ring opening of lactic acid and glycolic acid monomers and linkage via ester bonds (Equation 12.5). A common ratio of both monomers is 50:50, but other ratios may be used, which determine the physicochemical properties of the copolymer [26]:

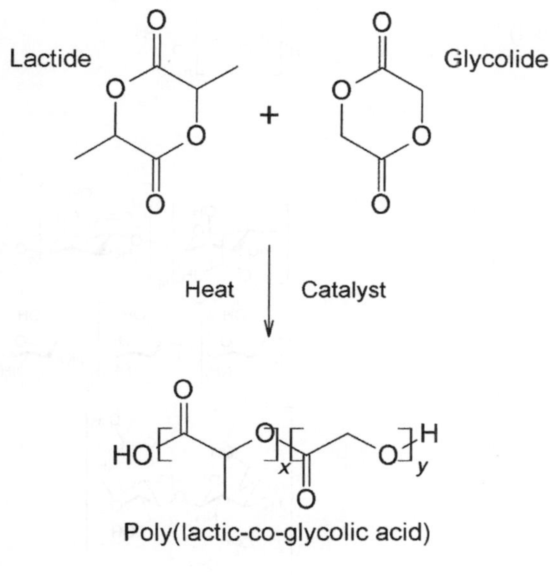

(12.5)

TABLE 12.4 Main Polymeric Materials Used in Drug Delivery Systems

Polymer	Structure
Polylactide or poly(lactic acid)	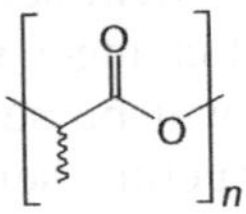
Poly(lactide-co-glycolide) or Poly(lactic-co-glycolic acid)	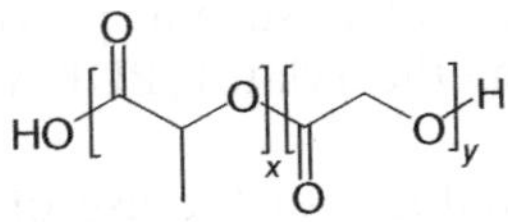
Cellulose and cellulose derivatives	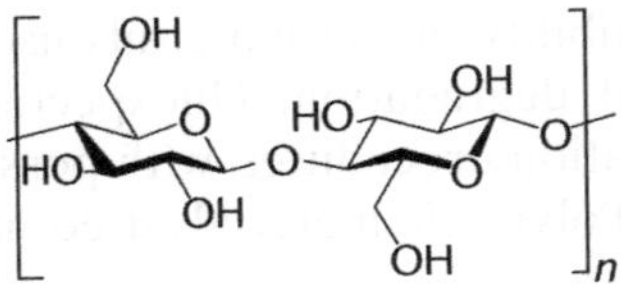
Starch	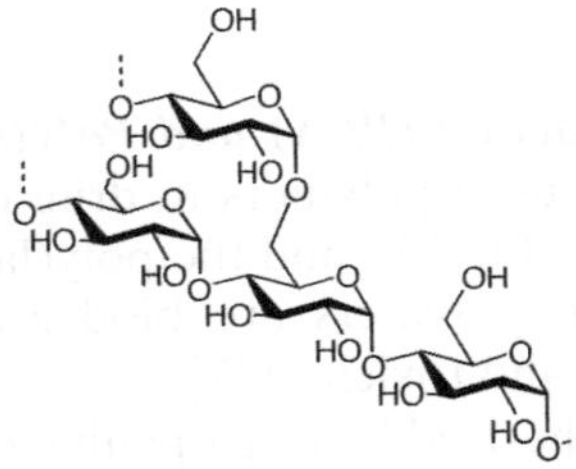
Gelatin	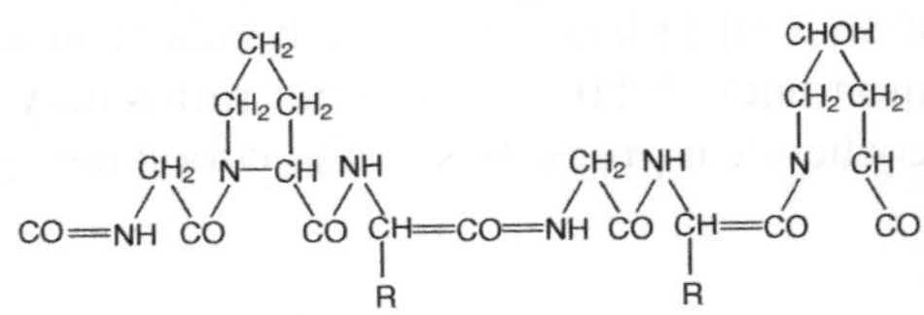
Poly(ethylene glycol)	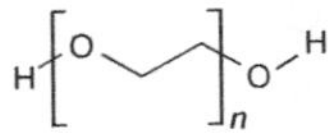
Collagen	–
Alginate	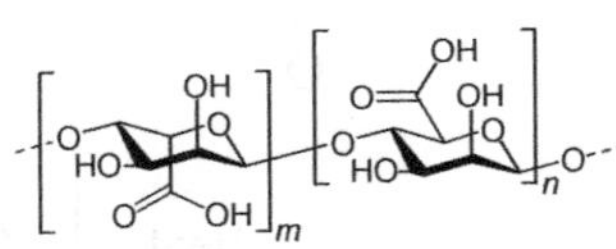
Chitosan	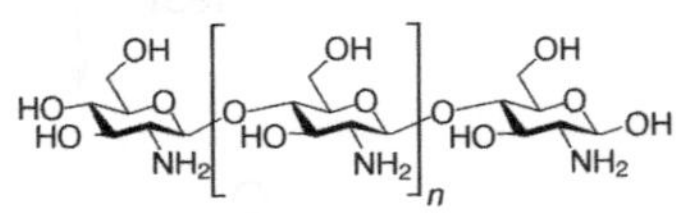
Chitin	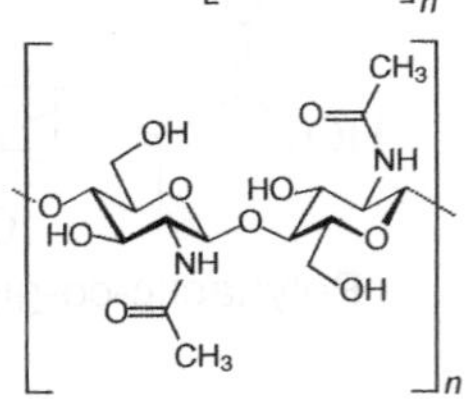

The degradation of PLGA occurs through hydrolysis of the ester linkage, which is responsible for its biodegradability:

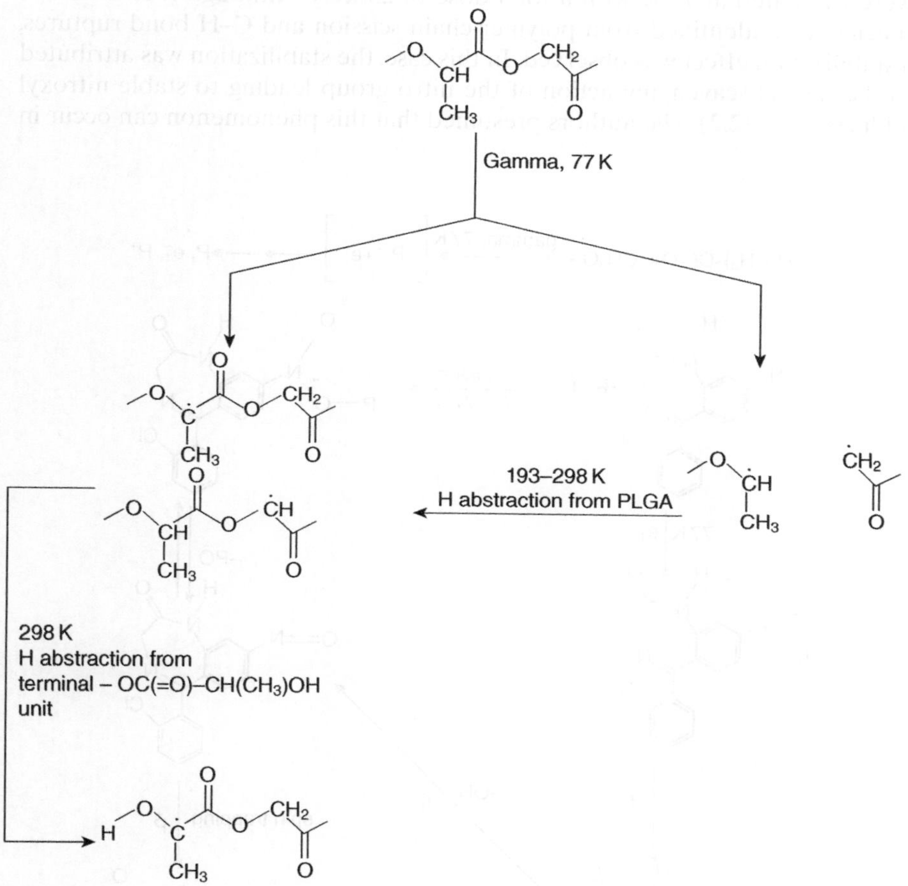

(12.6)

Polymer degradation also occurs during gamma-ray irradiation for terminal sterilization of the product, but in this case, the degradation is not owed to a hydrolysis reaction but to chain scission, where the radical cage effect plays a major role. Actually, studies have shown that the species generated arise from both chain scission and C–H bond ruptures (Equation 12.6 [28]). Moreover, different radiolysis mechanisms take place when PLGA or PLA are irradiated in the presence of a drug. A radiostabilization effect can occur in the mixed system due to the radical scavenging action of certain groups in the drug molecule. In terms of polymer stability, the radical scavenging properties of drugs may be worthwhile, but considering the entire product and depending on the drug, the formation of radiolysis species from drug decomposition can be even worse than the decomposition of the polymer itself.

Faucitano et al. [28] investigated the radiolytic behavior of the polymeric material itself in the presence of an encapsulated drug, clonazepam. Samples were irradiated at 77 K with a total dose of 25 kGy. Although four different species were identified from polymer-chain scission and C–H bond ruptures, a stabilization effect was observed. In this case, the stabilization was attributed to the radical scavenging action of the nitro group leading to stable nitroxyl adducts. (Fig. 12.2). The authors presumed that this phenomenon can occur in

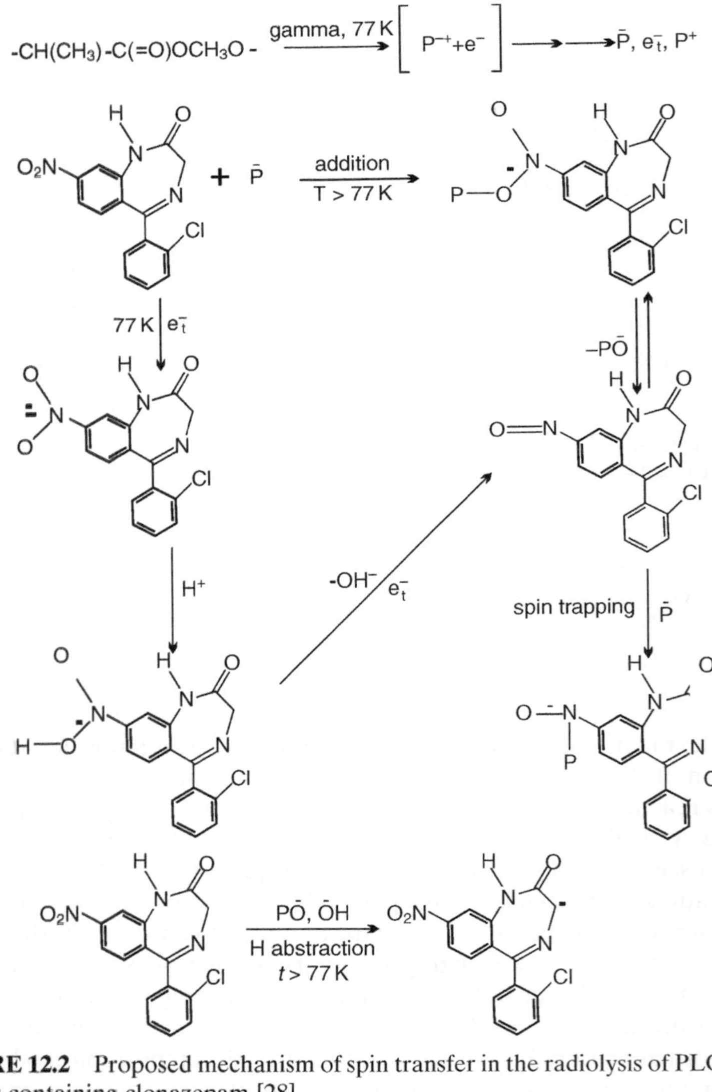

FIGURE 12.2 Proposed mechanism of spin transfer in the radiolysis of PLGA microspheres containing clonazepam [28].

pharmaceutical formulations based on components bearing nitro groups in their chemical structure.

The degradation of polymers can lead to the formation of low-molecular-weight species, which may include gases. In order to evaluate the possible release of gases from irradiated PLGA and PLA, Babanalbandi et al. [29] conducted radiation experiments (0–200 kGy) in evacuated and sealed glass ampoules containing the samples. Using gas chromatography (GC), CO, CO_2, CH_4, and C_2H_6 were identified as the major products formed for PLA and CO and CO_2 for PLGA. Figure 12.3 shows the gas yields with the radiation doses.

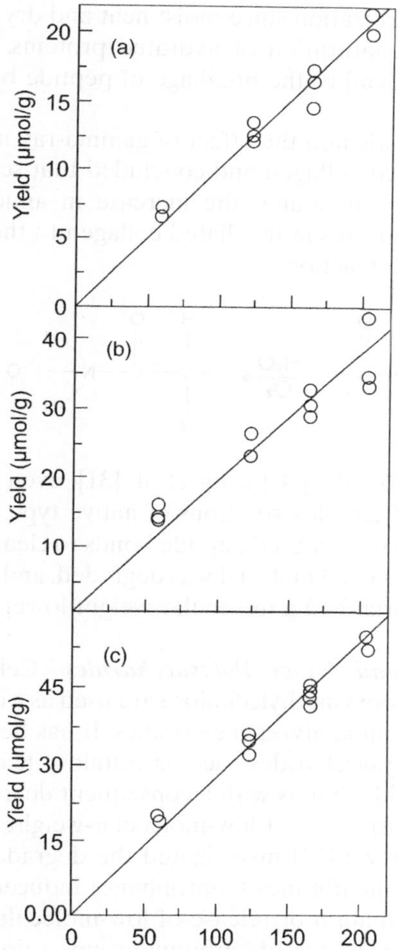

FIGURE 12.3 Plots of the gas yield for CO (a), CO_2 (b), and CH_4 (c) evolved on irradiation of L-PLA to different doses [29].

As the experiment was carried out under vacuum, the ester group (see Equation 12.6) is the only possible source for the production of CO and CO_2, so cleavage of the ester groups leading to the elimination of gases is the major process involved in the chain scission of the polymer. The amount of CH_4 released from irradiated PLA is less than 1.5% of the total, indicating that the susceptibility of the ester group toward degradation is much higher than that of the other chemical linkages in PLA.

12.1.3.2 Collagen The potential damage to collagen induced by gamma radiation is related to structural alterations and correlated decay of bioactivity. Although not ideal, radiation, along with EtO treatment, seems to be the option for collagen sterilization since moist heat and dry heat cannot be used due to the thermal denaturation of hydrated proteins. On the other hand, gamma radiation may lead to the breakage of peptide bonds and collapse of cross-linking.

Tyan et al. [30] investigated the effect of gamma-ray irradiation (from 5 to 40 kGy) on immobilized collagen and concluded that scission of the peptide linkage occurred. They attributed the increase in amide nitrogen and the detection of carbonyl groups in irradiated collagen to the breaking of –C–N– bonds according to the reaction

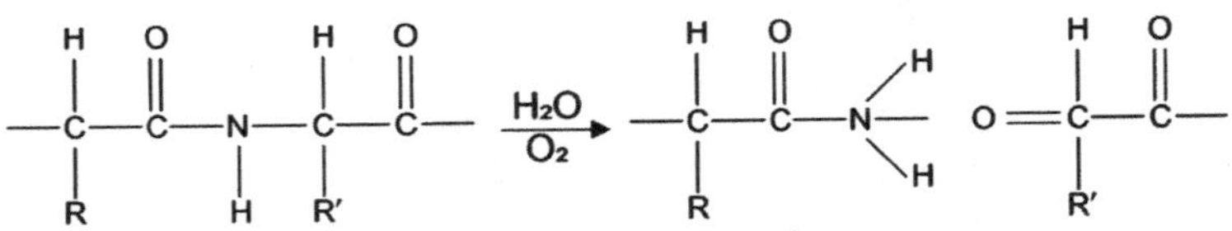

(12.7)

A similar result was found by Cheung et al. [31], even at dosages as low as 10 kGy. In all types of samples, solutions of native type II collagen as well as solid collagen, the radiation caused peptide bonds to cleave. At radiation doses of 10 kGy, 5% of the original material was degraded, and at 75 kGy, more than 40% of the total material had a molecular weight lower than an α-chain.

12.1.3.3 Cellulose and Other Polysaccharides Cellulose and cellulose derivatives such as carboxymethylcellulose are used as a carrier for bone grafts and to manufacture hemodialyzer membranes. It has been reported that ionizing radiation on polysaccharides such as cellulose leads to degradation by cleavage of the glycosidic bonds with a consequent decrease in the molecular weight and possible formation of low-molecular-weight species.

Kawano and Logarezzi [32] investigated the degradation of hemodialysis regenerated cellulose membranes (cuprophane) induced by X-ray. Although not evaluating the formation or release of low-molecular-weight compounds, they observed the appearance of carbonyl groups, scission of the glucosidic bond, and loss of OH groups in the membrane structure.

According to Kabanov [33], the decomposition of cellulose and of other polysaccharides is determined by the rupture of the pyranose cycle

TABLE 12.5 Gases and Other Compounds Formed by Decomposition of Cellulose after Gamma Irradiation [33]

H_2	Malic acid
CO	Fumaric acid
CO_2	Tartaric acid
Citric acid	Malonic acid
Succinic acid	Pyruvic acid
	Glyoxalic acid

accompanied by chain scission and the formation of CO_2, CO, and H_2 gases, as well as other compounds such as citric, succinic, and glyoxalic acids (see Table 12.5). The mechanism of their formation includes transformation of peroxide radicals, dehydration of radicals with β-elimination of water molecules, and emergence of allyl-type radicals.

It is believed that substituents play a protective role in the degradation of cellulose. In this case, polymer-chain scission is determined by the nature of the substituent and the degree of substitution. Apparently, the protective action of substituents is due to their ability to to draw the excitation energy or the primary radical ion of radiolysis to themselves. As would be expected, the presence of aromatic groups (benzyl, benzoyl, naphthoyl, etc.) in the cellulose is especially effective. This is probably what occurs in the case of chitin and chitosan. In chitin, the OH group in the C(2) position is replaced with the CH_3CONH group, which, according to Kabanov [33], substantially enhances the radiation resistance of this polymer when compared to cellulose. The same effect is observed in chitosan, where the OH group is substituted by a NH_2 group (the structure of these polymers is presented in Table 12.4).

The effect of radiation on chitosan was studied by Gryczka et al. [34], who confirmed that the higher stability of the polymer could be credited to the NH_2 group. Radiolysis of the NH_2 group led to the formation of stable nitroxyl radicals, but only when the radiation dose was over 100 kGy and the sample was irradiated in the presence of oxygen. The authors pointed out that other radiolysis products, such as hydroxylamine ethers, which might present a toxic effect, could be expected with this degradation route.

Zelinska et al. [35] also evaluated the radiolysis products of chitosan. The authors focused on gaseous products and measured the release of ammonia and hydrogen. They believed that mainly N–H, C–H, C–N, and glycoside bonds are cleaved as a result of irradiation. The first two types of scission lead to H_2 formation, via abstraction reactions ($RH + H \rightarrow H_2 + R^\circ$). The scission of C–N bonds is responsible for the formation of ammonia. The gas formation is dose dependent, as shown in Figure 12.4, and for the same radiation dose, the released amount of ammonia is much higher than the amount of hydrogen.

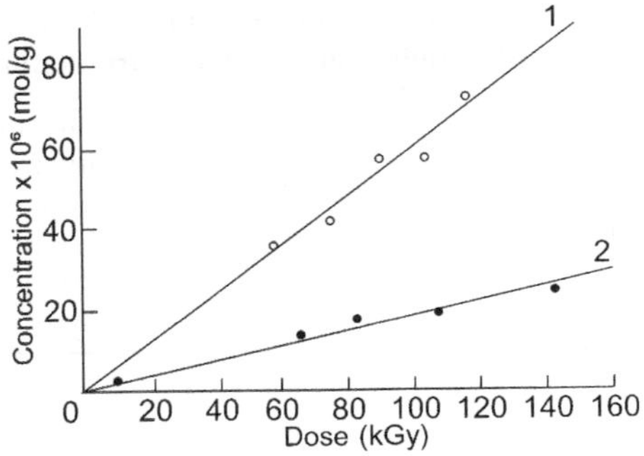

FIGURE 12.4 Dose dependence for the concentration of (1) ammonia and (2) hydrogen from an irradiated chitosan sample [35].

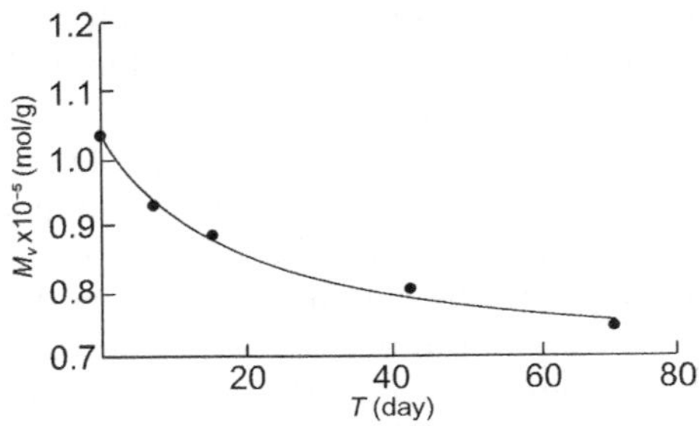

FIGURE 12.5 Time dependence for the viscosity-average molecular mass of chitosan after irradiation to a dose of 39 kGy [35].

Through the measurement of the molecular mass of the polymer 70 days after irradiation (39 kGy), the authors showed that degradation continued with subsequent formation of radicals and degradation products (Fig. 12.5).

12.1.3.4 Alginate Nagasawa et al. [36] studied the effect of gamma radiation on sodium alginate and evaluated the degradation process. Both an aqueous solution and solid sodium alginate were irradiated at doses ranging from 20 to 500 kGy, and the results showed that while the solid suffered direct oxidation, the indirect effect of water was significant in the degradation of alginate in aqueous solution. For both forms, chain scission occurred preferably, but the free radicals formed in water radiolysis accelerated this process, promoting a degradation 30 times higher when compared to the solid form, for the same radiation dose.

12.1.4 Radiolysis of Selected Nondrug Components

12.1.4.1 Residual Solvents Residual solvents, when present in the drug, may themselves suffer a radiolysis process or react with any other drug component due to the action of radiation, leading to the formation of contaminants. Barbarin et al. [37] evaluated the role of residual solvents in the formation of volatile radiolytic compounds after radiosterilization of certain drugs. After radiosterilization of three cephalosporins, several new compounds were identified in the irradiated drugs (Table 12.6). Some of them, such as carbon oxide sulfide and carbon disulfide, came from the degradation of the drug itself (cefotaxime and cefuroxime), whereas the formation of others required the presence of residual solvents, which were detected in the drug before irradiation. The results showed that only acetaldehyde directly came from the residual solvents (probably ethanol). The formation of the other compounds required the presence of both cefotaxime and residual solvents. In a following study conducted only with cefotaxime [38], among the eight products detected, only two (carbon oxide sulfide and carbon disulfide) were formed in the absence of the solvents. Figure 12.6 displays both chromatograms, from irradiated and nonirradiated cefotaxime samples.

12.1.4.2 Monosaccharides Sugars are frequently added to pharmaceutical formulations, which may undergo sterilization by ionizing radiation, leading to the formation of radiolysis products. The usual oxidation products of hexoses are glucuronic, gluconic, and arabonic acids, or products of ring scission such as formaldehyde or arabinose. The radiolysis of monosaccharides may also lead to the release of gases.

Shostenko et al. [39] evaluated the gaseous products of the radiolysis of D- and L- glucose, galactose, and mannose. They found hydrogen as the main gaseous product for all hexoses. Although other degradation products were not evaluated, electron paramagnetic resonance (EPR) spectra detected the formation of radicals, which were the same in all irradiated samples. In addition, the experiments allowed for the conclusion that the formed radicals were still present in the samples months after irradiation.

12.1.4.3 Starch Starch is one of the most used excipients in pharmaceutical formulations. It is used mainly in oral solid dosage forms as a filler, binder, or disintegrant, and being present in an irradiated drug, starch may suffer the effects of radiation. It has been shown that radiation causes fragmentation of the amylopectin leading to an increment in the amylase fraction. De Kerf et al. [40] evaluated the molecular weight distribution of cornstarch before and after irradiation (e-beam and X-ray, 50 kGy), and observed a decrease in the molecular weight of both amylase and amylopectin fractions, which was confirmed by the increase of the soluble fraction of the starch from 5.4% to 10.7%. Figure 12.7 shows the molecular weight distribution of cornstarch before and after X-ray irradiation determined by gel permeation chromatography (GPC). Although radiation has important effects on starch properties, it seems that they are related to the formation of low-molecular-weight species already present in starch.

TABLE 12.6 Volatile Impurities Found in Three Irradiated and Nonirradiated Cephalosporins [37]

Volatile Impurities Found in Nonirradiated Drugs	Cefotaxime	Cefuroxime	Cefazidime	Volatile Impurities Found in Irradiated Drugs	Cefotaxime	Cefuroxime	Cefazidime
Methanol	x	–	–	Carbon monoxide	x	x	–
Ethanol	x	x	–	Nitrogen oxide	–	x	–
Acetonitrile	x	–	–	Carbon oxide sulfide	x	x	x
Acetone	x	x	x	Carbon disulfide	x	–	–
Dichloromethane	x	–	–	Methanol	–	x	–
Isopropanol	x	–	–	Acetaldehyde	x	x	–
Acetic acid	x	–	–	Formic acid methyl ester	x	x	–
Tetrahydrofuran	x	–	–	Formic acid ethyl ester	x	x	–
Chloroform	x	x	–	Acetic acid methyl ester	x	–	–
Pyridine	–	–	x	Acetaldehyde *O*-methyl-oxime	x	x	–
Acetic acid ethyl ester	x	–	–				

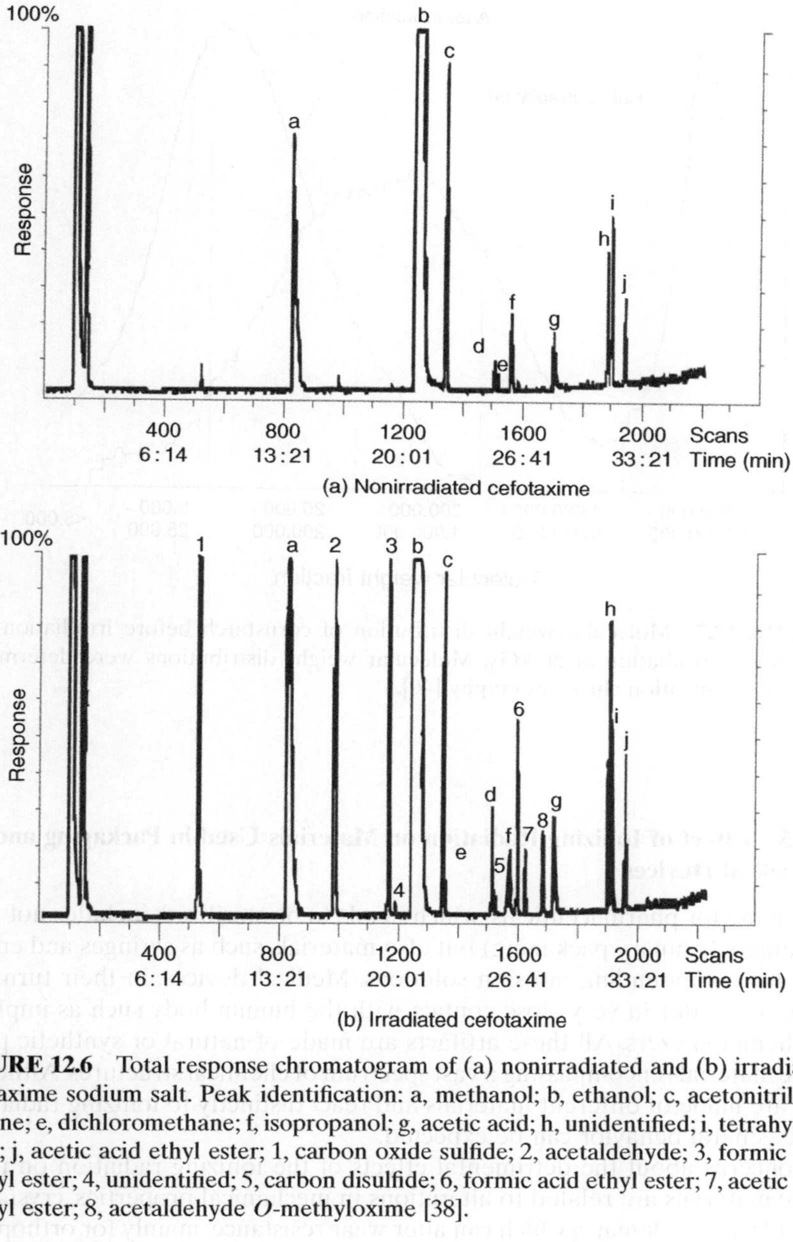

FIGURE 12.6 Total response chromatogram of (a) nonirradiated and (b) irradiated cefotaxime sodium salt. Peak identification: a, methanol; b, ethanol; c, acetonitrile; d, acetone; e, dichloromethane; f, isopropanol; g, acetic acid; h, unidentified; i, tetrahydrofuran; j, acetic acid ethyl ester; 1, carbon oxide sulfide; 2, acetaldehyde; 3, formic acid methyl ester; 4, unidentified; 5, carbon disulfide; 6, formic acid ethyl ester; 7, acetic acid methyl ester; 8, acetaldehyde *O*-methyloxime [38].

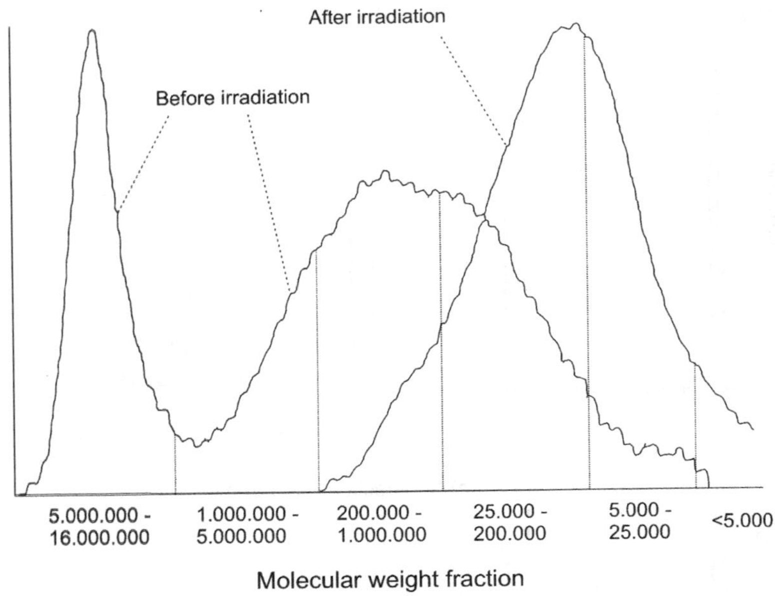

FIGURE 12.7 Molecular weight distribution of cornstarch before irradiation and after X-ray irradiation at 50 kGy. Molecular weight distributions were determined using gel permeation chromatography [40].

12.1.5 Effect of Ionizing Radiation on Materials Used in Packaging and in Medical Devices

Packaging for pharmaceuticals, which needs to be sterilized, includes not only containers (primary packaging) but also materials such as syringes and empty bags for compounding infusion solutions. Medical devices, in their turn, are those that enter in very close contact with the human body such as implants and hemodialyzers. All these artifacts are made of natural or synthetic polymeric materials encompassing a vast spectrum of chemical structures. Although they are made of different materials and react distinctly to ionizing radiation, some general behavior can be expected.

Concerns about the detrimental effects of the ionizing radiation on polymeric materials are related to alterations in mechanical properties, crystallinity, and surface damage, which can alter wear resistance, mainly for orthopedic implants. Although the degree of oxidation and the formation of free radicals have been extensively studied, there has been little study on the characterization of radiolysis products formed during sterilization processes.

The irradiation of polymeric materials, either natural or synthetic, with ionizing radiation leads to the formation of reactive intermediates, free radicals, ions, and atoms in excited states. These intermediates can undergo several reactions that result in hydrogen abstraction, rearrangements, and formation of new bonds. The degree of transformation depends on the structure of the polymer and the radiation conditions. Ionizing radiation on polymeric materials can promote either chain scission, which leads to the formation of low-molecular-weight species, or cross-linking with a consequent increase of the polymer's molecular mass. In fact, both reactions can occur simultaneously, but one is usually predominant, depending upon the specific polymer and additives involved. As a general rule, polymers containing aromatic rings are more resistant to radiation effects, while aliphatic polymers exhibit degrees of resistance depending upon their levels of instauration and substitution. The extent of radiation-induced change depends on the polymer type and the additives used to compound the polymer. Moreover, the radiation dose and environmental conditions play an important role in the formation of radiolysis products as well.

Radiation-induced cross-linking of polymers dominates under vacuum or in an inert atmosphere, while chain scission dominates in the presence of oxygen or air. Both reactions are random, generally proportional to, and dependent on dose rates and on the oxygen content of the atmosphere in which the polymer is irradiated. In an oxygen-rich atmosphere, radiation-induced degradation of both polymer base and additives is likely to occur, resulting in the formation of radiolysis products.

The stabilization of polymeric materials against high-energy radiation may be achieved through the use of antioxidants, UV, and heat stabilizers, which, in turn, also undergo degradation to some extent. Antioxidants as well as other additives used in manufacturing materials with a polymeric structure are displayed in Table A.2, Appendix. Radiolysis products, depending on the material, are much more a result of additive decomposition than originating from the polymer itself.

An important aspect regarding radiolysis products formed during gamma-ray or e-beam radiation is their later release. Due to the rigid structure, radiolysis products may be trapped inside the polymeric structure and be delivered later when the material enters into contact with an environment that favors their release. Buchalla et al. [41] evaluated radiolysis products in low-density polyethylene (LDPE) 7.5 months after irradiation (20 kGy), and the species found in the polymeric structure after this period were not present in the extracts obtained just after irradiation.

Yeom et al. [42] evaluated the oxidation profile of an ultra-high-molecular-weight polyethylene (UHMWPE) that underwent gamma radiation in air. On the polymer surface, peroxy radicals continued to react over years, promoting chain scission and generating carbonyls, mostly ketones. Peroxy radicals in the internal part of the polymer abstracted hydrogens and also formed hydroperoxides and ketones; however, due to the reduced concentration of oxygen, the formation of degradation products occurred to a lesser extent and for a shorter period of time.

Although ionizing radiation can cause damage in almost all types of polymeric materials, most of the studies dealing with this issue evaluate the extent of damage by measuring not the products of decomposition (which would be of interest for this book) but the alterations in the polymeric structure, such as chain extent, and the presence of certain groups which characterize oxidation processes. The use of vibrational spectroscopy or EPR spectroscopy enable the identification of free radicals derived from oxygen or OH° attack, or hydrogen abstraction from the polymeric chain. The generation of carbonyl groups (C=O) is one of the techniques used to show polymeric oxidant damage; however, the evaluation of released oxidation products is rarely carried out.

In the next sections, polymers used in medical devices and in primary packaging materials are discussed. They are presented together because of their similarity and because some of them are used in both areas, since the required inertness applies for both, in contrast to polymers used in drug delivery systems, which must be degradable in order to fulfill the desired delivery purpose.

Orthopedic implants and other medical devices that come into close contact with the human body are sterilized by either gas or high-energy radiation. One of the reasons is that autoclaving or dry heat sterilization may lead to intolerable dimensional changes of the material. Nonetheless, polymers are not free from the effects of ionizing radiation, particularly if the process is performed in air. Even packaging in an inert atmosphere does not warrant the total absence of oxygen in the polymer bulk; therefore, oxidation and the generation of radicals always occur. Table 12.7 lists the polymers most used to make implants and other medical devices and their standard resistance against radiation.

Primary packaging materials, conventional containers, syringes, bags, and infusion sets are usually made of typical synthetic polymers such as polyethylene (PE) and polypropylene (PP). Although they are radiation compatible, they may suffer degradation to some extent. Radiolysis products, once formed, may be leached when the container is filled, and delivered together with the fluid.

12.1.5.1 Ultra-High-Molecular-Weight Polyethylene (UHMWPE) Medical-grade UHMWPE is usually processed without any additives, which modifies the profile of radiolysis products, since one of the additives' functions is to prevent polymer oxidation.

Premnath et al. [43] published a comprehensive study on the oxidation of UHMWPE, discussing all aspects involved in the interaction of gamma radiation with this material. A list of the observed chemical reactions and species involved during and after polymer irradiation is given in Table 12.8. The authors confirmed that the physicochemical changes do not stop after irradiation but, rather, take place over a period lasting up to years due to the persistence of resonance-stabilized radicals (allyl radicals) and the low mobility of these species inside the polymer, especially in the crystalline regions. These

TABLE 12.7 Polymers Used in Implants, Medical Devices and Packaging, and Standard Resistance against Radiation

Polymer	Abbreviation	Use[a]	Radiation Stability[b]
Ultra-high-molecular-weight polyethylene	UHMWPE	Orthopedic implants	Good
Polyurethane	PU	Catheters, cardiovascular devices	Good
Silicone	–	Body implants, blood transfusion tubing, containers closures, sealing rings	Excellent
Polyamide (nylon)	PA	Nonabsorbable surgical sutures	Good
Poly(methyl methacrylate)	PMMA	Intraocular lenses	Good
Polytetrafluoroethylene	PTFE	Composite implants, nonabsorbable sutures	Poor
Poly(vinyl chloride)	PVC	Blood circuits, bags	Good
Polyethylene, high density and low density	HDPE and LDPE	Containers	Good
Polypropylene	PP	Syringes, surgical gloves	Poor
Polyethylene terephthalate	PET	Nonabsorbable surgical sutures	Good
Polystyrene	PS	Packaging	Excellent
Polysulfone	PSf	Hemodialyzer	Excellent
Ethylene vinyl acetate	EVA	Bags	Good
Multilayer materials	–	Containers	–
Cellulose	–	Hemodialyzer	Fair
Poly glycolic acid	PGA	Absorbable surgical sutures	Poor

Sources: [a]FDA; [b]Sterigenics (http://www.sterigenics.com).

long-lived allyl radicals are responsible for the long-term postirradiation oxidation in PE. The various steps involved in the oxidation of allyl radicals observed by the authors are listed in Table 12.9.

Costa et al. [44] investigated the mechanism and products originated from the oxidation process of UHMWPE induced by e-beam radiation. The study conducted in vacuum, in air, and in the presence of pure oxygen showed that, after irradiation at doses varying from 5 to 100 kGy, hydroperoxides, ketones, carboxylic acids, and alcohols were formed. Due to the way the study was conducted, the authors were able to measure the species formed after

TABLE 12.8 Chemical Reactions Involved during and after Irradiation of Polyethylene [43]

Generation of new species	Alkyl radicals Double bonds (unsaturation) Other radicals resulting from impurities/additives and gases within the polymer
Diffusion	Oxygen from the outside environment Alkyl radicals
Reactants and intermediates	Alkyl radicals Double bonds Allyl radicals Polyenyl radicals Other radicals resulting from impurities/additives and gases Oxygen from the outside environment
End products	Hydrogen gas Water CO and CO_2 Small molecular weight hydrocarbon gases Carbonyl groups—aldehydes and ketones Alcohol groups Branches and cross-links (includes -O-O- bridges) Esters and carboxyl acids Some backbone scission causes additional methyl groups

irradiation ceased. Figure 12.8 shows the increase in the hydroperoxide concentration as a function of radiation dose and storage time in air and at room temperature. For all irradiated doses, the hydroperoxide concentration continues to increase even after 15 days of storage. The authors measured not only the hydroperoxide concentration but also the concentration of other oxidation products for a sample irradiated in vacuum at 60 kGy. In Figure 12.9, it is possible to see that while the concentration of hydroperoxide and carboxylic acids stabilized after 40 days' postirradiation, the ketone concentration continued increasing even after 125 days of storage.

12.1.5.2 Polyurethane (PU) Owing to its good biocompatibility, PU is widely used in medical devices such as catheters and cardiovascular devices. Differently from polymers such as PE and PP, a large variety of raw materials have been employed to produce PUs [45]. These may contain different functional groups along with the urethane group (–NH–CO–O–), which sometimes is not even the principal group present.

PUs are known to be sensitive to sterilization processes, which can alter their physicochemical stability. Sterilization of PU, either by γ-ray or autoclaving, may lead to polymer degradation with the release of diisocyanates and their diamine degradation products: toluene diamine (TDA) and

TABLE 12.9 A Brief List of Possible Chemical Reactions Occurring during and after Irradiation of Linear Polyethylene [43]

1. Irradiation causes generation of new species	
$-CH_2-CH_2- \rightarrow -\dot{C}H-CH_2- + H^{\cdot}$	Mobile alkyl radicals
$-CH_2-CH_2- \rightarrow -CH=CH- + H_2$	Double bond
$-(CH_2-CH_2)_n- \rightarrow -(CH=CH)_n- + n\ H_2$	Conjugated double bonds
Other radicals like methyl radical or those from additives and dissolved gases	
2. Migration of mobile radicals and formation of relatively immobile allyl and polyenyl radicals	
$-\dot{C}H-CH_2- + H^{\cdot} \rightarrow -CH=CH- + H_2$	Double bond
$-(\dot{C}H-CH_2)_n- + H^{\cdot} \rightarrow -(CH=CH)_n- + n\ H_2$	Conjugated double bonds
$-\dot{C}H-CH_2- \rightarrow -CH_2-\dot{C}H-$	Intrachain migration of alkyl radicals
$-\dot{C}H-CH_2- + -CH_2-CH_2- \rightarrow -CH_2-CH_2- + -\dot{C}H-CH_2-$	Interchain migration of alkyl radicals
$-\dot{C}H-CH_2-(CH=CH)_n- \rightarrow -CH_2-\dot{C}H-(CH=CH)_n-$	Allyl and polyenyl radicals by intrachain migration
$-\dot{C}H-CH_2- + -CH_2-CH_2-(CH=CH)_n- \rightarrow -CH_2-\dot{C}H-(CH=CH)_n- + -CH_2-CH_2-$	Allyl and polyenyl radicals by interchain migration
3. Reactions among alkyl radicals and with oxygen (alkyl radicals are represented by $R^{\cdot}$)	
Note: the following reactions are a chain and until all radicals are terminated, they will continue	
$R^{\cdot} + O_2 \rightarrow ROO^{\cdot}$	Reversible step forming peroxy radicals
$ROO^{\cdot} + RH \rightarrow ROOH + R^{\cdot}$	Propagation
$ROOH \rightarrow RO^{\cdot} + HO^{\cdot}$	Hydroxyl radicals
$RO^{\cdot} + RH \rightarrow ROH + R^{\cdot}$	Propagation and formation of alcohol
$HO^{\cdot} + RH \rightarrow HOH + R^{\cdot}$	Propagation and formation of water
$R^{\cdot} + R^{\cdot} \rightarrow R-R$	Cross-linking and termination
$ROO^{\cdot} + ROO^{\cdot} \rightarrow O_2$	Cross-linking and termination
$R^{\cdot} + ROO^{\cdot} \rightarrow ROOR$	Cross-linking and termination
$-CH_2-CH(OOH)-CH_2- \rightarrow -CH_2-CHO + HO-CH_2-$	Scission and formation of aldehyde and alcohol
$-CH_2-CH(OOH)-CH_2- \rightarrow HOH + -CH_2-(C=O)-CH_2-$	Scission and formation of ketone and water
$-CH_2-CH(OO^{\cdot})-CH_2- \rightarrow CO_2 + -CH_3 + \dot{C}H_2-$	Scission and formation of CO_2
Also, other smaller radicals can also participate in termination; smaller-molecular-weight hydrocarbons and CO can also be formed	

(*Continued*)

TABLE 12.9 ***(Continued)***

4. Reactions of the allyl radicals	
$-CH_2-\dot{C}H-CH=CH-CH_2- + O_2$ (diffusion controlled) $\rightarrow -CH_2-C(OO^{\cdot})H-CH=CH-CH_2-$	
$-CH_2-C(OO^{\cdot})H-CH=CH-CH_2- \rightarrow -CH_2-C(COOH)H-\dot{C}=CH-CH_2-$	
$-CH_2-C(OO^{\cdot})H-CH=CH-CH_2- \rightarrow -CH_2-\dot{C}H-O-O-CH=CH-CH_2-$	
$-CH_2-C(COOH)H-\dot{C}=CH-CH_2- \rightarrow -CH_2-C(OH)H-(C=O)-\dot{C}H-CH_2-$	Alcohol and ketone carbonyl produced + an alkyl radical, which can participate in the chain reaction shown in the previous set of reactions
$-CH_2-\dot{C}H-O-O-CH=CH-CH_2- \rightarrow -CH_2-CHO + OHC-\dot{C}H-CH_2-$	Aldehyde carbonyl produced + chain scission + an alkyl radical produced, which can initiate a chain reaction

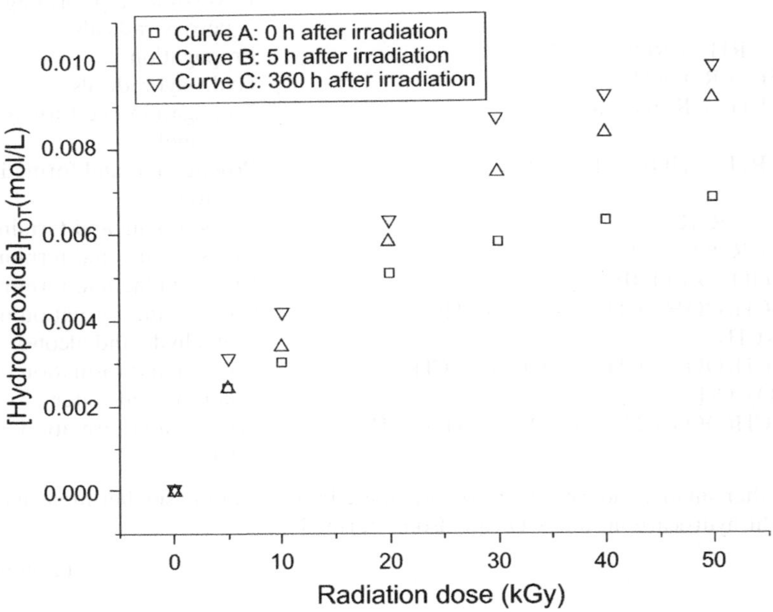

FIGURE 12.8 Total hydroperoxide concentration as a function of radiation dose and storage time in air at room temperature of a UHMWPE sample [44].

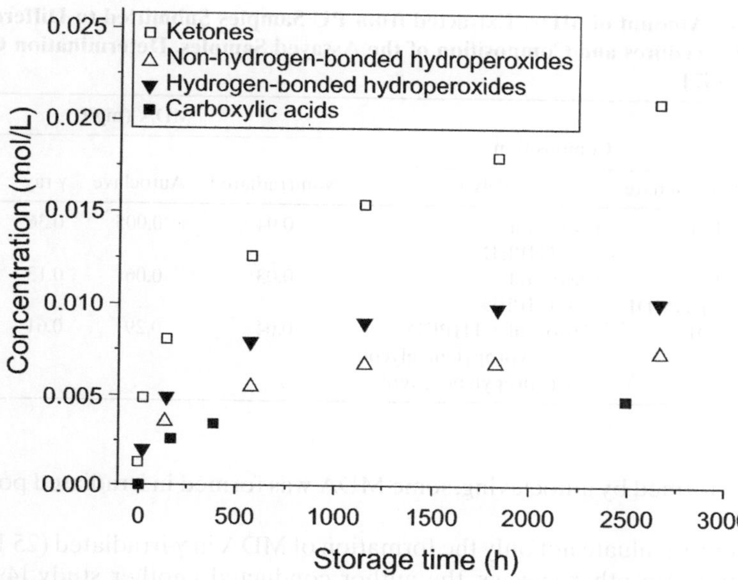

FIGURE 12.9 Oxidation products as a function of the storage time in a UHMWPE film irradiated in vacuum at 60 kGy and aged in air at room temperature [44].

4,4′-methylene dianiline (MDA). Although this release is attributed to the hydrolysis reaction, radiation may also lead to MDA formation due to chain scission. Concerns about the formation of TDA and MDA are related to their carcinogenicity and to significant noncarcinogenic health effects [46].

Hirata et al. [47] prepared different PUs to test the effect of sterilization procedures (γ-ray, EtO, and autoclaving) on the stability of polymeric materials. Samples were either synthesized in the laboratory or obtained from commercial dialyzers, where the diisocyanate components were 4,4′-methylene diphenyl-diisocyanate (MDI) and hexamethylene-diisocyanate (HMDI), and the polyols were castor oil and tetrakis (hydroxypropyl) ethylenediamine (THPED). The evaluation of MDA in methanolic extracts of sterilized and nonsterilized samples was carried out by high-performance liquid chromatography (HPLC) and the results are presented in Table 12.10. The three sterilization procedures led to MDA release and extraction. However, while a significant amount of MDA was extracted from γ-ray irradiated samples, autoclaving and EtO promoted the release of only a small amount. Moreover, in irradiated samples, the presence of polyols, other than castor oil and THPED, seemed to increase MDA formation. In a subsequent study, Shintani [48] compared the stability of two other PUs (chain-extended and non-chain-extended) concerning the formation of MDA after sterilization procedures (γ-radiation [25 kGy] and autoclaving). The author observed that while no

TABLE 12.10 Amount of MDA Extracted from PU Samples Submitted to Different Sterilization Procedures and Composition of the Assayed Samples. Determination Carried out by HPLC [47]

	Composition		MDA (mg/L)			
Sample	Diisocianate	Polyol	Nonirradiated	Autoclave	γ-ray	Ethylene Oxide
Dialyzer	MDI	Castor oil + THPED	0.04	0.005	0.36	0.05
Dialyzer	MDI + HMDI	Castor oil + THPED	0.03	0.06	0.13	0.10
Synthesized	MDI	Castor oil + THPED + polypropylene glycol + tripropylene glycol	0.04	0.29	0.61	0.02

MDA was formed by autoclaving, some MDA was formed in irradiated potting materials.

In order to evaluate not only the formation of MDA in γ-irradiated (25 kGy) samples but also other species, the author conducted another study [49]. In this case, only synthesized samples were used: chain-extended and non-chain-extended PU. Chain-extended PU was prepared by the reaction of MDI with polytetramethylene glycol (PTMG) and 1,4-butanediol (BU) as the chain extender. Non-chain-extended PU was prepared in the same way but without BU. Residual radicals were determined after 14 days and 6 months of irradiation by GPC, and as they did not significantly differ, the author concluded that there was little degradation after irradiation. Although the author did not present the amount of MDA released by each polymer sample, he commented that the more flexible the material (non-chain-extended), the greater the MDA formation. The author speculated that this effect was due to an increase in the polymer cross-linking by γ-ray radiation, which restricted MDA extraction from the internal part of the material. The GPC of methanolic extracts of irradiated samples also showed the extractable oligomers from PU. Figure 12.10 shows the chromatogram of the extraction from a chain-extended PU. The numbers displayed on the peaks are the molecular mass of the respective oligomers. In Table 12.11, the relative heights of the peaks in chromatograms obtained from different extracts are listed. The results showed that the extraction of oligomers increased with the radiation dose, and while the oligomer with mass 5400 was predominant in extracts from non-chain-extended PU, the oligomer with mass 9500 was predominant from chain-extended PU. The author believes that the peak with mass 9500 was the highest because it contained MDI. Moreover, higher amounts of oligomers were extracted from the chain-extended PU than from the non-chain-extended PU for the same radiation dose.

The degradation of a commercial PU (Tecoflex®) (structure showed in Fig. 12.11) containing practically the same constituents as the PU made for the

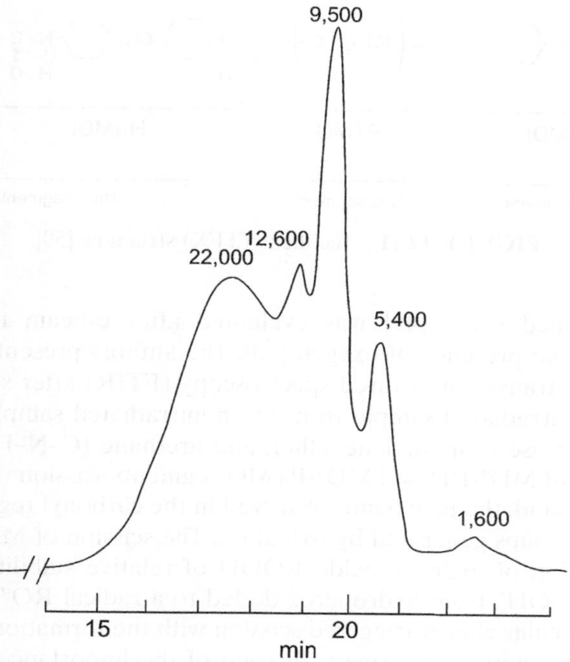

FIGURE 12.10 Gel permeation chromatography chromatogram of methanol extract from chain-extended thermoplastic PU detected by UV (290 nm) [49].

TABLE 12.11 Relative Peak Height of Gel Permeation Chromatograms of Methanol Extracts from γ-Ray Irradiated PU Collected at 290 nm [49]

Irradiation dose (kGy)	Molecular Weight 22,000	12,800	9500	5400	1600
Non-chain-extended PU					
0	17.4	6.6	32.8	100.0	4.6
25	20.8	5.7	23.6	100.0	4.7
50	41.2	7.9	29.4	100.0	4.4
75	62.5	5.0	36.3	100.0	17.5
100	112.6	11.5	43.7	100.0	4.6
Chain-extended PU					
0	72.1	54.4	205.9	100.0	20.6
25	125.0	35.7	207.1	100.0	10.7
50	324.1	41.4	248.3	100.0	12.1
75	348.0	52.0	232.0	100.0	7.2
100	465.9	63.6	281.8	100.0	13.6

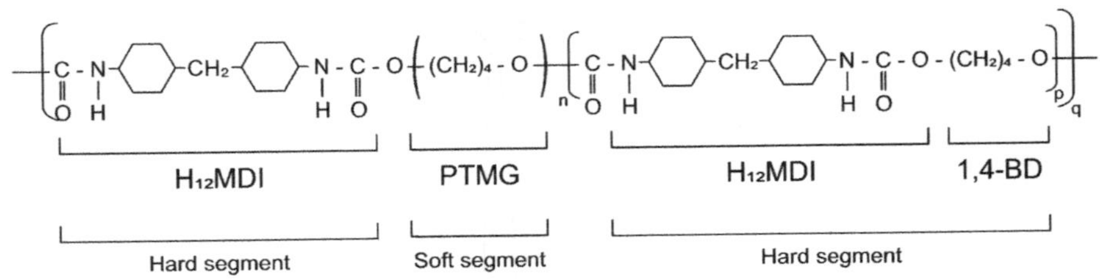

FIGURE 12.11 Tecoflex® (TFX) structure [50].

above mentioned study [49] was evaluated after e-beam radiation (25–1000 kGy) in the presence of oxygen [50]. The authors presented the results of the Fourier transform infrared spectroscopy (FTIR) after subtracting the spectra of the irradiated sample from the nonirradiated sample (Fig. 12.12). The large decrease of methylene, ether, and urethane (C–N–H bond) bands are indicative of MDI-BD and MDI-PTMG segments scission (see Fig. 12.11). On the other hand, the new bands observed in the carbonyl region have been attributed to groups generated by oxidation. The scission of MDI-PTMG led to the formation of hydroperoxide ROOH of relative stability (3388 cm^{-1}). The release of OH° from hydroperoxide led to a radical RO°, which, acting on macromolecular chains, triggered scission with the formation of carboxylic acids, alcohols, esters, and formats. In spite of the importance of MDA, the authors did not comment on its formation.

12.1.5.3 Silicone Upon radiation, silicones primarily undergo cross-linking with small amounts of degradation products as a result of chain scission. According to Palsule et al. [51], the degradation products are cyclic siloxanes of low molecular weight in extremely small amounts, trapped inside the matrix, and presenting no toxic effects. On the other hand, cross-linking results in the evolution of gases. The total gas yield should correspond to the extent of cross-linking occuring in the polymer with the formation of insaturation in the main chain. Mass spectrometric analysis of gases evolved from polydimethylsiloxane revealed these to be hydrogen, methane, and ethane [52]. Based on these occurrences, it is likely that scission occurred at the C–Si or C–H bonds, giving rise to methyl radicals.

Rogero et al. [53] also investigated the influence of ionizing radiation on silicone. The authors did not evaluate degradation products of irradiated samples, but they carried out a cytotoxicity test (ISO 10993-5) with extracts of irradiated samples. Silicone samples irradiated at 50 kGy presented the same toxicity behavior as the negative control; that is, the extracted products, if any, presented no cytotoxicity.

12.1.5.4 Polyamide (PA) Irradiated polyamide films show few compositional changes, mainly if they are aromatic polyamides, due to the stabilizing effect of the benzene ring. Nevertheless, aliphatic hydrocarbons (C9, C10)

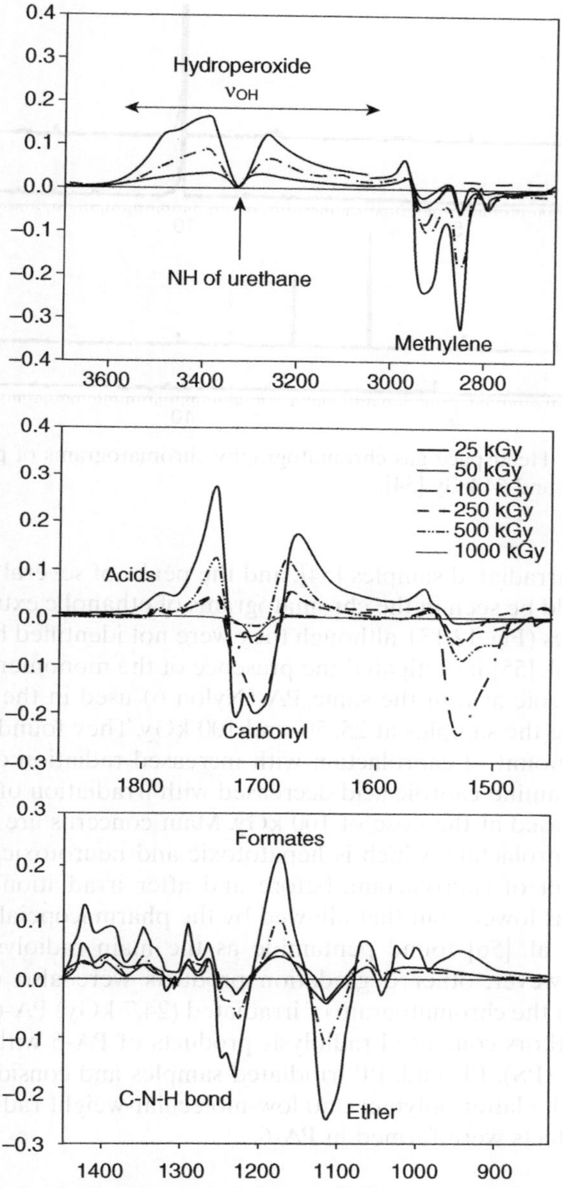

FIGURE 12.12 FTIR subtraction spectra of irradiated TFX as a function of the absorbed dose. Spectra are subtracted from nonirradiated film spectrum (positive bands correspond to new chemical groups, while negative bands correspond to disappearing ones) [50].

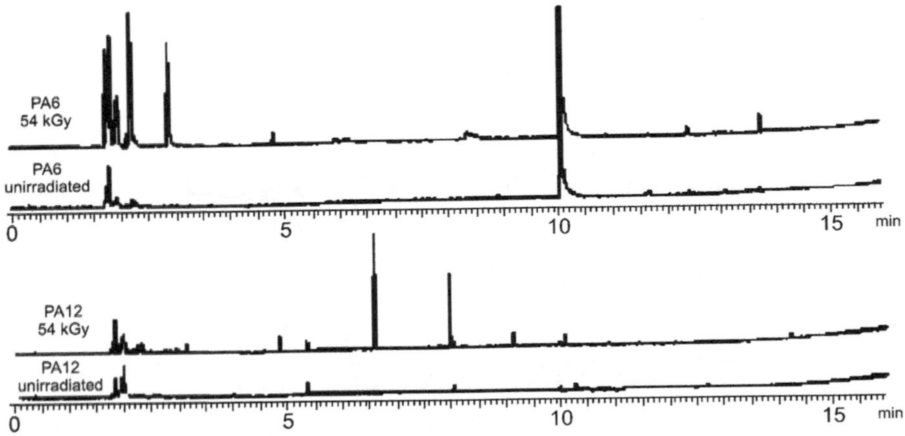

FIGURE 12.13 Headspace gas chromatography chromatograms of polyamide films irradiated with 0 and 54 kGy [54].

were found in irradiated samples [54], and the peaks of several other volatile substances could be seen in the chromatograms of ethanolic extracts of irradiated PA samples (Fig. 12.13), although they were not identified by the authors.

Yagoubi et al. [55] investigated the presence of the monomers caprolactam and amino caproic acid in the same PA (Nylon 6) used in the former study, after irradiating the samples at 25, 50, and 100 kGy. They found a decrease in the residual amount of caprolactam with increased radiation doses, whereas the amount of amino caproic acid decreased with irradiation of up to 50 kGy and then increased at the dose of 100 kGy. Main concerns are related to the presence of caprolactam, which is hepatotoxic and neurotoxic; however, the residual amount of caprolactam, before and after irradiation, according to the authors, was lower than that allowed by the pharmacopeial compendia.

Buchalla et al. [56] found pentamide as the main radiolysis product of polyamide; however, other degradation products were also observed in a selected part of the chromatogram of irradiated (24.7 kGy) PA-6 samples (Fig. 12.14). The authors compared radiolysis products of PA-6 with the products of polystyrene (PS), PE and, PP irradiated samples and considered that, differently from the latter polymers, no low-molecular-weight radiation-induced oxidation products were formed in PA-6.

12.1.5.5 Poly(methyl methacrylate) (PMMA) Due to its transparency and good biocompatibility, PMMA has been used in the manufacture of intraocular lenses. PMMA must be of high purity, but additives to prevent UV damage to the retina are usually added. Hydroxybenxophenone and hydroxybenzotriazole are typical UV radiation absorbers in PMMA lenses. As for all other polymers, the radiation may lead to either scission or cross-linking, and, in the case of PMMA, due to the presence of lateral side chains, scission occurs preferentially. Chain scission may be responsible for the formation of low-

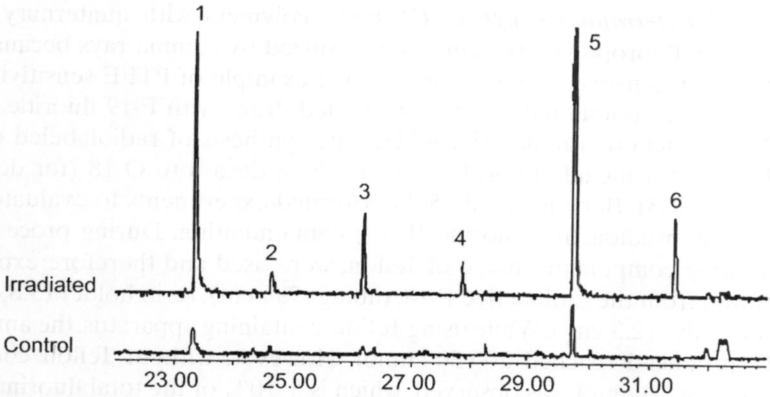

FIGURE 12.14 Polyamide-6, irradiated (24.7 kGy, top) versus nonirradiated control (bottom); detail (22–33 minutes) of fingerprints presented as extracted ion chromatograms for *m/z* (44 + 45 + 59). (1) Formamide, (2) acetamide, (3) propanamide, (4) butanamide, (5) pentanamide, (6) hexanamide. Experimental conditions: one pellet, 7 weeks' postirradiation [56].

TABLE 12.12 Data from the Subtraction of the FTIR Spectra of Gamma-Irradiated PMMA (25 kGy) and Nonirradiated PMMA [57]

Wave Number (cm^{-1})	Strength	Inference
3775	Weak positive	Gain of COO–H (hydroperoxide)
3583	Medium positive	Gain of free O–H
3258	Weak positive, broad	Gain of O–H, hydrogen bonded
1616	Strong negative	Loss of C=O
1468	Weak positive	Gain of C–OOH (hydroperoxide)
1277	Medium negative	Loss of C–O
1196	Medium negative	Loss of O–CH_3
898	Weak positive	Gain of CO–OH (hydroperoxide)

molecular-weight compounds, which can be released from the polymeric material when in contact with human body fluids.

Studies specifically on the species formed by scission of PMMA are not common; however, Pratt et al. [57] employed *size-exclusion chromatography* (SEC) and infrared (IR) spectroscopy to evaluate the effect of ionizing radiation on intraocular lenses made of PMMA (Vistacryl CQUV) and containing 2-[3-(2H-benzotriazol-2-yl)-4-hydroxyphenyl]ethyl methacrylate as a UV absorber. Gamma radiation (25 kGy) caused changes in both the polymer and the UV absorber, since the molecular weight was significantly reduced, and there was evidence of decarboxylation. Subtraction of IR spectra of irradiated and nonirradiated samples showed the gain or loss of several radicals. Table 12.12 presents the results of this estimation.

12.1.5.6 Polytetrafluoroethylene (PTFE) Polymers with quaternary carbons, such as fluoropolymers, cannot be sterilized by gamma rays because of their radiation sensitivity [58]. An interesting example of PTFE sensitivity to radiation is the contamination of F-18-labeled drugs with F-19 fluoride ions. As F-18 production is noncarrier added, the synthesis of radiolabeled compounds does not include fluoride ions, and F-18 decays to O-18 (for details, see Section 5.1.3), Berridge et al. [59] performed experiments to evaluate the sources of nonradioactive fluorine (F-19) contamination. During processing, the following components, made of Teflon, were used and therefore exposed to radiation from the radioactive F-18: tubing (790 cm), resin holder (3.8 cm^2), and slider valve (2.3 cm^2). While using Teflon-containing apparatus, the amount of F-19 produced reached 200-500 nmol. However, without Teflon components, only ca. 20 nmol was observed, which is 5–10% of the total fluorine and corresponds to the usual contamination from synthesis reagents. The results enabled the authors to state that Teflon radiolysis is the major source of F-19 in F-18-labeled drugs.

12.1.5.7 Polyvinyl Chloride (PVC) Flexible PVC accounts for a significant percentage of all plastic materials used in medical applications. Despite being a material of choice, PVC presents a significant drawback concerning its thermal stability. PVC easily undergoes thermal degradation, and its main reaction is known as dehydrochlorination, which leads to disruption of the C–Cl bonds in the polymeric chains, resulting in the evolution of hydrochloric acid (HCl) [60].

Manfredini et al. [61] investigated the thermal stability and the chemical durability upon ionizing radiation sterilization (25 kGy) of PVC-based blood circuits for hemodialysis. Degradation was monitored by measuring the amount of HCl evolved in a simulated hemodialysis procedure. Pure water recirculated in the sampled lines for 4 hours at a constant flow rate of 350 mL/min. The amount of HCl evolved as a consequence of dehydrochlorination was determined by ion chromatography. The highest measured amount of chlorine ions was 320 μg/L, which is within acceptable values (BP), probably due to the presence of stabilizers added to absorb the gas (see Table A.3 in the Appendix for these stabilizers).

Besides HCl, other decomposition products have been evaluated in irradiated PVC samples. Volatile radiolysis products originating from radiation sterilization (22.5 kGy) of PVC (Hyvin VQ 130 clear 116, organotin stabilized) were determined by gas chromatography–mass spectrometry (GC-MS) (Fig. 12.15). The four major products identified were octane, 1-octene, acetic acid 2-ethylexylester (peak 18), and 1-octanol (peak 13). The authors supposed that these products resulted not from the degradation of the polymer itself but from the stabilizers because none of the identified volatiles contained chlorine. However, after the elimination of HCl, PVC might also yield chlorine-free products.

Demertzis et al. [62] evaluated compositional changes in the solvent extract from PVC sheets (commercially available and complying with requirements

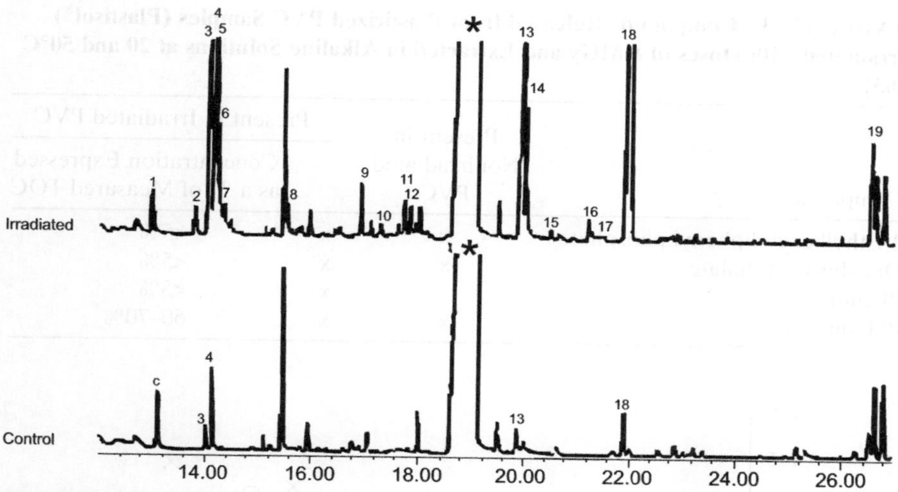

FIGURE 12.15 Radiolysis products of PVC, irradiated (22.5 kGy, top) versus nonirradiated control (bottom): (1) heptane, (c) unidentified, (2) 3-methylheptane, (3) 1-octene or isomer(s), (4) octane, (5) octane-isomer (?), (6) 2-octene, (7) 4-octene, (8) 3-heptanone, (9) 2-ethylhexanal, (10) 4-octanone or isomer, (11) 6-methyl-2-heptanone, (12) benzene derivative, (*) 2-ethyl-1-hexanol, (13) 1-octanol, (14) unidentified, (15) undecane, (16) pentylcyclopropane (?), (17) 2-ethylhexanoic acid, (18) acetic acid 2-ethylhexylester, (19) unidentified. Experimental conditions: three pellets, 2 months' postirradiation [56].

for coming into contact with foodstuff) after irradiation (44 kGy). In the chromatograms obtained by GC, 7 new peaks appeared, 10 peaks diminished or disappeared, and 5 peaks presented an increase, when compared to the control. The compounds related to the new peaks are presented in Table A.4, Appendix (the other compounds, not generated by radiation, are listed in Table A.3 in the Appendix, which displays compounds that are leachable and extractable from polymeric materials).

Colombani et al. [63] irradiated plasticized PVC samples (Plastisol®) with doses of 8 MGy and extracted the radiolysis products in alkaline solutions at 20 and 50°C. As the sample comprised only 27% pure PVC, the remainder being organic and inorganic additives, mainly di-isononyl phthalate, di-*n*-butyl phthalate, and 2-ethylhexyl diphenyl phosphate, the quantity of released compounds (Table 12.13) was much more related to the additives than to the PVC itself. Phthalic ions, formed by hydrolysis of phthalic esters, were leached in higher extent. Moreover, as the quantity of leachables decreases with the radiation dose (evaluated as total organic carbon [TOC] in Fig. 12.16), the authors considered that when plasticized PVC is irradiated, grafting reactions occurs, implying that radiolysis products are not leachable anymore.

TABLE 12.13 Compounds Released from Plasticized PVC Samples (Plastisol®) Irradiated with Doses of 8 MGy and Extracted in Alkaline Solutions at 20 and 50°C [63]

Compounds	Present in Nonirradiated PVC	Present in Irradiated PVC	
			Concentration Expressed as a % of Measured TOC
2-Ethylhexyl diphenyl phosphate	x	x	<5%
Di-*n*-butyl phthalate	x	x	<5%
Phenol		x	<5%
Phthalic ions	x	x	60–70%

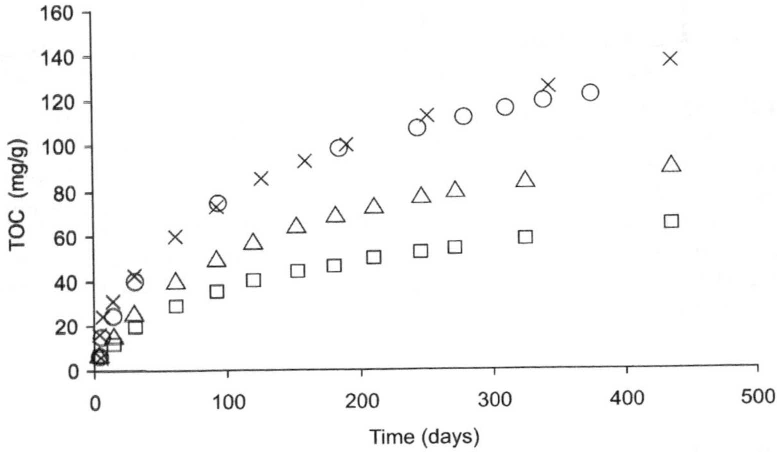

FIGURE 12.16 TOC measurements of unirradiated and irradiated plasticized PVC leached at 50°C. Leaching of unirradiated PVC (×), leaching of PVC irradiated to 2 MGy (○), leaching of PVC irradiated to 18 MGy (△), leaching of PVC irradiated to 26 MGy (□) [63].

The radiolysis products from irradiated PVC differ to a certain degree among the studies presented here probably because they have different origins and compositions. While a small amount of radiolysis products, mostly phthalic ions, was obtained from samples with greater amounts of additives [63], no formation of phthalic ions was observed in the other studies from PVC samples, although several other compounds were formed (see Table A.4, Appendix).

12.1.5.8 Polyethylene (PE) Buchalla et al. [41] analyzed irradiated (23.6 kGy) medical-grade high-density polyethylene (HDPE) (Lupolen 5031 L, BASF) by GC-MS. Major differences in the chromatograms of irradiated and nonirradiated samples were observed in the first 25 minutes, where several peaks, corresponding to more volatile products, were observed (Fig. 12.17).

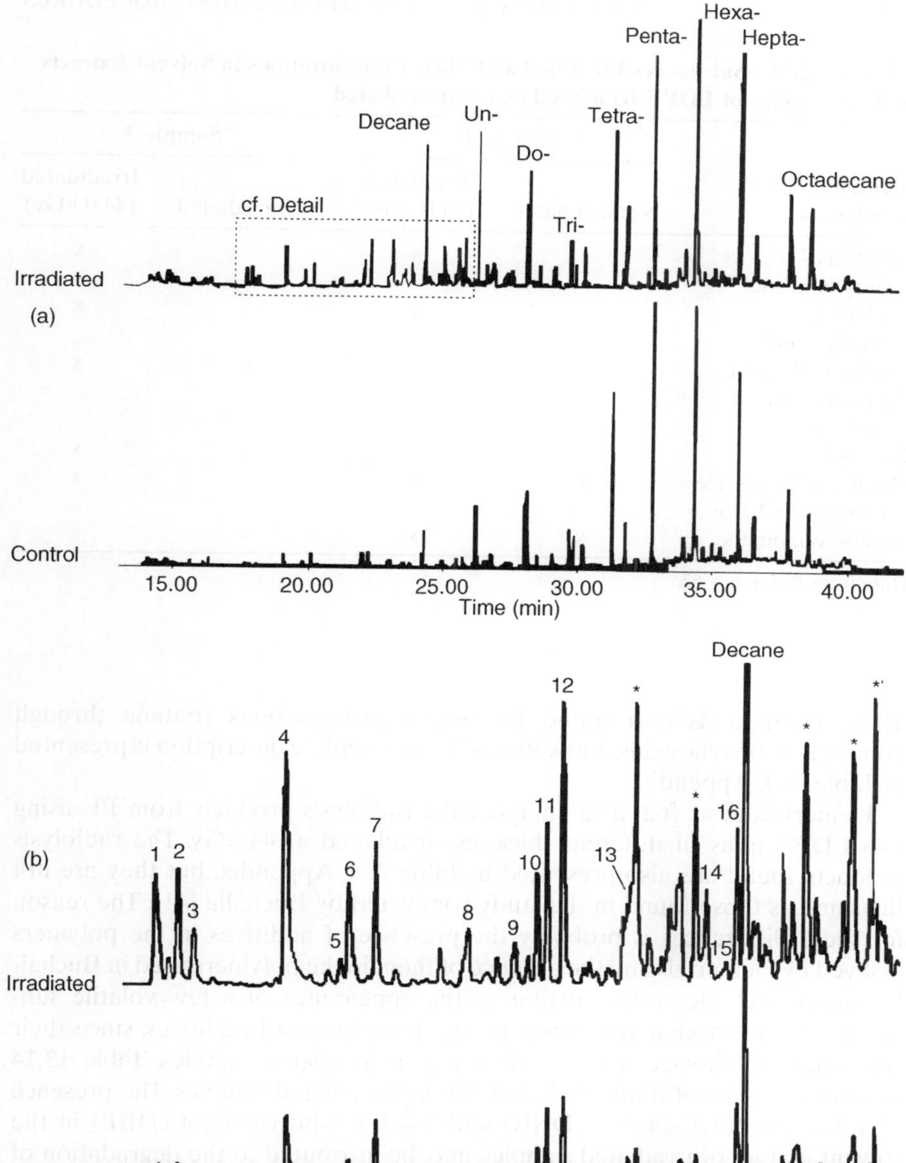

FIGURE 12.17 (a) Fingerprint chromatogram of a commercial LDPE film 7.5 months after irradiation with 20 kGy—irradiated (top) versus nonirradiated control (bottom). The main radiolysis products are detailed in (b). (b) Detail (17–26 minutes). Peak identification: (1) butanal, (2) 2-butanone, (3) heptane, (4) acetic acid, (5) 2-pentanone, (6) pentanal, (7) octane, (8) propanoic acid, (9) 3-hexanone (coelut.?), (10) 2-hexanone (11) hexanal, (12) nonane, (13) butanoic acid, (14) 3-heptanone, (15) 2-heptanone, (16) heptanal (coelut.). Peaks marked with an asterisk (*) are alkylbenzenes, a contamination possibly caused by printing inks on the film. Experimental conditions: sample: $3 \times (3 \times 100$ mm), 9 cm^2 of film [41].

TABLE 12.14 Substances Identified and Their Concentrations in Solvent Extracts of Two Samples of LDPE Irradiated and Nonirradiated

	Sample 1		Sample 2	
Identified Substances	Nonirradiated	Irradiated (44.0 kGy)	Nonirradiated	Irradidated (44.0 kGy)
1,3-Di-*tert*-butylbenzene		x		x
2,4-Di-*tert*-butylphenol		x		x
Erucic acid amide			x	x
Butanoic acid vinylester		x		
Irgafos 168			x	x
Oxidized Irgafos 168 or Irganox 1076	x	x	x	x
Several oligomers	x	x	x	x

Data from Reference 62.

These compounds correspond to several hydrocarbons (butane through nonane), carboxylic acids, and ketones. Their complete description is presented in Table A.4, Appendix.

Demertzis et al. [62] also analyzed the radiolysis products from PE using two LDPE films of different thickness irradiated at 44 kGy. The radiolysis products found are also presented in Table A.4, Appendix, but they are not the same as those found in the study conducted by Buchalla [41]. The reason for these differences is probably the presence of additives in the polymers assayed by Demertzis and the absence of them in the polymers used in Buchalla's study [41]. Demertzis attributed the appearance of a few volatile substances in the irradiated samples to the degradation of additives, since their concentration showed a strong decrease in irradiated samples. Table 12.14 shows the results of both irradiated and nonirradiated samples. The presence of 1,3-di-*tert*-butylbenzene (DBB) and 2,4-di-*tert*-butylphenol (DBP) in the solvent extract of irradiated samples may be attributed to the degradation of the antioxidants Irgafos 168 and Irganox 1076.

Stoffers et al. [54] specifically evaluated changes in the initial concentration of the stabilizers Irgafos 168 and Irganox 1076 in irradiated LDPE and HDPE. Samples were irradiated at 29 and 54 kGy, and it was observed that the higher the radiation dose, the lower the remaining stabilizer concentration. They identified the same degradation products reported by Demertzis et al. [62]: DBB and DBP; however, while the concentration of DBB increased with increasing radiation doses, DBP seemed to undergo further degradation as its concentration diminished.

The outcome of these two additives was also studied by Yagubi, but in PP syringes [64]. The experiment showed, however, that, due to the oxidizing effect of radiation, the phosphite group present in these stabilizers was converted into phosphate (see section below).

12.1.5.9 Polypropylene (PP) Products made of PP for the medical industry are generally single-use products requiring sterilization before use, and both gamma ray and e-beam may be used for this purpose. PP presents lower radiation resistance in comparison with PE due to its branched structure. In fact, this is evidenced in the great amount of branched molecular structures extracted from irradiated PP [62]. The compounds found in the extracts of PP samples (films) are presented in Table A.4 in the Appendix. The samples were γ-irradiated at 44 kGy, and extraction was carried out in CH_2Cl_2. The authors presented the results of irradiated and nonirradiated samples. Due to the way this chapter was organized, the compounds found in irradiated samples are shown in Table A.4, while those found in nonirradiated samples are in Table A.3 in the Appendix.

Figure 12.18 shows chromatograms of irradiated and nonirradiated medical-grade PP film where it is possible to see that the irradiated sample produced a much greater number of peaks. Considering that the measurement was

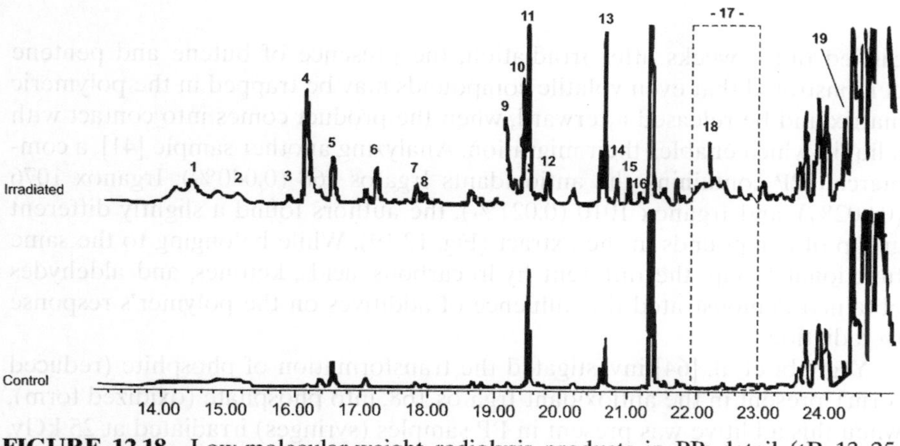

FIGURE 12.18 Low-molecular-weight radiolysis products in PP; detail (tR 13–25 minutes) of fingerprints, irradiated (24.4 kGy, top) versus nonirradiated control (bottom): (1) butene/isomers (?), (2) pentane (?), (3) 2-methylpentane, (4) 2-propanone, (5) 2-methyl-2-propanol, (6) 2-methyl-2-propenal, (7) unidentified, (8) 2-butanone, (9) acetic acid, (10) 3-methyl-2-cyclopenten-1-one, (11) 4-methylheptane, (12) 2-pentanone, (13) 2,4-dimethylheptane, (14) 2-hexanone, (15) 2-methyl-2-hepten-4-one (?), (16) propanoic acid, (17) 2,4-pentanedione (peak shape ?), (18) 4-methyl-3-penten-2-one (?), (19) 4-hydroxy-4-methyl-2-pentanone (?). Experimental conditions: one pellet, 6 weeks' postirradiation [56].

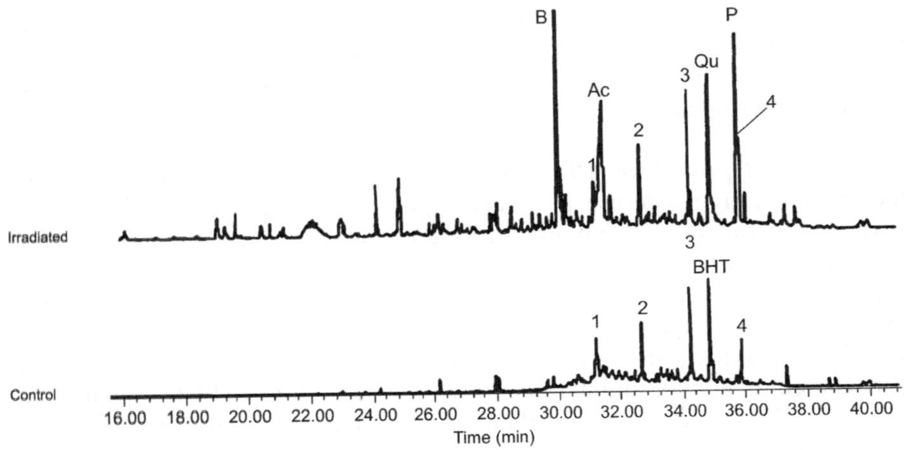

FIGURE 12.19 Fingerprint chromatograms of PP 6 days after irradiation with 26.6 kGy—irradiated (top) versus nonirradiated control (bottom) showing the three main radiolysis products (i.e., degradation products of phenol antioxidants): B, 1,3-bis-(1,1-dimethylethyl)-benzene; Qu, 2,6-bis-(1,1-dimethylethyl)-2,5-cyclohexadiene-1,4-dione [quinone]; P, 2,4-bis-(1,1-dimethylethyl)-phenol. Other peaks: BHT, butylhydroxytoluene; Ac, benzoic acid (contaminant?); 1, 2, 3, and 4, tetra-, penta-, hexa-, and heptadecane [41].

carried out 6 weeks after irradiation, the presence of butene and pentene demonstrated that even volatile compounds may be trapped in the polymeric matrix and be released afterward, when the product comes into contact with a liquid which enables their migration. Analyzing another sample [41], a commercial PP containing the antioxidants Irgafos 168 (0.040%), Irganox 1076 (0.002%), and Irganox 1010 (0.021%), the authors found a slightly different group of compounds in the extract (Fig. 12.19). While belonging to the same functional group, the different hydrocarbons, acids, ketones, and aldehydes obtained demonstrated the influence of additives on the polymer's response to radiation.

Yagoubi et al. [64] investigated the transformation of phosphite (reduced form), present in the antioxidant Irgafos 168, into phosphate (oxidized form), when this additive was present in PP samples (syringes) irradiated at 25 kGy, either by γ-ray or e-beam. In the chromatograms (Fig. 12.20) of the extracts, both compounds were identified: the reduced and the oxidized forms of the antioxidant Irgafos. The author attributed this reaction to oxygen and especially to the formation of hydroperoxide (ROOH) produced by the radiation in the bulk of the polymer, according to the reaction presented in Figure 12.21 [65]. It is interesting to note that high concentrations of both species were present in the nonirradiated sample. Although the peak of the phosphate compound decreased in both irradiated samples, there was a considerable

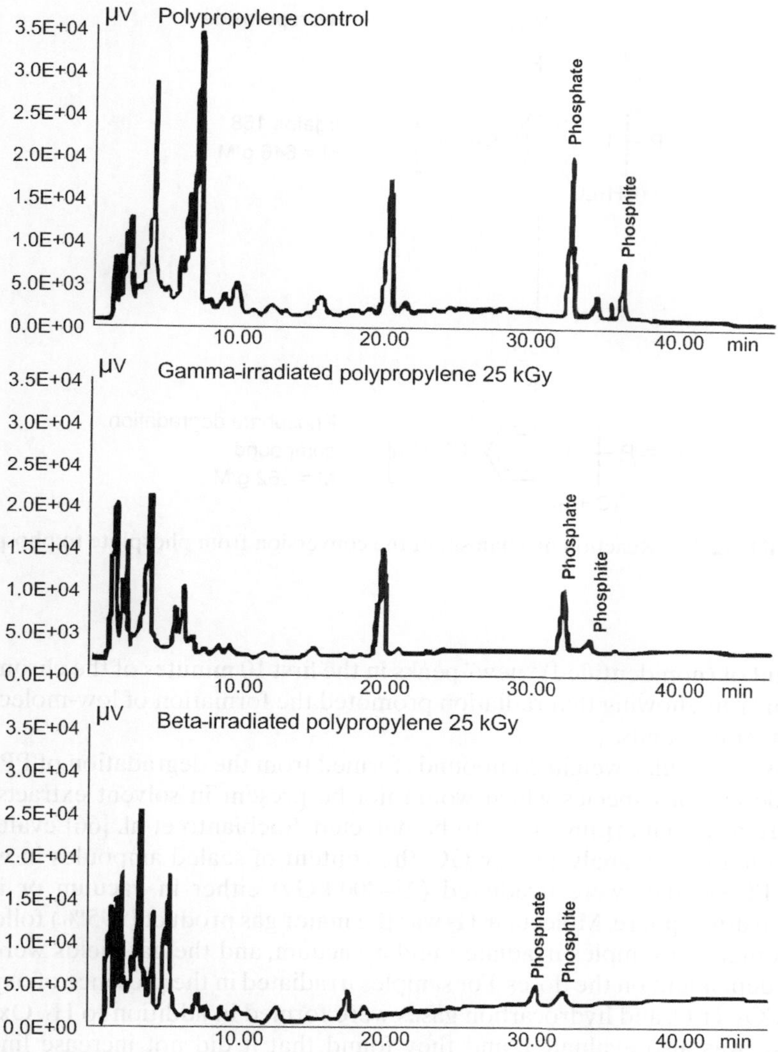

FIGURE 12.20 Chromatogram of extraction solutions obtained from PP control and γ-ray and e-beam (EB)-treated PP, at 25 kGy. Analytical conditions: Lichrospher column 100 RP-8 Select B, 250 × 4.6 mm ID, 5 μm (35°C), gradient elution varied during 35 minutes' interval. Phase A: *tetrahydrofuran* (THF), water, acetonitrile, methanol, 40:40:10:10, (v/v/v/v); phase B: THF, water, acetonitrile, 67.0:16.5:16.5, (v/v/v). Flow rate was set at 1 mL/min, λ = 280 nm [64].

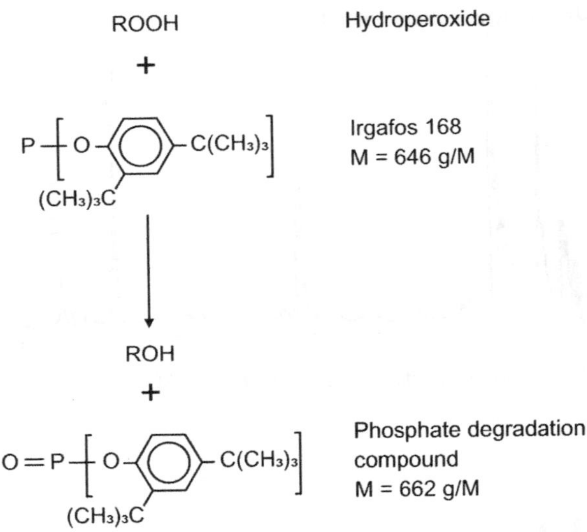

FIGURE 12.21 Reaction mechanism of the conversion from phosphite to phosphate [65].

amount of (nonidentified) "new" peaks in the first 10 minutes of the chromatographic run, showing that radiation promoted the formation of low-molecular-weight compounds.

Low-molecular-weight compounds formed from the degradation of PP may include gaseous species which would not be present in solvent extracts, but require a special arrangement to be detected. Soebianto et al. [66] evaluated gas formation by analyzing, by GC, the content of sealed ampoules in which pure PP samples were irradiated (25–300 kGy) either in vacuum or in an oxygen atmosphere. Molecular H_2 was the major gas product (>95%) followed by methane in samples irradiated under vacuum, and the gas yields were linearly dependent on the doses. For samples irradiated in the presence of oxygen, CO, CO_2, H_2O, and hydrocarbon gases were formed in addition to H_2. Oxygen uptake was also evaluated, and they found that it did not increase linearly with the radiation dose. The material balance showed that only a minor fraction (<1%) of the consumed O_2 was involved in the formation of gas products as CO, CO_2, or H_2O. It was rather converted into condensed products such as hydroperoxides, alcohols, and various kinds of carbonyl compounds. The authors also investigated the influence of additives on the degradation products of PP in the presence and in the absence of oxygen. Four different additives, Irganox 1010, butylhydroxytoluene (BHT), antirad, and Octahydrophenanthrene (OP), were separately inserted into the polymer in concentrations of 0.5 and 1.0% (m/m). In the presence of additives, O_2 uptake as well as gas evolution significantly decreased. As can be seen in Figure 12.22, the uptake was the lowest in the presence of BHT, although for all additives,

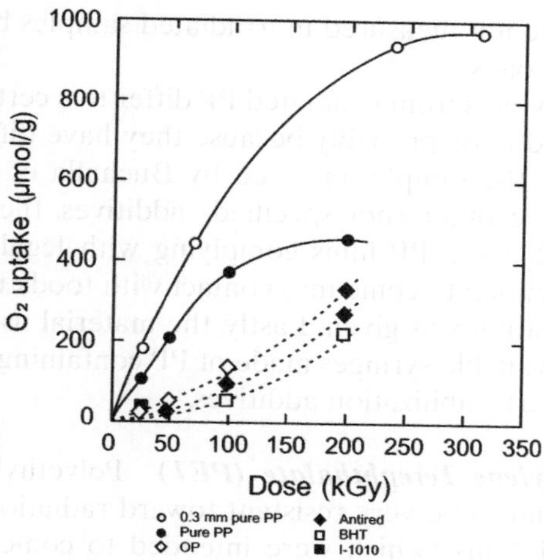

FIGURE 12.22 The O_2 uptake as a function of dose for the PP samples containing 1 wt % additives irradiated with 1 kGy/h. Initial O_2 pressure was 300 Torr (○) 0.3-mm-thick additive-free film (pure); other symbols are 0.6-mm films [66].

increased O_2 uptake was observed at higher doses, probably a consequence of the higher additive consumption. The gas analysis was also carried out after 6 months of storage. Pure PP samples showed approximately twofold increase of O_2 consumption and CO, CO_2, and CH_4 formation, whereas in the samples containing the additives, oxidation during storage was much less significant.

Fintzou et al. [67, 68] evaluated changes in the physicochemical and mechanical properties of PP syringes after gamma-ray and e-beam sterilization. Radiation doses of 30, 60, and 120 kGy were used and the procedure was carried out in the presence of air at room temperature. Radiolysis products, extracted with CH_2Cl_2 from small pieces of cut syringes, were characterized by GC-MS. The chromatograms of the extracts of irradiated and nonirradiated samples revealed the presence of 30 different compounds. The radiolysis products found in irradiated samples are listed in Table A.4, Appendix, while compounds present in nonirradiated samples and consequently not formed by action of radiation are listed in Table A.3, Appendix. The main classes of identified radiolysis products include hydrocarbons, acids, ketones and aldehydes, but almost 30% of the produced compounds contain chlorine atoms. Since radiation was conducted in air, the presence of acids, ketones, and aldehydes can be explained by the breaking of the branched structure of the hydrocarbon chain, producing persistent tertiary radicals able to react with atmospheric oxygen to build peroxide or peroxy radicals [62]. These intermediates decay to build oxygen-containing molecules, which were detected in the solvent extracts. However, the presence of chlorinated compounds is only possible if a chlorinated additive is present in the polymer. In fact, there are also compounds that were not measured in irradiated samples but were observed in the nonirradiated ones.

The radiolysis products from irradiated PP differ to a certain degree among the studies presented here, probably because they have different origins and compositions. While the sample analyzed by Buchalla et al. [41] contained Irgafos 168 and three other (not specified) additives, the samples used by Demertzis et al. [62] were PP films complying with legal requirements for plastic materials intended to come into contact with foodstuff, but no description of their composition was given. Lastly, the material used by Fintzou [68] was commercially available syringes made of PP containing 2% of LDPE and various (not specified) stabilization additives.

12.1.5.10 Polyethylene Terephthalate (PET) Polyethylene terephthalate (PET) film was found to be very resistant toward radiation. Demertzis et al. [62] submitted PET films (which were intended to come into contact with food) to gamma radiation (44 kGy) and found only a few substances, in low concentration, in the CH_2Cl_2 extracts. The only identified compound was di-isopropyl-ether.

Buchalla and Begley [69], using an LC-MS system, analyzed gamma-irradiated (25 and 50 kGy) PET samples and also found that this is a very radiation-resistant polymer. The concentration of the major compounds that were present in the acetone extract of nonirradiated samples did not change perceptibly in the extracts of irradiated samples. The main difference was the increase in the concentration of terephthalic acid ethylester from less than 1 mg/kg to ca. 2 mg/kg.

12.1.5.11 Polystyrene (PS) All styrenes are stabilized by benzene ring structure, so that PS is one of the most stable polymers against radiation. Just a few radiolysis products were found in extracts from irradiated PS and in so low concentration that they could not be fully identified. Only acetophenone, the oxidation product of the monomer styrene, was clearly identified by Demertzis et al. [62] in the CH_2Cl_2 extract of PS sheet irradiated at 44 kGy. Stoffers et al. [54] found no radiolysis products in the ethanolic extract from a 54-kGy irradiated PS sample. By contrast, Buchalla et al. [70] found, besides acetophenone, 1-phenylethanol, benzaldehyde, and phenol in the extract obtained from a PS sample (Styron 685D, Dow Chemical Co.) irradiated at 25 kGy. Figure 12.23 shows the complete chromatogram of the extract of an irradiated sample (0–45 minutes) and the detail between 16 and 25 minutes of irradiated and nonirradiated samples, where the peaks of the compounds mentioned above can be seen.

12.1.5.12 Polysulfone (PSf) PSf is a well-known material in the field of polymeric membranes, having an important application in dialyzers for blood purification. PSf presents an excellent stability against ionizing radiation due to the presence of the aromatic ring in its structure. In spite of that, some degradation occurred in PSf membranes irradiated at two different dose rates, 10.8 Gy/min (low dose) and 270.7 Gy/min (medium dose) [71]. Figure 12.24

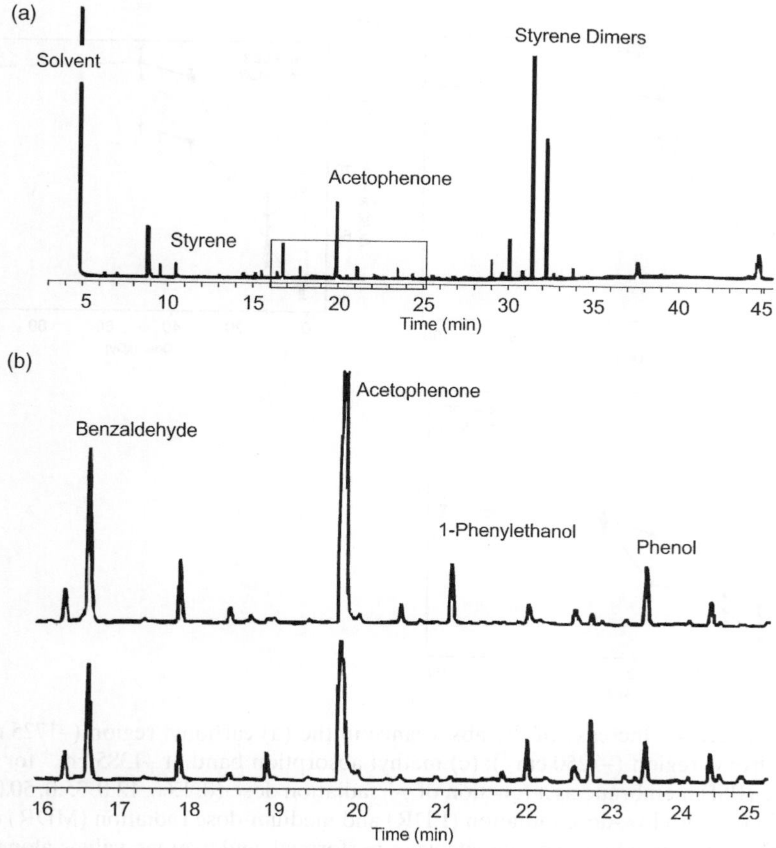

FIGURE 12.23 GC analysis of polystyrene extract. (a) PS foam sheet irradiated with 25 kGy, extract prepared by dissolution–precipitation. (b) Detail (16–25 minutes): polystyrene irradiated with 25 kGy (top) versus the nonirradiated control (bottom) [70].

shows the increase in the IR absorption bands at 3250, 1725, and 1385 cm^{-1}, which correspond to hydroxyl, carbonyl, and methyl groups, respectively. These observations indicated that γ-radiation in the presence of oxygen affected the PSf membranes by forming carbonyl groups and oxidizing the benzene ring in a dose-dependent manner. Carbonyl and hydroxyl groups result from the degradation of the diphenyethersulfone units (C–O and C–S bonds) in the backbone structure.

PSf degradation products were evaluated in another study [72], but in this case, the membranes were submitted to photodegradation under UV radiation. Although not as energetic as e-beam or γ-ray, the radiation was able to promote chain scission in the very first steps of irradiation. Low-molecular-weight compounds were extracted from irradiated samples by immersion of the membranes in methanol, and the IR spectra suggested that photolysis

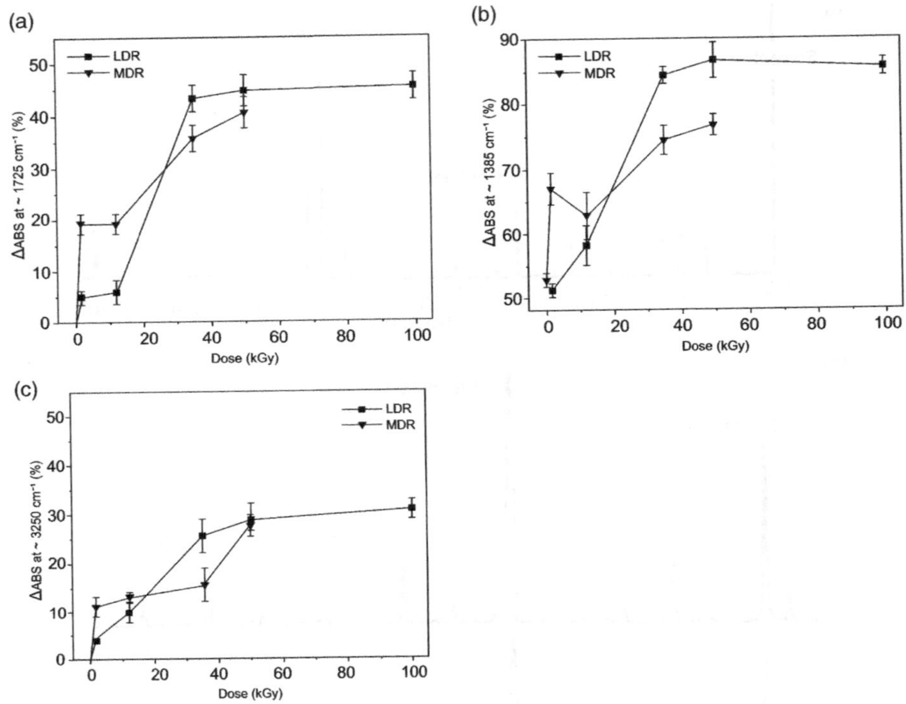

FIGURE 12.24 Increase of the absorbance in the (a) carbonyl region (~1725 cm^{-1}); (b) hydroxyl region (~3250 cm^{-1}); (c) methyl absorption band at ~1385 cm^{-1} for polysulfone (PSf) membrane as a function of γ-irradiation dose (0, 1.66, 12.0, 35.0, 50.0, and 100.0 kGy), with low-dose radiation (LDR) and medium-dose radiation (MDR) exposure. Three independent experiments were performed, and average values along with standard deviations are shown [71].

provoked C–O and C–S bond cleavages and reduction of the diphenylsulfone moiety, which is the same scission observed after γ-irradiation of the samples in the other study mentioned [71] (for PSf structure, see Table A.1, Appendix). The analysis of the extracts by HPLC allowed for the identification of muconic acid, hydroquinone, *o*-phenylphenol, 4-phenoxyphenol, and dibenzofuran among the degradation products. Figure 12.25 shows the chromatograms of the extracts of two different PSf membranes. The authors evaluated the formation of volatile products as well, and found that both CO_2 and SO_2 were formed during the photolysis of PSf membranes.

12.1.5.13 Ethylene Vinyl Acetate (EVA) Riganakos et al. [73] evaluated by GC-MS the radiolysis products of an ethylene vinyl acetate (EVA) film submitted to different radiation doses. Irradiation was carried out at room temperature and in the presence of air. In Figure 12.26, it is possible to see

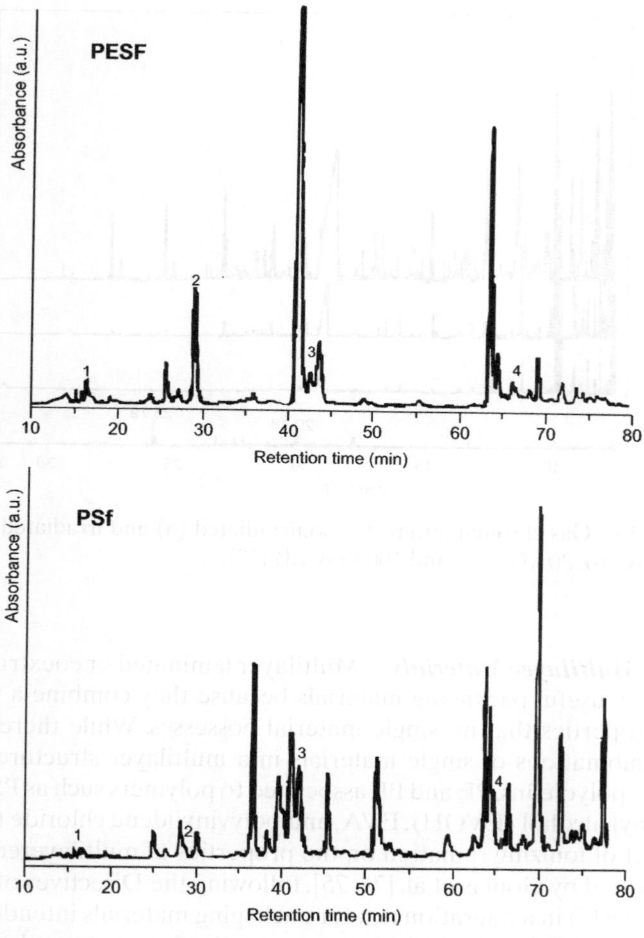

FIGURE 12.25 HPLC chromatogram of the methanolic solutions of the low-molecular-weight species extracted from polyethersulfone (PESF) and polysulfone (PSf) films irradiated 15 hours under vacuum. $\lambda_{detection}$ = 230 nm; gradient CH_3CN/H_2O acidified. Peak identification: (1) phenol, (2) 4-phenylphenol, (3) *o*-phenylphenol, and (4) dibenzofuran [72].

that the higher the radiation dose, the higher the number of radiolysis products. Although few peaks were observed in the chromatogram of the nonirradiated sample, 50 compounds were identified (these compounds are listed in Table A.3, Appendix). On the other hand, 102 compounds were recorded in irradiated samples and 65 of them were identified. Unsaturated hydrocarbons, ketones, and several methyl derivatives are among the identified compounds. Table A.4 in the Appendix lists the radiolysis products obtained from EVA polymers.

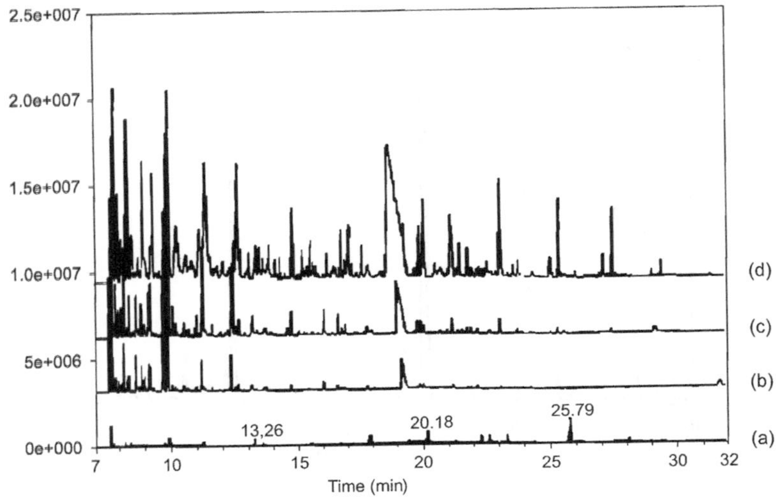

FIGURE 12.26 Gas chromatograms for nonirradiated (a) and irradiated samples of EVA at 5 kGy (b), 20 kGy (c), and 100 kGy (d) [73].

12.1.5.14 Multilayer Materials Multilayer laminated or coextruded structures are very useful packaging materials because they combine a number of desirable properties that no single material possesses. While there are many possible combinations of single materials in a multilayer structure, most are based on the polyolefins, PE and PP, associated to polymers such as PS, PET, PA, ethylene vinyl alcohol (EVOH), EVA, and polyvinylidene chloride (PVDC).

The effect of ionizing radiation on the properties of multilayered materials was investigated by Goulas et al. [74, 75], following the Directives of the European Union (EU) in a migration test for packaging materials intended to come in contact with foodstuffs. Although not exactly the same as the tests prescribed by pharmacopeial compendia, as it evaluates only the overall mass migration, the test gives an idea of the material behavior after being submitted to the action of ionizing radiation. Table 12.15 shows the overall migration values of different multilayer materials in function of the radiation dose and of the extraction solvent.

All results presented in Table 12.15 were significantly lower than the upper limit for overall migration set by EU for food grade plastic packaging materials of 10 mg/dm^2. There results, however, do not identify the radiolysis products and do not include volatile compounds since migration testing involves an evaporation step during which all volatile species are lost to the environment. The same research group conducted a study to evaluate volatile compounds produced during e-beam irradiation of multilayered packaging material, a PET/PE/EVOH coextruded structure [73]. Irradiation (100 kGy) was carried out at room temperature and in the presence of air, and the volatile compounds were identified by GC-MS. Out of 102 volatile compounds recorded from

TABLE 12.15 Overall Migration Values (mg/dm^2) into Distinct Solvents from Multilayer Films Irradiate and Nonirradiated after 10 Days at 40°C

Material	Distilled Water		3% Acetic Acid		Iso-octane	
	0 kGy	30 kGy	0 kGy	30 kGy	0 kGy	30 kGy
PP/EVOH/LDPE-LLDPE[a]	0.6	0.6	1.3	1.2	3.0	3.8
LDPE/EVOH/LDPE	0.9	0.8	1.0	0.8	2.5	2.7
Ionomer[b]/EVOH/LDPE	1.3	1.4	10.9	9.6	2.9	2.3
PA/LDPE	1.1	1.2	1.9	2.3	3.2	3.8
LDPE/PA/ionomer	1.7	2.0	16.4	12.8	5.5	4.6
PVC/HDPE	0.5	0.5	0.8	0.8	1.8	1.3
HDPE/PA	0.5	0.4	0.4	0.3	1.0	1.2

[a] Linear low-density polyethylene.
[b] Copolymer containing both nonionic repeat units, and a *small* amount (<15%) of ion containing repeat units.
Data from References 74 and 75.

irradiated samples, 80 were identified. Of course, several of these compounds were also present in the chromatogram of nonirradiated samples; however, aldehydes, ketones, and alcohols, which are products of an oxidative process triggered by ionizing radiation in the presence of oxygen, were mostly present in irradiated samples. These compounds are listed in Table A.4, Appendix.

12.2 HEAT STERILIZATION

Heat-labile products cannot be sterilized by heat. Drugs may degrade in the presence of heat or steam, and some plastics can melt or can be permanently strained. Therefore, heat-sensitive drugs and plastics should be sterilized by other processes. However, there are products containing heat-sensitive species that are sterilized by heat. One example of a heat-sensitive species is glucose, which has been observed to degrade to reactive carbonyl compounds during heat sterilization and subsequent storage [76]. Glucose degradation products (GDPs) have been identified as glyoxal, methylglyoxal, formaldehyde, 3-deoxyglucosone (3-DG), 3,4-dideoxyglucosone-3-ene (3,4-DGE), and 5-hydroxymethyl furaldehyde (5-HMF) [77].

Isotonic glucose solutions are used in intensive care units to hydrate patients. Glucose is also used as a component of peritoneal dialysis (PD) fluids. Conventional PD fluids contain high concentrations of glucose as an osmotic agent to promote the clearance of metabolic waste from the blood by dialysis across the peritoneal membrane. PD fluids need to be sterile and are commonly heat sterilized. Table 12.16 shows the concentration of GDPs found in different commercial infusion solutions containing glucose. Variations observed in the results presented in Table 12.16 are consequences of the differences in sterilization procedures and storage periods. The data in Table 12.16 show that the concentrations of GDPs are higher in bags in which the pH of the glucose

TABLE 12.16 Concentration of GDPs Found in Different Commercial Infusion Solutions and Peritoneal Dialysis Fluids Containing Glucose

Product	Glucose Conc. (%)	pH	GDPs Concentration (μmol/L)								Reference
			3,4-DGE	3-DG	5-HMF	GO	MGO	FuA	FoA	AcA	
Dianeal (Baxter) sample 1	1.36	–	–	224	–	3.5	4.0	–	–	–	[78]
Dianeal (Baxter) sample 2	2.27	–	–	335	–	6.2	5.1	–	–	–	
Dianeal (Baxter) sample 3	3.86	–	–	525	–	9.4	6.9	–	–	–	
Gambrosol (Gambro) sample 1	1.5	–	–	221	–	11.3	7.5	–	–	–	
Gambrosol (Gambro) sample 2	2.5	–	–	381	–	12.8	11.7	–	–	–	
Gambrosol (Gambro) sample 3	4.0	–	–	540	–	22.6	13.8	–	–	–	
Lockolys (Fresenius) sample 1	1.5	–	–	179	–	6.0	9.5	–	–	–	
Lockolys (Fresenius) sample 2	2.3	–	–	245	–	8.9	9.2	–	–	–	
Lockolys (Fresenius) sample 3	4.25	–	–	418	–	16.2	12.0	–	–	–	
PD fluid sample 1	2.5	5.3	13	175	5.4	–	–	–	6.4	292	[79]
PD fluid sample 2	2.27	5.0	19	213	15	–	–	–	6.0	173	
PD fluid sample 3	2.3	5.4	12	185	2.9	–	–	–	5.3	359	
PD fluid sample 4	50	3.1	0.5	29	19	–	–	–	2.3	<1.1	
PD fluid sample 5	5.82	4.2	11	178	30	–	–	–	3.4	5.5	
PD fluid sample 6	4.6	3.1	0.4	10	10	–	–	–	<1.7	1.5	
PD fluid sample 7	4.6	2.8	0.2	17	18	–	–	–	1.9	2.4	
PD fluid sample 8[a]	7.5	5.0	3.0	11	2.1	–	–	–	9.3	37	
PD fluid sample 1	7.5[a]	5.1	n.d.	7.5	2.9	n.d.	n.d.	n.d.	3.6	32.2	[80]
PD fluid 2	1.36	5.1	7.5	107.4	10.9	2.6	5.2	1.1	5.3	82.9	
PD fluid 3	2.27	5.1	11.3	166.8	16.7	4.1	6.9	1.7	6.8	114.3	
PD fluid 4	3.86	5.1	15.2	261.2	27.6	5.1	8.9	3.1	8.5	150.3	

PD fluid 5	3.75	4.1	9.9	69.1	29.0	2.3	n.d.	n.d.	n.d.	3.3	
PD fluid 6	6.26	4.1	14.3	93.3	44.8	4.4	n.d.	n.d.	3.4	3.2	
PD fluid 7	10.64	4.1	12.4	119.5	69.4	9.0	n.d.	n.d.	4.0	n.d.	
PD fluid 8	1.5	5.4	14.8	98.7	3.7	4.5	6.4	n.d.	5.0	136.2	
PD fluid 9	2.3	5.4	16.2	141.8	6.1	4.6	7.9	1.2	5.4	152.5	
PD fluid 10	4.25	5.4	18.1	234.7	12.7	8.8	9.9	1.6	7.4	170.9	
PD fluid 11	3.0	2.7	n.d.	10.2	7.0	n.d.	n.d.	n.d.	n.d.	n.d.	
PD fluid 12	4.6	2.7	n.d.	16.3	12.2	n.d.	n.d.	n.d.	n.d.	n.d.	
PD fluid 13	8.5	2.7	n.d.	29.9	24.1	n.d.	n.d.	n.d.	n.d.	n.d.	
PD fluid 14	3.0	3.0	n.d.	11.6	6.7	n.d.	n.d.	n.d.	n.d.	n.d.	
PD fluid 15	4.6	3.0	n.d.	17.6	9.4	1.7	n.d.	n.d.	n.d.	n.d.	
PD fluid 16	8.5	3.0	n.d.	31.5	16.3	3.9	n.d.	n.d.	n.d.	n.d.	
PD fluid 17	50	3.1	n.d.	12.6	11.8	n.d.	n.d.	n.d.	n.d.	n.d.	
PD fluid 18	50	3.1	n.d.	20.2	17.7	n.d.	n.d.	n.d.	n.d.	3.9	
PD fluid 19	50	3.1	n.d.	30.3	24.4	2.7	n.d.	n.d.	3.5	n.d.	
Glucose Fresenius 200	20	–	59	583	105	–	–	–	19	–	[81]
Glucose Fresenius 300	30	–	56	790	146	–	–	–	34	–	
Rehydrex (Fresenius)	2.5	–	22	141	2	–	–	–	17	–	
Glucose Baxter	10	–	50	400	42	–	–	–	10	–	
Glucose Baxter 25/50	2.5	–	22	123	2	–	–	–	10	–	
Glucose Baxter Viaflo	5	–	35	283	17	–	–	–	4	–	
Glucose Baxter Na40 K20	10	–	59	358	10	–	–	–	21	–	

[a] Glucose polymer. GO, glyoxal; MGO, methylglyoxal; FuA, 2-furaldehyde; FoA, formaldehyde; AcA, acetaldehyde.
n.d., not detected; –, data not furnished or not measured.

compartment is higher independent of the glucose concentration. In fact, the extent of glucose degradation is a result of the combination of pH and temperature.

Fifty percent glucose solutions with pH values ranging from 1 to 6 were heated, sterilized, and stored at 40°C for 30 days [79]. The concentrations of 3-DG, 3,4-DGE, and 5-HMF measured in these solutions after this period showed that the higher the pH, the higher the GDP concentration.These results confirmed that, in addition to time and temperature, pH influences the amount of GDPs formed in glucose solutions during sterilization. Moreover, once formed, the mixture of GDPs is not stable but changes with time as demonstrated by Erixon et al. [77]. They evaluated the concentration of 3,4-DGE, the most bioreactive among the GDPs, in laboratory-made PD fluids. The pH value of the fluids, containing 1.5% glucose and the other usual components, was adjusted to 5.5. The fluids were then subjected to heat sterilization at 121°C for 40 minutes. The samples were subsequently maintained at 25 and 40°C for 60 days. Aliquots were taken to measure 3,4-DGE, and the results demonstrate that immediately after heat sterilization, 125 μmol/L of 3,4-DGE was present. However, the concentration of 3,4-DGE rapidly decreased during the first week of storage. The 3,4-DGE decrease was faster at 40°C compared with 25°C. Furthermore, equilibrium was attained after about 1 week and 30 days for both temperatures, respectively. After 30 days of storage, the addition of 3,4-DGE to these fluids showed a similar decreasing profile during the following 30 days.

In another experiment, Erixon et al. [82] evaluated the behavior of pure solutions of 3,4-DGE, 3-DG, and 5-HMF that were not subjected to sterilization but were stored at 40°C for 30 days. The results showed that the incubation of 100 μmol/L of 3,4-DGE resulted in the production of 36 μmol/L 3-DG and 4 μmol/L 5-HMF, with a decrease in the 3,4-DGE concentration of 79 μmol/L. In 3-DG solution, 3,4-DGE and 5-HMF were also produced, whereas the concentration of 5-HMF remained stable during the entire incubation period. The authors estimated that 50% of 3,4-DGE was converted into 3-DG and the other 50% was converted into another substance that was neither 5-HMF nor any other known aldehyde.

To summarize, infusion solutions containing glucose should have low pH values between 2 and 3 and should be stored at low temperatures during shelf life (below 20°C) to avoid production of GDPs in high extent. Considering the behavior of 3,4-DGE, 30 days after sterilization is a reasonable period of time to allow its concentration to decrease [83].

12.3 RESIDUALS FROM CHEMICAL DISINFECTION AND STERILIZATION AGENTS

The most frequent contaminant in medical devices sterilized or disinfected with chemical agents is the residual of the agent. Residuals of such chemicals

TABLE 12.17 Toxicity of Sterilizing and Disinfecting Agents

Parameter	LD_{50} (rat) mg/kg Body Weight	LC_{50} (rat) mg/m^3
Etylene oxide	72	5029 (1 h)
Ethylene glycol	4700	>200 (4 h)
Ethylene chlorohydrin	71	290 (4 h)
Peracetic acid	1540	590 (1 h)
Hydrogen peroxide	6667	2000 (4 h)
Formaldehyde	100	578 (4 h)
Paraformaldehyde	800	1070 (4 h)
Glutaraldehyde	134	480 (4 h)

Sources: Health Protection Agency-UK (hpa.org.uk), Oxford University: http://msds.chem.ox.ac.uk.

in materials that come into close contact with patient's skin or mucous membranes may pose health risks because they are used due to their toxic action on living organisms. Table 12.17 displays the toxicity levels of sterilants and disinfectants and on some subproducts formed as a result of their utilization.

The process of sterilizing using chemical agents requires the placement of the devices, wrapped or unwrapped, into a compartment where a chemical is delivered in the form of a gas. After an appropriate period of contact, poststerilization aeration is necessary to eliminate the residuals of the chemical agent. Contamination will occur if the chemical agent remains adsorbed onto the device's surface or if the gas penetrates into the device or container and the subsequent aeration is not satisfactory to eliminate the residue entirely. In fact, the ability to penetrate is an important feature of a sterilizing agent because the extent of its action depends on this feature. Therefore, depending on the item to be sterilized, poststerilization aeration used for residual elimination has to be in accordance with the penetrating ability of the chemical agent.

Nevertheless, the permeation of gases through packaging materials can be a problem if the item to be sterilized or disinfected contains a drug product sensitive to the chemical agent. The permeation of gases occurs through plastic and elastomeric materials, and the extent of permeation is determined by the nature of the gas and the structure of the polymeric material. Because drug stability depends on the insulating property of the packaging material, the permeation of gases is a well-defined parameter. Table 12.18 lists the resistance against gas permeation of selected packaging materials.

12.3.1 Ethylene Oxide (EtO)

EtO is a highly reactive chemical agent mainly employed for sterilizing disposable medical devices. The compatibility of EtO with a wide range of materials

TABLE 12.18 Permeability Coefficient of Selected Gases across Common Polymeric Materials

Polymer	T (°C)	Permeability ($P \times 10^{13}$) Oxygen	Carbonic gas	Water
Low-density polyethylene	25	2.2	9.5	68
High-density polyethylene	25	0.30	0.27	9.0
Polytetrafuorethylene	25	3.17	10.0	13.4
Polypropylene	30	1.7	6.9	51.0
Polyurethane	–	–	–	–
Polyvinyl chloride unplasticized	25	0.034	0.12-	206
Polyvinyl chloride plasticized	25	0.68	2.40	4120
Ethylene vinyl acetate	25	1.53	12.8	968.2
Poly(methyl methacrylate)	23–34	0.116	–	480
Polystyrene	25	2.0	7.9	1350
Polycarbonate	25	1.4	–	1400
Polyethylene terephthalate	30	0.019	–	130
Polyester	23–25	0.014	–	114
Silicone	25	389	1594	–
Nylon 6	20–30	0.0285	0.066	0.139
Polybutadiene	25	14.2	104	–
Butyl rubber	23–25	0.977	3.89	157
Natural rubber	25	17.6	115	2290
Cellulose	25	0.0016	0.00353	18900
Cellulose acetate	25–30	0.585	17.3	4130
Ethyl cellulose	25–30	11.0	84.8	6700
Cellulose nitrate	25–30	1.46	1.59	4720

Adapted from Reference 5.
Permeability (P) has the dimension: (quantity of permeant) × (film thickness)/(area) × (time) × (pressure drop across the film). The SI unit of permeability coefficient is $cm^3 \times cm/cm^2 \times s \times Pa$.

and its capacity of penetration make it an adequate sterilant for the majority of heat and moisture-sensitive medical products. Sterilization is carried out at low temperatures (37–60°C) in the presence of water vapor (40–80% relative humidity). EtO has a characteristic sweet odor of ether and is readily soluble in water, ethanol, and several other organic solvents. EtO may be absorbed by numerous materials and may remain unchanged. Consequently, EtO might be present as a residual in sterilized devices.

EtO is produced by the direct oxidation of ethylene with oxygen (air) using the following silver-based catalyzed reaction:

$$CH_2=CH_2+O_2 \rightarrow 2(CH_2CH_2)O. \quad (12.8)$$

Because EtO reacts with water to form a series of ethylene glycols (EtGs), EtG may be a concomitant residue in EtO sterilization:

$$(CH_2CH_2)O + H_2O \rightarrow HO\text{–}CH_2CH_2\text{–}OH \qquad (12.9)$$

$$2(CH_2CH_2)O + H_2O \rightarrow HO\text{–}CH_2CH_2\text{–}O\text{–}CH_2CH_2\text{–}OH \qquad (12.10)$$

$$3(CH_2CH_2)O + H_2O \rightarrow HO\text{–}CH_2CH_2\text{–}O\text{–}CH_2CH_2\text{–}O\text{–}CH_2CH_2\text{–}OH \qquad (12.11)$$

$$n(CH_2CH_2)O + H_2O \rightarrow HO\text{–}(\text{–}CH_2CH_2\text{–}O\text{–})_n\text{–}H. \qquad (12.12)$$

If the reaction occurs in the presence of chloride ions, ethylene chlorohydrin (2-chloroethanol) (EtC) may also form:

$$(CH_2CH_2)O + H_2O(Cl^-) \rightarrow HO\text{–}CH_2CH_2\text{–}Cl. \qquad (12.13)$$

Although EtG does not pose severe hazards, EtC does, which is even more toxic than the EtO itself (see Table 12.17).

Thus, the residue limits for EtC are established in regulatory standards for EtO sterilization. The current allowable limits for EtO residues in medical devices are set by the Association for the Advancement of Medical Instrumentation (AAMI)-Industrial Ethylene Oxide Sterilization for Medical Devices, which is equivalent to the ISO 10993-7 [85].

Table 12.19 lists residues of EtO in different types of materials. The levels of EtO residue vary greatly among the samples because the permeation of gases into polymeric materials depends on features of these materials and aeration conditions were not the same. Nevertheless, these data show that higher levels of EtO residues were found in PVC, silicone, and methacrylate polymers, and the aeration times of these polymers must be longer than those of other polymers. Additionally, wall thickness also influenced the residue levels in PVC tubing from the same manufacturer that have been sterilized all together [87]. The thicker the tubing, the higher the EtO residue levels after 24 and 48 hours of aeration.

Differences in the levels of EtO residues among the same polymeric material were also observed between two samples of PU [93]. A study designed to investigate the influence of resterilization on the EtO residuals included two PUs of different crystallinity. The more crystalline PU 75 retained at least 400 times more EtO than the less crystalline PU 80.

Wang et al. [94] established a correlation between the residual of EtO in a sterilized device and the amount of EtO inside the packaging bag. EtO residues in a PVC infusion set and the gaseous space inside the packaging bag were measured for a period of 16 days after sterilization. The results in Figure 12.27 show that the EtO residue in the infusion set decreased exponentially

TABLE 12.19 Ethylene Oxide Residuals in Different Materials after Sterilization and Aeration Procedures

Sample	Sterilization Conditions				Aeration Procedure		EtO Residue	Reference
	EtO Conc.	RH (%)	Temp. (°C)	Exposure (h)	Temp. (°C)	Time (h)	(mg/g)	
Polyethylene (company A)	12%	50	57	5	25	0	3.20	[86]
						24	0.18	
Polyethylene (company B)	12%	50	57	5	25	0	2.20	
						24	0.40	
Polyethylene (company C)	12%	50	57	5	25	0	0.50	
						24	0.15	
PVC endotracheal tube 1/16 in.	12%	50	54	2	Ambient	0	11,000	[87]
						24	5,000	
						48	3,000	
PVC endotracheal tube 1/32 in.	12%	50	54	2	Ambient	0	14,000	
						24	3,000	
						48	2,000	
PVC endotracheal tube 1/64 in.	12%	50	54	2	Ambient	0	19,000	
						24	1,000	
						48	800	
PVC tubing	1200 mg/L	50	60	2		0	18,519	[88]
PVC tubing	1200 mg/L	50	60	3		0	26,570	
PVC tubing	1200 mg/L	50	60	3		20	14,712	
Methacrylate powder	n.i.	n.i.	n.i.	n.i.	n.i.	1	>200 ppm	[89]
						24	>200	
						2 days	>100	
						8 days	>100	
						22 days	65–70	
						27 days	10–12	
						35 days	2–3	
Silicone tubing	1200 mg/L	n.i.	50	1	23	24	4 ppm	[90]
PVC tubing	n.i.	n.i.	n.i.	n.i.	n.i.	4	18.91	[91]
						24	4.31	
						48	2.10	
						96	1.81	
Natural rubber (sample A)	n.i.	n.i.	n.i.	n.i.	n.i.	0.5	6.01	
						24	<0.4	
						48	<0.4	
Natural rubber (sample B)	n.i.	n.i.	n.i.	n.i.	n.i.	0.5	5.86	
						24	<0.4	
						48	<0.4	
Látex rubber	n.i.	n.i.	n.i.	n.i.	n.i.	0.5	<0.4	
						24	<0.4	
						48	<0.4	

TABLE 12.19 (*Continued*)

	Sterilization Conditions				Aeration Procedure		EtO Residue	
Sample	EtO Conc.	RH (%)	Temp. (°C)	Exposure (h)	Temp. (°C)	Time (h)	(mg/g)	Reference
Cuprophane	n.i.	n.i.	n.i.	n.i.	n.i.	3	<0.4	
						24	<0.4	
						48	<0.4	
						96	<0.4	
PVC tubing 1.6-mm wall thickness	n.i.	n.i.	n.i.	n.i.	23	0	77,700	[92]
						2	47,400	
						6.5	28,000	
						24	8,820	
						72	999	
						168	21	
						240	2.40	

n.i., not informed.

over time and was positively correlated to the concentration of gaseous EtO in the package. Due to the equilibrium that exists between adsorption and desorption, the removal of EtO from the device depends on the rate at which it crosses the wrapping material.

Martins et al. [95] evaluated the residues of EtO, EtG, and EtC in perfusion cannulas (made of PVC) sterilized with 450 mg/L EtO for 2 hours at 45°C according to the ISO 10993-7 method. The data listed in Table 12.20 show the contaminant residuals before aeration and the aeration time necessary to lower the levels of these residues to those set by the ISO 10993-7. EtG residues were also evaluated to abide by Brazilian standards [96]. The results showed that EtG and EtC residues are much lower and are more easily removed than the EtO residues. Additionally, the results showed the aeration times for identical devices made of the same polymer varied when their dimensions changed.

Because the reuse of medical devices is a current praxis and their reprocessing may demand sterilization, Lucas et al. [93] investigated the influence of resterilization on EtO residues. Strips of PU and nylon 66 and catheters made of PU were tested after the first and fifth resterilization cycles. In all cases, a 12-hour poststerilization aeration period at 132°C and a 5-day period at room temperature were adopted between resterilizations. The results in Figure 12.28 show that the number of sterilization cycles had little influence on the amount of EtO residues in these polymers. The depth of EtO penetration and the residue remaining were also studied. One of the catheters contained embedded wires (solid) and the other contained removable wires (hollow), which were removed before sterilization. The amount of residue in the solid catheter was measured to be at least 10 times higher than that in the hollow catheter.

Bone grafts are among the materials sterilized by EtO. Demineralized bone matrices have been used in a variety of clinical applications, including oral,

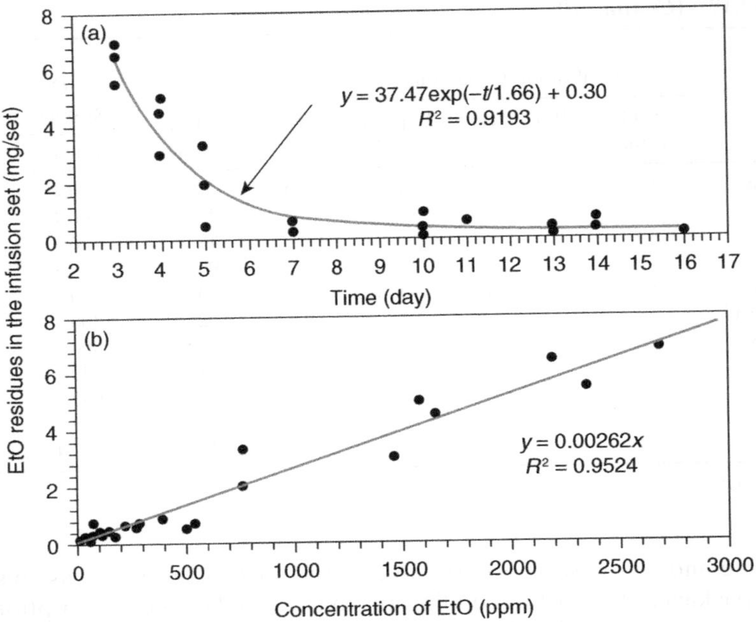

FIGURE 12.27 In the infusion set of brand A, (a) the EtO residues variation along with stock time and (b) the correlation between EtO residues and the concentration of EtO gas within the packaging bag [94].

TABLE 12.20 Ethylene Oxide, Ethylene Glycol, and Ethylene Chlorohydrin Residues in Cannulas of Different Dimensions (after Sterilization Cycle and before Aeration) and Aeration Time Necessary for the Residues to Lower to Standard Levels[a]

	EtO		EtG		EtC	
Sample Dimension	Residue (ppm)	Time for Residue <250 ppm (h)	Residue (ppm)	Time for Residue <250 ppm (h)	Residue (ppm)	Time for Residue <25 ppm (h)
12.7 mm × 50 cm	765.41	26	163.42	<1	16.88	<1
12/15.3 mm × 37.5 cm	363.28	27	183.65	<1	7.38	<1
12.0/17.0 mm × 37.5 cm	409.43	20	189.37	<1	8.20	<1
12.0 mm × 40.0 cm	658.51	27	283.13	<1	14.77	<1
8.0 mm × 28 cm	564.64	25	105.22	<1	6.71	<1
6.7 mm × 28 cm	330.64	20	137.53	<1	9.15	<1
7.3 mm × 28 cm	563.71	28	70.97	<1	4.18	<1
8.0 mm × 28 cm	459.92	19	94.05	<1	4.42	<1

[a] Aeration at 35°C and 26 air changes per hour.
Adapted from Reference 95.

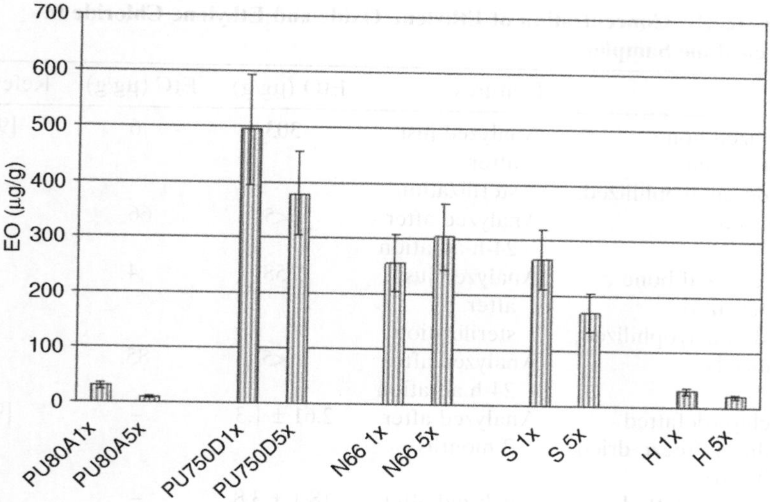

FIGURE 12.28 Resterilization effect. Materials were resterilized with EtO one (1×) or five (5×) times. There was no significant difference in residue levels between sterilizing once or five times for the polymers and devices used in this study. PU, polyurethane; N66, nylon; S, solid catheter; H, hollow catheter [93].

craniofacial, and spinal procedures. Sterilization is a necessary practice to reduce the risk of disease transmission. Although EtO has been shown to be an effective sterilant for bone tissue, the generation of EtC during the EtO sterilization process has been implicated in the failure of these grafts by inducing a localized inflammatory response. The blood residue (not completely removed) in the bone matrix provides the chloride ions to form EtC.

Lomas et al. [97] investigated the presence of EtO and EtC in mineralized and demineralized bone matrices after sterilization. Both mineralized and demineralized bone retained detectable amounts of EtO immediately following sterilization. The demineralized bone retained approximately eight times more EtO than mineralized bone (Table 12.21). However, subsequent aeration successfully eliminated EtO traces. In contrast, low levels of EtC were measured in the samples immediately after sterilization. However, these levels increased after aeration. The remaining EtO is potentially converted into EtC during the aeration time. The authors speculated that demineralized bone, a more open, protein scaffold as opposed to the solid matrix of mineralized bone, absorbed more EtO, which was then released at a slower rate.

Arizono obtained similar results with defatted bone samples (Reference 98 in Table 12.21). Although samples analyzed 2 months after sterilization contained EtO residues as low as those found in samples by Lomas 24 hours after aeration, differences were observed among the samples. In freeze-dried samples, residues were seven times lower than those in only defatted samples.

TABLE 12.21 Concentration of Ethylene Oxide and Ethylene Chloride in Sterilized Bone Samples

Sample	Features	EtO (μg/g)	EtC (μg/g)	Reference
Mineralized bone (pasteurized, powdered, lyophilized, sterilized)	Analyzed just after sterilization	303	6	[97]
	Analyzed after 24-h aeration	<5	66	
Demineralized bone (pasteurized, powdered, lyophilized, sterilized)	Analyzed just after sterilization	2358	4	
	Analyzed after 24-h aeration	<5	85	
Bone chip (defatted sterilized, freeze-dried, sterilized)	Analyzed after 2 months	2.61 ± 1.3	–	[98]
Bone chip (defatted sterilized, sterilized)	Analyzed after 2 months	18.1 ± 3.8	–	
Bone chip removed from 0 to 3 mm from the surface	Defatted and sterilized EtO measured after 24 h	1850 ± 77	–	
Bone chip removed from 6 to 9 mm from the surface	Defatted and sterilized EtO measured after 24 h	1080 ± 100	–	

Arizono also measured the EtO concentration at two levels of depth in bone matrices. From 6 to 9 mm from the surface, 1850 ± 77 μg/g of EtO was measured. In contrast, from 0 to 3 mm, the EtO concentration was 1080 ± 100 μg/g. EtO has a strong ability to penetrate. Thus, the deeper EtO penetrates, the slower is its release.

12.3.2 Peracetic Acid and Hydrogen Peroxide

PA or peroxyacetic acid is employed as a chemical sterilization agent for medical devices, including isolators for the safe handling of hazardous drugs and reused hemodialyzers. PA is an organic peroxide, containing the functional group ROOR′. Similar to all peroxides, PA presents strong oxidizing properties and is highly corrosive. PA is also used in commercial mixtures with hydrogen peroxide.

PA is produced by the reaction of acetic acid with hydrogen peroxide:

$$H_2O_2 + CH_3CO_2H \rightarrow CH_3CO_3H + H_2O. \qquad (12.14)$$

It decomposes into either acetic acid and oxygen or acetic acid and hydrogen peroxide as shown in the following equations:

$$CH_3CO_3H + H_2O \rightarrow H_2O_2 + CH_3CO_2H \tag{12.15}$$

$$2CH_3CO_3H \rightarrow 2CH_3CO_2H + O_2. \tag{12.16}$$

Hydrogen peroxide is mainly used in the surface sterilization of flexible and rigid isolators, production filling lines, biological safety cabinets, and cleanrooms.

Hydrogen peroxide is produced by the Riedl–Pfleiderer process in which 2-ethyl-9,10-dihydroxyanthracene ($C_{16}H_{16}O_2$) undergoes autoxidation using oxygen from the air:

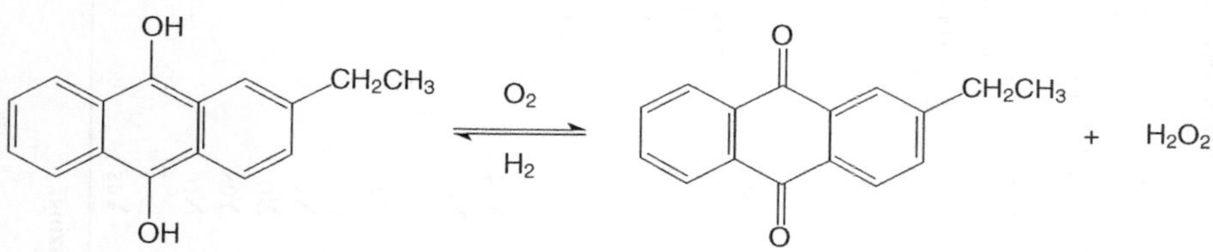

(12.17)

Hydrogen peroxide cannot be used with highly absorptive materials such as cellulosic materials, nylon, and certain types of rubber. Additionally, PP and polyester should be not sterilized with hydrogen peroxide because residues will be retained.

The permeation capability of PA and H_2O_2 through some plastics may be a concern when formulations should be locally prepared and sterilized. The potential risk arising from this event is drug degradation due to the oxidizing properties of these agents. Additionally, changes in pH can occur due to the formation of acetic acid by the hydrolysis of PA (Equation 12.15). Permeation was observed during the sterilization of infusion bags used for cytotoxic preparations, parenteral fluids, and ambulatory pumps. Ambulatory pumps are disposable elastomeric reservoirs used either in hospitals or in home settings for the infusion of cytotoxic drugs, antibiotics, and pain relief medication. These pumps are made of different plastic materials such as silicone, PVC, and PET, with a drug reservoir of approximately 100 mL. Bags and pumps may be sterilized wrapped or unwrapped. Tests on bags filled with water for injection (WFI) before sterilization and different types of pumps filled with 5% glucose after sterilization showed that the permeation of both PA and H_2O_2 is possible. The results presented in Table 12.22 demonstrate that permeation is associated with plastic composition. There was no permeation of PA detected in both multilayered and LDPE bags and pumps made of silicone. On the other hand, PA and hydrogen peroxide were detected in bags and pumps made of PVC, EVA, and polyisoprene. PVC and EVA are known to be permeable for steam,

TABLE 12.22 Peracetic and Hydrogen Peroxide Permeated into Wrapped and Unwrapped Infusion Bags and Ambulatory Pumps after Peracetic Acid Sterilization Procedure

Sample	Materials	Precleaning	Wrapped	Peracetic Acid Mean (n = 6) (mg/L)	Hydrogen Peroxide Mean (n = 6) (mg/L)	Reference
Macroflex®	PVC	–	Yes	0.071 ± 0.040	–	[99]
		50% Isopropanol		0.008 ± 0.004	–	
		50% Isopropanol + 0.35% peracetic acid		0.008 ± 0.006	–	
Clearflex®	Three-layer: PP + PA + PE (inner)	–	Yes	n.d.	–	
		50% Isopropanol		n.d.	–	
		50% Isopropanol + 0.35% peracetic		n.d.	–	
Ecoflac®	LDPE	50% Isopropanol	No	0.005 ± 0.004	–	
		50% Isopropanol + 0.35% peracetic		0.003 ± 0.002	–	
Nutripoche®	EVA	50% Isopropanol	No	0.014 ± 0.001	–	
		50% Isopropanol + 0.35% peracetic		0.016 ± 0.004	–	
Aminomix®	Two-layer: PA + PE	50% Isopropanol	No	0.002 ± 0.002	–	
		50% Isopropanol + 0.35% peracetic		0.007 ± 0.003	–	
Accufuser®	Silicone + PET	–	No	n.d.	0.52–2.35	[100]
		–	Yes	n.d.	0–0.5	
Infusor®	Polyisoprene + PET glycol modified	–	No	n.d.	4.64–23.13	
		–	Yes	n.d.	0–0.5	
Surefuser®	Silicone	–	No	n.d.	n.d.	
		–	No	n.d.	n.d.	
Ultraflow®	PVC	–	No	5–10	1.84–6.65	
		–	Yes	n.d.	0–0.5	
Easypump®	Elastomeric material + PVC protection	–	No	0–5	0.56–2.23	
		–	Yes	n.d.	n.d.	

Bags were filled with water before sterilizing and pumps with 5% glucose after sterilizing.
n.d., not detected; -, not measured.

TABLE 12.23 Amount of H_2O_2 in Sterilized Materials after 2 and 12 Hours' Aeration. H_2O_2 Extraction: 24 Hours' Submersion in Water [101]

Material	H_2O_2 (μg/piece)	
	2-h Aeration	12-h Aeration
Polystyrene (PS)	2.2 ± 0.0	0.23 ± 0.03
Polyurethane (PU)	8.3 ± 0.4	0.19 ± 0.16
Silicone (Si)	0.24 ± 0.06	0.09 ± 0.04
Silicone polyurethane blend (Si/PU)	18.0 ± 0.4	0.09 ± 0.02
Polyethylene (PE)	0.18 ± 0.0	n.t.
Fluorosilicone acrylate	1.4 ± 0.2	0.08 ± 0.01
Poly(methyl metacrylate) (PMMA)	5.1 ± 0.3	1.5 ± 0.0
Poly(2-hydroxyethyl metacrylate) (HEMA)	186 ± 7	98 ± 15

n.t., not tested.

oxygen, and carbon dioxide (see Table 12.18). The tests were also designed to evaluate precleaning and the influence of wrapping. Although precleaning did not alter permeation, unwrapped devices presented higher levels of contamination. Bags supplied with an outer bag (Macoflex and Clearflex) presented lower levels of PA than those from unwrapped bags that were sterilized (Ecoflac, Nutripoche, and Aminomix). Additionally, hydrogen peroxide dropped to less than 0.5 mg/L when pumps were sterilized wrapped.

Ikarashi et al. [101] evaluated the residue of hydrogen peroxide remaining in different plastic materials (0.5-mm films). After vapor phase hydrogen peroxide sterilization was conducted, samples were aerated by introducing air into the chamber at a flow rate of 500 L/min for 2 and 12 hours. After 24 hours of extraction in water, hydrogen peroxide was measured. The polymers behaved differently. After 2 hours of aeration, poly(2-hydroxyethyl metacrylate (HEMA), PU, Si/PU, and PMMA still retained H_2O_2 in large amounts, whereas Si and PE did not (Table 12.23). After 12 hours of aeration, the amount of H_2O_2 residue decreased to less than 0.2 μg for most polymers other than both metacrylates. The penetration and absorption of H_2O_2 may potentially be enhanced in metacrylate derived polymers. As a result, these residues are more difficult to remove.

12.3.3 Formaldehyde

Formaldehyde is a colorless gas with a characteristic pungent odor. The chemical sterilant is commercialized as a saturated water solution called formalin, which contains approximately 37% formaldehyde in mass. A small amount of stabilizer, usually methanol, is added to limit the oxidation and polymerization. Typical commercial-grade formalin contains 10–12% methanol.

Formaldehyde is industrially produced via the catalytic oxidation of methanol. The most common catalysts used in this reaction are silver metal or a mixture of iron with molybdenum or vanadium oxides. Methanol and oxygen react at ca. 250–400°C to produce formaldehyde according to the following equation:

$$2CH_3OH + O_2 \rightarrow 2CH_2O + 2H_2O. \tag{12.18}$$

An aqueous solution of formaldehyde consists of methylene glycol, generated by its reaction with water, and polymeric forms of methylene glycol, known as paraformaldehyde, as shown in the following equations:

$$CH_2O + H_2O \rightarrow HOCH_2OH, \tag{12.19}$$

$$nCH_2 = O \rightarrow [-CH_2-O-CH_2-O-CH_2-O-]_n, \tag{12.20}$$

or

$$nCH_2 = O + H_2O \rightarrow HO[-CH_2-O-CH_2-O-CH_2-]_nOH, \tag{12.21}$$

where n may vary from 8 to 100.

According to regulatory standards (European Standards EN 14180) [102], the formaldehyde residues after low-temperature steam formaldehyde (LTSF) sterilization are evaluated by measuring the amount of formaldehyde on a standard filter paper (placed in the sterilization chamber with the items to be sterilized), which may not exceed 200 μg. These standards were created because formaldehyde interacts differently with various types of polymers and other materials that are used in medical devices. Porous materials, such as textiles and paper, have been shown to absorb a considerably higher amount of formaldehyde than plastics, glass, or metals. Filter paper is used as an indicator because it is considered the "worst case" regarding formaldehyde absorption. Thus, if the limit is not exceeded in the filter indicator, it will not be exceeded in other materials. A comparative list of formaldehyde residues on different materials is shown in Table 12.24.

Kanemitsu et al. [103] investigated the residuals of formaldehyde on the filter indicator using different conditions, such as temperature and the number of washing pulses after the sterilization procedure. The results showed that even below the level established in the norm (200 μg), formaldehyde residues were present in the filter paper. Temperature played a significant role in the amount of formaldehyde retrieved. The mean value of residual formaldehyde on filter paper was 181 μg at a sterilization temperature of 50°C and decreased to 32 μg at 80°C (Table 12.25). The same authors [104] demonstrated that certain polymeric materials, after being submitted to the same sterilization program, may retain higher amounts of formaldehyde than in the filter paper. Among the materials investigated, PU and polyamide 6 contained notably high amounts of formaldehyde, whereas the other materials residual contained amounts similar to that of filter paper (Table 12.26). The authors also

TABLE 12.24 Approximate Formaldehyde Residual Figures Compared to Filter Paper after Low-Temperature Steam Formaldehyde Sterilization under the Same Conditions[a]

Material	Approximate Residual Figures Compared to Filter Paper[b]
Filter paper	1.0
Polyester fibers	1.3
Cotton linen	0.65
Polyamide 6	0.8
Polyethylene	0.4
Polypropylene	0.1
Polymethylmetacrylate	0
Cellulose acetate	0.15
Butyl rubber	1.1
Aluminum foil	0
Copper foil	0
Stainless steel	0.1

[a] According to the EN 14180.
[b] Amount material/amount filter paper.

TABLE 12.25 Formaldehyde Residue Absorbed in Filter Indicators after Low-Temperature Steam Formaldehyde Sterilization Carried Out at Different Temperatures and Washing Pulses [103]

Program	Temperature (°C)	Washing Pulse	Formaldehyde Residue (μg)
1	80	20	31.9 ± 13.6
2	65	50	56.3 ± 11.4
3	60	60	54.9 ± 7.4
4	55	60	82.2 ± 10.9
5	50	60	181.0 ± 23.4

evaluated residues retained in medical equipment sterilized with LTSF. Equipment (Table 12.27) was double-packed in sterilization bags. After completing the sterilization cycle, they were submersed in a 0.2 M NaOH solution for 72 hours at room temperature for residue desorption (European Norm test). The stainless steel tweezers did not present any formaldehyde residue. However, the equipment made of EVA and PVC presented residues exceeding the European Standards. These results demonstrated that retention is influenced by surface conditions, material permeability, and material affinity for formaldehyde.

Polymeric formaldehyde (paraformaldehyde) is also present in the residuals of formaldehyde. Paraformaldehyde can form on wet surfaces as a result

TABLE 12.26 Amount of Formaldehyde Found in the Plastic Materials and the Ratio to Amount Retained by the Filter Indicator [104]

Material	Dimension Thickness (mm)	Formaldehyde Residue (μg)	Ratio Sample/ Filter Paper
Soft polyvinyl chloride	0.5	14.8	0.3
Low-density polyethylene	0.5	1.5	0
Polypropylene	0.5	7.2	0.1
Polycarbonate	0.5	26.6	0.5
Polyurethane	1.0	866.0	15.2
Polystyrene	1.0	15.7	0.3
Silicone	0.5	74.5	1.3
Polymethylmetacrylate	0.5	4.8	0.1
Polytetrafluoroethylene	0.5	−0.7	0
Polyethylene terephthalate	0.5	6.3	0.1
Polyamide 6	0.5	1243.6	21.9
Polysulfone	1.0	47.5	0.8
Acrylonitrile butadiene styrene	1.0	25.7	0.5
Polyacetal	0.5	119.6	2.1
Natural rubber	0.5	172.1	3.0
Butyl rubber	1.0	43.9	0.8
Filter paper	0.22	56.9	1.0

TABLE 12.27 Amount of Formaldehyde Found in Different Medical Equipment Sterilized with LTSF [104]

Equipment	Material	Weight (g)	Formaldehyde Residue (μg)
Breathing circuit	Ethylene vinyl acetate	176	260
Anesthesia circuit	Ethylene vinyl acetate	196	240
Oxygen tubing	Soft polyvinyl chloride	54	594
Airway tube	Soft polyvinyl chloride	20	56
Tweezers	Stainless steel	22	0

of a reaction between formaldehyde and water (Equation 12.21), implying that the wettability of a material, which is a measure of its hydrophobic/hydrophilic nature, is important in determining the formation of residuals. Therefore, hydrophilic polymers can be expected to retain more residues than hydrophobic polymers. During aeration for residual elimination, paraformaldehyde dehydrates and forms a precipitate on the surface, which is even more difficult to remove. Table 12.28 shows formaldehyde residues in different materials measured 13 days after sterilization. The residuals were higher on EVOH and very low (no residuals) on HDPE, confirming that wettability is an important parameter in formaldehyde retention.

TABLE 12.28 Formaldehyde Residues on Different Polymeric Materials 13 Days after Sterilization [105]

Polymer	Formaldehyde Residue (μg/cm²)	Polymer	Formaldehyde Residue (μg/cm²)
Ethylene vinyl alcohol	2.94	High-density polyethylene	0.00
Natural rubber	0.74	Polyethylene terephthalate	0.09
Polyamide 6 (film)	0.37	PVC plasticized (25% di-2-ethylhexylphthalate [DEHP])	0.02
Poly(vinylidene chloride)	0.07		

The influence of these properties is more explicit in the rebound release in dialyzers disinfected with formaldehyde for reuse. Rebound release is the delayed liberation of residual amounts of germicide into an infusion, indicating the binding of the disinfectant or sterilant to the material. This effect is mainly observed with dialyzers due to the way they are utilized and reused, which requires disinfection.

Rebound release is illustrated in Figure 12.29 [106], where the formaldehyde concentration in saline emerging from both blood and dialysate lines of a hollow fiber dialyzer was evaluated during the rinsing period. Although the concentration of formaldehyde decreased to approximately 2 μg/mL in both lines, a peak of more than 40 μg/mL in the blood line occurred after clamping off for needle insertion, even when the line had been rinsed with 2 L saline. These data suggest that some formaldehyde binds to the plastic material of the dialyzer during the sterilization process with slow dissociation during the washout phase.

The amount of formaldehyde remaining in different parts of a dialyzer after the washout phase was also evaluated. Lewis et al. [106] performed an experiment that showed that among the tested parts, the membrane retained the highest amount of formaldehyde.

The rebound release of formaldehyde and hydrogen peroxide from reused polysulfone and cellulose acetate membranes dialyzers was evaluated [107]. After dialysis, the blood compartment of the dialyzer was rinsed with 500 mL saline solution. Subsequently, both compartments were cleaned with purified water at 500 mL/min for 15 minutes. The compartments were then filled with the chemical agent and were left to stand for 48 hours. After this period, the dialyzers were rinsed again with saline solution and dialysate at a flow rate of 500 mL/min until disinfectant residuals became undetectable (40 minutes for formaldehyde and 20 minutes for hydrogen peroxide). Rebound release was monitored by recirculating 230 mL of saline solution in the blood

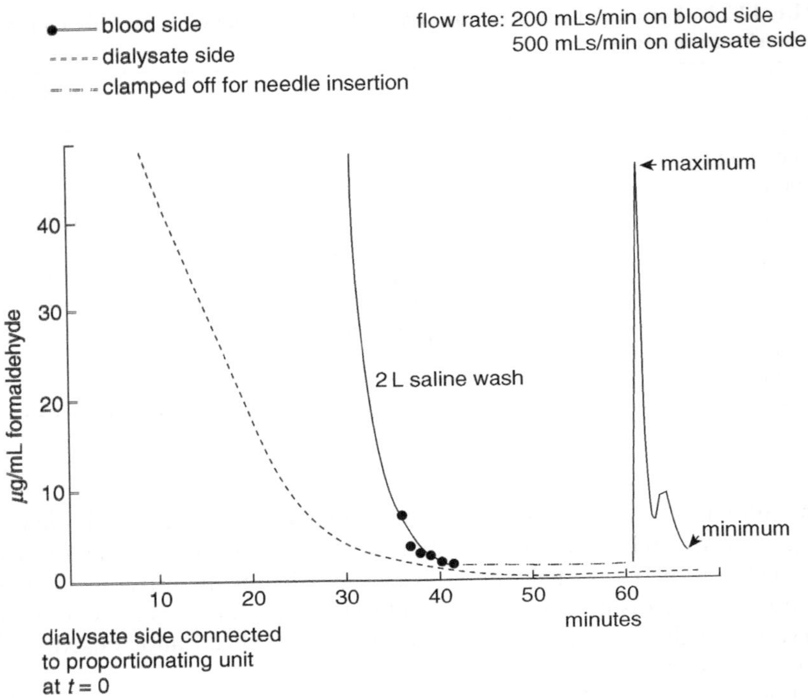

FIGURE 12.29 Formaldehyde concentration in saline emerging from the venous line of a Tri-EX hollow fiber dialyzer sterilized with formalin for reuse, during the original rinsing technique [106].

compartment for 240 minutes and collecting aliquots every 30 minutes to determine the amount of disinfectants. The results in Table 12.29 show that the release of formaldehyde still increases after 240 minutes of saline recirculation for both membranes. However, the concentration of formaldehyde is at least five times higher with cellulose acetate than with polysulfone. In contrast, the same amount of hydrogen peroxide was released from both membranes, which remained the same during the entire period of recirculation.

Rebound release of formaldehyde and the disinfectants sodium hypochlorite and Renalin® was measured 30 minutes after rinsing the dialyzers [108, 109]. Four different materials and new and reused (three times) dialyzers were tested. The results (Table 12.30) indicate that although different membranes retained different amounts of disinfectants, the rebound release of formaldehyde is significantly higher than that of the two other disinfectants. The reuse of dialyzers appears to affect the rebound of formaldehyde in many ways among different membranes. Cellulose acetate and cuprophane rebound release decreased by almost two-thirds after three reuses, whereas it increased more than five times for PMMA [109].

TABLE 12.29 Rebound Release of Formaldehyde and Hydrogen Peroxide from Cellulose Acetate and Polysulfone Membranes in 230 mL Saline during 240 Minutes' Recirculation in the Blood Compartment of Dialyzers [107]

	Cellulose Acetate		Polysulfone	
Time (min)	Formaldehyde (ppm)	Hydrogen Peroxide (ppm)	Formaldehyde (ppm)	Hydrogen Peroxide (ppm)
30	27.7	0.5	4.7	0.5
60	36.1	–	6.4	0.5
120	45.3	0.5	8.4	0.5
180	54.0	–	9.6	0.5
240	55.3	0.5	10.1	0.5

$n = 4$.

TABLE 12.30 Rebound Release of Disinfectants from Different Membranes

	Disinfectant (mg/L)					
Membrane	Formaldehyde 2%	Formaldehyde 4%	Sodium Hypochlorite 0.5%	Renalin 3.5%	Renalin 0.5%	Ref.
Cellulose acetate	–	3–11	–	–	–	[108]
Polyacrylonitrile	–	5–6	≤1	–	0.5–2	
Polysulfone	–	–	–	0.5–10	–	
Cuprophane	–	–	–	2–5	0.8	
Cellulose acetate (new)	98	227	–	–	–	[109]
Cellulose acetate (reused)	37	88	–	–	–	
PMMA (new)	12	47	–	–	–	
PMMA (reused)	160	280	–	–	–	
Cuprophane (new)	63	151	–	–	–	
Cuprophane (reused)	14	55	–	–	–	

Concentration recovered in 100 mL collected after 30 minutes' rinsing
Renalin: stabilized mixture of 4.5% peracetic acid, 6% acetic acid, and 28% hydrogen peroxide.

CONCLUDING REMARKS

Temperature and oxygen are important factors for the formation of radiolysis products in drugs and polymeric materials. Vacuum- and cryo-radiation as well as irradiation of drugs in solid state minimize oxidative degradation reactions. Irradiation of aqueous solutions should account for the "extra" power of water radiolysis products and free radicals, with either reducing or oxidizing properties.

The action of radiation on the backbone structure of carbon chain polymers leads to the formation of carbonyl radicals, which react to build oxygen-containing molecules: acids, aldehydes, and ketones. Even when carrying out the irradiation in vacuum, some radicals are persistent and may later, when the product comes in contact with air, react to build these oxygen-containing species.

Polymers with aromatic rings are more resistant to radiolysis than the equivalent aliphatics with the same functions due to the stabilizing effect of the benzene ring structure. This is valid whether or not the aromatic group is directly in the chain backbone. Thus, both PSs, with a pendant aromatic group, and polyimides, with an aromatic group, directly in the polymer backbone, are relatively resistant to high radiation doses.

In crystalline polymers such as PE, the effects of radiation may not stop after the process finishes but rather may persist for years. Very low diffusion rates of reactive species (radiolysis radicals) in the solid media (implants, which are bulky structures) lead to very slow reaction rates. The more crystalline the polymer, the higher the probability of postirradiation reactions. Consequently, irradiation and subsequent storage may be inappropriate.

Radiolysis products may be trapped in the polymeric matrix during the irradiation procedure and be released afterward when the material comes in contact with a favorable environment such as a drug solution, in the case of a syringe, or the fluids of the human body, in the case of an implant.

Heat and radiation may either lead to decomposition of the drug product or other formulation constituent. Moreover, heat and radiation may act on the container material, causing its decomposition and the formation of undesirable contaminant species. Polymeric materials exhibit a different resistance against heat, radiation, and chemical agents for sterilization.

Chemical agents are effective sterilants or disinfectants for items that are not able to withstand heat or radiation. The actions of these agents are related to their capability to penetrate different materials, which also depends on the material itself. Residual contamination is associated with the extent of the absorption of the chemical agent into the material. Therefore, aeration is essential for the effective removal of residuals, adaptation of aeration time to the material, and piece size and shape.

REFERENCES

1. Centers for Disease Control and Prevention. Guideline for Disinfection and Sterilization in Healthcare Facilities, Rutala WA, Weber DJ, 2008. Available at http://www.cdc.gov/hicpac/pdf/guidelines/Disinfection_Nov_2008.pdf (accessed November 2011).
2. Richter HW. Radiation chemistry: principles and applications. In: *Photochemistry and Radiation Chemistry*, Wishart J (ed.). Advances in Chemistry. American Chemical Society, Washington, DC, 1998, pp. 5–24.

3. Basly JP, Duroux JL, Bernard M. Gamma radiation induced effects on metronidazole. Int J Pharm 139, 219–221, 1996.
4. Horsch P, Bigler L, Altorfer HR. Influence of radiation sterilization on the stability of trifluorothymidine. Int J Pharm 222, 205–215, 2001.
5. Fauconnet AL, Basly JP, Bernard M. Gamma radiation induced effects on isoproterenol. Int J Pharm 144, 123–125, 1996.
6. Basly JP, Longy I, Bernard M. Radiation sterilization of formoterol. Pharm Res 14, 810–814, 1997.
7. Basly JP, Bernard M. Radiosterilization dosimetry by ESR spectroscopy: ritodrine hydrochloride and comparison with other sympathomimetics. Int J Pharm 149, 85–91, 1997.
8. Naskrent M, Mielcarek J. The effect of ionizing radiation of 1,4-dihydropyridine derivatives in the solid state. Spectrochim Acta [A] 67, 251–255, 2007.
9. Colak S, Korkmaz M. Investigation of radiosterilization and dosimetric features of sulfacetamide sodium. J Pharm Biomed Anal 36, 791–798, 2004.
10. Katusin-Razem B, Hamitouche K, Maltar-Strmecki N, Kos K, Pucić I, Britvić-Budicin S, Ražem D. Radiation sterilization of ketoprofen. Radiat Phys Chem 73, 111–116, 2005.
11. Colak S, Korkmaz M. Spectroscopic features of radiolytic intermediates induced in gamma irradiated sulfatiazole: an ESR study. Int J Pharm 285, 1–11, 2004.
12. Varshney L, Dodke PB. Radiation effect studies on anticancer drugs, cyclophosphamide and doxorubicin for radiation sterilization. Radiat Phys Chem 71, 1103–1111, 2004.
13. Basly JP, Duroux JL, Bernard M. Gamma irradiation sterilization of orciprenaline and fenoterol. Int J Pharm 142, 125–128, 1996.
14. Montanari L, Cilurzo F, Conti B, Genta I, Groppo A, Valvo L, Faucitano A, Buttafava A. Gamma irradiation effects and EPR investigation on poly(lactide-co-glycolide) microspheres containing bupivacaine. J Control Release 57, 427–433, 2002.
15. Basly JP, Duroux JL, Bernard M. The effect of gamma radiation on the degradation of Salbutamol. J Pharm Biomed Anal 15, 1137–1141, 1997.
16. Basly JP, Longy I, Bernard M. Radiation effects on dopamine and norepinephrine. Pharm Res 14, 1192–1196, 1997.
17. Vas G, Alquier L, Maryanoff CA, Cohen J, Reed G. Investigation of mass-balance issue in e-beam sterilized paclitaxel eluting coronary stents by SPME/GC-MS. J Pharm Biomed Anal 48, 568–572, 2008.
18. El-Bagory IM. Protective effect of scavengers and surfactants on gamma irradiated cortisone acetate aqueous solutions. J Drug Deliv Sci Technol 17, 437–442, 2007.
19. Maquille A, Slegers C, Habib JL, Tilquin B. Electron beam and gamma radiolysis of solid-state metoclopramide. Pharm Res 23, 1343–1349, 2006.
20. Kryczka T, Marciniec B, Popielarz-Brzezinska M, Bero M, Kasperczyk J, Dobrzyński P, Kazimierczuk Z, Grieb P. Effect of gamma-irradiation on cladribine and cladribine-containing biodegradable copolymers. J Control Release 89, 447–456, 2003.

21. Moyne P, Botella A, Peyrouset A, Rey L. Sterilization of injectable drugs solutions by irradiation. Radiat Phys Chem 6, 703–704, 2002.
22. Friess W, Schlapp M. Sterilization of gentamicin containing collagen/PLGA microparticle composites. Eur J Pharm Biopharm 63, 176–187, 2006.
23. Soriano I, Martin AY, Evora C, Sánchez E. Biodegradable implantable fluconazole delivery rods designed for the treatment of fungal osteomyelitis: influence of gamma sterilization. J Biomed Mater Res A 77, 632–638, 2006.
24. Mohr D, Wolff M, Kissel T. Gamma irradiation for terminal sterilization of 17 beta-estradiol loaded poly-(D,L-lactide-co-glycolide) microparticles. J Control Release 61, 203–217, 1999.
25. Volland C, Wolff M, Kissel T. The influence of terminal gamma-sterilization on captopril containing poly(D,L-lactide-co-glycolide) microspheres. J Control Release 31, 293–305, 1994.
26. Uchegbu IF, Schatzlein AG (eds.). *Polymers in Drug Delivery*. Taylor & Francis, Boca Raton, FL, 2006.
27. Anderson JM, Shive MS. Biodegradation and biocompatibility of PLA and PLGA microspheres. Adv Drug Deliv Rev 28, 2–24, 1997.
28. Faucitano A, Buttafava A, Montanari L, Cilurzo F, Conti B, Genta I, Valvo L. Radiation-induced free radical reactions in polymer/drug system for controlled release: an EPR investigation. Radiat Phys Chem 67, 61–72, 2003.
29. Babanalbandi A, Hill DJT, Whittaker AK. Volatile products and new polymer structures formed on Co-60 γ-radiolysis of poly(lactic acid) and poly(glycolic acid). Polym Degrad Stab 58, 203–214, 1997.
30. Tyan YC, Liao JD, Lin SP, Chen CC. The study of the sterilization effect of gamma ray irradiation of immobilized collagen polypropylene nonwoven fabric surfaces. J Biomed Mater Res A 67A(3), 1033–1043, 2003.
31. Cheung DT, Perelman N, Tong D, Nimni ME. The effect of γ-irradiation on collagen molecules isolated α-chains, and cross-linked native fibers. J Biomed Mater Res 24, 581–589, 1990.
32. Kawano Y, Logarezzi AJM. X-ray induced degradation of regenerated cellulose membrane films. Polym Degrad Stab 50, 125–130, 1995.
33. Kabanov VYa, Feldman VI, Ershov BG, Polikarpov AI, Kiryukhin DP, Apel PYu. Radiation chemistry of polymers. High Energy Chem 43, 1–18, 2009.
34. Gryczka U, Dondi D, Chmielewsky AG, Migdal W, Buttafava A, Faucitano A. The mechanism of chitosan degradation by gamma and e-beam irradiation. Radiat Phys Chem 78, 543–548, 2009.
35. Zelinska K, Shostenko AG, Truszkowski S. Radiology of chitosan. High Energy Chem 43(6), 445–448, 2009.
36. Nagasawa N, Mitomo H, Yoshii F, Kume T. Radiation-induced degradation of sodium alginate. Polym Degrad Stab 69, 279–285, 2000.
37. Barbarin N, Crucq AS, Tilquin B. Study of volatile compounds from the radiosterilization of solid cephalosporins. Radiat Phys Chem 48(6), 787–794, 1996.
38. Barbarin N, Rollmann B, Tilquin B. Role of residual solvents in the formation of volatile compounds after radiosterilisation of cefotaxime. Int J Pharm 178(2), 203–212, 1999.

39. Shostenko AG, Truszkowski S, Bukhal'skii A, Churkin AA, Dobrov VI. Radiolysis of L- and D-monosaccharides. High Energy Chem 37, 205–208, 2003.

40. De Kerf M, Mondelaers W, Lahorte P, Vervaet C, Remon JP. Characterisation and disintegration properties of irradiated starch. Int J Pharm 221(1–2), 69–76, 2001.

41. Buchalla R, Boess C, Bogl KW. Analysis of volatile radiolysis products in gamma-irradiated LDPE and polypropylene films by thermal desorption-gas chromatography-mass spectrometry. Appl Radiat Isot 52(2), 251–269, 2000.

42. Yeom B, Yu YJ, McKellop HA, Salovey R. Profile of oxidation in irradiated polyethylene. J Polym Sci Part A 36(2), 329–339, 1998.

43. Premnath V, Harris WH, Jasty M, Merril EW. Gamma sterilization of UHMWPE articular implants: an analysis of the oxidation problem. Biomaterials 17, 1741–1753, 1996.

44. Costa L, Carpentieri I, Bracco P. Post electron-beam irradiation oxidation of orthopaedic UHMWPE. Polym Degrad Stab 93, 1695–1703, 2008.

45. Feldman D, Barbalata A. *Synthetic Polymers*. Chapman & Hall, London, 1996, pp. 272–317.

46. Lewandowski TA, Hayes AW, Beck BD. Risk evaluation of occupational exposure to methylene dianiline and toluene diamine in polyurethane foam. Hum Exp Toxicol 24(12), 655–662, 2005.

47. Hirata N, Matsumoto Ki, Inishita T, Takenaka Y, Suma Y, Shintani H. Gamma-ray irradiation, autoclave and ethylene-oxide sterilization to thermosetting polyurethane—sterilization to polyurethane. Radiat Phys Chem 46, 377–382, 1995.

48. Shintani H. Formation and elution of toxic compounds from sterilized medical products—methylenedianiline formation in polyurethane. J Biomater Appl 10(1), 23–58, 1995.

49. Shintani H. Formation and elution of toxic compounds from sterilized medical products: toxic compound formation from irradiated products. Radiat Phys Chem 47, 139–148, 1996.

50. Guignot C, Betz N, Legendre B, Le Moel A, Yagoubi N. Degradation of segmented poly(etherurethane) Tecoflex® induced by electron beam irradiation: characterization and evaluation. Nucl Instrum Methods Phys Res B 185, 100–107, 2001.

51. Palsule AS, Clarson SJ, Widenhouse CW. Gamma irradiation of silicones. J Inorg Organomet Polym 18, 207–221, 2009.

52. Miller AA. Radiation chemistry of polydimethylsiloxane. J Am Chem Soc 82, 3519, 1960.

53. Rogero SO, Souza JS, Alário D, Jr., Lopérgolo L, Lugao AB. Silicone cross-linked by ionizing radiation as potential polymeric matrix for drug delivery. Nucl Instrum Methods Phys Res B 236, 521–525, 2005.

54. Stoffers NH, Linssen JPH, Franz R, Welle F. Migration and sensory evaluation of irradiated polymers. Radiat Phys Chem 71, 203–206, 2004.

55. Yagoubi N, Baillet A, Pellerin F, Ferrier D. Physico-chemical behaviour of β irradiated plastic materials currently used as packagings and medical products. Nucl Instrum Methods Phys Res B 105, 340–344, 1995.

56. Buchalla R, Boess C, Bogl KW. Characterization of volatile radiolysis products in radiation-sterilized plastics by thermal desorption-gas chromatography-mass

spectrometry: screening of six medical polymers. Radiat Phys Chem 56(3), 353–367, 1999.

57. Pratt CM, Barton S, McGonigle E, Kishi M, Foot PJS. Effects of ionising radiation on poly(methyl methacrylate) used in intra-ocular lenses. Polym Degrad Stab 91, 2315–2317, 2006.
58. Kowalski JB, Morrissey RF. Sterilization of implants. In: *Biomaterials Science*, Ratner BD, Hoffman AS, Schoen FJ, Lemons JE (eds.). Academic Press, San Diego, CA, 1996, pp. 415–420.
59. Berridge MS, Apana SM, Hersh JM. Teflon radiolysis as the major source of carrier in fluorine-18. J Labelled Comp Radiopharm 59, 543–548, 2009.
60. Hill DJT, O'Donnell JH, Pomery PJ, Sangter DF. Radiation chemical yields: g-values. In: *Polymer Handbook*, 3rd ed. Brandrup J, Immergut EH (eds.). Wiley-Interscience, New York, 1989, pp. II/387–II/397.
61. Manfredini M, Bodecchi LM, Marchetti A. Thermal stability and chemical durability of PVC-based biomedical devices. J Appl Polym Sci 102, 5378–5387, 2006.
62. Demertzis PG, Franz R, Welle F. The effects of gamma-irradiation on compositional changes in plastic packaging films. Packag Technol Sci 12(3), 119–130, 1999.
63. Colombani J, Herbette G, Rossi C, Joussot-Dubien C, Labed V, Gilardi T. Leaching of plasticized PVC: effect of irradiation. J Appl Polym Sci 112, 1372–1377, 2009.
64. Yagoubi N, Peron R, Legendre B, Grossiord JL, Ferrier D. Gamma and electron beam radiation induced physico-chemical modifications of poly(propylene). Nucl Instrum Methods Phys Res B 151, 247–254, 1999.
65. Yagoubi N, Boucherie P, Ferrier D. Radiosterilization and stem autoclaving sterilization effects on phosphate antioxidant stability. Nucl Instrum Methods Phys Res B 131, 398–404, 1997.
66. Soebianto YS, Kusuhata I, Katsumura Y, Ishigure K, Kubo J, Kudoh H, Seguchi T. Degradation of polypropylene under gamma irradiation: protection effect of additives. Polym Degrad Stab 50, 203–210, 1995.
67. Fintzou AT, Badeka AV, Kontominas MG, Riganakos KA. Changes in physicochemical and mechanical properties of γ-irradiated polypropylene syringes as a function of irradiation dose. Radiat Phys Chem 75, 87–97, 2006.
68. Fintzou AT, Badeka AV, Kontominas MG, Stahl MR, Riganakos KA. Changes in physicochemical and mechanical properties of electron-beam irradiated polypropylene syringes as a function of irradiation dose. Radiat Phys Chem 76, 97–107, 2007.
69. Buchalla R, Begley TH. Characterization of gamma-irradiated polyethylene terephthalate by liquid-chromatography-mass-spectrometry (LC-MS) with atmospheric-pressure chemical ionization (APCI). Radiat Phys Chem 75(1), 129–137, 2006.
70. Buchalla R, Begley TH, Morehouse KM. Analysis of low-molecular weight radiolysis products in extracts of gamma-irradiated polymers by gas chromatography and high-performance liquid chromatography. Radiat Phys Chem 63, 837–840, 2002.
71. Rupiasih NN, Vidyasagar PB. Comparative study of the effect of low and medium dose rate of γ-irradiation on microporous polysulfone membrane using spectroscopic and imaging techniques. Polym Degrad Stab 93, 1300–1307, 2008.

72. Rivaton A, Gardette JL. Photodegradation of polyethersulfone and polysulfone. Polym Degrad Stab 66, 385–403, 1999.

73. Riganakos KA, Koller WD, Ehlermann DAE, Bauer B, Kontominas MG. Effects of ionizing radiation on properties of monolayer and multilayer flexible food packaging materials. Radiat Phys Chem 54(5), 527–540, 1999.

74. Goulas AE, Riganakos KA, Kontominas MG. Effect of ionizing radiation on physicochemical and mechanical properties of commercial multilayer coextruded flexible plastics packaging materials. Radiat Phys Chem 68(5), 865–872, 2003.

75. Goulas AE, Riganakos KA, Kontominas MG. Effect of ionizing radiation on physicochemical and mechanical properties of commercial monolayer and multi-layer semirigid plastics packaging materials. Radiat Phys Chem 69(5), 411–417, 2004.

76. Kjellstrand P, Erixon M, Wieslander A, Lindén T, Martinson E. Temperature: the single most important factor for degradation of glucose fluids during storage. Perit Dial Int 24(4), 385–391, 2004.

77. Erixon M, Lindén T, Kjellstrand P, Carlsson O, Ernebrant M, Forsbäck G, Wieslander A, Jönsson JA. PD fluids contain high concentrations of cytotoxic GDPs directly after sterilization. Perit Dial Int 24(4), 392–398, 2004.

78. Schalkwijk CG, Posthuma N, Ten Brink HJ, Ter Wee PM, Teerlink T. Induction of 1,2-dicarbonyl compounds, intermediates in the formation of advanced glycation end-products, during heat-sterilization of glucose-based peritoneal dialysis fluids. Perit Dial Int 19(4), 325–333, 1999.

79. Erixon M, Wieslander A, Lindén T, Carlsson O, Forsbäck G, Svensson E, Jönsson JA, Kjellstrand P. How to avoid glucose degradation products in peritoneal dialysis fluids. Perit Dial Int 26(4), 490–497, 2006.

80. Frischmann M, Spitzer J, Fünfrocken M, Mittelmaier S, Deckert M, Fichert T, Pischetsrieder M. Development and validation of an HPLC method to quantify 3,4-dideoxyglucosone-3-ene in peritoneal dialysis fluids. Biomed Chromatogr 23(8), 843–851, 2009.

81. Bryland A, Broman M, Erixon M, Klarin B, Lindén T, Friberg H, Wieslander A, Kjellstrand P, Ronco C, Carlsson O, Godaly G. Infusion fluids contain harmful glucose degradation products. Intensive Care Med 36(7), 1213–1220, 2010.

82. Erixon M, Wieslander A, Lindén T, Carlsson O, Forsbäck G, Svensson E, Jönsson JA, Kjellstrand P. Take care in how you store your PD fluids: actual temperature determines the balance between reactive and non-reactive GDPs. Perit Dial Int 25(6), 583–590, 2005.

83. Zimmeck T, Tauer A, Fuenfrocken M, Pischetsrieder M. How to reduce 3-deoxyglucosone and acetaldehyde in peritoneal dialysis fluids. Perit Dial Int 22(3), 350–356, 2002.

84. Brandrup J, Immergut EH (eds.). *Polymer Handbook*, 3rd ed. Wiley & Sons, New York, 1989,

85. ANSI/AAMI/ISO 10993-7; Biological Evaluation of Medical Devices—Part 7: Ethylene Oxide Sterilization Residuals, 2008.

86. O'Leary RK, Watkins WD, Guess WL. Comparative chemical and toxicological evaluation of residual ethylene oxide in sterilized plastics. J Pharm Sci 58(8), 1007–1010, 1969.

87. Stetson JB, Whitbourne JE, Eastman C. Ethylene oxide degassing of rubber and plastic materials. Anesthesiology 44(2), 174–180, 1976.
88. Mogenhan JA, Whitbourne JE, Ernst RR. Determination of ethylene oxide in surgical materials by vacuum extraction and gas chromatography. J Pharm Sci 60(2), 222–224, 1971.
89. Zagar LA. Determination of residual ethylene oxide in methyl methacrylate polymer powders by GLC. J Pharm Sci 61(11), 1803, 1801.
90. White JD, Bradley TJ. Residual ethylene oxide in gas-sterilized medical-grade silicones. J Pharm Sci 62(10), 1634–1637, 1973.
91. Jones AB. In vitro evaluation of hemolytic and cell culture toxicity potential of residual ethylene oxide in selected medical materials. J Biomed Mater Res 13(2), 207–216, 1979.
92. Matthews IP, Gibson C, Samuel AH. Enhancement of the kinetics of the aeration of ethylene oxide sterilized polymers using microwave radiation. J Biomed Mater Res 23(2), 143–156, 1989.
93. Lucas AD, Merritt K, Hitchins VM, Woods TO, McNamee SG, Lyle DB, Brown SA. Residual ethylene oxide in medical devices and device material. J Biomed Mater Res B Appl Biomater 66(2), 548–552, 2003.
94. Wang Y, Shen C, Li J, Wang H, Wang H, Jiang H, Chu Y. Thermal desorption extraction proton transfer reaction mass spectrometer (TDE-PTR-MS) for rapid determination of residual solvent and sterilant in disposable medical devices. J Pharm Biomed Anal 55(5), 1213–1217, 2011.
95. Martins AMS, Penna TCV, Diais F, Godoi I, Nogaroto S. Determinação dos níveis de óxido de etileno, etilenocloridrina e etilenoglicol em cânulas de perfusão após processo de esterilização. Rev Lat Tecnol Extracorpórea 10(4), 2003. Available at http://perfline.com/revista/volume10/v10n4/v10n4-02.html.
96. Agencia Nacional de Vigilância Sanitária—ANVISA. Portaria 482 de 16 de abril de 1999–Ministério da Saúde, Brazil 1999.
97. Lomas RJ, Gillan HL, Matthews JB, Ingham E, Kearney JN. An evaluation of the capacity of differently prepared demineralised bone matrices (DBM) and toxic residuals of ethylene oxide (EtOx) to provoke an inflammatory response in vitro. Biomaterials 22(9), 913–921, 2001.
98. Arizono T, Iwamoto Y, Okuyama K, Sugioka Y. Ethylene oxide sterilization of bone grafts: residual gas concentration and fibroblast toxicity. Acta Orthop Scand 65(6), 640–642, 1994.
99. Escalup L, Crauste-Manciet S, Richard L, Brossard D. Chemical sterilization with peracetic acid: permeation risks inside parenteral infusion fluid. PDA J Pharm Sci Technol 55(1), 24–32, 2001.
100. Bourguignon C, Grain A, Schlatter J, Vermerie N. Permeation risks with peracetic acid and hydrogen peroxide sterilizing agent inside ambulatory pumps. PDA J Pharm Sci Technol 59(3), 200–205, 2005.
101. Ikarashi Y, Tsuchiya T, Nakamura A. Cytotoxicity of medical materials sterilized with vapour-phase hydrogen peroxide. Biomaterials 16(3), 177–183, 1995.
102. European Standards. CSN EN 14180+A2—Sterilizers for medical purposes—Low temperature steam and formaldehyde sterilizers—Requirements and testing, 2009.

103. Kanemitsu K, Kunishima H, Imasaka T, Ishikawa S, Harigae H, Yamato S, Hirayama Y, Kaku M. Evaluation of a low-temperature steam and formaldehyde sterilizer. J Hosp Infect 55(1), 47–52, 2003.

104. Kanemitsu K, Kunishima H, Saga T, Harigae H, Imasaka T, Hirayama Y, Kaku M. Residual formaldehyde on plastic materials and medical equipment following low-temperature steam and formaldehyde sterilization. J Hosp Infect 59(4), 361–364, 2005.

105. Vink P. Residual formaldehyde in steam—formaldehyde sterilized materials. Biomaterials 7, 221–227, 1986.

106. Lewis KJ, Ward MK, Kerr DNS. Residual formaldehyde in dialyzers: quantity, location, and the effect of different methods of rinsing. Artif Organs 5(3), 269–277, 1981.

107. Tanaka N, Tanaka N, Fujisawa T, Daimon T, Fujiwara K, Yamamoto M, Abe T. The use of electrolyzed solutions for the cleaning and disinfecting of dialyzers. Artif Organs 24(12), 921–928, 2000.

108. Stragier A, Wenderickx D, Jadoul M. Rinsing time and disinfectant release of reused dialyzers: comparison of formaldehyde, hypochlorite, warexin, and Renalin. Am J Kidney Dis 26(3), 549–553, 1995.

109. Hakim RM, Friedrich RA, Lowrie EG. Formaldehyde kinetics and bacteriology in dialyzers. Kidney Int 28(6), 936–943, 1985.

13

BIOTECHNOLOGICAL PRODUCTS

Biopharmaceuticals are pharmaceuticals derived from life forms. These substances can be obtained through different processes, including extraction from natural sources and production by microorganisms or cell cultures either *in vitro* or *in vivo*. Biopharmaceuticals are either specific proteins, such as enzymes and antibodies, which are used as medication because of their particular function in the body, or they are living microorganisms, such as viruses and bacteria, which, after attenuation, can be used for therapeutic or *in vivo* diagnostic purposes.

Products such as albumin and human growth hormone are both biopharmaceuticals; however, whereas albumin is obtained from plasma (human or bovine) through fractionation, human growth hormone is produced by recombinant DNA (rDNA) technology. While small drug molecules are synthesized through chemical reactions, the larger protein molecules are created in living cells, bacteria, yeast, animals, or plants.

Biotechnological rDNA technology in the field of biopharmaceuticals encompasses the following applications:

- therapeutic agents
- vaccines

Sources of Contamination in Medicinal Products and Medical Devices,
First Edition. Denise Bohrer.
© 2013 by John Wiley & Sons, Inc. Published 2013 by John Wiley & Sons, Inc.

- gene therapy
- diagnostic agents.

The production requires distinct steps [1]:

- incorporation of specific genes into select host cells using a vector that can carry these genes (DNA fragments)
- cultivation of modified host cells through cell culture or fermentation processes to produce large quantities of the desired gene product
- separation and purification of cell cultures or fermentation processes to obtain the desired product.

Although several different routes may be followed, the most widely employed hosts for cloning DNA fragments are the bacterium *Escherichia coli* and mammalian cells of Chinese hamster ovary (CHO).

Figure 13.1 displays a general flowchart for the production of biopharmaceuticals using the rDNA technology, the main raw materials used in each step, and the potential impurities and contaminants associated with the process. Because of the nature of the host cells, the need for the addition of different substances in each step of production associated with the fragility of the target products makes separation and purification the most challenging steps when considering the final product quality.

When working with biological products, the usual impurities associated with manufacturing processes and product instability are sources of contaminants. Protein degradation may be responsible for the generation of undesirable species, and protein aggregation and unfolding may lead to product potency reduction and the formation of species that cause adverse reactions. In addition, product-related impurities are specific to each product type, and their presence depends on the manufacture and purification processes.

Because of the origin in living organism, the most significant impurities and contaminants in biopharmaceuticals are hostile species, such as viruses and endotoxins. When working with genetically modified products, molecules originating from the host cells, namely, DNA and host cell proteins (HCPs), are also important impurities. Although mutual impurities, HCPs are unique to their respective hosts. Recombinant proteins produced by *E. coli* may be expressed directly in the cytoplasm or secreted into the periplasm, whereas in mammalian cells, they are secreted into the culture medium. Each of the different expression modalities and techniques can produce a different population of HCP [2, 3]. The European Pharmacopoeia does not establish qualitative nor quantitative limits for HCP and residual DNA in the monographs of products generated by rDNA technology (Table 13.1). However, in the FDA document "Points to Consider in the Manufacture and Testing of Monoclonal Antibody Products for Human Use" [4], screening for retroviruses and adventitious viruses is advised, and a limit of 10 pg/dose of DNA is suggested.

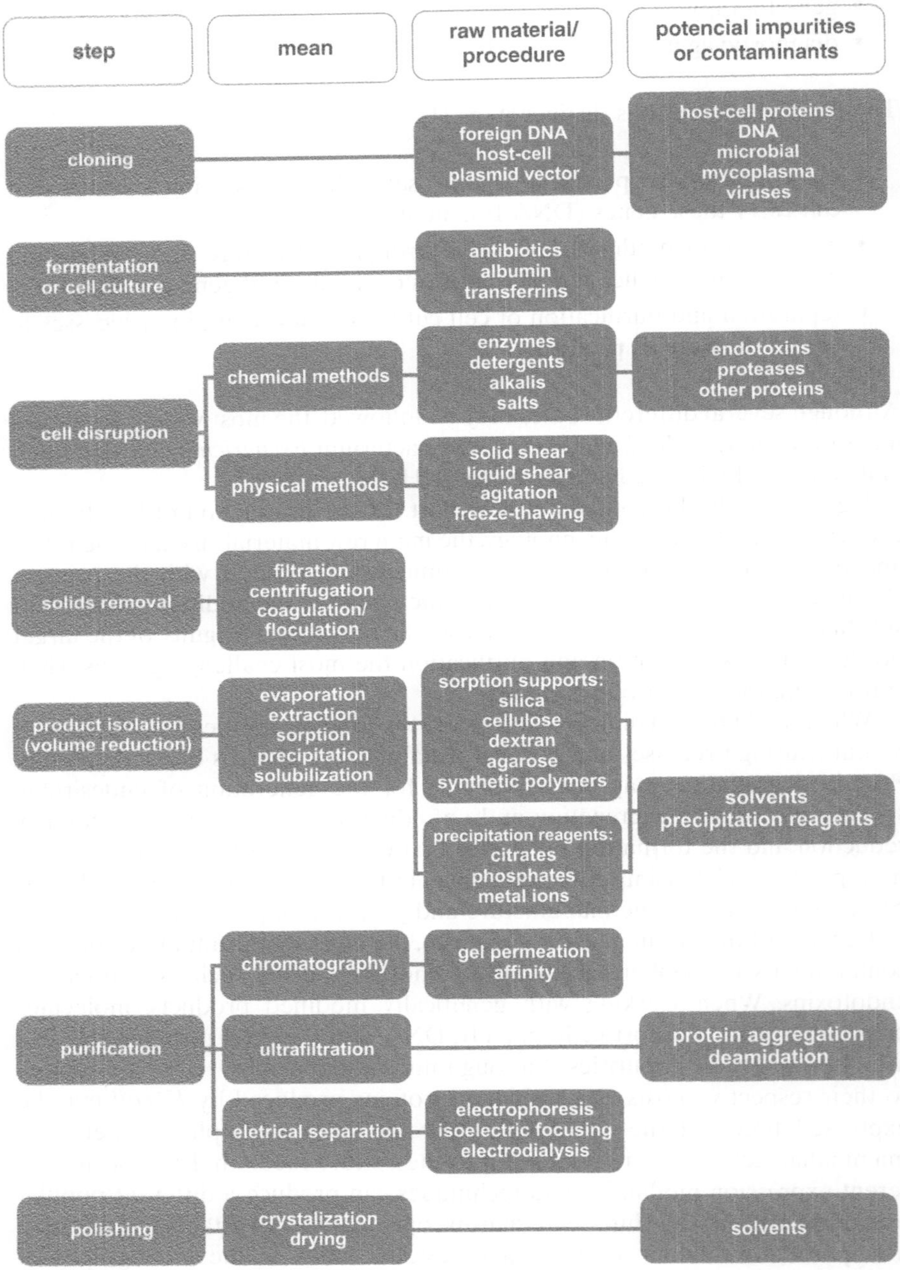

FIGURE 13.1 General flowchart for the production of biopharmaceutical using the rDNA technology.

TABLE 13.1 Impurities in Products of Recombinant DNA Technology of the European Pharmacopeia

Product	Host Cells	Impurity				
		Host cell-derived proteins	Vector-derived DNA	Endotoxins	Higher molecular mass substances or related proteins	Others
Erythropoietin	Rodent cells	Limit approved by the competent authority	Limit approved by the competent authority	20 IU in 100,000 IU EPO	<2% of the EPO content[a]	–
Insulin lispro	n.i.	Limit approved by the competent authority	–	<10 IU/mg	<0.25% of the insulin content[a]	<1% A21 desamido insulin lispro <0.5% Any other impurity
Insulin aspart	n.i.	Limit approved by the competent authority	–	<10 IU/mg	<0.5% of the insulin content[a]	<1% B28isoAsp insulin aspart <1.5% Any other impurity
Interferon alfa-2	Bacteria	Limit approved by the competent authority	Limit approved by the competent authority	100 IU/1.0 mg protein	<5% of the interferon content[b]	–
Interferon gamma-2	Bacteria	Limit approved by the competent authority	Limit approved by the competent authority	5 IU/20 × 10^6 IU interferon	<2% of the interferon content[a]	<10% Deamidated and oxidized forms, <3% heterodimers, <0.2 mol norleucine /mol interferon
Coagulation factor VIII	Mammalian cell	–	–	3 IU/100 IU factor VIII	–	–
Hepatitis B vaccine	Yeast or mammalian cells	<10 pg of DNA in a quantity equivalent to a single-dose vaccine	–	–	<5% of the hepatitis B surface antigen	–

[a] Determined by size-exclusion chromatography. Sum of the areas of the peaks with retention time less than the principal peak (percentage of the total).
[b] Determined by liquid chromatography.
n.i., not informed.

13.1 DNA AND HCP RESIDUALS

The host produces HCPs during culture or fermentation. These proteins are not the target proteins and can cause an immune response in patients. The HCP composition is extremely heterogeneous and depends on the host cell and the protein product.

Table 13.2 shows the residual DNA and HCP in process streams of recombinant protein products expressed in *E. coli* and other cells. In the studies using *E. coli*, cells were grown to express the protein of interest as insoluble inclusion bodies (IBs) within the cells. The IBs were isolated by repeated homogenization and centrifugation, and the IB slurry was solubilized with urea. Refolding was conducted before or after the solution was loaded on several chromatographic phase columns. The products were eluted, and fractions were collected and analyzed for DNA and HCPs. Similar processes were used for the clarification and purification of the target species in other studies. Depth filtration combined with size exclusion chromatography was employed by Nayak et al. [11] for the purification of equine influenza virus, while Stein and Kiesewetter [13] utilized an optimized pH for cation exchange antibody purification. The reduction of HCP to less than 100 μg/g was achieved by Toueille et al. [12], exploiting a combination of both ionic and hydrophobic chromatography, and Forcic et al. [10] purified the rubella virus using monolithic chromatographic supports.

13.2 VIRUSES AND MYCOPLASMA

Viral contamination may arise from the original source of the cell lines or from the adventitious introduction of viruses during the production process. Viruses may be present in the master cell bank (MCB) as a result of the derivation of cells from infected animals, the use of viruses to establish the cell line, the use of contaminated reagents such as serum components, and contamination during handling. Adventitious viruses can be introduced into the final product through the use of contaminated reagents, the use of a virus for the induction of expression of specific genes, and contamination during handling.

Kerr and Nims [14] presented an overview on viral contaminants detected in biopharmaceutical bulk harvested samples at a biotech industry over a period of 10 years. Viruses were detected only in processes involving CHO cells. The findings confirmed that, although viruses are of great concern for biopharmaceuticals manufactured using mammalian cells, contamination rates are low. In this period of time, only three viruses had been isolated from the CHO cell manufacturing process on more than one occasion: mouse minute virus (MMV), reovirus (REO), and the bunyavirus Cache Valley. Sources were attributed to the nonhomogeneous contamination of raw materials, rodent contamination in the case of MMV, and bovine serum used at some point during the manufacturing process.

TABLE 13.2 Quantification of Residual DNA and HCPs in Purified Recombinant Proteins

Sample	Agent	Medium	DNA (pg/mL)	HCP (μg/g Drug Protein) or μg/mL	Reference
Monoclonal antibody	–	–	12.4 ± 1.0	–	[5]
Vaccine sample 1	Malaria	*E. coli*	–	0.362 μg/g	[6]
Vaccine sample 2	Malaria	*E. coli*	–	1.112 μg/g	[6]
Vaccine pool 1	Cancer	*E. coli*	6160	160 μg/mL	[7]
Vaccine pool 2	Cancer	*E. coli*	5220	10 μg/mL	[7]
Vaccine	Human respiratory syncytial virus	*E. coli*	–	200 μg/mL	[8]
Plasminogen activator	Thrombotic disorder	*E. coli*	–	<1 ng/mg	[9]
Vaccine	Rubella	Human lung fibroblasts	600 pg	0 mg	[10]
Vaccine	Equine influenza virus	Madin–Darby canine kidney	23.500	9 μg/mL	[11]
Monoclonal antibody		CHO cell culture	–	<100 μg/g	[12]
Monoclonal antibody		NS/0 hybridoma cell culture	–	84 ng/mg	[13]

Another investigation on an adventitious agent contaminant in a CHO cell-derived MCB resulted in the identification of the viral nature of the contaminant [15]. The contaminant was an equine rhinitis virus, whose source was not originated from the MCB, but the equine serum added to the cell growth medium.

In a research conducted by Kozasa et al. [16], of 49 samples of commercial fetal bovine serum(FBS) from 10 different manufacturers in seven countries, 57.1% presented pestiviral genes, and two samples (4.1%) contained infectious bovine viral diarrhea virus (BVDV). BVDV is a common pathogen that contaminates bovine-derived materials, such as FBS, which is used in cell cultures. The results indicated that the parent cow may be infected during pregnancy, causing the fetuses to also be infected.

The inactivation of viruses may be attained by gamma irradiation of the frozen sera. Nims et al. [17] reviewed this topic and concluded that viruses, as well as mollicutes, may effectively be inactivated by gamma irradiation. However, while mollicutes (acholeplasma and mycoplasmas) appear to be readily inactivated by the radiation doses typically employed (25–40 kGy), for viruses, the data indicate that inactivation depends on the size of the target virus. The authors showed that larger viruses were more effectively inactivated than smaller viruses. Table 13.3 displays the $\log_{10}$ reduction in titer of mollicutes and viruses in frozen bovine serum after gamma irradiation. The same radiation dose was much more effective for infectious bovine rhinotracheitis virus (120- to 200-nm particle size) than for porcine circovirus (17 nm).

13.3 ENDOTOXIN

Endotoxin is a lipopolysaccharide (LPS) present in the cell wall of most gram-negative bacteria, including *E. coli*. Endotoxins are not secreted but are released only when the cells are disrupted, that is, when the bacterium dies. After release, the endotoxin is capable of producing a toxic manifestation within the host. Endotoxin comprises approximately 3% of the bacterial cell lysate constituents [18] (Fig. 13.2) and is found in large quantities in *E. coli* and in other cell cultures (Table 13.4) [19].

The usual techniques for endotoxin removal are ultrafiltration, which is not applicable to protein purification, and ion-exchange chromatography, which may also not be applicable to proteins because of their own affinity with the stationary phase. Liu et al. [20] compared three methods for endotoxin removal from *E. coli* recombinant protein purification. Recombinant proteins, creatinine kinase isozymes (rCK-BB, rCK-MB, rCK-MM), cardiac troponin I (rcTnI), and recombinant myoglobin were purified from *E. coli* lysates and IB using three different methods: polymyxin B and histidine affinity chromatography and Triton X-114 phase separation. Endotoxin levels were determined in purified proteins using the limulus amebocyte lysate (LAL) gel-clot assay. The results presented in Table 13.5 show that endotoxin levels in recombinant

TABLE 13.3 Efficacy of Gamma Irradiation for Mollicute and Virus Inactivation in Frozen Serum [17]

Microorganism	Family	Particle Size (nm)	Irradiation Dose (kGy)	Log_{10} Reduction in Titer
Acholeplasma laidlawii	–	–	12.5–13.4	9.62
			17.0–17.8	>10.30
			7	6.00
			15	>8.0
Mycoplasma orale	–	–	26–34	≥6.6
Mycoplasma pneumoniae	–	–	26–34	≥6.9
Mycoplasma hyorhinis	–	–	26–34	≥6.3
Infectious bovine rhinotracheitis	Herpesvirus	120–200	25–35	≥4.7
Parainfluenza type 3	Paramyxovirus	100–200	25–35	≥7.1
Feline leukemia virus	Retrovirus	80–110	25–35	3.3
			20	3.0–4.0
			20	2.0–3.0
Cache Valley virus	Bunyavirus	80–100	26–34	≥5.4
Parainfluenza type 3	Reovirus	60–80	26–24	≥7.1
Bluetongue virus	Reovirus	60–80	25–35	3.3
			25	3.5–4.0
Bovine viral diarrhea virus	Flavivirus	50–70	25–35	≥4.3
			20	4
Simian virus	Polyomavirus	40–50	26–34	1.4
Mouse minute virus	Parvovirus	18–24	25–35	3.8
Porcine circovirus	Circovirus	17	30	1.0
			45	1.0

proteins from *E. coli* could be as high as 500,000 EU/mg protein, and the most effective method of purification was Triton X-114 phase separation. However, even the best method was not sufficiently effective to reach endotoxin levels prescribed by pharmacopeias for the protein drugs listed in Table 13.1.

Reichelt et al. [21] combined affinity chromatography with Triton X-114 in the washing step to remove endotoxins from recombinant proteins purified

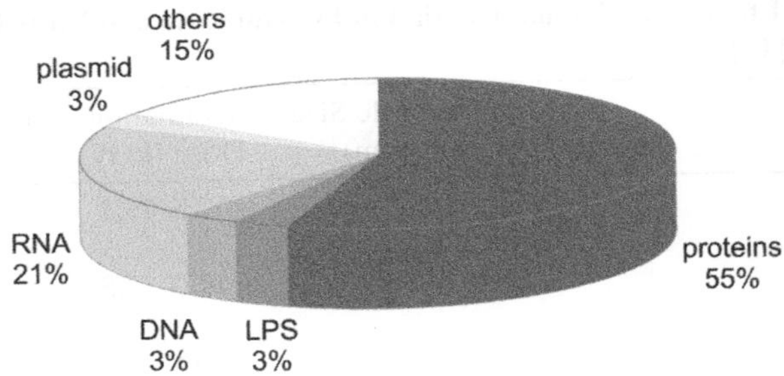

FIGURE 13.2 Composition of bacterial cell lysate (adapted from Reference 18).

TABLE 13.4 Concentration of Endotoxin in Different Protein Solutions of Different Origins [19]

Source	Solution	Endotoxin (EU/mL)
Protein from high cell density culture of *E. coli* TG1:pγFGFB	Supernatant after homogenization	>2,000,000
Protein from shaking flask cell cultures of *E. coli*	Culture filtrate	70,000–500,000
Murine IgG_1 from cell culture	Culture filtrate	≤100
Commercial bovine serum albumin (supplier 1)	1 mg/mL of lyophilized protein	50
Commercial bovine serum albumin (supplier 2)	1 mg/mL of lyophilized protein	0.5

TABLE 13.5 Endotoxin Levels in Proteins Produced by *E. coli* Fermentation and Endotoxin Levels after Purification Procedures [20]

Protein	Endotoxin Level before Purification (EU/mg protein)	Endotoxin Level after Purification (EU/mg protein)		
		Polymyxin B	Histidine	Triton X-114
rCK-MB	5,000–10,000	250–1000	–	25.0–125.0
rCK-MM	100,000–500,000	250–1000	–	5.0–25.0
rCK-BB	100,000–500,000	–	500–1000	5.0–25.0
rMyo	5,000–10,000	5,000–10,000	2,500–5,000	2.5–25.0
rcTnI	5,000–10,000	–	–	2.5–25.0

–, protein binds to the affinity column.

TABLE 13.6 Endotoxin Levels in Proteins after Purification with and without Triton X-114 in the Washing Step [21]

		Endotoxin (EU/mg)	
Protein	Purification Conditions	Without Triton X-114	With Triton X-114
Human TAR RNA binding protein	Denatured	>3,000	0.20 ± 0.05
Transcription activator Tax protein from human T-cell	Native	>16,000	3.5 ± 0.5
cAMP receptor protein SyCRP	Native	>34,000	1.0 ± 0.5
LysR-type transcription regulator NdhR	Denatured	>2,500	0.36 ± 0.03
Oxal 1 from *Saccharomyces cerevisae*	Native	>20,000	4.0 ± 0.5
Oxal 1 from *Neurospora crassa*	Native	>15,000	2.5 ± 0.5

from *E. coli* and achieved better results. Table 13.6 shows the endotoxin content of various recombinant proteins. The purification consisted of Ni-NTA or glutathione-Sepharose column under native or denaturing conditions with and without 0.1% of Triton X-114 in the washing step. The results confirm the efficacy of the nonionic detergent in removing more than 99% of the endotoxin contaminants.

Porcine vaccines produced in Japan were assayed for endotoxin contamination using the kinetic turbidimetric LAL test [22]. Different vaccines and different batches of the same vaccine were evaluated. In only 9 of the 39 samples, endotoxins were not detected (Fig. 13.3). For positive cases, the levels varied from less than 0.1 EU/mL to more than 60 EU/mL. Higher contamination levels were found in viral vaccines and in combined vaccines, although in three viral vaccine samples, endotoxin was not detected. These results show that the level of endotoxin may vary from sample to sample and from lot to lot. A similar result was found among reagents used for the isolation and purification of human pancreatic islets. In Table 13.7, the results showed that in different lots of type P collagenase produced by the same manufacturer, endotoxin levels varied from 208 to 7688 EU/mL. By contrast, all other reagents presented low levels of endotoxin. Low levels of endotoxin were also presented in blood products. The results in Table 13.8 were obtained with the kinetic turbidimetric LAL test.

The clearance of DNA, HCPs, and endotoxins at the different stages of recombinant protein purification was evaluated during downstream process optimization for monoclonal antibody [25], promegapoietin [26], human granulocyte colony-stimulating factor (rhG-CSF) [27], and interferon γ (INF-γ) [28] expression in *E. coli*. Although expressed in the same host cell, the impurity levels present enormous variation among the different proteins and processes

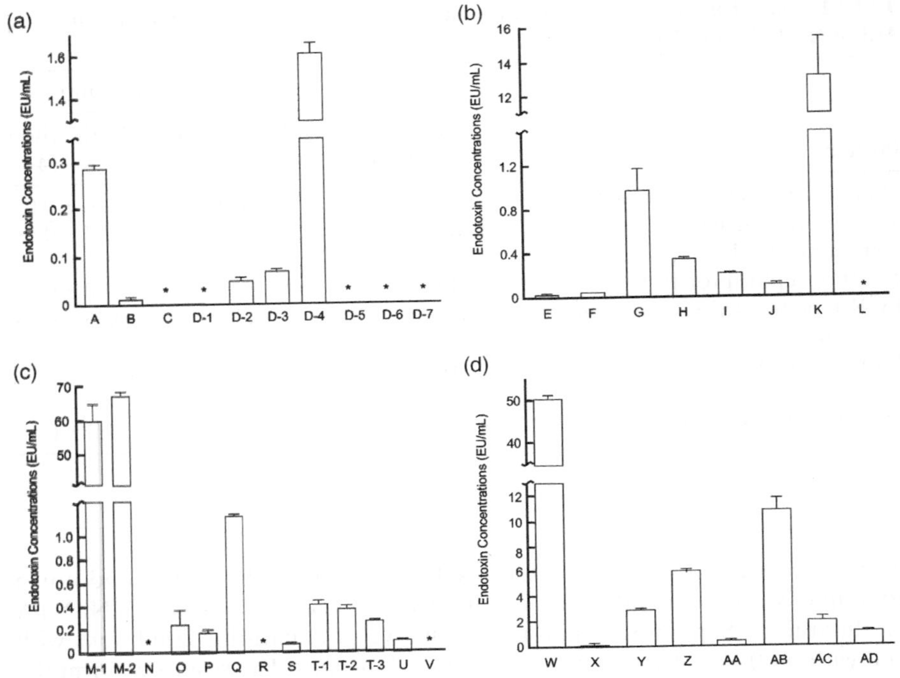

FIGURE 13.3 Quantification with LAL ES of contaminating endotoxin in commercial porcine vaccines. (a) *M. hyopneumoniae* vaccines, (b) gram-positive bacterial vaccines, (c) viral vaccines, and (d) combined gram-positive bacterial and viral vaccines. Each sample is expressed as A to AD, and different batches of the same product are expressed as D-1 to D-7, M-1 to M-2, and T-1 to T-3. Inactivated vaccines are A to D, L, M, and O. *: below the limit of detection [22].

(Table 13.9). Purification steps included ion exchange, adsorption, chelating, or affinity chromatography. Nevertheless, other separation techniques, such as dialysis and membrane filtration, were employed as well. Protein purification was not attained in only one step, and even though efficient, methods were not able to remove all impurities. The impurity levels at the end of the each process were, in most cases, compatible with the specifications (International Conference on Harmonisation (ICH), pharmacopeias) for rDNA therapy products.

13.4 PROTEIN DEGRADATION

Protein degradation is caused by physical or chemical changes in the protein structure. These changes include deamidation, oxidation, hydrolysis, and

TABLE 13.7 Endotoxin Contaminations of Reagents Used during Isolation and Purification of Human Pancreatic Islets [23]

Reagent Type	*N*	Endotoxin (EU/mL)
Liberase	6	10.1 ± 4.0
Collagenase, Type P (lot 43)	6	2683.8 ± 1264.8
Collagenase, Type P (lot 38)	6	7687.7 ± 3785.9
Collagenase, Type P (lot 33)	6	4354.3 ± 1581.5
Collagenase, Type P (lot 14557520)	6	433.4 ± 296.2
Collagenase, Type P (lot 24)	3	539.2 ± 198.2
Collagenase, Type P (lot 9)	3	207.7 ± 186.5
Collagenase, Type P (lot 89)	3	1290.7 ± 1,053.5
Collagenase, Serva (lot 80106)	6	607.5 ± 101.3
Collagenase, Sigma, Type V (lot 32H6802)	6	518.5 ± 207.3
Hank's balanced salt solution	2	<0.03
Eurocollins (solution for storage of organs)	2	<0.03
Fetal calf serum	2	<0.03
Human serum	2	<0.25
Human serum albumin	2	<0.125
CMRL-1066 (cell culture medium)	2	<0.03
Aim-V (serum-free cell culture medium)	2	<0.25
RPMI-1640 (cell culture medium)	2	<0.03
Ficoll (sterile medium for isolation of human lymphocyte)	6	68.9 ± 36.6
Optiprep (sterile and endotoxin-tested solution of iodixanol)	6	<0.25
Nicoprep (sterile and endotoxin-tested medium for separation of human monocytes)	6	<0.125
Distilled water	2	1

changes in carbohydrate structure and glycosylation. Deamidation results in the conversion of asparagine to a mixture of isoaspartate and aspartate. Oxidation occurs in the side chains of the amino acids, histidine, methionine, cysteine, tryptophan, and tyrosine, and is accelerated in the presence of air, elevated temperatures, and metal traces. Figure 13.4 shows these common protein degradation pathways.

Kothari et al. [30] evaluated degradation products formed during storage of the recombinant human parathyroid hormone. They found that the hormone degraded because of not only oxidation but also deamidation and peptide bond cleavage. Figure 13.5 displays chromatograms of the drug product after storage for 6 months at 5 and 25°C. Although practically the same peaks were observed in the chromatogram of the drug stored at both temperatures, they are higher in the drug stored at 25°C. These peaks, listed in Table 13.10, were identified by gel electrophoresis and mass spectrometry and correspond to protein degradation products.

TABLE 13.8 Endotoxin Levels in Blood Products [24]

Product	Number of Batches	Endotoxin (EU/mL)
Polyethylene glycol-treated normal human immunoglobulin	3	0.062
Polyethylene glycol-treated human anti-HBs immunoglobulin	3	0.013
Polyethylene glycol-treated human anti-tetanus immunoglobulin	3	0.035
pH 4-treated normal human immunoglobulin (acidic)	3	0.066
pH 4-treated normal human immunoglobulin	3	0.041
Ion-exchange resin-treated normal human immunoglobulin	3	0.026
Sulfonated normal human immunoglobulin	3	0.036
Pepsin-treated normal human immunoglobulin	1	0.016
Normal human immunoglobulin[a]	3	0.064
	2	0.038
Human anti-HBs immunoglobulin[a]	3	0.063
	2	0.077
Human anti-tetanus immunoglobulin	2	0.054
Human anti-D(Rho) immunoglobulin	3	0.047
Normal human immunoglobulin with histamine	3	0.061
Human haptoglobin	4	0.012
Concentrated human blood coagulation factor VIII[a]	3	0.013
	1	0.367
	3	0.009
	3	0.008
Concentrated human blood coagulation factor IX	4	0.011
Concentrated human blood coagulation factor IX	4	0.011
Human fibrinogen	4	0.023
Human activated protein C concentrate	3	0.014

[a] More than one manufacturer.

13.5 PROTEIN AGGREGATION

Protein aggregation is a common phenomenon that results from protein instability and that may occur during fermentation, refolding, purification, formulation, and storage. Aggregates can be soluble or insoluble, covalent or noncovalent bind, and the process can be reversible or irreversible. The protein aggregate is an impurity in the drug product. Temperature is the most critical parameter affecting protein aggregation; higher temperatures increase the rate of protein aggregation by increasing both the frequency of molecular collision and hydrophobic interactions. Medium pH is also a critical factor since pH strongly affects the type and distribution of charges in proteins. Protein

TABLE 13.9 Dawnstream Processes for Purification and Recovery of Recombinant Proteins and Examples of the Efficacy of Each Step

Protein	Host Cell	Raw Material	Cell Debris Removal	Purification Steps					Reference
Monoclonal antibody	*E. coli*		Centrifugation/ filtration	Q-sepharose	Ni-chelating	Q-sepharose	Hydroxyapatite	Q-sepharose	[25]
Endotoxin (EU/mg)			24,000,000	731,000	1871	0.2	–	<0.2	
DNA (pg/mg)			–	–	–	–	–	39.8	
HCP (ng/mg)			4,000,000	85,000	960	300	–	18	
Promegapoietin	*E. coli*	Raw material	Filtration	Dialysis	Dialysis				[26]
Endotoxin (EU/50 mL)		5.95×10^8	–	1.79×10^7	3.30×10^6				
DNA (mg/ 50 mL)		203 ± 9.3	64 ± 6.0	–	–				
HCP (mg/50 mL)		1400	640	33	170				
Granulocyte colony-stimulating factor	*E. coli*	Cell mass	Centrifugation	Washing	Diafiltration	Sephadex G-25			[27]
Endotoxin (EU)		1600	150	40	>1	>1			
DNA (pg/mL)		1250	975	250	115	85			
HCP (mg)		2436	1013	22	19	0			
Interferon	*E. coli*	Cell lysis	Washing	Butyl-sepharose	Q-sepharose	Refolding	SP-sepharose		[28]
Endotoxin (EU)		900	520	15	15	15	<1		
DNA (pg/mL)		800	800	800	100	100	80		

Q-sepharose: anion-exchange resin; Ni-chelating: nickel chelating resin; Sephadex G-25: gel filtration medium; butyl-sepharose: hydrophobic resin; SP-sepharose: cation exchange resin.

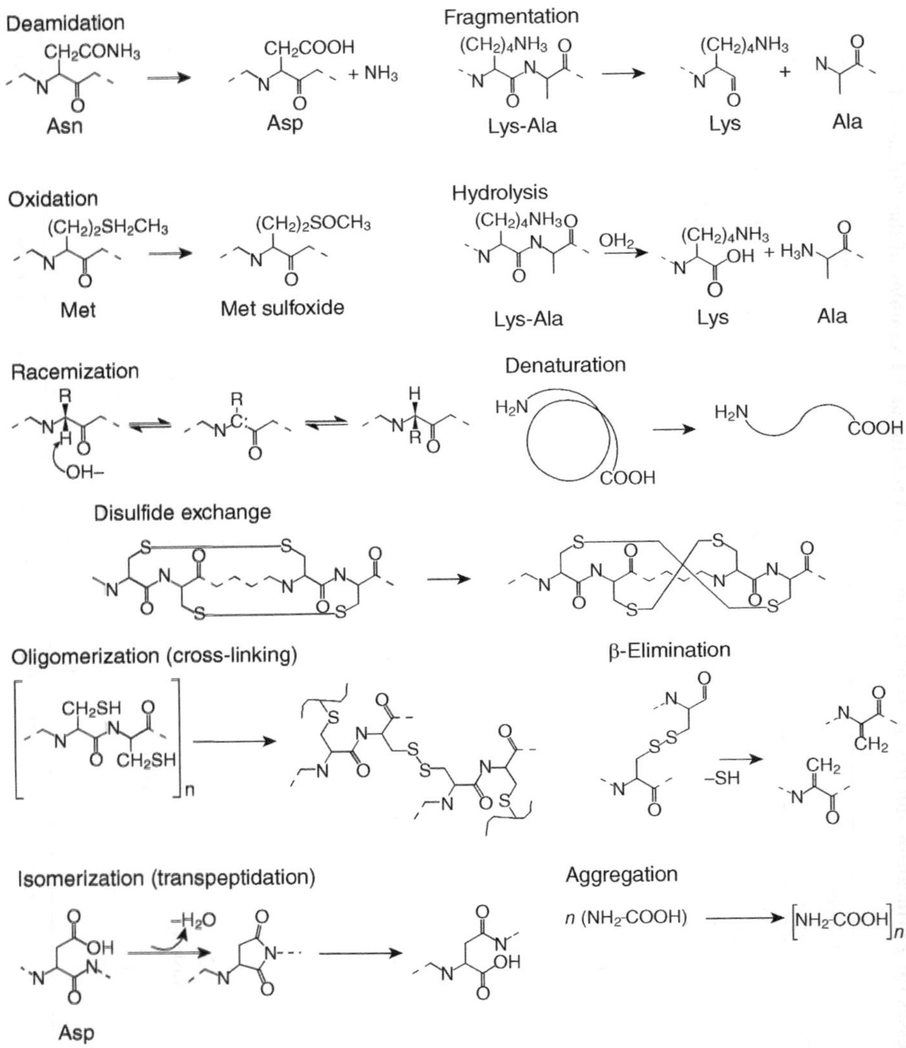

FIGURE 13.4 Pathways of physical and chemical changes that may alter protein product quality [29].

concentration is another important factor because a critical concentration may be required to form the initial nucleus for the aggregation process. Reconstitution, freeze-thawing, and freeze-drying may also play a role in protein aggregation. Table 13.11 lists the influence of these and other factors on the aggregation of selected drug proteins.

Silva et al. [32] examined the structural integrity of preparations of interferon alfa-2a (IFN-α2a) and interferon alfa-2b (IFN-α2b), which had been

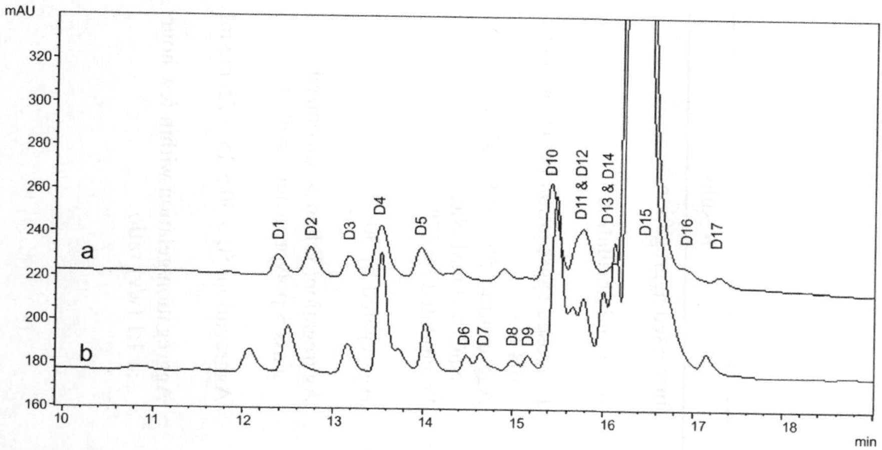

FIGURE 13.5 Reverse-phase high-performance liquid chromatography (RP-HPLC) chromatogram of drug product (recombinant human parathyroid hormone) stored for 6 months at (a) 5 ± 3°C and (b) 25 ± 2°C [30].

TABLE 13.10 Identification of the Peaks Observed in the Chromatograms in Figure 13.5 [30]

Peak	Identification
D1	Truncated but position not determined
D2	1–16 (cleavage at 16)
D3	17–34 (cleavage at 16)
D4	Cleavage at 33 + oxidation
D5	Dioxy (Met 8, Met 18)
D6	Not responding as a protein peak
D7	Cleavage at 33 + oxidation
D8	Oxidation at Met 8
D9	Not responding as a protein peak
D10	Cleavage at 33 + oxidation at Met 8
D11 and 12	(1–30) cleavage at 30 and deamidation
D13	Deamidation (Asn 15)
D14	Deamidation (Asn 33)
D15	rHu-PTH (1–34)
D16	Succinimide intermediate
D17	Insufficient protein amount to be identified

subjected to stress conditions, namely, storage at elevated temperature (37°C) for 9 weeks. The sodium dodecyl sulfate polyacrylamide gel electrophoresis (SDS-PAGE) analysis of the samples stored at 37 and −70°C showed that high-molecular-weight bands were detected at 37°C for both preparations, revealing the occurrence of protein aggregation at higher temperature

TABLE 13.11 External Factors Affecting Protein Aggregation

Factors	Protein	Protein Formulation	Study Conditions	Results
Temperature	IL-1β	0.1 mg/mL in acetate, pH 5.0	Incubation at 39–60°C	Increased aggregation
	Bovine/porcine insulin	30 mg/mL, pH 2.5	Incubation at 35–75°C	Increased fibrillation
pH	IL-1β	0.1 mg/mL	Incubation at pH 2–9, 30°C	Increased aggregation below or above pH 6
Ionic strength	RNase A	5 mg/mL in 100 mM citrate, pH 3.0	Incubation at 75°C	Aggregation: 500 mM NaCl > 250 mM NaCl > 0 mM NaCl
Shaking	hGH	0.5 mg/mL in phosphate, pH 7.4	Vortexing for 1 min	Precipitated by 67%
Shearing	hGH	2 mg/mL in phosphate, pH 7.4	Exposed to a shear of 10^5	Aggregated by 30%
Hydrophobic surface	Bovine insulin	0.6 mg/mL in PBS, pH 7.4	Agitation at 37°C with different beads	Aggregation: Teflon > siliconized glass > polypropylene > glass
Protein concentration	Bovine insulin	7.5 to 50 mg/mL, pH 2.5	Incubation at 21°C	Aggregation: 50 > 30 > 15 > 7.5 mg/mL
Organic solvent	β-Lactoglobulin	50 mg/mL at pH 7–9	Addition of ethanol at 25°C	Aggregation/gelation within few hours at 1:1 (v/v) ratio

Metal ions	Deoxyribonuclease I	~18 mg/mL in 150 mM NaCl	Incubation at 40°C for 30 days	Complete aggregation but not visible aggregation in the presence of 1 mM $CaCl_2$
Protein source	Bovine insulin	15 mg/mL, pH 2.1	Incubation at 21°C	Lot A > lot B
Pressure	H^+-ATPase	0.05 mg/mL in 3 mM $MgCl_2$, 5% glycerol, and 50 mM Tris-HCl, pH 7.0	Increasing pressure from 0 to 2.3 bar	Increased aggregation with increasing pressure
Freeze-thawing	Hemoglobin	50 mg/mL in PBS, pH 7.4	Freeze-thawing 0, 3, and 5 times	Aggregation: 5 > 3 > 0 times
Freeze-thawing	Bovine IgG	0.5–8.0 mg/mL in 0.22 M NaCl and 10 mM phosphate, pH 7.1	Freeze-drying	Increased formation of insoluble aggregated with decreasing protein concentration
Spray-drying or	rhMAb	10 mg/mL	Spray-drying	Aggregated by 5.6%
Spray-freeze drying	Bovine serum albumin (BSA)	20 mg/mL	Spray-freeze drying	Particle size dependent aggregation
Reconstitution	Keratinocyte growth factor (KFG)	0.4–0.65 mg/mL in 10 mM phosphate, pH 7.0	Reconstitution with water or 0.5% heparin	Aggregation: water > heparin

Data collected by Wang [31].

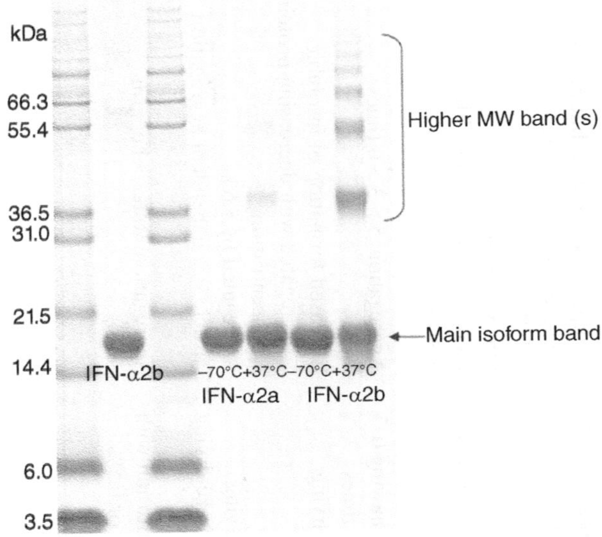

FIGURE 13.6 Selected SDS-PAGE gel in 4–12% reducing conditions for interferon preparations, IFN-α2a and IFN-α2b, after storage at −70 and +37°C for 9 weeks (4-μg loading). Molecular weight (MW) ladder (kDa) is Marker 12 (Invitrogen Ltd.) [32].

(Fig. 13.6). IFN-α2b presented a large portion of aggregates, and the band profile suggested that associations existed corresponding to dimmers and trimers, and indicated that it is less stable than IFN-α2a.

CONCLUDING REMARKS

Biopharmaceuticals, obtained either by rDNA technology or by other technologies, are originated from living organisms and are therefore subject to harmful contaminants. Viruses, mycoplasmas, and endotoxins are the most obvious endangered species, but others, which originate from the cell culture, DNA and proteins, may also pose risks if present in the final drug. Challenges dealing with recombinant proteins are to manage satisfactorily all purification steps and to keep protein integrity in the final product. They may aggregate, denature, or become fragmented, introducing new entities in the final product, which may behave as toxicants, even after distribution, during shelf life.

REFERENCES

1. *Ullmann's Biotechnology and Biochemical Engineering*. Wiley-VCH Verlag GmbH, Weinheim, 2007.

2. Wang X, Hunter AK, Mozier NM. Host cell proteins in biologics development: identification, quantitation and risk assessment. Biotechnol Bioeng 103(3), 446–458, 2009.
3. Krawitz DC, Forrest W, Moreno GT, Kittleson J, Champion KM. Proteomic studies support the use of multi-product immunoassays to monitor host cell protein impurities. Proteomics 6(1), 94–110, 2006.
4. Center for Biologics Evaluation and Research. FDA. Points to Consider: Manufacture and Testing of Monoclonal Antibody Products for Human Use, Rockville, MD, 1997.
5. Cai H, Gu X, Scanlan MS, Lively CR. Development of a quantitative PCR assay for residual mouse DNA and comparison of four sample purification methods for DNA isolation. J Pharm Biomed Anal 55(1), 71–77, 2011.
6. Zhu D, Saul AJ, Miles AP. A quantitative slot blot assay for host cell protein impurities in recombinant proteins expressed in *E. coli*. J Immunol Methods 306(1-2), 40–50, 2005.
7. Rathore AS, Sobacke SE, Kocot TJ, Morgan DR, Dufield RL, Mozier NM. Analysis for residual host cell proteins and DNA in process streams of a recombinant protein product expressed in *Escherichia coli* cells. J Pharm Biomed Anal 32(6), 1199–1211, 2003.
8. Dagouassat N, Haeuw JF, Robillard V, Damien F, Libon C, Corvaïa N, Lawny F, Nguyen TN, Bonnefoy JY, Beck A. Development of a quantitative assay for residual host cell proteins in a recombinant subunit vaccine against human respiratory syncytial virus. J Immunol Methods 251(1-2), 151–159, 2001.
9. Wan M, Wang Y, Rabideau S, Moreadith R, Schrimsher J, Conn G. An enzyme-linked immunosorbent assay for host cell protein contaminants in recombinant PEGylated staphylokinase mutant SY161. J Pharm Biomed Anal 28(5), 953–963, 2002.
10. Forcic D, Brgles M, Ivancic-Jelecki J, Santak M, Halassy B, Barut M, Jug R, Markušić M, Strancar A. Concentration and purification of rubella virus using monolithic chromatographic support. J Chromatogr B 879(13-14), 981–986, 2011.
11. Nayak DP, Lehmann S, Reichl U. Downstream processing of MDCK cell-derived equine influenza virus. J Chromatogr B 823(2), 75–81, 2005.
12. Toueille M, Uzel A, Depoisier JF, Gantier R. Designing new monoclonal antibody purification processes using mixed-mode chromatography sorbents. J Chromatogr B 879(13-14), 836–843, 2011.
13. Stein A, Kiesewetter A. Cation exchange chromatography in antibody purification: pH screening for optimised binding and HCP removal. J Chromatogr B 848(1), 151–158, 2007.
14. Kerr A, Nims R. Adventitious viruses detected in biopharmaceutical bulk harvest samples over a 10 year period. PDA J Pharm Sci Technol 64(5), 481–485, 2010.
15. Chen D, Nims R, Dusing S, Miller P, Luo W, Quertinmont M, Parekh B, Poorbaugh J, Boose JA, Atkinson EM. Root cause investigation of a viral contamination incident occurred during master cell bank (MCB) testing and characterization—a case study. Biologicals 36(6), 393–402, 2008.
16. Kozasa T, Aoki H, Nakajima N, Fukusho A, Ishimaru M, Nakamura S. Methods to select suitable fetal bovine serum for use in quality control assays for the detection of adventitious viruses from biological products. Biological 39, 242–248, 2011.

17. Nims RW, Gauvin G, Plavsic M. Gamma irradiation of animal sera for inactivation of viruses and mollicutes—a review. Biologicals 39(6), 370–377, 2011.
18. Stadler J, Lemmens R, Nyhammar T. Plasmid DNA purification. J Gene Med 6(1 Suppl), S54–S66, 2004.
19. Petsch D, Anspach FB. Endotoxin removal from protein solutions. J Biotechnol 76(2-3), 97–119, 2000.
20. Liu S, Tobias R, McClure S, Styba G, Shi Q, Jackowski G. Removal of endotoxin from recombinant protein preparations. Clin Biochem 30(6), 455–463, 2006.
21. Reichelt P, Schwarz C, Donzeau M. Single step protocol to purify recombinant proteins with low endotoxin contents. Protein Expr Purif 46(2), 483–488, 2006.
22. Ogikubo Y, Norimatsu M, Noda K, Takahashi J, Inotsume M, Tsuchiya M, Tamura Y. Evaluation of the bacterial endotoxin test for quantification of endotoxin contamination of porcine vaccines. Biologicals 32(2), 88–93, 2004.
23. Linetsky E, Inverardi L, Kenyon NS, Alejandro R, Ricordi C. Endotoxin contamination of reagents used during isolation and purification of human pancreatic islets. Transplant Proc 30(2), 345–346, 1998.
24. Ochiai M, Yamamoto A, Naito S, Maeyama J, Masumi A, Hamaguchi I, Horiuchi Y, Yamaguchi K. Applicability of bacterial endotoxins test to various blood products by the use of endotoxin-specific lysates. Biologicals 38(6), 629–636, 2010.
25. Premsukh A, Lavoie JM, Cizeau J, Entwistle J, MacDonald GC. Development of a GMP Phase III purification process for VB4-845, an immunotoxin expressed in *E. coli* using high cell density fermentation. Protein Expr Purif 78(1), 27–37, 2011.
26. Ledung E, Eriksson PO, Oscarsson S. A strategic crossflow filtration methodology for the initial purification of promegapoietin from inclusion bodies. J Biotechnol 141(1-2), 64–72, 2009.
27. Dasari VKR, Are D, Joginapally VR, Mangamoori LN, Adibhatla KSBR. Optimization of the downstream process for high recovery of rhG-CSF from inclusion bodies expressed in *Escherichia coli*. Process Biochem 43(5), 566–575, 2008.
28. Mohammadian-Mosaabadi J, Naderi-Manesh H, Maghsoudi N, Nassiri-Khalili MA, Masoumian MR, Malek-Sabet N. Improving purification of recombinant human interferon gamma expressed in *Escherichia coli*; effect of removal of impurity on the process yield. Protein Expr Purif 51(2), 147–156, 2007.
29. DiPaolo B, Pennetti A, Nugent L, Venkat K. Monitoring impurities in biopharmaceuticals produced by recombinant technology. Pharm Sci Technol Today 2(2), 70–82, 1999.
30. Kothari R, Kumar V, Jena R, Tunga R, Tunga BS. Modes of degradation and impurity characterization in rhPTH (1–34) during stability studies. PDA J Pharm Sci Technol 65, 348–362, 2011.
31. Wang W. Protein aggregation and its inhibition in biopharmaceutics. Int J Pharm 289(1-2), 1–30, 2005.
32. Silva MM, Lamarre B, Cerasoli E, Rakowska P, Hills A, Bailey MJ, Wheeler JX, Burns CJ, Gaines-Das RE, Jones C, Robinson CJ. Physicochemical and biological assays for quality control of biopharmaceuticals: interferon alpha-2 case study. Biologicals 36(6), 383–392, 2008.

Appendix

POLYMERIC MATERIALS: COMPONENTS, ADDITIVES, EXTRACTABLES, AND DEGRADATION PRODUCTS

This appendix is meant to be a "special chapter" on polymeric materials. Because containers, closures, and several different devices for medical use are made of polymeric materials, and one specific polymeric material can find several different uses, polymers appear in various parts of the book. So as not to be repetitive, general information on polymeric materials is brought together in this Appendix.

Table A.1 lists usual polymers, the monomeric units, and their uses. In Table A.2, catalysts and usual additives for each polymer are presented. Table A.3 lists extractable substances from polymeric materials by the action of solvents, either as prescribed by pharmacopeial compendia or following an author's strategy. The action of drugs, vehicles, or body fluids on promoting the leaching of any material component is discussed in the respective chapter. For example, extractables from polysulfone into water or ethanol are presented in Table A.3; on the other hand, the release of contaminants from hemodialysis devices into serum samples is dealt with in Chapter 9, "Medical Devices" under Section 9.2, Extracorporeal Circuits. Finally, Table A.4 presents species formed and extracted from polymeric materials that were submitted to radiation for sterilization purposes. This table is linked to Section 12.1, Residuals from Radiation Sterilization.

Sources of Contamination in Medicinal Products and Medical Devices, First Edition. Denise Bohrer.
© 2013 by John Wiley & Sons, Inc. Published 2013 by John Wiley & Sons, Inc.

TABLE A.1 Natural and Synthetic Polymers, Their Structural Units, and Use in the Pharmaceutical Industry

Polymer	Abbreviation	Structure	Use
Low-density polyethylene	LDPE	$-(CH_2-CH_2)_n-$	Containers Syringes
High-density polyethylene	HDPE	$-(CH_2-CH_2)_n-$	Containers Syringes
Ultra-high-molecular-weight polyethylene	UHMWPE	$-(CH_2-CH_2)_n-$	Implants
Polytetrafuorethylene	PTFE	$-(CF_2-CF_2)_n-$	Implants Sutures
Polypropylene	PP	$-(CH_2-CH(CH_3)-CH_2)_n-$	Containers Syringes Hemodialyzers Sutures
Polyurethane	PU	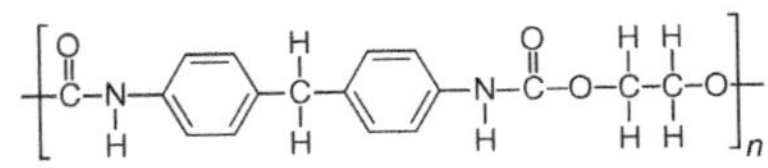	Implants Hemodialyzers Catheters
Polyvinyl chloride	PVC	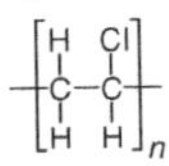	Bags Infusion sets Catheters Gloves
Ethylene vinyl acetate	EVA	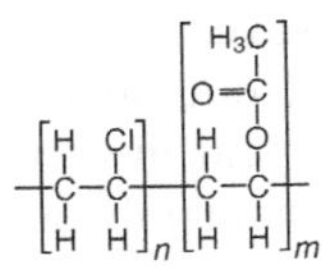	Bags Infusion sets
Poly(methyl methacrylate)	PMMA	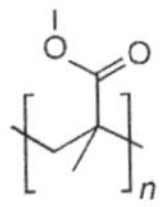	Membranes Hemodialyzers Tissue substitute
Polystyrene	PS	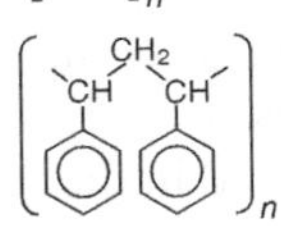	Hemodialyzers
Polyamide	PA	$-(NH-(CH_2)_5-CO)_n$	Multilayer packaging material Filters Sutures
Polysulfone	PSf	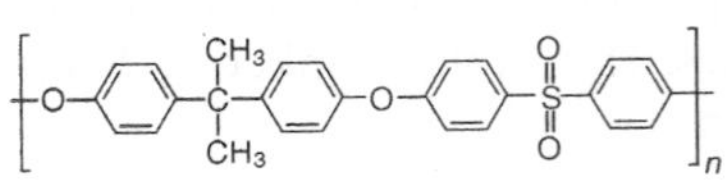	Filters
Polyethylene terephthalate	PET	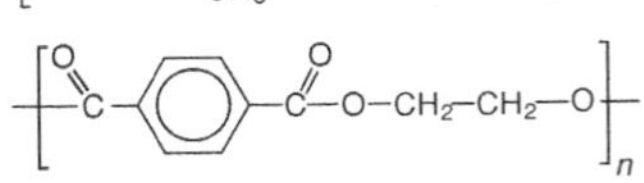	Containers Sutures
Silicone	–	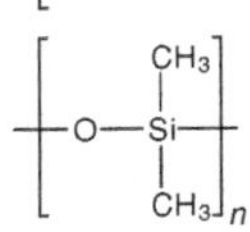	Tubing Implants

TABLE A.1 (*Continued*)

Polymer	Abbreviation	Structure	Use
Polycarbonate	PC	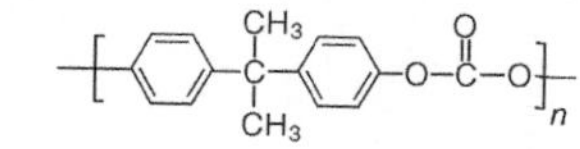	Hemodialyzers Burette
Polyvinylidene fluoride	PVDF	$-(CH_2-CF_2)_n-$	Membrane filters
Poly(lactic-co-glycolic) acid	PLGA	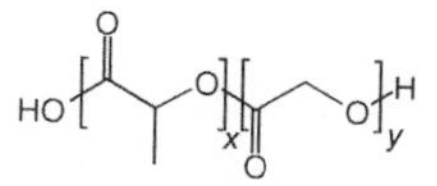	Drug encapsulation
Polylactic acid	PLA	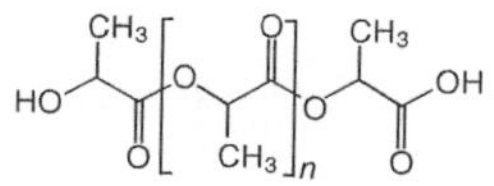	Drug encapsulation Tissue substitute
Cellulose	–	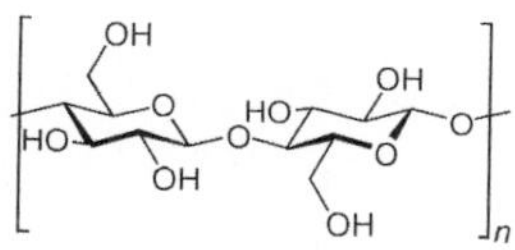	Membranes Tissue substitute
Alginate	–	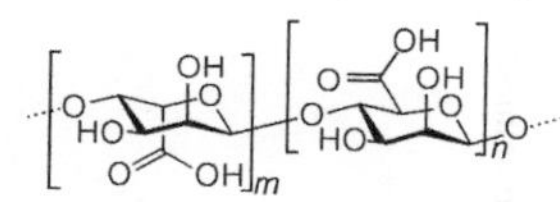	Tissue substitute

Polymeric materials seem to be the most prominent source of exogenous contamination in pharmaceutical products, either as containers, delivery systems, or medical devices. Since the universe of polymeric materials and their additives is vast, the characterization of all extractables and leachables is probably an unreachable task. Table A.3 of this appendix alone lists more than 200 different substances.

Not only the nature of the polymeric material itself but also specific interactions that can occur between the material components and the formulation components contribute to this. Two opposite examples are the leaching of disposable syringe components into an erythropoietin formulation and the delivery of polyvinyl chloride (PVC) additives into the air in a tubing of respiratory therapy. While the inactive ingredient polysorbate 80 of EPREX was able to leach components of the rubber syringe stopper in an amount sufficient to cause immunogenicity in patients when administered subcutaneously, several compounds were found in a sample as simple as the air delivered by PVC tubing used in respiratory therapy. In the first case, a specific and unexpected reaction occurred, and in the second, there was no solvent, drug, or other "inactive" ingredient to be found responsible for the event.

TABLE A.2 Process Components and Functional Additives Used in Common Plastic Polymers

	Polymerization Process			Functional Additives					
Polymer	Solvent/Emulsifier	Catalyst	Initiator	Antioxidants	Stabilizers	Plasticizers	Lubricants	Flame retardants	Fillers
LDPE	Benzene Water		Azoisobutyronitrile Benzoyl peroxide Oxygen	Hindered phenols Phosphites Thioethers Oxamides Hydrazine	Nickel complexes 2-hydroxybenzophenone		Waxes	Chlorinated paraffins Sb_2O_3	Chalk Silane Titanate Carbon Black Talc
HDPE	Liquid hydrocarbon such as cyclohexane Methanol or ethanol or isopropanol + HCl for catalyst deactivation and solubilization	Zigler process: organometallic compounds, halides, oxyhalides, alkoxy, acetylacetonyl, or phenyl with Ti, V, Cr, Mo, Hf,or Zr Phillips process: Cr_2O_3 + silica + alumina Indiana process: Mo_2O_3, TiO_2, or ZrO_2 + Na or Ca hydrides + alumina Unipol process: Cr_2O_3 or $TiCl_4$,or bis-tri-phenylsilyl chromate or Chromocene $(C_5H_5)_2Cr$ + silica		Hindered phenols Phosphites Thioethers Oxamides Hydrazine	Hindered amines (HALS)		Waxes	Brominated aromatics Sb_2O_3	Chalk Talc

UHMWPE	Same as for HDPE	Zigler-type catalyst (same as for HDPE)		α-tocopherol			Waxes		Chalk Talc
PTFE	Water		Ammonium persulfate Benzoyl peroxide Hydrogen peroxide						Graphite
PP	Naphtha or C4–C6 paraffins Methanol + HCl for catalyst deactivation and solubilization	$TiCl_3$ + triethyl or tributyl aluminum or aluminum diethyl monochloride or metallocene-based: L_2MX_2 L: cyclopentadienyl, indenyl, fluorenyl ligands M: Ti, Zr, Hf X: halogen, alkyl, phenyl, benzyl, trimethylsilyl		Hindered phenols Hydroperoxides Phosphites	2-Hydroxybenzophenone 2-Hydroxyphenyl-benzotriazole Hindered amines (HALS) 4-Hydroxybenzoate		Waxes	Tetrabromobisphenol Sb_2O_3	Talk $CaCO_3$
PU		KOH Tertiary amines Tin, Mercury, and bismuth compounds	Glycols	Hindered phenols Aromatic amines Phenothiazene	Hindered amines (HALS) Phenolic antioxidants			Melamine Tris (chloroethyl)phosphate Chlorinated paraffins	Talc
PVC	Sodium persulfate Sodium dodecylbenzene sulfonate Fatty alcohols (C16–C18)		Azoisobutyronitrile Benzoyl peroxide Acetyl cyclohexane sulfonyl peroxide Peroxidicarbonates Lauroyl peroxide	Organic phosphites	Lead soap Organo tin compounds Ca-Zn soaps Thiourea derivatives Indol derivatives Diketones Epoxy esters Oxalanilide	Phthalates Adipates Trimellitates Citrates	Stearic acid 12-Hydroxystearic acid Glycerol-monostearate	$Al(OH)_3$ Sb_2O_3	Chalk Talc

(*Continued*)

TABLE A.2 (*Continued*)

	Polymerization Process			Functional Additives					
Polymer	Solvent/Emulsifier	Catalyst	Initiator	Antioxidants	Stabilizers	Plasticizers	Lubricants	Flame retardants	Fillers
PMMA	Water		Azoisobutyronitrile Benzoyl peroxide		2-hydroxyphenyl-benzotriazole Hindered amines (HALS) Oxalanilide	Phthalates Adipates		Halogenated polyphosphonate	Graphite
PS	Ethyl benzene		Benzoyl peroxide *t*-butyl hydroperoxide	Phenolic antioxidants Dilauryl ester of thiodipropionic acid	Hindered amines (HALS)	Dibutyl phthalate	Butylstearate Glycerol-monostearate Fatty acid esters paraffin	Hexabromocyclododecane Dicumyl peroxide	
PA	Water Acetic acid	Caprolactam magnesium bromide Sodium caprolactam	Metal hydrides Metal amides Alkali metals Bis-acyllactams	Hindered phenols Phosphites		Phthalates sulfonamides	Pentaerythriol esters of fatty acids Amide wax	Brominated aromatics Sb_2O_3 Ethylene-bis(tetrabromophthalimide)	
PET		Sb_2O_3 Co acetate Mn acetate					Pentaerythriol esters of fatty acids Amide wax		
PC	Pyridine Chloroform Methylene dichloride tetrachloroethane	Hydroxides or hydrides of alkaline metals and zinc or Antimony oxides Quaternary ammonium salts		Hindered phenols Phosphites	Oxalanilide		Pentaerythriol esters of fatty acids Amide wax	Tetrabromobisphenol Sb_2O_3	

Sources: Feldman D, Barbalata A. *Synthetic Polymers, Technology, Properties, Application*. Chapman & Hall, London, 1996; *Ullman's Encyclopedia of Industrial Chemistry*, 5th ed., VCH, Weinheim, 1993.

TABLE A.3 Organic Extractables from Polymers with Pharmaceutical Use

Polymer	Sample	Extraction Conditions	Compounds Extracted	Reference
LDPE	Film	95% Ethanol	Irgaphos 168 Irganox 1076	[1]
LDPE	Film	CH_2Cl_2 at 60°C, 24 h	Oxidized Irgaphos 168 Oxidized Irganox 1076 Erucic acid amide	[2]
LDPE	Film	Direct thermal desorption at 80°C, 30 min	Hexane Heptane Octane Acetone trans-1-Butyl-2-methyl-cyclopropane Nonane Decane Hexanal Undecane 1-[1-Cyclohexen-1-yl]-ethanone 4-Hydroxy-4-methyl-pentanone Dodecane 2,3-Dehydro-4-methyl-furan Dimethylphenol 1,3,5,7-Cyclooctatetraene Octanal Tridecane 6-Methyl-5-hepten-2-one Tetradecane 2,3,6-Trimethyl-1,5-heptadiene Acetic acid Pentadecane Propanoic acid 2,2-Dimethylpropanoic acid Hexadecane 3,5,5-Trimethyl-2-cyclopenten-1-one Butanoic acid Acetophenone Pentanoic acid 3-Methyl-2-butanoic acid Hexanoic acid 6,10-Dimethyl-5,9-undecadien-2-one Butyl hydroxytoluene Heptanoic acid Octanoic acid	[3]
HDPE		95% Ethanol	Irgaphos 168 Irganox 1076	[1]
HDPE	Pellets	Direct thermal desorption at 160°C	Alkanes (octane through eicosane) 1-Alkenes (decene, 10, 12, 14, 16 18 C octadecene) 1,4-Benzenedicarboxylic acid dimethyl ester	[4]

(*Continued*)

TABLE A.3 (*Continued*)

Polymer	Sample	Extraction Conditions	Compounds Extracted	Reference
PP	Film	CH_2Cl_2 at 60°C, 24 h	Irgaphos 168 Oxidized Irgaphos 168 Oxidized Irganox 1076 Acid amide Erucic acid amid 2,4-Di-*tert*-butylphenol	[2]
PP	Syringe	CH_2Cl_2 at 60°C, 24 h	2-Ethyl phenol 1-(2-Hydroxymethyl)-4-hydroxy-2,2,6,6-tetramethylpiperidine succinic acid (HALS) 8-Acetoxy-2-chloro-5,6,7,8-tetrahydroquiniline Palmitic acid Stearic acid Irgaphos 168	[5]
PP	Pellets	Direct thermal desorption at 160°C	2-Methyl-2-propanol 4-Methylheptane 2,4-Dimethylheptane	[4]
PP	Tubing	Ethyl acetate at 22°C, 5 min	Bis(2,2,6,6-tetramethyl-4-piperidyl) sebacate (Tinuvin 770)	[6]
PU	Dialyzer	Methanol at room temperature, 1 week	4,4′-Methylenedianiline	[7]
PU	Breast implant	Methanol 37°C	2,4-Dimethyl-6-t-butylphenol Polyol	[8]
	Breast implant	Phosphate buffer pH 7.4 at 37°C	2,4-Toluenediamine 2,6-Toluenediamine	[9]
	Breast implant		2,4-Toluenediamine 2,6-Toluenediamine Toluene diisocyanate Polydiethyleneglycol adipate Polydimethylsiloxanes 2,3-Dimethyl-6-t-butylphenol	[10]
PVC	Sheet	Acetone/ ethanol 1:1 at 60°C, 24 h	2-Propyl-1-pentanol Mercapto acetic acid ethylester 2-Ethyl-4-methyl-1,3-dioxolane 4-Hydroxy-4-methyl-2-pentanone 3-Methylheptyl acetate 3-Ethoxy-3-methyl-2-butanone Nonanoic acid ethylester n-Alkane n-Alkane acid ethylester	[2]
PVC	Pellets	Direct thermal desorption at 160°C	Octane 1-Octene 1-Octanol Acetic acid 2-Ethyl-1-hexanol	[4]

TABLE A.3 (*Continued*)

Polymer	Sample	Extraction Conditions	Compounds Extracted	Reference
PVC	Feeding tube	CCl_4/MeOH Soxhlet extraction	Butylhydroxytoluene Hexanedioic acid, mono(2-ethylhexyl) ester Bis (2-ethylhexyl) adipate Di(cyclohexyl) phthalate Di(isooctyl) phthalate Methyl pentyl adipate Methyl 2-ethylhexyl adipate Hexadecanoic acid, methyl ester Octadecanoic acid, methyl ester	[11]
PVC	Suction catheter/ funnel	CCl_4/MeOH Soxhlet extraction	Butylhydroxytoluene Hexanedioic acid, mono(2-ethylhexyl) ester Bis (2-ethylhexyl) adipate Di(cyclohexyl) phthalate Di(isooctyl) phthalate Methyl pentyl adipate Methyl 2-ethylhexyl adipate Hexadecanoic acid, methyl ester Octadecanoic acid, methyl ester Hexanediodic acid, ethyl methyl ester.	[11]
PVC	Feeding tubing	Water at 110°C, 70 days	Adipic acid 1,4-Butanediol Monomers and oligomers of polyadipates Hexadecanoic acid Octadecanoic acid	[12]
PVC	Dialysis tubing	Direct thermal desorption at 120°C, 10 min	Cyclohaxanone Butylhydroxytoluene Diethylhexyl phthalate 2-Ethylhexanoic acid	[13]
EVA	Film	Direct thermal desorption at 80°C, 30 min	Hexane Heptane Octane 1-Octene Nonane 3-Methylbutanal Methyl isobutyrate Decane Toluene Hexanal Undecane 3-Hepten-2-one Ethylbenzene p-Xylene 1,3-Dimethylbenzene 1-Butanol Heptanal D,L-limonene	[3]

(*Continued*)

TABLE A.3 (*Continued*)

Polymer	Sample	Extraction Conditions	Compounds Extracted	Reference
			3-Octanol 1,3,5-Trimethylbenzene Octanal Tridecane 6-Methyl-5-hepten-2-one Nonanal Tetradecane 1-(Ethenyloxy)-octadecane 9-Methyl-nonadecane Acetic acid cis-5-Methyl-2-(1-methylethyl)-cyclohexanone 2-Propyl-1-pentanol Pentadecane 3,7-Dimethyl-1,6-octadien-3-ol Propanoic acid Hexadecane 3,5,5-trimethyl-2-cyclopenten-1-one Butanoic acid 5-Methyl-2-(1-m-ethylethyl)-cyclohexanol Heptadecane Pentanoic acid Naphthalene 1,3-Butanediol Hexanoic acid 6,10-Dimethyl-5,9-undecadien-2-one Propylene carbonate Benzyl alcohol Butylhydroxytoluene 1-Phenylethanol Heptanoic acid Octanoic acid Dimethylphthalate	
PS		95% Ethanol	Irganox 1076	[1]
PS	Sheet	n-Hexane at 90°C, 24 h	–	[2]
PS	pellets	Direct thermal desorption at 160°C	Benzaldehyde Phenol Acetophenone 2-Phenylpropenal	[4]
PS		Sunflower, olive or palm oil at 37°C, 60 days	Butylhydroxytoluene Butylated hydroxyanisole	[14]
PA	Film	Acetone at 60°C, 24 h	3,3,5-Trimethyl-2-cyclohexen-1-one Lauroic acid lactame	[2]
PA	Pellets	Direct thermal desorption at 160°C	Formamide Pentanamide	[4]
PC	Hemodialyzer	0.9% NaCl room temperature	Bisphenol A	[15]

TABLE A.3 (*Continued*)

Polymer	Sample	Extraction Conditions	Compounds Extracted	Reference
PET	film	Hexafluoro-iso-propanol at 60°C, 24 h Isopropanol at 60°C, 24 h	–	[2]
Multilayered	Polyolefin with one nylon-6 layer	40/60 Ethanol water at 70°C, 24 h (three sequential extraction with three fresh portion of extraction solution)	Caplolactam oligomers with $n = 1$ to 9 Caplolactam hydrolyzed dimer Caplolactam hydrolyzed trimer	[16]
Very low-density PE Ethyl-vinyl-alcohol Polyurethane	Laminate	Water at 70°C, 3 days	Caprolactam Erucamide p-toluenesulfonamide Caproic acid Caprylic acid 1,4-Dioxacyclotetradecane-5,14-dione 1,8-dioxacyclotetradecane-5,7-dione 3,3-dimethyl-1,5-dioxacyclopentadecane-6,15-dione 3,3-Dimethyl-1,5-dioxacycloundecane-6,11-dione 7,9-Di-t-butyl-1-oxaspiro[4.5] deca-6,9-diene-2,8-dione 3-(3′,5′-Di-t-butyl-4′-hydroxyphenyl) propanoic acid 3-(3′,5′-Di-t-butyl-1′-hydroxy 4′-oxacyclohexa-2′,5′-dienyl) propanoic acid	[17]
Multilayer PETG/PVDC/PE Glycol modified PET PVDC polyvinylidene chloride HDPE PU adhesive between PVDC and PE	Laminate	Soxhlet extraction Ethanol, 24 h	Pentadecane Nonanedioic acid, dimethyl ester Heptadecane Octadecane Butyl, 2-methylpropyl phthalate Eiocosane Henicosane Docosane Bis(2-ethylhexyl)adipate Tricosane Tetracosane Pentacosane Bis(2-ethylhexyl)phthalate Hexacosane Heptacosane 13-Docosenamide (erucyl amide) 2,2′-[(1-methylethylidene)bis(4,1-phenyleneoxymethylene)] bisoxirane Hexacosanoic acid, ethyl ester	[18]

(*Continued*)

TABLE A.3 (*Continued*)

Polymer	Sample	Extraction Conditions	Compounds Extracted	Reference
Multilayer PET/PE/EVOH/ PE	Film	Direct thermal desorption at 80°C, 30 min	Hexane Heptane Octane trans-1,2-Dimethyl-cyclohexane cis-1-Butyl-2-methyl-cyclopropane trans-1-Butyl-2-methyl-cyclopropane 4-Methyl-3-heptane Nonane 3,5-Dimethyloctane Tetramethyloctane Decane Toluene 5-(1-Methylpropyl)-nonane Undecane Ethylbenzene 1,3-Dimethylbenzene 1-Butanol 1-Ethylmethylbenzene 1,2-Dimethylbenzene Dodecane Propylbenzene 1-Ethyl-3-methylbenzene Cyclodecane 1,3,5,7-Trimethylbenzene 1-Ethyl-4-methylbenzene 1-Methyl-2-(1-methylethyl)-benzene 1,2,4-Trimethylbenzene Tridecane 1-Methyl-3-propylbenzene 1,4-Diethylbenzene 1-Ethyl-3,5-dimethylbenzene 1,2-Diethylbenzene a-Methylstyrene 2,5-Dimethylbenzaldehyde 2-Ethyl-1,4-dimethyl-benzene 1-Methyl-2-(1-methylethyl)-benzene Tetradecane 1-(Ethenyloxy)-octadecane Cyclotetradecane Acetic acid 2-Propyl-1-pentanol Pentadecane 1,7,7-Trimethyl-bicyclo- [1, 2, 2] heptane-2-one Hexadecane Cyclohexadecane Menthol 2-Methoxy-1-phenyl-ethanone Trichloromethylbenzene Naphthalene	[3]

TABLE A.3 (*Continued*)

Polymer	Sample	Extraction Conditions	Compounds Extracted	Reference
			Hexanoic acid	
			6,10-Dimethyl-5,9-undecadien-2-one	
			Butylhydroxytoluene	
			Phenol	
			Octanoic acid	
			Nonanoic acid	
			2,4-Bis-(1,1-dimethylethyl)-phenol	
Rubber	Sealing gasket of aerosol valve	Soxhlet Methylene chloride	2,6-Di-t-butyl-4-bromomethyl phenol	[18]
			Dimethyl terephthalate	
			Dibutyl fumarate	
			3-Ethyl-3-methylpentanedioic acid	
			Diphenyl amine	
			4,6-Dimethyl-2-(2-methyl cyclohexyl) phenol	
			Palmitic acid	
			Phenanthrene	
			Dibutyl phthalate	
			Bis-(3-t-butyl-2-hydroxy-5-methyl-phenyl)methane	
			Stearic acid	
			N,N-Dimethyl-9-octadecenoamide	
			2-Propen-1-one,1-(methylphenyl)-3-phenyl oxime	
			Fluoranthene	
			Hexadecanoamide	
			Bis-(2-ethylhexyl)phthalate	
			Pyrene	
			Triphenylphosphine oxide	
Rubber (type not informed)	Closure	Acetonitrile, refluxed, 8 h	4-(1,1-Dimethyl-propyl)-phenol	[19]
			Sulfur	
			2,6-Di-tert-butyl- [1, 4]benzoquinone	
			Furan-2-yl-(5-hydroxymethyl-furan-2-yl)-methanol	
			2-Bromo-4-(1,1-dimethylpropyl)-phenol	
Rubber (carbon black nitrile based elastomer)	Closure	2-Propanol refluxed, 7 h	Acenaphtylene	[20]
			Phenanthrene	
			Anthracene	
			Fluoranthene	
			Pyrene	

TABLE A.4 Degradation Susceptibility to Ionizing Irradiation of Natural and Synthetic Polymers with Pharmaceutical Use and Degradation Products

Polymer	Sample	Irradiation Conditions	Degradation Products	Reference
LDPE	Film	44.0 kGy Atmospheric conditions Room temperature	1,3-Di-*tert*-butylbenzene 2,4-Di-*tert*-butylphenol Butanoic acid vinylester Acid amide Irgaphos 168 Oxidized Irgaphos 168 Oxidized Irganox 1076 Erucic acid amide	[2]
LDPE	Film	100 kGy Atmospheric conditions Room temperature	Hexane 1-Hexene 3-Methylhexane Heptane 2-Heptene Octane Acetone 3-Ethyl-4-methyl-1-pentene trans-1-Butyl-2-methyl-cyclopropane 3-Ethylheptane Nonane Benzol Decane 3-Hexanone Hexanal Undecane Ethylbenzol 1-[1-Cyclohexen-1-yl]-ethanone 4-Hydroxy-4-methyl-pentanone 3-Heptanone Dodecane 2,3-Dehydro-4-methyl-furan Dimethylphenol 3-Methyl-1-butanol 5-Methyl-3-heptanone 1,3,5,7-Cyclooctatetraene 2-Octanone Octanal Tridecane 2-Methyl-4-octanone 6-Methyl-5-hepten-2-one 5-Methyl-1-heptene Hexyl formate Tetradecane 2,3,6-Trimethyl-1,5-heptadiene Acetic acid Pentadecane Propanoic acid Octyl formate 2,2-Dimethylpropanoic acid	[3]

TABLE A.4 (*Continued*)

Polymer	Sample	Irradiation Conditions	Degradation Products	Reference
			Hexadecane	
			3,5,5-Trimethyl-2-cyclopenten-1-one	
			Butanoic acid	
			Acetophenone	
			Pentanoic acid	
			3-Methyl-2-butanoic acid	
			Hexanoic acid	
			Heptanoic acid	
			Octanoic acid	
HDPE	Pellets	23.6 kGy	Hydrocarbons (butane through eicosane)	[4]
			2-Methyl-2-propanol	
			3-Heptanone	
			Butanoic acid	
			2,2-Dimethyl propanoic acid	
			Pentanoic acid	
PP	Syringes	30 kGy	1,3-Dichloro-2-propanone	[5]
			3,3,3-Rrichloro-2-methyl-1-propene	
			2-Ethyl phenol	
			3,4-Dimethyl phenol	
			1-(2-Hydroxymethyl)-4-hydroxy-2,2,6,6-tetramethylpiperidine succinic acid (HALS)	
			1,3-Di-tert-butyl-benzene	
			4-tert-Butyl-2-chlorophenol	
			3,5-Dimethylbenzoic acid	
			Di-tert-butyl-chloro-benzene	
			2-Chloro-6-(1,1 dimethylethyl) phenol	
			2,4-Di-tert-butylphenol	
			3,5-Bis-(1,1 dimethylethyl) phenol	
			8-Acetoxy-2-chloro-5,6,7,8-tetrahydroquiniline	
			Tetraethyldichlorobenzene	
			Palmitic acid	
			Stearic acid	
			Irgaphos 168	
PP	Film	44.0 kGy Atmospheric conditions Room temperature	Acetanhydride	[2]
			3-Methyl-butanoic acid	
			Acetic acid-(1-ethylhexyl)ester	
			Octanoic acid	
			3-Methyl-4-methylen-hexan-2-one	
			Hexadecanol or octadecanol	
			Several oligomers	
			Erucic acid amide	
			Oxidized Irgaphos 168	
			Oxidized Irganox 1076	
			1,3-Di-tert-butyl-benzene	

(*Continued*)

TABLE A.4 (*Continued*)

Polymer	Sample	Irradiation Conditions	Degradation Products	Reference
PP	Pellets	24.4 kGy	Butene	[4]
			Pentane	
			2-Methylpentane	
			2-Propanone	
			2-Methyl-2-propanol	
			2-Methyl-2-propenal	
			2-Butanone	
			Acetic acid	
			3-Methyl-2-cyclopenten-1-one	
			4-Methylheptane	
			2-Pentanone	
			2,4-Dimethylheptane	
			2-Hexanone	
			2-Methyl-2-hepten-4-one	
			Propanoic acid	
			2,4-Pentanedione	
			4-Methyl-3-penten-2-one	
			4-Hydroxy-4-methyl-2-pentanone	
Polyurethane		100 kGy	4,4′-Methylenedianiline	[7]
		4 MGy Atmospheric conditions Room temperature	2-Ethylhexyl diphenyl phosphate	[21]
			Di-*n*-butyl phthalate	
			Phenol	
			Phthalic ions	
PVC	Pellets	22.5 kGy	Heptane	[4]
			3-Methylheptane	
			1-Octane	
			Octane	
			2-Octene	
			4-Octene	
			3-Heptanone	
			2-Ethylhexanal	
			4-Octanone	
			6-Methyl-2-heptanone	
			2-Ethyl-1-hexanol	
			1-Octanol	
			Undecane	
			Pentylcyclopropanone	
			Ethylhexanoic acid	
			Acetic acid 2-ethylhexylester	
			Acetic acid	
PVC	Sheet	44.0 kGy Atmospheric conditions Room temperature	4-Hydroxy-4-methyl-2-pentanone	[2]
			5-Hexen-2-one	
			1-Ethoxy-2-heptanone	
			Methoxy acetaldehyde diethyl acetale	
			2-Propyl-1-pentanol	
			Diethoxy acetic acid ethylester	
			3-Methylheptyl acetate	
			Nonanoic acid ethylester	
			n-Alkane	
			n-Alkane acid ethylester	

TABLE A.4 (*Continued*)

Polymer	Sample	Irradiation Conditions	Degradation Products	Reference
EVA	Film	100 kGy Atmospheric conditions Room temperature	Hexane 1-Hexene 3-Methylhexane Heptane trans-1,2-Dimethyl-cyclopentane 3-Heptene Octane 1-Octene 3-Ethyl-4-methyl-1-pentene Nonane 3-Methylbutanal Methyl isobutyrate Decane Toluene Hexanal Undecane 3-Hepten-2-one Ethylbenzene p-Xylene 1,3-Dimethylbenzene 1-Butanol 3-Heptanone Heptanal D,L-limonene 2,3-Dimethyl-2-hexanol 3-Octanol 3-Methyl-1-butanol 1,3,5-Trimethylbenzene 2-Octanone Octanal Tridecane 3-Heptanol 6-Methyl-5-hepten-2-one Hexyl formate 4-Hydroxy-4-methyl-pentanone Nonanal Tetradecane 1-(Ethenyloxy)-octadecane 9-Methyl-nonadecane Acetic acid cis-5-Methyl-2-(1-methylethyl)-cyclohexanone 2-Propyl-1-pentanol 2-ethyl-1-hexanol Pentadecane Benzaldehyde 3,7-Dimethyl-1,6-octadien-3-ol Propanoic acid	[3]

(*Continued*)

TABLE A.4 (*Continued*)

Polymer	Sample	Irradiation Conditions	Degradation Products	Reference
			Octyl formate Dimethyl-propanedioic acid Hexadecane 3,5,5-Trimethyl-2-cyclopenten-1-one Butanoic acid 5-Methyl-2-(1-m-ethylethyl)-cyclohexanol Nonanol 2-Methoxy-1-phenyl-ethanone Heptadecane Pentanoic acid Naphthalene 1,3-Butanediol Hexanoic acid 6,10-Dimethyl-5,9-undecadien-2-one Propylene carbonate Benzyl alcohol Heptanoic acid Octanoic acid	
PS	Sheet	44.0 kGy Atmospheric conditions Room temperature	Acetophenone Carbon acid	[2]
PS	Pellets	25.5 kGy	Benzaldehyde Phenol Acetophenone Phenylacetaldehyde 1-Phenylethanol 2-Phenylpropenal	[4]
PA	Film	44.0 kGy Atmospheric conditions Room temperature	Lauroic acid lactame	[2]
PA	Pellets	24.7 kGy	Formamide Acetamide Propanamide Butanamide Pentanamide Hexanamide	[4]
PSf	Films	UV radiation	Muconic acid Hydroquinone 2,2′-Phenylphenol 4-Phenoxyphenol Dibenzofuran	[22]
PET	Film	44.0 kGy Atmospheric conditions Room temperature	Di-iso-propylether	[2]

TABLE A.4 (*Continued*)

Polymer	Sample	Irradiation Conditions	Degradation Products	Reference
Multilayer PET/PE/EVOH/PE	Laminate	100 kGy Atmospheric conditions Room temperature	Hexane	[3]
			Heptane	
			3,4,5-Trimethylheptane	
			Octane	
			trans-1,2-Dimethyl-cyclohexane	
			cis-1-Butyl-2-methyl-cyclopropane	
			4-Methyloctane	
			trans-1-Butyl-2-methyl-cyclopropane	
			3-Ethylheptane	
			4-Methyl-3-heptane	
			Nonane	
			3,5-Dimethyloctane	
			2-Nonene	
			4-Methylnonane	
			2,6,7-Trimethyldecane	
			Tetramethyloctane	
			Pentanal	
			Decane	
			2,6-Dimethylundecane	
			Toluene	
			3-Hexanone	
			5-(1-Methylpropyl)-nonane	
			Hexanal	
			Undecane	
			Ethylbenzene	
			1,3-Dimethylbenzene	
			1-Butanol	
			3-Heptanone	
			1-Ethylmethylbenzene	
			2-Heptanone	
			Dodecane	
			Propylbenzene	
			1-Ethyl-3-methylbenzene	
			tert-Butylbenzene	
			Cyclodecane	
			3-Methyl-1-butanol	
			1,3,5-Trimethylbenzene	
			1,3,5,7-Cyclooctatetraene	
			1-Ethyl-4-methylbenzene	
			1-Methyl-2-(1-methylethyl)-benzene	
			1,2,4-Trimethylbenzene	
			2-Octanone	
			Octanal	
			Tridecane	
			1-Methyl-3-propylbenzene	
			1,4-Diethylbenzene	
			1-Ethyl-3,5-dimethylbenzene	
			1,2-Diethylbenzene	
			a-Methylstyrene	

(*Continued*)

TABLE A.4 ***(Continued)***

Polymer	Sample	Irradiation Conditions	Degradation Products	Reference
			6-Methyl-5-hepten-2-one	
			Hexyl formate	
			2-Ethyl-1,4-dimethyl-benzene	
			1-Methyl-2-(1-methylethyl)-benzene	
			Tetradecane	
			1,3-Bis-(1,1-dimethylethyl)-benzene	
			1-(Ethenyloxy)-octadecane	
			Cyclotetradecane	
			Heptanol	
			Acetic acid	
			2-Propyl-1-pentanol	
			Pentadecane	
			1,7,7-Trimethyl-bicyclo- [1, 2, 2] heptane-2-one	
			Propanoic acid	
			Octyl formate	
			2,2-Dimethyl-propanoic acid	
			Hexadecane	
			Butanoic acid	
			Cyclohexadecane	
			Menthol	
			2-Methoxy-1-phenyl-ethanone	
			Trichloromethylbenzene	
			Pentanoic acid	
			Naphthalene	
			Dimethylbenzenemethanol	
			Hexanoic acid	
			Heptanoic acid	
			Phenol	
			Octanoic acid	
			Nonanoic acid	
			2,4-Bis-(1,1-dimethylethyl)-phenol	
Multilayer PETG/PVDC/PE	Laminate	2.7-3.2 MRad	Pentadecane	[18]
Glycol modified PET			Nonanedioic acid, dimethyl ester	
PVDC polyvinylidene chloride			Nonanedioic acid, monomethyl ester	
HDPE			Nonanedioic acid, diethyl ester	
PU adhesive between PVDC and PE			Heptadecane	
			Octadecane	
			Nonanedioic acid, ethyl, 2-hydroxyethyl ester	
			Butyl, 2-methylpropyl phthalate	
			Ethyl palmitate	
			Eiocosane	
			Henicosane	
			Docosane	
			Bis(2-ethylhexyl)adipate	
			Tricosane	
			Tetracosane	

TABLE A.4 (*Continued*)

Polymer	Sample	Irradiation Conditions	Degradation Products	Reference
			Pentacosane	
			Bis(2-ethylhexyl)phthalate	
			Hexacosane	
			Heptacosane	
			13-Docosenamide (erucyl amide)	
			Tetracosanoic acid, ethyl ester	
			2,2′-[(1-Methylethylidene)bis(4,1-phenyleneoxymethylene)] bisoxirane	
			Hexacosanoic acid, ethyl ester	
Cellulose	Hemodialysis membrane	Not informed	CO, CO_2, H_2	[23]
			Citric acid	
			Succinic acid	
			Malic acid	
			Fumaric acid	
			Tartaric acid	
			Malonic acid	
			Pyruvic acid	
			Glyoxalic acid	

The potential leaching of contaminants due to the physicochemical nature of the polymeric material and formulation ingredients is an authentic problem and must be a matter of concern when evaluating the safety of a new product. Several examples, beyond the well-known leaching of DEHP into lipophilic formulations, substantiate how vital this safety appraisal is.

REFERENCES

1. Stoffers NH, Linssen JPH, Franz R, Welle F. Migration and sensory evaluation of irradiated polymers. Radiat Phys Chem 71, 203–206, 2004.
2. Demertzis PG, Franz R, Welle F. The effects of gamma-irradiation on compositional changes in plastic packaging films. Packag Technol Sci 12(3), 119–130, 1999.
3. Riganakos KA, Koller WD, Ehlermann DAE, Bauer B, Kontominas MG. Effects of ionizing radiation on properties of monolayer and multilayer flexible food packaging materials. Radiat Phys Chem 54(5), 527–540, 1999.
4. Buchalla R, Boess C, Bogl KW. Characterization of volatile radiolysis products in radiation-sterilized plastics by thermal desorption-gas chromatography-mass spectrometry: screening of six medical polymers. Radiat Phys Chem 56(3), 353–367, 1999.
5. Fintzou AT, Badeka AV, Kontominas MG, Stahl MR, Riganakos KA. Changes in physicochemical and mechanical properties of electron-beam irradiated polypropylene syringes as a function of irradiation dose. Radiat Phys Chem 76, 97–107, 2007.

6. Glossmann H, Hering S, Savchenko A, Berger W, Friedrich K, Garcia ML, Goetz MA, Liesch JM, Zink DL, Kaczorowski GJ. A light stabilizer (Tinuvin 770) that elutes from polypropylene plastic tubes is a potent L-type Ca(2+)-channel blocker. Proc Natl Acad Sci U S A 90(20), 9523–9527, 1993.
7. Shintani H. Formation and elution of toxic compounds from sterilized medical products: toxic compound formation from irradiated products. Radiat Phys Chem 47, 139–148, 1996.
8. Daka JN, Chawla AS. Release of chemicals from polyurethane foam in the Même breast implant. Biomater Artif Cells Immobilization Biotechnol 21(1), 23–46, 1993.
9. Luu HMD, Biles J, White KD. Characterization of polyesterurethane degradation products. J Appl Biomater 5(1), 1–7, 1994.
10. Benoit FM. Degradation of polyurethane foams used in the même breast implant. J Biomed Mater Res 27(10), 1341–1348, 1993.
11. Wang Q, Storm BK. Separation and analysis of low molecular weight plasticizers in poly(vinyl chloride) tubes. Polym Test 24(3), 290–300, 2005.
12. Wang Q, Storm BK. Migration of additives from poly(vinyl chloride) (PVC) tubes into aqueous media. Macromol Symp 225, 191–203, 2005.
13. Wahl HG, Hoffmann A, Haring HU, Liebich HM. Identification of plasticizers in medical products by a combined direct thermodesorption-cooled injection system and gas chromatography-mass spectrometry. J Chromatogr A 847(1–2), 1–7, 1999.
14. Tawfik MS, Huyghebaert A. Interaction of packaging materials and vegetable oils: oil stability. Food Chem 64(4), 451–459, 1999.
15. Haishima Y, Hayashi Y, Yagami T, Nakamura A. Elution of bisphenol-A from hemodialyzers consisting of polycarbonate and polysulfone resins. J Biomed Mater Res 58(2), 209–215, 2001.
16. Jenke DR, Poss M, Sadain S, Story J, Smith W, Reiber D. Identification of caprolactam oligomers and related compounds in aqueous extracts of nylon-6. J Appl Polym Sci 95(5), 1262–1274, 2005.
17. Jenke DR, Jene JM, Poss M, Story J, Tsilipetros T, Odufu A, Terbush W. Accumulation of extractables in buffer solutions from a polyolefin plastic container. Int J Pharm 297(1–2), 120–133, 2005.
18. Kim-Kang H, Gilbert SG. Isolation and identification of potential migrants in gamma-irradiated plastic laminates by using GC/MS and GC/IR. Appl Spectrosc 45(4), 572–580, 1991.
19. Zhang F, Chang A, Karaisz K, Feng R, Cai J. Structural identification of extractables from rubber closures used for pre-filled semisolid drug applicator by chromatography, mass spectrometry, and organic synthesis. J Pharm Biomed Anal 34(5), 841–849, 2004.
20. Paskiet DM. Strategy for determining extractables from rubber packaging materials in drug products. PDA J Pharm Sci Technol 51(6), 248–251, 1997.
21. Colombani J, Herbette G, Rossi C, Joussot-Dubien C, Labed V, Gilardi T. Leaching of Plasticized PVC: effect of irradiation. J Appl Polym Sci 112, 1372–1377, 2009.
22. Rivaton A, Gardette JL. Photodegradation of polyethersulfone and polysulfone. Polym Degrad Stab 66, 385–403, 1999.
23. Kabanov VYa, Feldman VI, Ershov BG, Polikarpov AI, Kiryukhin DP, Apel PYu. Radiation chemistry of polymers. High Energy Chem 43, 1–18, 2009.

INDEX

Note: Page numbers in **bold** refer to tables; page numbers in *italics* refer to figures.

Sources of Contamination in Medicinal Products and Medical Devices, First Edition. Denise Bohrer.
© 2013 by John Wiley & Sons, Inc. Published 2013 by John Wiley & Sons, Inc.

Printed and bound by CPI Group (UK) Ltd, Croydon, CR0 4YY

16/09/2026

14963417-0005